COLD WAR ANTI-SUBMARINE WARFARE

81
NAVY

COLD WAR ANTI-SUBMARINE WARFARE

NORMAN FRIEDMAN

NAVAL INSTITUTE PRESS
Annapolis, Maryland

Frontispiece: F-1: The ASW team of the late 1950s: Task Group Alfa at sea. Alfa was an experimental hunter-killer group built around an ASW carrier, in this case USS *Randolph*. Her flight deck shows HSS-1 Seabat helicopters (later designated SH-34) and a twin-rotor Kaman HUK-1 plane guard. Farther aft are S2F Trackers and, at right, an AD-5W Skyraider with the big radome of an APS-20 radar. In the foreground to the right is a *Fletcher*-class DDE (probably *Cony*), with a *Fletcher*-class destroyer astern, and another *Fletcher* DDE farther astern. At left are two more *Fletcher* DDEs (probably *Bache* leading *Beale*). The devices on the sterns of the two Guppies in the foreground may be noisemakers. The helicopter in the foreground is an HSS-1. Overhead are a land-based P2V Neptune, four carrier-based S2F Trackers, and an AD-5W Skyraider with its massive APS-20 radome under it. *Randolph* replaced the first carrier assigned to the Task Group, *Valley Forge*, late in 1959. USNI

Naval Institute Press
291 Wood Road
Annapolis, MD 21402

Library of Congress Cataloging-in-Publication Data
Names: Friedman, Norman, author
Title: Cold War anti-submarine warfare / Norman Friedman.
Description: Annapolis, Maryland : Naval Institute Press, 2025. | Includes bibliographical references and index.
Identifiers: LCCN 2025011025 | ISBN 9781682478578 hardcover
Subjects: LCSH: Anti-submarine warfare—History—20th century | Submarines (Ships)—Design and construction—History—20th century | Cold War | United States. Navy—History—20th century | Soviet Union. Voenno-Morskoĭ Flot—History
Classification: LCC V214 .F75 2025 | DDC 359.9/3097309045—dc23/eng/20250416
LC record available at https://lccn.loc.gov/2025011025

♾ Print editions meet the requirements of ANSI/NISO z39.48-1992 (Permanence of Paper).
Printed in the United States of America.

33 32 31 30 29 28 27 26 25 9 8 7 6 5 4 3 2 1
First printing

CONTENTS

List of Acronyms . . . vii

1 The Cold War and ASW . . . 1

2 Finding Submarines: Sound in the Sea . . . 20

3 Legacy . . . 36

4 New Kinds of Submarine . . . 71

5 Facing the Postwar Nightmare . . . 99

6 The Crisis and Its Solution . . . 133

7 The Mobilization Decade: The Fifties . . . 150

8 The Nuclear Revolution . . . 184

9 Fighting a New Soviet Submarine Generation . . . 217

10 End-Game . . . 254

Acknowledgments . . . 285

Appendices

A. Surface Ship and Aircraft Sonars . . . 287

B. Submarine Sonars . . . 313

C. Torpedoes . . . 331

Notes . . . 361

Bibliography . . . 397

Index . . . 399

ACRONYMS

AAW	anti-air warfare
ACNO	Assistant Chief of Naval Operations
ADAR	Air-Deployed Active Receiver
ADAWS	Action Data Automated Weapons System
AEC	Atomic Energy Commission
AEW	airborne early warning
AGC	automatic gain control
AGI	intelligence-gathering ship
AIC	Action Information Center
AIDA	acoustic intelligence program
ALICE	adaptive line-canceler and enhancer
ALWT	Advanced Lightweight Torpedo
ARL	Admiralty Research Laboratory
ASAP	Anti-Submarine Attack Plotter
ASCAC	ASW Classification and Analysis Center
ASCL	Advanced Sonobuoy Communications Link
ASDevLant	ASW Development Group Atlantic
ASO	acoustic-sensor operator
ASPECT	Acoustic Short-Pulse Echo Classification Technique
ASROC	rocket boost for depth charge or torpedo
ASTOR	Anti-Submarine Mk 45 Torpedo
ASWORG	ASW Operations Research Group
ASWSCCS	ASW Ship Command and Control System
ATACO	air-tactical control officer
ATERE	Airship Towed Echo-Ranging Equipment
ATO	airborne tactical officer
ATT	automatic target training
AUWE	Admiralty Underwater Weapons Establishment
BBG	missle-armed battleship (concept)
BDI	bearing deviation indicator
BRASS	Bottom Reflected Active Sonar System
BT	bathythermograph
BuOrd	Bureau of Ordnance
BuShips	Bureau of Ships
CDS	Comprehensive Display System
CODAR	correlation detection
COMINT	communications intelligence
CONOPS	concept of operations
COTP	CO Tactical Plot
CRT	cathode-ray tube (computer display)
CUW	Committee on Undersea Warfare
CVS	ASW carrier
DASH	drone antisubmarine helicopter
DA/SW	Admiralty Anti-Submarine Warfare Division
DATAR	Digital Automatic Tracking And Remoting
DCG	Doppler-controlled gain
DDE	Escort Destroyer
DDR&E	Defense Directorate of Research and Engineering
DE	Destroyer Escort (FMR and WGT are versions)
DER	destroyer escort radar picket
DEXTOR	deep experimental torpedo

DIMUS	digital multi-beam sonar
DMCC	Dual Maintenance of Close Contact
DME	distance-measuring equipment
DMWD	Department of Miscellaneous Weapon Development (British)
DNC	Director of Naval Construction
DSB	Defense Science Board (US)
DTSD	Director of the Tactical and Staff Duties Division, British
ECCM	electronic counter-countermeasures (anti-jam)
ECM	electronic countermeasures
EDR	effective detection range
ERA	Extended-Range ASROC
ERAPS	Expendable Reliable Acoustic Path Sonobuoy
ESM	electronic support measures
ETASS	Escort Towed Array Sensor System
E2R	enhanced echo-ranging
EW	electronic warfare
EWO	electronic warfare officer
FDS	Fixed Distributed System
FLIR	forward-looking infrared
FMR	Fairbanks-Morse Reversing (type of DE)
FOSM	Flag Officer Submarines (UK)
FRAM	Fleet Rehabilitation and Modernization
FSED	full-scale engineering development
FY	Fiscal Year (cited as FY 48, etc.)
FYDP	Five-Year Defense Plan, U.S.
GCCS	Government Code and Cipher School (British)
GCHQ	communications intelligence headquarters (British)
GHG	Gruppen Horch-Geraet, "group listening device"
GIUK Gap	Greenland-Iceland-UK Gap, the Gap
GNAT	German Navy Acoustic Torpedo
GUPPY	Greater Underwater Propulsive Power
HE	hydrophone effect
HF/DF	high-frequency direction-finding
HLA	horizontal line array
HTP	high-test hydrogen peroxide
HUK	hunter-killer
HUSL	Harvard Undersea Sound Laboratory
IETASS	Interim Escort Towed Array
IFF	identification friend or foe
IR	infrared
ITASS	Interim Towed Array Surveillance System
IUSS	Integrated Undersea Surveillance System
JIC	British Joint Intelligence Committee
LAAV	Light Airborne ASW Vehicle
LAMPS	Light Airborne Multipurpose System
LCS	low-cost sonobuoy
LFA	Low Frequency Adjunct
LIDAR	laser radar
LOFAR	low-frequency analysis and ranging, a fake designation
LORA	long-range addition
LRO	Long-Range Objectives Group
LUSC	Lightweight Undersea Components
LUT	German pattern-running torpedo
MAD	magnetic anomaly detection
MARK	maintainability and reliability kit
MATCH	Medium Range Torpedo-Carrying Helicopter
MCAR	Multi-Channel Aural Reception
MCJR	multi-channel Jezabel repeater
MIP	modest improvement program
MSS	Moored Surveillance system
MTB	maintenance of true bearing

NADC	Naval Air Development Center
NAVFAC	naval facility
NAVSTIC	ONI Scientific and Technical Intelligence Center
NEARTIP	near-term improvement program (Mk 46)
NID	Naval Intelligence Division (UK)
NIDAR	cover name for non-acoustic technique
NOL	Naval Ordnance Laboratory
NOTS	Naval Ordnance Test Station
NDRC	National Defense Research Committee (US)
NRL	Naval Research Laboratory
NSIA	National Security Industrial Association (US)
NTDS	Naval Tactical Data System
NTU	New Threat Upgrade
NUSL	Naval Underwater Sound Lab
ODN	own Doppler nullifier
ODT	omni-directional transmission
OEG	Operational Evaluation Group
OKM	German high command
ONI	Office of Naval Intelligence
ONR	Office of Naval Research
OpDevFor	Operations Development Force
Op-95	ASW office
OSD	Office of the Secretary of Defense
OSP	Off-Shore Purchases
OSRD	Office of Scientific Research and Development (US)
OTC	officer in tactical command
PADLOC	Passive-Active Detection and Localization
PAIR	Performance and Integration Refit
POMCUS	pre-positioned overseas materiel configured to unit sets
PPI	plan position indicator
PUFFS	Passive Underwater Fire-control Feasibility Study
QHB	a scanning sonar
RAP	reliable acoustic path
RAPLOC	rapid passive localization
RAT	rocket-assisted torpedo
RDSS	Rapidly Deployable Surveillance System
RDT	rotationally directed transmission
REMRO	remote-radar operator
RFP	radio finger printing
RORSAT	radar satellite
SAC	Strategic Air Command
SACEUR	Supreme Allied Commander Europe
SACLANT	Supreme Allied Commander for the Atlantic
SCB	Ship Characteristics Board
SCEPS	Stored Chemical Energy Power System
SDT	steered directional transmission
SESCO	secure underwater communications system
SHAB	steerable hull-array beam-former
SIOP	Single Integrated [Strategic] Operational Plan
SITTAR	Submarine Integrated Tactical Towed Array
SLBM	ballistic missile
SLOC	sea lines of communication
SLOT	submarine-launched one-way transmission
SO	sensor operator
SOFAR	sound fixing and ranging
SOKS	*Sistema Obnaruzheniya Kil'vaternogo Sleda* (Soviet non-acoustic sensors)
SOR	Specific Operational Requirement
SOSS	Soviet Ocean Surveillance System
SOSUS	sound surveillance system
SOW	Standoff Weapon
SPA	submarine patrol area (UK)

SRS	sonobuoy reference system
SSAP	Sonar Signal Acoustic Processors
SSBN	ballistic missile submarine (nuclear)
SSMP	Surface Ship Sonar Modification Program
SSN	nuclear-powered submarine
SSPA	SOSUS Probability Area
STASS	Submarine Towed Array System
SUBROC	submarine rocket
SURTASS	surveillance towed array sensor system
TACCO	tactical coordinator
TACTLASS	tactical towed array sensor system
T-AGOS	towed array surveillance ship
TASS	towed array sensor system
TCS	tactical support center
TE	turbo-electric
TF	task force
TMA	target motion analysis
TRAM	test, reliability, and maintenance
TSOR	Tentative Specific Operational Requirement
TUBA	acoustic intelligence collection towed array
TVD	Soviet oceanic theater of operations
TVG	time variation of gain
UAV	unmanned aerial vehicle
UEP	underwater electric potential
UHF	ultra-high frequency
UPTIDE	United Pacific Fleet Project for Tactical Improvement and Data Extraction
USGW	underwater-launched antiship missiles (Britain)
UWC	Undersea Warfare Conference
VDS	variable-depth sonar
VLA	vertical linear array
VLA	Vertically Launched ASROC
VLAD	vertical line array DIFAR
VP	patrol plane squadron
VS	carrier ASW squadron
VTOL	vertical takeoff and landing
WAA	wide aperture array
WAP	weapons alternative processor
WATU	Western Approaches Tactical Unit
WGT	Westinghouse geared turbine (type of DE)
ZP	blimp squadron

1
THE COLD WAR AND ASW

ABOUT 30 YEARS AGO the Cold War abruptly ended when the Soviet Union disintegrated. The stricken Russian economy, the rump of the Soviet state, could barely support even a deeply diminished submarine force; much of the Cold War Soviet fleet was soon scrapped. Thirty years is about the length of a senior naval officer's career, which means that very few of the world's active naval officers have memories of the character of Cold War ASW.

Now, however, it seems that Cold War submarine threats are back. Cold War forms of ASW—generally with different technology—seem relevant again. This book looks at the kinds of problems we are once again likely to face. It reminds us that ASW is a multidimensional problem, the domain of air and surface forces as well as of our own submarines. History also reminds us that ASW can have multiple meanings. Is it primarily about destroying an enemy submarine force? About frustrating an enemy attack on shipping? About frustrating enemy attacks on our fleet? About frustrating an enemy's strategic submarines? How are the different aspects of ASW related?

This is not to say that submarine threats vanished with the Soviet Union, but rather that the deep-water threat permeating the world's oceans was replaced by local threats largely confined to littoral areas. Littoral ASW is very different from Cold War ASW, not least because the latter typically involved diesel-electric rather than nuclear submarines and shallow rather than deep water. For example, a diesel submarine can bottom in shallow water, but a nuclear submarine risks clogging her condensers with mud and being put out of action.

The enemy: a Soviet Foxtrot-class submarine photographed in 1982. DIA PHOTO VIA USNI

Instincts built up to combat one kind of submarine and one kind of threat need to be revisited if we return to the earlier kind of threat.

Antisubmarine warfare was inescapably bound up with the Cold War. It was widely assumed, for example, that a Western army fighting for its life in central

The prize: the shipping connecting the New World to the Old, sustaining NATO and its army in Europe. This 24-ship convoy was photographed off Newfoundland on 28 July 1942, but an early Cold War convoy would have been similar. The question in the late Cold War was whether the Soviets planned to wield their massive submarine fleet against vital Allied shipping. NHHC

Europe could survive only thanks to supply across the Atlantic. That supply line had to be protected.

Among the important Western equalizers were nuclear-armed carrier battle groups, which were expected to face Soviet submarines. Both sides became intensely interested in strategic ASW. The way antisubmarine warfare featured in Western thinking changed considerably during the Cold War. Those changing tactics and strategies are likely to carry important lessons for the future.

The Soviets had built the world's most numerous submarine fleet before World War II. Then, in 1945, the Red Army captured revolutionary German submarine technology. It was generally assumed that the Soviets would soon build large numbers of German-type submarines. The Western ASW fleets that had won the Battle of the Atlantic were not well suited to the new underwater threat. The Western Allies knew that they had needed great numbers of ships and aircraft to win. The great early postwar question was how they could amass sufficient numbers of the more capable ships and aircraft needed to deal with the expected more capable Soviet submarines. Initially it seemed that the West would have breathing space; in 1947 the British Joint Intelligence Committee estimated that the Soviets would not be ready to fight for a decade. The outbreak of war in Korea in 1950 was thus a terrible shock. It seemed to many to be the beginning of a new world war. NATO became a military alliance with military and naval planners and international commanders such as the Supreme Allied Commander for the Atlantic (SACLANT). Alliance members were encouraged to build up their fleets and naval air arms to provide the numbers that would be needed to fight a new Battle of the Atlantic.

ADM Sergei Gorshkov, the driving force behind the modern Soviet Navy, later pointed to the disproportionate Allied force that had been necessary to defeat U-boats in World War II. Could Soviet investment in submarines unbalance the West, perhaps reducing the threat the Soviets felt from U.S. aircraft carriers? One wonders if the Chinese see things similarly now.

This history is dominated by the U.S. Navy and the Royal Navy. In 1945, and for many years afterward, they were the West's principal sea powers. The Royal Navy led in the Battle of the Atlantic, and in 1945 was by far the most experienced (and advanced) ASW navy in the world. In numbers of ships and in

Massed escorts that had survived World War II: destroyer escorts off San Diego, 14 October 1948.
RICK DAVIS

postwar resources, however, it was dwarfed by the U.S. Navy. The two had cooperated closely during the war. Well before there was any formal postwar connection between them, they exchanged extensive information on their approaches to ASW. Beginning in 1947, for example, British officers attended the annual classified U.S. Navy Undersea Warfare Conferences. British groups were welcomed at U.S. Navy research centers closed to other foreigners. Before NATO was created, the U.S. Navy promoted agreements to exchange new sensors and weapons for trials.

The U.S. Navy and the Royal Navy retained large numbers of ships built during World War II and modernized many of them to serve through the 1960s. Eventually, however, the urgent question became how to fight an oceanic ASW battle with much smaller numbers. Naval programs naturally competed with those for air forces and armies. In both Britain and the United States, the defense budgets of the mid-1950s were dominated by investments in bomber deterrence and nuclear weapons. Armies also became considerably more expensive as they became increasingly mechanized. The competition between naval and ground force investment was evident in both the United States and Britain, particularly in the 1970s. The U.S. Carter administration wanted to balance its budget by cutting naval forces while building up ground forces in Europe. The British reached a similar conclusion slightly later. Both navies argued that they offered invaluable flexibility. In the U.S. case the need for such flexibility was demonstrated by crises in the late 1970s in Iran and Afghanistan. The parallel British experience was the Falklands Conflict.

During World War II the Royal Canadian Navy contributed a large ASW force which in 1945 was (numerically) the third largest navy in the world. Much of the tonnage involved was rapidly disposed of after the war, but Canada's wartime alliance with the United States survived. Particularly in the 1950s the Royal Canadian Navy made substantial contributions to Western ASW, including important technology. Canada and the United States worked together in the strategic defense of North America, including ASW directed against Soviet missile submarines. The Royal Canadian Navy became a partner in the SOSUS surveillance system. The close relationship between the American, British, and Canadian ASW forces was sometimes described as the "ABC Partnership." Australia and New Zealand also enjoyed especially close relationships with the United States and Britain, but they were less intensely involved in ASW.

The other two major Western naval powers, France and Italy, lost much of their fleets during World War II, and their postwar military recovery favored their armies and air forces. More important for this history, both France and Italy had large Communist parties controlled by the Soviets. Many of the most important labor unions in both countries, moreover, were communist controlled. Indeed, the French air minister had been a Communist before the war. The U.S. and British navies were thus very reluctant to share crucial ASW secrets. This security issue probably explains why fixed SOSUS arrays were never built in the Mediterranean. In the mid-1950s the French were particularly resentful of that restricted access.

The Royal Netherlands Navy had an important Cold War ASW role as well but seems not to have participated in developing ASW concepts, technology, or tactics comparable to those of the ABC Partnership. For that reason I have not highlighted its Cold War ASW development.

There were important differences between the Western Cold War navies. The U.S. Navy included both ships (with their aircraft) and land-based maritime aircraft. It managed to gain a share in the U.S. strategic nuclear force, developing both carrier-based bombers and submarine-launched missiles. By way of contrast, the Royal Navy controlled only ships and their associated aircraft. Land-based ASW aircraft were owned by the Royal Air Force and subject to that service's preference for spending its money on fighters and bombers. Interservice rivalry had ASW implications. Until the British adopted the Polaris system in the 1960s, the Royal Navy was largely shut out of British nuclear striking power, and it naturally concentrated on ASW and other forms of trade protection. It emphasized the value of convoy operations while RAF Coastal Command emphasized offensive air ASW.

Canada's defense was organized like Britain's in that the RCAF operated land-based maritime aircraft. However, Canada abandoned strategic air power after 1945, to the benefit of Canadian air ASW. The revived French navy was organized more like the U.S. Navy. It retained its prewar control of land-based maritime aircraft, and it gained a nuclear strike role when France began developing nuclear weapons. The Royal Netherlands Navy was also organized similarly to the U.S. Navy, operating land- as well as carrier-based aircraft. The revived Italian navy operated no organic aircraft beyond shipboard helicopters, though after the Cold War it operated carrier-based fighter/attack aircraft.

Soviet submarines were only one of the postwar threats to Western shipping and Western naval forces. From the mid-1950s the Soviets maintained a strong naval air force armed with antiship missiles. Soviet surface warships could break up convoys, as German heavy ships had demonstrated during World War II. Mining was expected to be a serious threat to European ports and a lesser but still significant threat to U.S. ports. A new generation of particularly sensitive magnetic mines, introduced by the Germans during World War II, demanded the construction of new minesweepers for NATO navies that were ton for ton at least as expensive as submarines. The minehunters that followed them were even more expensive. The cost of these programs necessarily impinged on escort construction.

ASW was always a multi-dimensional form of warfare. For much of World War II, for example, maritime patrol aircraft sank more U-boats than surface ships did; but surface ships directly protected convoys. The ships and aircraft often worked together, as in groups built around escort carriers. The initial postwar ASW crisis arose largely because new technology had badly devalued aircraft, placing a much greater burden on surface ships. Submarines were brought much more prominently into ASW because the situation was so dire. Later, maritime patrol aircraft regained much of their former prominence with the advent of SOSUS and long-range sonobuoys.

THE COLD WAR

Virtually from the 1917 Revolution on, the Soviets claimed that the West was inherently hostile because their mere existence would inspire revolution in the capitalist West, which would inevitably seek to destroy the USSR. This apparently defensive posture could justify what the Soviets might describe as preemptive attacks (but the West would consider aggression). The World War II alliance between the Soviets and the United States and the United Kingdom generated hope in the West that the USSR, badly damaged by World War II, would cooperate with the Western powers to prevent any new war. This hope was shattered early in 1946 when Stalin announced that the Soviet Union regarded the West as a permanent enemy.

It was not clear when—or whether—Stalin would convert his announcement into military action. Some believed that Stalin would not risk war and that a firm stance on the part of the West might suffice. That was the context of the strategy of containment proposed by Cold War strategist George Kennan in 1946: if war could be staved off, the internal strains inevitably generated by the Soviet system would destroy the USSR. But Stalin might choose to fight. Unlike the Western powers, the Soviets apparently did not completely demobilize after the war. In 1947, based on knowledge and assumptions of the degree to which the Soviet Union had been damaged by World War II, the British Joint Intelligence Committee (JIC), which seems to have been the leading Western analytic agency of the time, estimated that Stalin would be entirely unable to fight until 1952. For the next five years his ability to attack the West would grow, with 1957 being the "year of maximum danger." The term recalled a 1934 British estimate that Germany would need five years to prepare for war, making 1939—very accurately—the year of maximum danger. This timeline was adopted by the U.S. government.

World War II had introduced such radically new weapons, including fast submarines, that most of the equipment produced during that time seemed already obsolescent at best. The JIC's assumption made it possible for the British, whose country had also been badly damaged by World War II, to concentrate initially on research, rather than on producing new weapons reflecting wartime lessons. They decided to forgo "interim" weapons in favor of "ultimate" weapons they could have in 1956, in time to deal with the anticipated war in 1957. The U.S. approach seems to have been similar. In both countries much of the limited money available went into nuclear weapons (the British government having decided that Britain could not retain Great Power status without them).

The Soviets' enmity was obvious in occupied Germany and also in Iran, which the Soviets and the two Western allies had occupied (the Soviets left only under U.S. pressure). At first the U.S. government thought that U.S. occupation troops in Germany could soon be withdrawn from Europe, leaving the British to maintain order. This was fantasy: Americans did not realize the extent to which World War II had crippled Britain. In 1947 the British told the U.S. government that they could no longer afford to support the Greeks, who were fighting Soviet-backed Communists. In response, President Harry S. Truman announced that the United States would back governments trying to protect themselves against the Soviets.

Through 1948 the Soviets turned the countries their army had occupied in central Europe into satellites they completely dominated. Their rigging of the 1947 Hungarian election finally convinced the French, who had seen the Soviet Union as a bulwark against a revived Germany, that the Soviets were a threat.

The U.S. government realized that the Soviet threat was more than merely military. In 1946–47 European economic recovery was failing. Stalin might overrun Western Europe by a combination of subversion and popular disenchantment with governments unable to revive their economies. The United States pursued European economic recovery via the Marshall Plan and offered aid to all European governments, but Stalin vetoed participation by those he now controlled. When Czechoslovakia, which had a coalition government not entirely dominated by Communists, tried to participate in the Marshall Plan, Stalin ordered a coup that eliminated the non-Communists. He also demonstrated his hostility to the West in Germany, culminating in the unsuccessful blockade of West Berlin in 1948–49. In view of these events the United States decided to rearm. In 1948 Congress passed the first peacetime draft in American history.

Military assistance to the Europeans complemented the Marshall program. In 1948 the British sponsored the Western European Union to integrate the region's defense. The United States and Canada joined the union in 1949, creating the North Atlantic Treaty Organization (NATO). The North Atlantic was at the core of the organization because North America was its most productive territory. Should war with the Soviet Union erupt, supply across the North Atlantic would be crucial, as it had been during World War II. The main barrier to such resupply would be the huge Soviet submarine fleet.

NATO was not initially a military alliance, and no U.S. troops were deployed to Europe; it was an effort to strengthen Western European countries so that they could deter (or resist) a Soviet attack. The United States transferred much of the army equipment held in storage after 1945, as well as many aircraft, to NATO. The object was to create a military force sufficient to slow a Soviet invasion of Western Europe long enough for the United States to stop the Soviets with nuclear weapons.[1] The United States also sponsored the revival of European arms production with Off Shore Purchases (OSP) and with the Mutual Defense Assistance Program. The transfer of U.S. warships that might be needed badly for ASW, however, became contentious, to the point that the 1948 U.S. ASW Conference wanted the Navy to reject a proposal to earmark 44 of its best destroyer escorts (*Buckley* class) for transfer.[2]

Confronted by a Congress unwilling to raise taxes, President Truman pointed out the Marshall Plan's value as a security measure. Money that might otherwise have been allocated solely to U.S. rearmament instead had to pay for the Marshall Plan and for assistance to NATO countries. As a consequence, U.S. rearmament was drastically cut in 1949. Among the victims was the Navy's supercarrier *United States*, which was considered partly an ASW asset because its aircraft could attack Soviet submarines "at source" (in their bases). Plans to build new escorts were dropped. U.S. Navy antisubmarine forces were heavily affected—personnel were axed, operating funds were deeply cut, and ships were laid up.

The Truman administration knew that it was taking a chance but was betting that the U.S. nuclear deterrent would prevent any Soviet attack. Through his spies, Stalin was almost certainly aware that the United States actually had very few nuclear weapons. The U.S. deterrent began to melt when Stalin's scientists exploded a prototype nuclear bomb in August 1949. The U.S. government had been aware that the Soviets were scrambling to build a bomb but assumed that the program was still some years away from completion. The U.S. estimate of the "year of maximum danger" was reassessed and connected with a calculation of how long it would take the Soviets to build enough bombs to disable the United States. In 1947 it had been estimated that Stalin would not have his first bomb until 1952. Now that he had beaten that estimate by three years, the U.S. government advanced the year of maximum danger to 1954.

Because the United States did not yet have enough nuclear weapons to disable the Soviet Union in 1950, Stalin had a window of opportunity to attack in Asia. With his approval, in June 1950 North Korea invaded South Korea. Given their previous fears of Soviet aggression, the Americans and British read the invasion as a possible step toward World War III and mobilized. The U.S. mobilization not only built America's current forces, it also corrected military deficiencies (including naval ones) and developed a new generation of sensors, ships and airplanes, and weapons. The momentum generated by the Korean War mobilization carried the United States through the 1950s. Britain, in contrast, had barely recovered from the enormous destruction of World War II, not just physical but also financial, and was compelled to put interim weapons into production while continuing work on the ultimate ones. The stress of mobilization necessitated the first in a series of defense reviews, which culminated in the 1957 review that reshaped the Royal Navy and the other British services.

Given the postwar emphasis on building a nuclear arsenal, the only part of the pre-Korea U.S. rearmament program *not* deeply cut was a major expansion in nuclear weapon production, including obtaining sufficient enriched uranium. The consequences would become obvious in the 1950s' era of "nuclear plenty." Uranium stockpiling had important ASW consequences. Not only did it provide nuclear material for weapons, it also provided the weapons-grade, highly enriched uranium for the reactors that propelled the nuclear navy, particularly its submarines.

For the first time in peacetime, the United States deployed a combat army—rather than an occupying force—abroad, into West Germany.[3] NATO became a military alliance with designated regional commanders, including SACEUR (Supreme Allied Commander Europe) and SACLANT (Supreme Allied Commander Atlantic, appointed in 1952). As understood in 1950, the main features of World War III would be a Soviet invasion of Western Europe coupled with a new battle for the Atlantic. Transatlantic resupply could enable NATO, which outproduced the Soviet Union, to make its weight felt—if NATO forces could survive long enough.

Huge numbers of escorts would be needed. In 1950, before the outbreak of war in Korea, the U.S. Navy envisaged a wartime total of 212 destroyers and destroyer escorts for ocean escort, backed by another 102 if Soviet submarines reached U.S. waters. Ninety more were required for fast-carrier escort and 17 for forward-area Pacific patrol stations.[4] Each convoy would need five to eight escorts, but it seemed that only three (if that) would be available on the outbreak of war.

NATO developed specific force goals, including numbers of escorts.[5] To meet the goals, the U.S. Navy assisted in the reconstruction of the French and, to a lesser extent, the Dutch and Italian navies, partly by financing some of their initial escort programs as OSP under the FY 52 and later programs. Their U.S. financing was reflected in the U.S. hull numbers initially assigned to them. Although the periodically revised NATO force goals were never met, they compelled NATO navies to declare which resources they expected to provide to the alliance in wartime. Navies could argue to their governments that forces promised to SACLANT should not be cut.

The first SACEUR, GEN Dwight D. Eisenhower, emphasized the need for carrier strikes on the flanks of an advancing Soviet army to slow it down. Thus, from the outset, NATO ASW had two complementary aspects: support of carrier strike forces and protection of transatlantic (and complementary coastal) shipping.

The newly created SACLANT made attack at source (using nuclear weapons) the primary role of U.S. carriers, recognizing, however, that before beginning a war the Soviets might well surge many of their submarines into the North Atlantic. Attack at source would be pointless until these submarines had returned to base to refuel and to replenish their weapons and stores. In these terms the key geographic feature of the North Atlantic was the Greenland-Iceland-UK Gap (GIUK Gap) between submarine bases in the Soviet Union and their likely patrol areas in the open North Atlantic. NATO ASW increasingly envisaged a barrier in the Gap. The NATO carrier force in the Atlantic, the Strike Fleet Atlantic, could help defend Greenland, Iceland, and Norway against initial Soviet attacks and thus help keep the Gap more or less closed.

NATO would have two carrier strike forces, one in the North Atlantic and one in the Mediterranean (the U.S. Sixth Fleet and some Allied carriers). Strike Fleet Atlantic would be formed at the beginning of a war. In classical naval terms, the Strike Fleet was the NATO battle fleet that would prevent the Soviets from sweeping away NATO convoys. Many Soviet submarines would be occupied trying to attack the Strike Fleet, so the fleet could be viewed as a blocking force in the North Atlantic. In the first big NATO naval exercise, Mainbrace, the carriers demonstrated their ability to support NATO troops in Denmark and Norway. British naval observers were concerned that Americans on board their carrier *Midway* considered the war won when they managed to launch simulated nuclear strikes.[6] They doubted the carriers would survive Soviet submarine attacks and considered convoy a much more important ASW measure.[7]

The Mediterranean force would operate continuously in peacetime. It became the main tactical nuclear support for NATO forces in the south, hence a major Soviet target. The threat it faced grew as the

Soviets deployed a Mediterranean fleet (the Fifth Eskadra) from 1967 on.

The threat to the carriers was at least two-dimensional, because the Soviets had a large naval bomber force in addition to their submarines. By the late 1950s that force was armed with antiship missiles. A 1960 U.S. Navy report named Tu-16 "Badgers" armed with AS-1 standoff missiles as the most serious threat to U.S. carriers. The Badgers were probably considered the greater threat because the Soviets did not yet have many nuclear submarines, and their diesel submarines certainly could not keep up with the carriers. The Soviets soon put antiship missiles on board surface ships and submarines, creating a three-dimensional threat.

The NATO view evolved to envisage a two-phase war. The first phase would begin with a massive (but indecisive) exchange of nuclear strikes. In the war's second phase, "broken-backed war," the survivors of the first phase would struggle to stay alive. Control of sea communications would be essential during this phase; without it the survivors of the first phase could not be fed, let alone kept fighting. NATO would fight to maintain sea control during the broken-backed phase. Skeptics argued that there might not even be an initial nuclear phase, and sea control would be essential throughout a war. Some pointed out that SACLANT war plans tended to overemphasize the first phase, with little reference to the broken-backed phase. The Royal Navy tended to emphasize broken-backed war in internal British defense debates. The British objected to war plans that emphasized the initial strikes and neglected what might follow.

The 1954 SACLANT Emergency Defense Plan again stressed the carrier offensive, particularly against submarine bases. By this time overall NATO strategy was changing. Initial U.S. hopes that the European allies would build an army powerful enough to stop a massive Soviet attack were frustrated as European governments felt pressure to cut defense spending when tensions relaxed after Stalin died in 1953. The U.S. government argued that the nuclear deterrent America provided had to be backed by an effective ground force. For a time it appeared that that a smaller (i.e., affordable) army could be an effective shield for Western Europe, tactical nuclear weapons being the associated sword. This idea was gradually abandoned as the Soviets deployed a substantial tactical nuclear force of their own, although it later returned in modified form. At least in theory, strategy for a war in Europe determined key factors for SACLANT such as the likely duration of a war and the degree to which the NATO army needed reinforcement across the Atlantic.

The 1956 SACLANT Emergency Defense Plan no longer separated the two phases of a possible war. It emphasized a "forward defense" that would confine Soviet submarines to areas from which they could not threaten shipping in the Atlantic—and from which they could not emerge to threaten North America with sea-based missiles. SACLANT wanted to create forward barriers that would impede the Soviets' Northern and Baltic fleets.[8] Against that was the general assumption that war would begin with a deliberate, premeditated Soviet attack on NATO. In that case the Soviets would send submarines to sea beforehand, when barrier units would not be allowed to attack them as they transited. The only defense against submarines already at sea would be defended convoys. SACLANT thought it would be necessary to begin convoy operations at the outbreak of war. The convoys would be backed by carrier air attacks at source.[9] Unfortunately, the convoy strategy required large numbers of modern escorts, and those did not exist.

NATO defense planning was complicated by the outbreak of a war for independence in the French colony of Algeria in 1956. France had just lost Indochina, but Algeria, with its large French population (albeit dwarfed by the Arab population, which was behind the revolution), was considered far more important. The French considered Algeria an integral part of metropolitan France. They tried and failed to convince the NATO Council that Algeria came under the alliance pledge that "an attack on one is an attack on all." The French army was the largest in NATO, but to fight in Algeria the French withdrew troops that had been committed to NATO.

The French believed that Gamel Abdel Nasser of Egypt was the Algerians' principal supporter. The British (or at least Prime Minister Anthony Eden) considered Nasser a mortal threat to their position in the Middle East. On that basis they and the French combined with the Israelis (who were threatened by Egyptian-backed Arab commandos) to attack Egypt at Suez in November 1956. The result was a crisis for NATO. The United States backed Nasser in (futile) hopes of maintaining Third World credibility, wielding both the Sixth Fleet and U.S. economic power. NATO was seriously weakened by the conflict. The French began their long slide to leaving the military part of NATO and concluded that they needed a fully independent military force backed by nuclear weapons. That decision shaped the French navy in the 1960s and ended France's interest in mass-producing specialized convoy escorts. The British lost much of their influence in the Middle East as their client governments, most prominently that of Iraq, were forced out by Nasser-backed nationalists. Unlike the French, the British opted to maintain their ties to the United States. Their failure at Suez (to win before the United States could block them) helped justify sweeping defense cuts and a new defense orientation.

The Suez crisis was also Khrushchev's first attempt to use his new nuclear missiles to pressure the West. It is not clear how effective his missile rattling was, but it certainly demonstrated that NATO's ideas about using nuclear weapons to counter Soviet hordes might not be very realistic. Many NATO Europeans concluded that a rough nuclear balance between NATO and the Soviet Union was emerging, hence that a European war might be less and less probable. A new NATO strategy announced in May 1957 supported continued emphasis on wartime shipping protection. A stalemate in the initial phase of the war might be broken by "NATO naval forces, still organized and relatively intact, with mobility, atomic fire power and capable of sustained operations at sea."

As the balance of power in Europe was changing, the U.S. Navy became vitally interested in strategic ASW. In the late 1940s and early 1950s the Navy began to develop submarine-launched missiles, which were soon assigned nuclear warheads. The Navy correctly

Transformation: a Regulus land-attack missile on board USS *Tunny*, 26 August 1954. The Soviets were expected (correctly) to follow suit. The advent of expected Soviet submarine-launched land-attack missiles changed U.S. ASW priorities and caused a rift with the British. NHHC

Expectation fulfilled: the sail of a Soviet Zulu V submarine shows two ballistic missile launching tubes. USNI

assumed that the Soviets were doing the same.[10] In 1954 the British defense attaché in Washington reported with dismay that some in the U.S. Navy were becoming intensely interested in measures to be taken against the expected Soviet missile submarines (which did not yet exist).[11] Any such measures would detract from SACLANT forces available to protect Atlantic convoys. The attaché thought it bizarre that the threat of a few missiles fired at coastal towns could exercise so much leverage. Presumably he did not appreciate how important those coastal towns (such as New York and Washington) were. The British were not impressed by the submarine missile threat because they were already within range of a massive Soviet medium bomber fleet and, increasingly, within range of Soviet land-based missiles.

Moreover, the attaché did not realize that Soviet missile submarines would bypass the increasingly expensive U.S. strategic warning system, which guaranteed, at least in theory, that Strategic Air Command bombers could get into the air ahead of any Soviet bomber attack. Submarine-launched missiles might evade detection altogether, and at the least would offer very little warning time. Throughout the Cold War, Soviet missile submarines remained an important focus of U.S. ASW efforts. In 1967, for example, the U.S. Navy asked the British to cooperate in providing SOSUS facilities on the far side of the Atlantic, arguing that the arrays would make it difficult for Soviet missile submarines to hide there.

RETHINKING NUCLEAR WAR

Given its other costly defense burdens, including the continuing effort to develop an atomic bomb (and means of delivering it), Britain could not afford to mobilize for the Korean War. Maintaining security in the remaining British colonies, which involved the maintenance of a peacetime draft, was proving costly: in 1950 Britain was fighting a hot counterinsurgency war in Malaya, and another was about to break out in Kenya.

On 1 November 1952 the United States exploded the world's first thermonuclear device, which was about a thousand times as powerful as the early atomic bombs. The Soviets exploded their own thermonuclear bomb (of a much less powerful but still devastating type) on 12 August 1953. Broken-backed war was no longer a possibility, and its economic burden could be dismissed. The British Chiefs of Staff now proposed an affordable national strategy based on deterrence rather than nuclear warfighting. First priority should go to measures concerned with the Cold War. Second priority should go to deterring a global (hot) war. Third priority should go to measures designed to fight the hot war should it break out.[12]

The Admiralty, led by First Sea Lord Admiral Mountbatten, soon argued for a more nuanced view. The East-West struggle would move to the much safer Third World. The Admiralty distinguished between wars of different "temperatures": Hot, Warm, and Cold. Hot War was the global fight that seemed increasingly unlikely. Warm War was the increasingly important war on the periphery, in British parlance war "East of Suez." Cold War was the ongoing struggle, in which navies participated by their visible presence. Defence Minister Duncan Sandys accepted this reasoning. The Royal Navy kept the carriers and cruisers it needed to operate in the Warm War environment and show the flag in the Cold War environment.

The Royal Navy had retained a large reserve of war-built escorts to fight a protracted broken-backed war. Maintaining these ships imposed considerable costs. If Hot War were discounted, they were no longer worth keeping. Similarly, plans to build large numbers of low-end ASW escorts, useful for wartime ASW but not for any major Cold War or Warm War duty, were dropped. The emphasis shifted to the needs of a fleet facing a Third World opponent.

Although the United States weathered mobilization much better, when he entered office in 1953 President Dwight D. Eisenhower was determined to limit the burden of largely non-nuclear U.S. mobilization. He proclaimed the "New Look" (a phrase otherwise associated with French fashions) national strategy emphasizing nuclear deterrence. The threat of nuclear escalation would preclude war in Europe (and permit a deep reduction of the expensive U.S. Army). Eisenhower's policy of threatening the Soviets with massive retaliation was popular in Europe because it seemed to guarantee against invasion by the Soviets.

Given nuclear plenty, the Air Force would deter the Soviets from contemplating general war, and conflict would move to the Eurasian periphery (as typified by the ongoing war in French Indochina). The nuclear-armed U.S. Navy would deter such adventures by threatening strikes at "a time and place of our choosing." By 1955 U.S. Navy long-term planners had adopted these ideas. They were less and less interested in building the huge escort force that SACLANT had envisaged.

However, in the late 1950s the U.S. Navy also envisaged what it called a limited-war scenario: a major war that did not escalate to nuclear attacks. That possibility justified continued interest in protecting shipping in the North Atlantic, though not investment in convoy escorts. The U.S. Navy was increasingly interested in attack at source and mining as a way of blunting any Soviet submarine offensive, while recognizing that this type of offensive could not immediately affect submarines the Soviets had deployed into shipping areas prior to the outbreak of war.[13] Significant shipping losses would have to be tolerated until that Soviet submarine force had been destroyed. That would take about three months. It is not clear to what extent the Royal Navy adopted this strategy.

The limited wars in peripheral areas envisaged by the U.S. Navy and the Royal Navy had ASW implications. The Soviets might covertly support a local fight with their submarines,[14] as German and Italian submarines had done during the Spanish Civil War of 1936–39. Soviet client states might use Soviet-supplied submarines. The Soviets transferred Whiskey-class submarines to Egypt beginning in 1957, after Suez, and to Indonesia beginning in 1959 (initially via Poland).

THE BURDEN OF LIMITED WAR

The East-West conflict shifted toward the Third World. After a number of limited interventions in places like Kuwait, in 1964 the British deployed carrier strike groups to intervene in the "confrontation" between Malaysia and Indonesia. The confrontation was fought on a much larger scale than earlier British interventions. Although the British and their allies won, the cost of the operation led the British government to abandon operations East of Suez and ultimately to concentrate on the NATO central European scenario. The Royal Navy faced further cutbacks as the cost of ground and air forces to fight in central Europe escalated.

The U.S. Navy intervened in the Middle East in 1967, 1970 (in Jordan), and 1973. Given the strain of the simultaneous war in Vietnam, in 1973 there were fears that the Sixth Fleet would have been unable to face down the missile firepower of the Soviet Fifth Eskadra in the eastern Mediterranean. The financial burden imposed by the Vietnam War made it impossible to replace the older ships the war wore out—as the Soviet fleet grew. The cost of fighting the war also cut back a wide variety of ASW weapon and sensor programs.

The French navy faced similar problems. The war in Indochina took a toll on the French warships that had survived World War II. France was able to begin to rebuild its fleet during the Indochina War thanks largely to massive U.S. assistance. The decline of that assistance coincided roughly with the outbreak of the war in Algeria, which roughly marked the beginning of the U.S. and British turn away from the initial NATO view of future war. France meanwhile invested heavily in the program to guarantee its own independence by creating independent nuclear forces. To the extent that the French continued to invest in ASW, it was to support their nuclear-oriented carrier striking force rather than escort ships suitable for a future convoy war.

The other focus of French naval policy was the French Union, which replaced the French Empire, mainly in Africa. The French Union mission did justify building a few ships which in wartime might be convoy escorts, and it provided further justification for the emerging carrier-centric French fleet.

THE STRATEGIC SHIFT

By the late 1950s, questions were being raised about massive retaliation. What if the Soviets made a limited attack in Europe and their own growing tactical nuclear arsenal deterred NATO from escalating? In the late 1950s some U.S. Army officers argued publicly that Eisenhower's strategy of massive retaliation was too risky. In 1960 John F. Kennedy adopted their argument. As president, Kennedy ordered an army buildup. He could argue that tensions over Berlin in 1961 made his point. To some extent he seems to have been interested in building an army to take on the Soviets in peripheral "wars of national liberation" (as a senator he had been a strong advocate of South Vietnam during the early years of North Vietnam's attacks). For example, in 1962 the Kennedy administration planned to place a newly formed army division in the Philippines specifically to deal with a possible war in Southeast Asia.

Given his new national strategy, President Kennedy's administration tried to change NATO strategy to emphasize an initial non-nuclear phase of a war. The Europeans were unenthusiastic. They knew what a "robust" conventional defense was: World War II revisited. In the administration's view, NATO already had sufficient forces to fight off anything short of a full-scale Soviet attack, but their preferred stated policy was massive retaliation. They argued that the Soviets would not risk even a small attack if it ended in their complete destruction, as Eisenhower had argued. Too, massive retaliation tied the United States into European defense to an extent that a more graduated policy did not. The French were particularly unhappy with the new idea, which accepted tacitly that in wartime the NATO army would fall back toward the French frontier with Germany. NATO did not formally adopt a new strategy until the French left the military part of the alliance in 1969.[15]

To demonstrate its commitment to an initial non-nuclear phase of a European war, the Kennedy administration created the capability to rapidly reinforce U.S. forces in Europe by air. Equipment and supplies would normally come by sea, but that would be too slow in the event of a surprise Soviet attack. The United States therefore placed prepositioned equipment and stores in special European POMCUS (Prepositioning of Materiel Configured in Unit Sets) warehouses.[16] In annual REFORGER exercises U.S. troops were flown in to activate prepositioned equipment. The tacit assumption was that the initial phase of the war would not last very long, hence that resupply by sea might not be vital. POMCUS was sufficient for a week of non-nuclear combat. The Soviets certainly imagined that any future war would be short, but so did the Germans in both 1914 and in 1939.

The flexible strategy required NATO to create a shield force that could resist a limited Soviet attack, preventing the Soviets from simply marching into Western Europe. The delays these forces could impose on a Soviet advance into Europe would give the alliance time both to mobilize and to convince the Soviets to stop before it was too late. It seems to have taken some time for NATO naval strategy to catch up with the new strategic idea.[17] NATO navies (particularly the Royal Navy) argued that NATO had to demonstrate the ability both to protect Allied shipping and to inflict great enough losses on the Soviets at sea that NATO land and air forces could sustain the defense of central Europe. In 1969, NATO strategists assumed that the non-nuclear phase of a war might last no longer than a week, after which the alliance would be compelled to escalate in order to convince the Soviets to stop.[18] This estimate was still being quoted in the late 1970s.

Much depended on how much warning NATO would enjoy. In the 1950s and the early 1960s, planners assumed that it would take at least a month for the Soviets to assemble a sufficient ground force to attack the West. For example, they had to move most of their heavy armor forward by rail on special flatcars ("war flats"). The possibility of a much more sudden attack became more plausible in 1968, when the Soviets and their allies managed to surprise the Czechs with an attack (in the event, largely bloodless) preparation for which was concealed by a major exercise. In the 1970s it was apparently assumed that it would take the Soviets about a month to prepare a full-scale offensive in Europe, and that NATO would probably become aware of preparations after about a week. However, there was also the possibility of a surprise attack with no more than a few days' warning.

A great deal depended on whether the Soviets surged submarines into the Atlantic as part of the run-up to an attack. A detected surge would be a war warning. Through the 1950s the U.S. Navy developed the ability to detect and track Soviet submarines in transit areas leading to the likely wartime patrol areas in the North Atlantic, mainly through the development of SOSUS. Unfortunately, SOSUS coverage of likely Soviet submarine patrol areas was far less complete, such that SOSUS would likely lose track of surging submarines before they were attackable. Once war broke out, most Soviet submarines would be sunk approaching or leaving their bases, so it would take considerable time to destroy them all. Classic convoy tactics could limit losses at the outset, but probably would not destroy enough Soviet submarines.

The NATO scenario was extensively analyzed by the powerful Office of the Secretary of Defense (OSD) and the armed services, and each viewed it very differently. For the Army and the Air Force, the defense of Western Europe was their central task. The Navy considered a ground war in Europe extremely unlikely and thought the Eurasian periphery was likely to continue to be a realistic focus. The Navy's view was that the fleet had to be shaped to deal with the most likely scenarios, attention being paid to how that fleet would be used in the less likely European case. Through the 1970s, intelligence analysis increasingly showed that the Soviets assumed that their army would rapidly defeat NATO on land and they would thus focus on protecting their homeland against strategic submarines and nuclear-armed carrier battle groups. In that case, the Soviets seemed unlikely to devote much effort to cutting the lifeline between North America and Europe. The question in the late 1970s was how much surplus effort the Soviets might devote to this task. The Navy generally assumed that the Soviets would mainly use diesel-electric attack submarines, which had been designed for that role in the 1950s.

Until the late 1960s the Royal Navy's explicit East of Suez role justified a power projection fleet similar in shape to (though much smaller than) the U.S. Navy's. When that role was formally dropped, the Royal Navy was ordered to concentrate on NATO missions. The protection of sea lines of communication (SLOCs) became the Royal Navy's central mission, the only other major ones being protection of the national deterrent (the ballistic missile submarines) and an amphibious role in support of Norway in the event of war. Like the U.S. Navy, the Royal Navy was well aware that a big European war was far less likely than a wide variety of peripheral wars; the logic of the 1950s was still valid. The question was how it could preserve the ability to carry out vital national tasks in such crises despite pressure to optimize its ability to fight the unlikely NATO war.

For the U.S. Navy, the NATO-only problem became acute when Jimmy Carter became president in 1977. Carter considered a NATO war in Europe the determining factor in U.S. defense; other crises could all be handled using "soft power." Upon becoming president he asked for a study of the likely course of a war in Europe. The resulting report estimated, surprisingly, that if properly supplied, NATO could fight a conventional war for several months. To the Carter administration, the only important U.S. naval mission was thus to safeguard the transatlantic supply route that would keep the NATO armies alive. The portion of the U.S. Navy serving on the Eurasian periphery could be cut to balance the budget while efforts were concentrated on Europe.

SACLANT commander ADM Isaac ("Ike") Kidd wrote in 1977 that resupply would be essential, and that convoys would have to fight through strong Soviet opposition.[19] According to Kidd, 1 million men and 11 million tons of supplies would have to be transported, the latter by sea. Kidd's view differed radically from the Soviets' expressed emphasis on defending against Western naval forces poised to attack their strategic submarines, carriers, and amphibious groups. Attacks on Western shipping would come a distant second, if that. However, in a protracted crisis or war, interdiction of Western SLOCs might become rather important to the Soviets—as would SLOC defense to NATO.[20]

In 1978 a British naval intelligence officer explained the disparity in views. Naturally the Soviets concentrated on defending their homeland, while SACLANT worked on his most difficult problem—protracted reinforcement of a NATO army fighting in central Europe. Neither possibility could be ruled out altogether.[21] Military planners had to take into account a range of possibilities reflecting not only what the enemy might be expected to do (or might believe it could do) but other potential actions as well. If for some reason the initial Soviet offensive failed, Western navies could expect to find themselves fighting a protracted shipping protection war. In the face of nuclear deterrence, the Soviets might never decide to attack Western Europe en masse. However, they might be drawn into a protracted crisis.

President Carter's OSD favored investment in SLOC protection. The U.S. Navy argued that buying ASW assets such as patrol aircraft and frigates was not enough. Offensive forces such as carrier battle groups had an essential sea control role. At the least they were the covering force needed to protect escorts and convoys from Soviet naval threats such as missile-armed bombers. It was usual to characterize carriers as power projection forces and escorts as sea control forces. The Navy increasingly argued that power projection and sea control were parts of a single unified whole. The fleet that was effective on the Eurasian periphery would be effective in a NATO war. It could and should do more than solve the SLOC problem. Offensive naval forces could affect the land battle in Europe by operating on the flanks of an advancing Soviet army. This was much the view Eisenhower had held as SACEUR. The Navy's argument became the basis for the 1980s Maritime Strategy.

The decentralized U.S. decision-making process made it possible for the U.S. Navy to make its case despite the Carter administration's policy. Part of the Navy's argument was that the new microcomputers becoming widely available in the West (but not in the Soviet Union) were revolutionizing its capability to deal with Soviet naval attackers. Ironically, the Carter administration relied on microcomputers to solve the problem of non-nuclear defense against massed Soviet land forces but did not recognize the consequences of the same technology for naval warfare. The Soviets took the West's advantage with the new technology seriously. In 1984 Marshal Nikolai V. Ogarkov, the outgoing chief of the Soviet General Staff, said that the Soviets had to match the new "reconnaissance-strike complexes" if they were to retain the ability to fight a European war.[22]

The late 1970s demonstrated that the United States could not limit itself to the NATO problem, as Iran was transformed from ally to enemy and the Soviets invaded Afghanistan.

The Royal Navy had a much more difficult problem because it faced a more centralized administration that concentrated on a NATO scenario. The Royal Navy argued vainly that its surface fleet could help head off a crisis by visibly demonstrating its determination and capability against Soviet surface ships—something submarines could not do without actually attacking. The surface fleet was also needed to meet remaining British responsibilities outside NATO Europe. Through the late 1970s the Royal Navy argued forcefully for its wartime SLOC protection role, despite considerable evidence that the Soviets did not contemplate early action against NATO SLOCs. As finances tightened, the British government sought economies which that analysis seemed to justify; hence the deep cuts in the Royal Navy surface force ordered by Minister of Defence John Nott in 1981. The following year the Falklands Conflict proved that the Royal Navy was right: a force shaped for combat (but not prewar deterrence) in the big NATO war was not the best for dealing with other situations.

The U.S. Navy was well aware of the Royal Navy's struggle. It can be argued that its own support for the SLOC defense role was shaped in part by the need to back the British so that the Royal Navy would be able to fight alongside it in likely scenarios. British surface warships were likely to make contributions in the smaller peripheral wars that were far more likely to occur than the big NATO war in Europe.

An important difference between British and U.S. naval thinking at this time was that the British emphasized the need to deal with the growing Soviet surface fleet. The U.S. Navy had a large strike carrier force that could be turned to that role; the British had almost completely abandoned their equivalent. The Royal Navy emphasized the antiship role of its nuclear attack submarines, seeking to arm them with underwater-launched antiship missiles (USGW in their parlance) at the expense of some torpedoes. They sponsored development of a submarine-launched version of the U.S. Harpoon, which the U.S. Navy adopted.

Neither navy seems to have been aware of the emphasis the Soviets were placing on the protection of the bastions in which Soviet ballistic missile submarines operated. The bastion idea was certainly known; a Barents Sea bastion figured, for example, in a U.S. 1976 scenario. In that scenario, its protection did not absorb the bulk of Soviet naval resources. It appears that the extent of the Soviets' obsession with such protection gradually became evident from the late 1970s onward, offering the West invaluable naval leverage that dominated the Cold War end-game.

SOVIET THINKING

We have only limited evidence of the Soviets' Cold War naval thinking or planning, so what follows is necessarily somewhat speculative. We know that before World War II, a defense-minded Soviet "young school" had argued against an "old school" that wanted the Soviet Union to build a battle fleet (or fleets) to contend for sea control, at the least in contiguous seas. In the 1930s Stalin chose large ships over submarines—but he also approved a very large submarine-building program, including construction of the Gorkiy submarine yard.

Stalin approved the initial postwar Soviet building program in September 1945. A Soviet account of the crucial meeting gives little sense of what (if any) naval strategy he had in mind.[23] The Soviets had had the world's most numerous submarine fleet in 1939, and Stalin may simply have envisaged replacing it.[24] Much of the prewar fleet consisted of small (M-class) submarines intended mainly for the Far East to counter the Imperial Japanese Navy.[25] Stalin's rebuilt submarine fleet incorporated the new technology the Soviets had captured from the Germans, which promised to revolutionize submarine warfare and make Western ASW fleets largely obsolete.[26] Stalin stated that he wanted first to build a defensive fleet and then a dominant oceanic fleet. Some analysts have suggested that the numerous new W-class medium submarines were primarily defensive, the larger Z-class submarines being the only Soviet submarines considered suitable for an antishipping offensive.

A successful open-ocean submarine campaign required some means of ocean surveillance coupled with communication with the submarines. It is difficult to be sure of the intent of the observed Soviet developments. For example, the Soviets built many large HF/DF arrays (Krugs), but whether they were intended primarily to track Western strike groups or Western convoys (or, more likely, both) could never be established.

It appears that after Stalin died, Marshal Georgi Zhukov, the new defense minister, decided that in a European war the Soviet submarine force would interdict Western shipping.[27] Stalin's successor, Nikita Khrushchev, initially supported Zhukov.[28] Soviet naval chief ADM Nikolai Kuznetzov was excluded from a fall 1955 meeting at Sevastopol that set the future course of Soviet naval development. Khrushchev declared that for the future, submarines and shore-based naval aviation should be the basis of the navy. Zhukov agreed and emphasized attacks against SLOCs. The conference decided that naval nuclear weapons were to be adopted and that the main role of the navy would be to disrupt enemy SLOCs. Having recently witnessed the destruction of the old cruiser *Krasny Kavkaz* by air-launched guided missiles, Khrushchev concluded that surface ships were doomed. Kuznetzov was soon removed from office and replaced by Admiral Gorshkov, who initially agreed with Khrushchev but managed to convince him to allow a few surface ships to be built. In 1957 ADM Vladimir Andreyev explained the wartime role of Soviet submarines as SLOC interdiction.[29] The numbers envisaged were never reached, perhaps because Soviet policy changed radically within a few years.

Khrushchev later wrote that he could not sleep after seeing film of the first Soviet hydrogen-bomb test. Then he realized that the bomb would deter the West; the global war previously considered inevitable might never happen. Like the Americans and the British, he understood that the East-West conflict would shift to the Third World. To make sure that happened, he would emphasize nuclear/missile forces over conventional ones. Khrushchev proclaimed a Revolution in Military Affairs emphasizing nuclear weapons and missiles—and nuclear deterrence. He feared that any direct conflict could escalate uncontrollably. However, he had to deal with China's attempt to take over the world communist

Hunter: a Soviet November (Project 627A) nuclear attack submarine intended to destroy nuclear-armed Western aircraft carriers. She was photographed in the Gulf of Mexico 30 July 1969. NHHC

movement. He supported the North Vietnamese in 1959 only because they shrewdly threatened to support the Chinese in their ongoing fight with the Soviets if he did not.

Given Khrushchev's new emphasis on nuclear weapons and missiles, the large, expensive army in Eastern Europe could be cut back (perhaps also a necessary reaction to the demographic echoes of World War II and the prewar purges). To enforce his changes, Khrushchev suspended the 1956–60 Five-Year Plan and introduced a Seven-Year Plan (1959–65). In its original form, the Seven-Year Plan called for construction of no fewer than 421 nuclear submarines and 132 diesel submarines, figures never actually approached.[30]

Westerners generally attribute Khrushchev's fall in 1964 to the failure of his attempt to place strategic missiles in Cuba, but it seems likelier that he fell afoul of those who had been hurt by his disruption of economic (i.e., military) planning. In the Soviet system, industry was tied to politics. Industry managers gained their positions for political reasons. They enjoyed valuable bonuses based on how much they produced. The drastic change Khrushchev envisioned would inevitably cut production, hence their bonuses. Thus, any attempt to change production goals had political implications. Khrushchev's successor, Leonid Brezhnev, proclaimed the "stability of cadres," which reassured factory managers that they need not worry about their jobs or sudden changes in their production quotas.[31] Several Soviet industrial experts said after the end of the Cold War that the Soviet military industrial base became self-sustaining, and justifications for programs became irrelevant. One Russian naval historian explained Admiral Gorshkov's expansion of the Soviet fleet in just this way, arguing that anyone in his position would have achieved much the same results.[32]

The character of the Soviet economy was little understood in the West. It was relatively sluggish, so that the lag between a decision—to design quieter submarines, for example—and its implementation might take a surprisingly long time. Too, the Soviet system was poorly adapted to maintain or modernize equipment.[33] One of the great surprises of post–Cold War Russia was that vital systems still in service, such as ASW aircraft, had never been updated, as would have been standard practice in the West. The rigidity enforced by industrial planning probably explains how slow the Soviets were to silence their submarines in the 1970s.

Khrushchev's insistence on the primacy of nuclear weapons survived, however, to the point where the Soviets were generally concerned that crises such as the 1973 Middle East War not be allowed to escalate to the nuclear level. Khrushchev's logic set the character of the oceanic part of the Soviet navy. If war came, Western nuclear forces, including strike carriers, would be the primary targets. It took time for the change to be understood in the West. Western intelligence gained an idea of what was happening when COL Oleg Penkovskiy provided copies of articles in the Top Secret version of the Soviet official publication *Voenny Mysl'* (Military Thought). The journal's editors raised vital topics connected with Khrushchev's revolution in turn, inviting articles by highly placed authors before debate was closed. Naval articles (translations of which are now declassified) made it clear that the new priority targets were nuclear-armed strike carriers. One article pointed to the need for good timing, so that missiles bombers fired at the carriers would not obliterate submarines making torpedo attacks. Penkovskiy's arrest in 1963 cut off Westerners' knowledge of the further development of Soviet naval thinking.

The cost of the anticarrier force and the nuclear submarine land-attack force was enormous. Construction of submarines suited to an antishipping campaign declined, but previous programs had already produced large numbers of submarines that seemed well suited to just such a campaign.

After the United States deployed Polaris submarines, the Soviets made efforts to hunt them. That proved both extremely expensive and, apparently, ineffective. The Polaris hunt was also the first role really appreciated by the political leadership. That is why so many Soviet ships and other naval projects were given ASW designations.

Equalizer: a U.S. strategic submarine. From about 1960 on, Soviet naval thinking was increasingly dominated by the need to counter U.S. and other Western strategic submarines while defending their equivalent. It took considerable time for Western naval leaders to accept that the Soviets considered the defense of their strategic submarine force as vital as destroying Western strategic submarines. This is the first U.S. ballistic-missile submarine, USS *George Washington*. The Soviets' problem was further complicated by the emergence of British and French underwater deterrents. USNI

As the Soviets developed their own force of strategic submarines armed with long-range weapons, they came to see them as an essential strategic reserve. This idea was apparently first mooted in public by the creator of the modern Soviet navy, ADM Sergei Gorshkov, in his 1973 book. The protection of that strategic reserve helped justify investment in the Soviet navy; Gorshkov explained that he needed his surface fleet to maintain the "combat stability" of his submarines. Presumably much the same argument applied to Soviet naval air strike forces.

Sensitive intelligence indicating the extent of the resources the Soviets thought would be needed to protect their ballistic missile submarines (nuclear) (SSBN) during a major war was a great surprise for the U.S. Navy in the late 1970s and early 1980s. Presumably the Soviets were aware of U.S. and British success in trailing their SSBNs. They could not expect their SSBNs to survive in the open sea, so the submarines would instead shelter in protected bastions in the White Sea and the Sea of Okhotsk. This in itself was not a surprise; the concentration of SSBNs in the bastions had been observed. What seems to have been shocking was the extent to which the bastions would be protected. It seemed that bastion protection would absorb so much of the Soviet fleet as to preclude other operations, such as attacks on the transatlantic SLOCs. This revelation did not, of course, predict what the Soviets might do in a local war, for example in the Mediterranean; but it did indicate that continuing pressure on the bastions would likely keep the Soviets from using their fleet in any more effective way in the event of a protracted NATO war.

Threats to the bastions included U.S. carrier strike forces. Dealing with them could justify local offensive action out to a considerable distance. Soviet naval doctrine of the 1970s envisaged sufficient sea control to keep Western naval forces far from the Soviet coast. That tallied with their interest in precluding Western attacks, particularly nuclear ones. By the early 1980s the Soviets' goal was to maintain sea control or sea denial out to 2,000 km (about 1,080 nm).[34] Such control would have interdicted SLOCs

The prize: a Soviet Delta I class SSBN (Project 667B) designed to operate from within a defended bastion. The Cold War ASW end-game exploited the Soviets' obsession with protecting the bastions while attacking Western naval nuclear forces.

connecting the United States to Europe and to the Far East. It seemed likely that the Soviets would feel compelled to extend the sea denial boundary out to 3,000 km as the United States put Tomahawk cruise missiles in service. Although the Soviets seem not to have been interested in interdicting the SLOCs per se, exerting sea control and sea denial over the areas of interest would likely have involved them in attacks on SLOC shipping.

2

FINDING SUBMARINES

Sound in the Sea

SOUND SEEMS TO TRAVEL EASILY through water, which is why acoustics dominates ASW. But using sound as the basis of ASW also presents problems that have prompted efforts to develop non-acoustic alternatives. During and after the Cold War, many hoped that devices in aircraft could render the ocean "transparent." Nothing of the sort happened. In the 1970s the U.S. Air Force claimed that the Soviets had devised wide-area non-acoustic devices that would render the survivability of U.S. strategic submarines questionable. Exploration of these claims, mainly by the CIA (because the Navy was likely to be biased), turned up no such devices. Eventually the Air Force admitted that it had spread the rumor mainly to support its own program for a land-mobile MX missile. The only non-acoustic sensor still in widespread use is magnetic anomaly detection (MAD). MAD is effective only at limited ranges, but it has the advantage of greater certainty than acoustics because it detects only masses of metal. MAD was first used during World War II (the United States also deployed a shipboard towed magnetic submarine detector, which was unsuccessful).

During World War I the Allies developed and deployed passive sonars (hydrophones), which had some limited success in detecting submarines but much less in holding contact with them. The only means of detection—that is, of deciding that what an operator heard was a real target—was the operator's ear, so hydrophones operated at low (audible) frequencies. Among other problems, they could be neutralized if the target submarine became silent, for example by lying on the bottom. The solution was to reflect a recognizable pulse of sound (a ping) off the submarine. Initial experiments showed that low-frequency (audible) sonic echoes (at the strength then available) were useless, so by 1918 the French, the British, and the Americans were all working at higher (inaudible) frequencies, often described as supersonic.

The beginning of U.S. sonar: this oil-filled sphere contained the flat-faced transducer (projector) of the QC-1 sonar aboard the destroyer *Monahan*, 4 August 1937. RICK DAVIS

Using high frequencies eliminated much of the background noise, and it also enabled a relatively small device to produce a fairly narrow beam. Echo-ranging devices typically applied a high-frequency electrical signal to a piezoelectric crystal, causing it to vibrate in the water and thus to send out a high-frequency ping. The returning echo hitting the crystal produced a weak signal that could be electronically amplified. Success thus depended not only on choosing the right crystal but also on the sophistication of the vacuum tube amplifier employed to make the returning echo audible.

In October 1918 the Allies held a conference in Paris to discuss their experiments and results. None had an operational device. The French and the Americans were ahead in vacuum tube development, but the British seem to have been ahead overall. At the end of the war, the British and Americans persisted in their efforts, with the British in the lead. The first experimental British set was tested immediately after the 1918 Armistice.

The British development organization was called A/S, so the Royal Navy turned those letters into "asdic" to create a pronounceable cover name.[1] The U.S. Navy initially called its active underwater sound devices "supersonics," but by 1940 it had switched to "sonar" (SOund Navigation And Ranging); the contemporary radio pinging system became "radar." Asdic and sonar enabled a hunter to maintain contact with a fleeing U-boat and to attack and reattack, hence to gain a reasonable chance of sinking the submarine. The prewar Admiralty seems to have greatly overestimated how well that could be done, based on extensive exercises between asdic-equipped ships and evading submarines.

The British, and later the Americans, discovered that attenuation (the loss of signal strength with distance) was reduced at lower frequencies—an important point in postwar sonar/asdic development.[2] Thus in 1929 the Royal Navy reduced its standard asdic frequency range from 21–31 kHz to 14–26 kHz, later settling on 14–22 kHz. Research went no further because the navy was unwilling to increase transducer size to maintain a tight "searchlight" beam (typically about 15 degrees wide) at lower frequencies. Too, at lower frequencies there was more background noise. After numerous trials, late in 1937 the British classified water conditions as good if the maximum range of an asdic set was about 2,000 yds, moderate for 500 yds, and bad if range could not reach 500 yds.

A British Type 144 searchlight transducer is shown in its sonar dome, some of the sides having been removed. It emitted a beam in one direction. Material at the right inside the dome provided some insulation from the noise of the ship's machinery and propellers. AUSTRALIAN WAR MEMORIAL/PAUL MARKS AND DR. JOSEF STRACZEK

Sonar frequency limited the reliable range of wartime sets to about 1,500 yds, though such sonars sometimes reached out considerably farther. The British used quartz transducers. They stockpiled suitable quartz (which had to be imported) for assembly into the large number of simple asdics they expected to install on board trawlers and other extemporized ASW craft in wartime.[3] By the late 1930s the British considered war imminent, so they avoided refinements, which would come too late. The U.S. Navy began with quartz (the QA series) but soon switched to Rochelle salt (the QB series) and the more powerful magnetostriction transducers (the QC series).[4]

Prewar and wartime sets employed a "searchlight" projector to send out a single beam (typically 12–15 degrees wide) that was tilted down in order to detect a submerged submarine. The operator sent out a pulse, listened for an echo, and if nothing was heard he then turned the beam to look in another direction. The beam was too wide to give a precise bearing. The operator would turn the beam back and forth, seeing where the submarine echo vanished on either side. The estimated submarine bearing lay midway between the bearings where the echo vanished.

Although the beam was about 15 degrees wide, it was claimed in 1939 that bearing could be measured with 2 degrees accuracy. Typically, on hearing a ping a wartime operator trained his beam in 5-degree steps until he could no longer hear an echo. Then he stepped back toward the contact in 2.5-degree steps and across it in 5-degree steps. When the extent of the echo area was established, the operator adjusted the step angle to obtain three echoes for each crossing of the target bearing, after which he would classify the contact as "submarine," "possible submarine," or "non-sub." Operators were to hold the contact and find its cut-ons continually to keep the sonar projector trained on any submarine detected.[5]

In early British installations (which were the most advanced at the time), the training control for the asdic transducer was in an asdic office or bridge hut. Oscillator bearings were repeated on the captain's bearing instrument, which was used as a basis for plotting estimated target bearings. This plot enabled the captain to bring the ship onto a collision course with the target as a first approximation to the desired attack course. Work on this problem began in 1938, but early in World War II the British had not yet developed a systematic method to deduce the attacking course.

The chemical range recorder, introduced in 1934, was in effect the first asdic display.[6] The recorder's chemically treated paper (hence the name) advanced at a set rate. An electric stylus passed across the moving paper at a rate proportional to the speed of sound in water. The stylus marked the paper each time the set received a sound, flying back periodically to record sounds received after the next ping. As used by the U.S. Navy, the device had two range scales, for long- and short-range keying (i.e., ping repetition): 3,700 yds and 1,500 yds. Each division on the paper represented 100 yds. A dark mark or a space (depending on whether the loudspeaker was in the same circuit) on the left side of the paper indicated the outgoing ping. Received sound was indicated to the right of this mark; for example, reverberation would appear as a line. An echo appeared as a darker line, its length proportional to the duration of the echo. An operator looking down the paper could see how the range of the echo varied. He had to disentangle the echo of a submarine from extraneous echoes, such as those from wakes and reefs. The recorder would also pick up miscellaneous noises, such as those from propellers, from other ships' sonars, and from some fish.

The slope of a line through the echo marks on the paper gave the rate at which the attacking ship was approaching the submarine. An operator could predict when the ship would pass over the submarine, hence when to release depth charges. At least the U.S. Navy version of the recorder analyzer had a transparent plotter bar that could be lined up with the echo marks. Lines on the bar could be used to find firing time, with a correction for the time that would elapse between the decision to fire and the moment when depth charges or Hedgehog or Mousetrap projectiles struck the water. For depth charge attacks there was an additional correction for the distance between the

The Royal Navy found that enclosing a sonar (asdic) transducer in a water-filled dome reduced flow noise and made it possible to operate at higher speeds. This is the retractable sonar dome of the destroyer escort *Griswold*, 5 May 1944. The dome was retractable so that the ship could be dry-docked. RICK DAVIS

sonar projector and the stern of the ship. In the case of ahead-thrown projectiles, an allowance was made for time of flight. For example, Hedgehog dead time (between the decision to fire and firing) was 2.5 seconds, and time of flight was 7.5 seconds, for a total of 10 seconds between the decision to fire and the moment projectiles struck the water. Sinking time was additional, depending on the target's depth. Projectiles would reach terminal velocity below 100 ft. In the case of Hedgehog, that was 25.2 ft/sec. Hedgehog projectiles would reach 100 ft in 4 seconds, and 600 ft in 23.8 seconds. During the attack, the conning officer would announce estimated target depth and the recorder operator would set it on the machine. Typically, only two settings would be used—normal (to 300 ft) and deep (greater than 300 ft). Time to sink to normal depth was 7.5 seconds, a setting that would normally be correct to 10 yds when the target was above 300 ft. For deeper targets corrections had to be made for slant range.

The British discovered that installing a free-flooding dome around a sonar head reduced flow noise and thus made it possible for an escort to use asdic at higher speeds (ultimately 25 kts). The dome could not be longer than the distance between dry dock blocks (55 in.) or deeper than the blocks (about 54 in.). By 1934 British destroyers equipped with asdic were being fitted with retractable domes. The wartime U.S. Navy adopted this idea. Prior to adopting British sonar domes, the U.S. Navy enclosed its searchlight sonars in spherical rubber domes.

The result of the search for an asdic better adapted to attack led to the introduction of the Type 144 in June 1942.[7] The new asdic used the same searchlight transducer as the prewar sets, but its inboard components were new: a bearing recorder and associated automatic training, and an improved amplifier with automatic volume control. The transducer was stabilized in bearing. Automatic training was intended to simplify the operator's task. The transducer trained automatically in 5-degree steps over an 80-degree arc to each side of the ship. The operator set it going, and he registered an echo by closing a key (the echo push). In an attack, the operator stepped the beam back and forth across the target to determine its bearing (as in the earlier sonars), closing a key each time he detected an echo.

This set introduced a bearing recorder. Unlike the range recorder, it was energized only when the operator closed his key at the bearings to either side of the target. The recorder thus showed two series of dots, against which lines of light could be set. Their slope gave the crossing rate of the submarine. A predictor built into the set gave the appropriate gun bearing (for a Hedgehog) or collision course (for depth charges). Both the range recorder and the bearing recorder transmitted to repeaters on the bridge. The captain's instrument also repeated the gun or collision course bearing. Standard postwar British surface ship asdics were developed versions of Type 144: first the experimental Type 160X in 1948, then the operational Type 164 in 1950.

Submarines capable of diving deeper than 300 ft presented a further problem. The Germans found that by going deep, U-boats could confound escorts. ASW weapons took longer to reach them, giving them more scope for escape. Deeper diving exacerbated an existing problem. Sonar beams tilted down at a moderate angle. As an attacking ship approached, a submarine passed out of the tilted beam, and contact was lost, typically at about a third of the submarine's depth. The deeper the submarine, the worse the problem.[8]

The British devised countermeasures. The Q attachment produced a narrow vertical wedge-shaped beam about 3 degrees wide in the horizontal, mounted beneath the main transducer. It extended coverage down from the usual 10-degree depression to 45 and later 60 degrees. It was intended mainly to maintain contact with a U-boat at close range. Type 147, however, created a depressible beam, broad in bearing (about 60 degrees) but of limited depth (5 degrees), which could be depressed to 45 degrees.[9] It always faced forward, covering 30 degrees to each side of a ship. Type 147 was intended specifically to work with the Squid ASW mortar. Maximum depth was initially 880 ft. In addition to its value in adjusting lead angles and setting depth for depth charging, Type 147B was considered a valuable aid to classifying a target by distinguishing a real submarine from a knuckle in the water or a German bubble decoy (a *Pillenwerfer*).

The British provided 50 Type 147s to the U.S. Navy pending production of a U.S. equivalent (which did not materialize in wartime).[10] The modified destroyer escorts can presumably be recognized by their extended bridge fronts (to accommodate the additional sonar gear in their sound room just below the open bridge). A U.S. scanning depth-determining sonar (QDA) was intended to replace the British Type 147. Like the Type 147, it produced a beam that was broad in bearing but narrow in elevation. A combination of QDA and the scanning QHB, described as an integrated sonar, was the early postwar ideal. It was displaced by the lower-frequency SQS-4, which could determine depth by itself by phasing the two different windings on each of its staves.

When the United States entered the war, it was understood that echo-ranging equipment was needed in quantity, hence that instead of developing new sonars, the effort should concentrate on improving existing ones. American scientists recognized that sonar was valuable mainly for maintaining contact with a submarine, so the wartime Harvard Underwater Sound Lab (HUSL) developed a bearing deviation indicator (BDI), often called a split-beam sonar. The sonar transducer produced two slightly overlapping beams whose echoes were displayed on a cathode ray tube, telling the operator whether the target was moving right or left. Given BDI, it was possible to keep the sonar projector trained on a target (ATT, automatic target training). ATT was developed but not put into production because it required too much monitoring by the sonar operator; BDI was considered good enough.

Another improvement was maintenance of true bearing (MTB), which linked the training motor of the sonar transducer to the ship's gyro compass. The transducer could therefore be held on a desired bearing while a ship maneuvered during an attack. Automatic gain control (AGC) was introduced to compensate for increased transmission loss at greater ranges. The gain required to distinguish a weak echo from a distant target would make local reverberation too loud. Gain control was considered particularly important for plan position indicator (PPI) systems. Doppler-controlled gain (DCG) enhanced an echo showing a Doppler shift, which indicated that the target was moving. The pinging ship was also moving, creating another Doppler shift in an echo. ODN (own-Doppler nullifier) was intended to solve this problem. Incorporation of DCG and ODN was in progress as the war ended. QGA and QGB both incorporated BDI and MTB. They had new magnetostriction transducers, which were more powerful than the crystals.

Although the human ear was considered the most reliable means of detecting an echo (when it was translated into the audible range), a visual indication on a CRT (with a persistent screen) was considered an important aid to detection. U.S. wartime systems employed an A-scan, a visual representation that indicated the range of the target. It was considered an important aid to detection of a weak signal. For tracking and attacking, the British chemical recorder was considered ideal.

U.S. sonar scientists emphasized the need to make sonars operator-friendly; they packaged the usual sonar components into consoles developed by the New London laboratory and the Key West Sound School. In the sonar stacks or racks that were previously standard, range and bearing indicators had been widely separated from the main operating controls. There was no direct means of indicating the direction in which the sonar projector should turn to stay on a target. An operator had to refer to both the direction of the projector in the ship and the ship's heading. It was difficult to understand a tactical situation as a ship turned. Initially the hope was simply to incorporate improved circuits in the standard rack: MTB, BDI, and TGV (time variation of gain), but it was soon evident that a completely redesigned console would provide a seated operator with easier access to controls and would also facilitate maintenance. In the new system (QGB), the console incorporated a chemical range recorder with a second pen to indicate right-to-left target movements (BDI). A single long-persistence CRT could be switched between displaying the echo pattern and right-to-left indication. A wheel at the side of the console trained the searchlight projector.

Despite all the work on sonar aids to attack, the main World War II ASW tactic was to hold down a submarine that had revealed itself by attacking. The submarine might be destroyed by repeated depth-charging or held down to exhaustion of her battery. Asdic made it impossible for the submarine to escape by going quiet, as she could have done when hunted by hydrophone. Submariners learned that to escape they had to create false targets, for example by turning radically to produce a "knuckle" in the water that was about the same size as a submarine. They could also turn to present a smaller sonar target. Wakes, wrecks, and eddies all produced echoes that might confuse an asdic operator. Late in the war the U.S. and German navies produced sonar decoys—initially devices emitting bubbles. The hunter's asdic pings made it possible for a submarine to estimate its pursuer's position. Having created a false target, a submarine might glide away with motors stopped, then withdraw at slow speed.

THE NEXT STEP

In 1943 the Germans began work on a much faster and more agile U-boat, the Type XXI. Existing searchlight sonars were not likely to track the new targets because operators could not turn them fast enough. Too, while the ship attacked a contact, the sonar operator could not search for any others, or even hear the noise of other targets or of torpedoes. It was difficult for the operator to maintain contact on an evasive target. It was difficult to regain a lost contact, because a new search would likely be laborious. The need for a search system that would rapidly scan through 360 degrees and present what it detected on a radar-like PPI had already been recognized. An unexpected benefit was that the sonar could pick up noise, such as torpedo sounds, from all directions more or less simultaneously. Both a target and its wake could be seen at the same time, considerably helping the conning officer making an attack. Even if the operator failed to see a target when it first appeared, he was likely to see it when it was illuminated by a later ping.

The U.S. scanning sonar system, Rotoscope, was developed at the HUSL at New London.[11] The sonar emitted a ping in all directions (omnidirectional transmission, or ODT). The receiver was a cylindrical array of vertical staves of transducers. Directional receiving beams were created in sequence by applying the appropriate phases to a group of adjacent receivers, using a spinning commutator that scanned rapidly through all directions. The relevant phases were obtained by moving a contact over the surface of a small-scale replica of the multistave sonar cylinder. The scanning system had the vital advantage of simplicity, but it paid for that by receiving only a fraction of the energy transmitted in any one direction. Each direction (beam) received only a fraction of the energy reflected by a target, an effect called scanning loss.

For example, a scanning sonar with 10-degree beams had to scan through all 360 degrees (36 beams). At best it would receive only ⅓₆th of a returning ping. In effect the moving receiving beam spent only a short time staring in any one direction. The emitted pulse had to be at least as long as the whole scanning cycle. Any echo picked up by the rotating beam was displayed as a spot on the PPI display. The display was arranged so that spots were farther out as time passed, that is, as they represented echoes from more distant objects.

Range resolution was set by the scanning cycle. For example, a sonar might scan at 30 cycles/sec. It would take a second for a ping to go out to a target 800 yds away and then return to the sonar. Because the sonar would be unable to distinguish ranges within one scan, in this case range resolution would be about 27 yds. For example, a sonar with 36 staves would create 36 successive receiving beams, each of which would receive no more than 1/36th of the energy reflected off a target. The postwar standard QHB scanning sonar operated at about 20–28 kHz, with a source level

The lower the sound frequency, the less water absorbs it. Postwar U.S. sonars operated at progressively lower frequencies, hence at longer wavelengths, and hence needed larger transducers. This is the largest on a surface ship, the dome for the 3.5 kHz sonar on board the destroyer *Kidd*. She is shown at the Litton yard prior to launch, 28 August 1979. NHHC

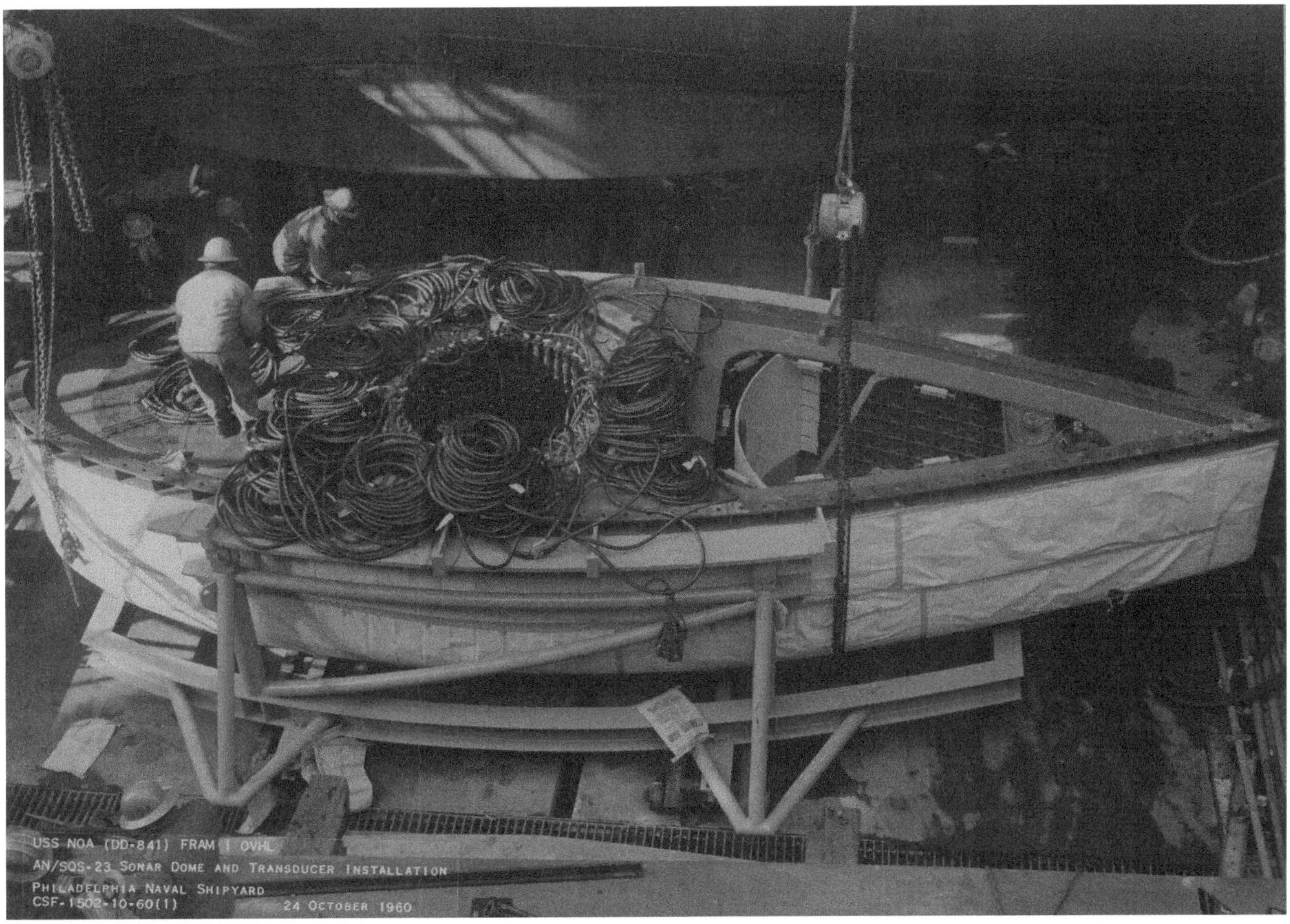

Modern sonar domes incorporate baffling to absorb the sound of a ship's propellers and machinery. This is the dome of an SQS-23 sonar prior to installation on the FRAM I destroyer USS *Noa* (DD 841) at the Philadelphia Naval Shipyard, 24 October 1960. The transducer is already in place; it is the assembly of vertical staves in the circular cavity. Extraneous sound was also reduced by hull damping treatment and by Prairie-Masker silencing using air, both to create a barrier to water transmission of ship machinery and propeller noise, and to reduce propeller noise directly. The British Nightshirt was another propeller air-silencing device. USNI

of 209 dB. It was nearly ready for production at the end of the war, but with the postwar drawdown of R & D, prototypes were not ready for ship trials until 1947. Among early U.S. countermeasures to new kinds of submarines was widespread replacement of searchlight sonars by QHB, which could fit in much the same space; installations in destroyers began in 1948. QHB became the basis for most U.S. Cold War active surface ship sonars.

QHB scanned at 30 cycles/sec, creating a 15 × 15 degree beam keyed for ranges of 1,500 and 3,750 yds. It operated at 25.5 kHz, but in QHBd a field change reduced frequency to 20 kHz. Pulses were 35 msec long so that the beam fully scanned within each pulse. QHBa increased pulse power by 50 percent. QHBc introduced a faster scan (150 cycles/sec).

The standard U.S. postwar hull sonars (SQS-4 and SQS-23 series) were lower-frequency equivalents of QHB. They exploited an additional virtue of scanning sonar that was not suspected when it was designed. A lower frequency (longer wavelength) made possible a much greater sonar range. It would have been difficult to build much-lower-frequency spotlight sonars, because their transducers would have been unwieldy.

The British later claimed that several of their asdic developers had unsuccessfully proposed scanning sonars before the war.[12] Rotation speed was limited by the strength of the scanning element; higher scanning speed could be achieved using electronic rotation. Work began late in 1943 on an electronically scanned Type 172 sonar as part of the British response to fast submarines. It used 72 magnetostriction staves connected through beam-forming networks similar to those the Germans used in their passive sonars. In contrast to QHB, the networks were always in place, but like QHB the sonar sampled beams in sequence (1,000 cycles/sec). The pulse rate corresponded to a range of 2,500 yds. As in the Americans' scanning sonars, the display was a PPI. Each stave held the energy of a received pulse on a rectifier; the rectifiers were scanned at high speed by an electronic switch and the results presented on a PPI. This technique stored the whole length of the pulse received, in contrast to the U.S. system, in which each stave was activated only briefly as the scanner rotated. A late 1948 trial showed that the British sonar maintained contact better than the U.S. QHB due to its rapid scan, which made for a much cleaner presentation on its PPI. However, the British sonar developer said that QHB with a cleaner display would have been adequate.

The British later solved the scanning loss problem by introducing preformed beams. Instead of creating beams as a commutator turned, a preformed-beam sonar listened in all directions at once, its beams formed by delay lines permanently connected to all its transducers. The drawback was that the delay lines were cumbersome, so that a preformed-beam sonar could not produce many beams (this was solved once digital technology was introduced).

Comparison between an improved Type 166 searchlight sonar and QHB (HMS *Helmsdale* vs. USS *Meredith*) convinced the British to develop preformed beams, beginning in 1949. A U.S. sonar scientist independently conceived preformed beams in 1951. Preforming made it possible to build a scanning sonar with an irregularly shaped transducer, such as the flat array of the first British preformed-beam sonar, Type 177. The digital technique was introduced in the 1970s; current sonars are typically described as digital multibeam sonars (DIMUS).

Immediately after the war the British improved their wartime Type 144 as Type 164. A secondary passive set was replaced by the depth-determining Type 147F, which was effective down to 1,500 ft. The forward transducer would find the range and bearing of the target; the after one was for attack using a weapon (Squid and its successor, Limbo). The frigate version was Type 165. With the new 100 in. dome, it was hoped that Type 166 (improved Type 164) would be able to detect a submarine at 25 kts and to support an attack at 28 kts. The searchlight Type 164/165 was acceptable when coupled with a fire control sonar (Type 170) that could lock onto a submarine and track it.

Initial experiments with a tracker used split beams to determine bearing or depth, but in 1943 the British asdic developers proposed a four-part transducer, two side by side for bearing and two one above the other for depth. A U.S. QC transducer was modified to prove this "four-square" concept. Although hardly a scanning sonar, this tracking sonar offered some of the same benefits against a fast submarine crossing the path of an escort. Once locked on, however, it could not detect or track any other submarine. Compared with a scanning sonar, four-square was much simpler and more compact.

An experimental transducer was ready for tests in November 1945. It became the fire-control asdic for the postwar Limbo ASW mortar. This Type 170 was first tested in late February and early March 1947 and demonstrated better accuracy than the current combination of Type 144 and Type 147. Quantity production began in 1952, acceptance trials of a production set following in June 1954. Type 170 became the first British postwar sonar, followed by the preformed-beam Types 177 and 184.

UNDERWATER ACOUSTICS

Early experimenters with asdic and sonar seem to have been nearly completely ignorant of the properties of the sea as a medium.[13] Sound speed varies with temperature, salinity, and pressure, affecting the behavior of the beams on which detection depends. In the 1920s the British began to appreciate that asdic beams in water were subject to refraction (bending due to variation in the speed of sound) and reflection. British experiments in the Mediterranean in 1930 demonstrated the existence of a warm surface layer of lower density than the water below it. Mixing due to wind and wave action maintained a constant temperature in the surface layer, which was called simply "the layer" or "the surface duct." The constant temperature and minor variation in water pressure made for constant sound velocity in the layer. Sound was typically trapped in the layer, in which it could travel considerable distances, greater than those in the layer below. Sound trapped in the surface layer could not penetrate much below it, so that a submarine below the layer usually could not be detected.

In the winter North Atlantic, the layer can be as much as 600 ft deep, which is why World War II ASW was often conducted as though sonar paths were not bent at all. A submarine caught by sonar might escape by diving below the layer, but had to go just above crush depth. Modern submarines can dive considerably deeper.

Below the surface layer ocean temperature decreases rapidly with depth; the depth at which that sudden decrease begins is called the thermocline. In calm waters the sun might warm the surface of the sea, creating a very shallow duct above the usual surface duct. Under some conditions of calm weather and sun (which warmed the surface without mixing surface water with deeper, colder water) surface ship sonars could not penetrate very effectively, hence could not detect most submarines.[14] This was the basis of the "afternoon effect." It also turned out that sound in the layer below the shallow layer could bounce off the bottom of the shallow layer and also off the top of the thermocline. Thus, a submarine might be detectable at some ranges but not at others.

The distribution of the layers is much affected by factors such as north–south flows, as for example, when warm water from the tropics meets denser colder water from the Arctic regions. Major currents, such as the Gulf Stream, further complicate matters. In winter beyond a latitude of about 50 degrees north or south, there is a very deep isothermal region. In contrast, the main thermocline approaches the surface in areas near the equator, and submarines can submerge below this layer.

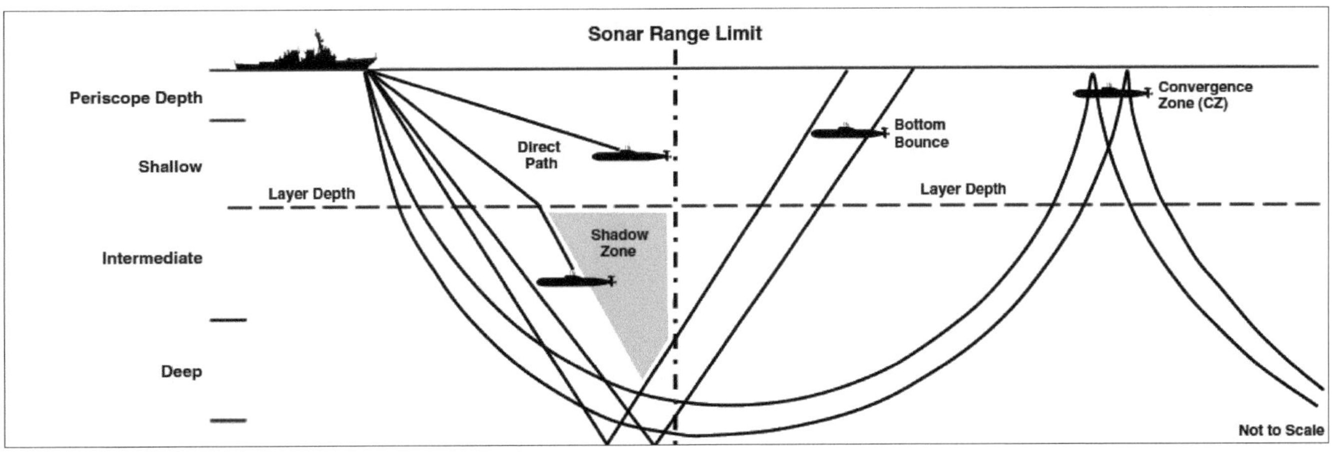

Sonar propagation paths, not shown to scale. The path to the convergence zone bends back up at the deep sound channel (not indicated here except that it is deep). Note the shadow zone (due to refraction, the bending of the sonar beam) below the layer, which drastically reduces direct-path sonar range to a submarine below the layer. The line showing the limit of sonar range is the limit of direct-path range, as a target can be seen in a considerably more distant convergence zone. However, the width of the convergence zone (and of the bottom-bounce zone) makes convergence zone range ambiguous, typically within a 5-nm-wide area. In theory, bottom bounce enables a ship to detect a submarine outside direct-path range but inside convergence zone range. Note that more distant convergence zones occur at multiples of the inner convergence zone: for a zone at 35 nm range, they occur at 70 and 105 nm and beyond. In the Atlantic in winter, the layer is about 600 ft deep, and the deep channel is about 4,000 ft deep. The seabed, which would reflect a bottom-bounce ping, is generally much deeper. ADMIRALTY TRILOGY GROUP

These variations are why graphs of the speed of sound at various depths, which are central to ASW, vary so much with location and time of year. Layers can be formed by fresh water pouring out of major rivers, like the St. Lawrence in Canada and the rivers that feed the Baltic Sea.

Below the thermocline, sound velocity falls slowly because deeper water is cooler. At great depths it begins to rise again due to increasing pressure. In deep enough water there is a depth at which sound velocity is at a minimum, increasing at greater depths. The layer of minimum sound velocity is the "deep channel." Sound bends away from regions of higher velocity, so sound traveling up from the deep channel will bend back down into it, and sound traveling down from it will be refracted back up into it. Thus, sound in the deep channel is effectively trapped. The deep channel also traps a fraction of sound generated above it, because that sound spreads in all directions. Sound trapped in the deep channel spreads cylindrically (in two dimensions) rather than spherically (in three dimensions) like shallower sound. That is, sound intensity declines at a rate proportional to distance rather than to the square of distance—an enormous difference. This phenomenon became extremely important for postwar ASW, though it was discovered as early as 1944. Its first application was to enable downed airmen to signal via what was then called the SOFAR (SOund Fixing And Ranging) channel using practice depth charges. Differences in time of arrival at different receiving stations would locate the airmen. Several SOFAR detection stations were established after World War II. An inverse form of SOFAR, RAFOS, was proposed for postwar Regulus missile submarine navigation, with the explosions at different stations being precisely timed.

Another phenomenon associated with the deep channel is the convergence zone. Sound spreading down from a shallow source, such as a sonar ping, is bent up toward the surface in a circular zone around the source called the convergence zone. This is evident only if the sonar is operating at a low enough frequency that its pings are not absorbed well before they can be bent around. The distance to the convergence zone depends on the depth of the deep channel. In the Atlantic, the first convergence zone is typically 5 nm wide, 30 or 35 nm from the pinging ship. Additional zones are spaced at intervals of about 30 or 35 nm. The pinging sonar cannot distinguish distances within the convergence zone. In the Mediterranean, where the deep channel may be as shallow as 200–400 ft, the convergence zone is only 18–20 nm from a sonar, so a higher-frequency (inherently shorter-range) sonar may be able to exploit this phenomenon.[15] The convergence zone was predicted in 1945 but not exploited until much later.[16]

The convergence zone was first demonstrated by a U.S. submarine balanced silently on a density layer, listening while a surface ship dropped practice depth bombs. The submarine heard them at ranges of 35, 70, and 105 nm. No convergence zone could form if the surface temperature was too high, or in relatively shallow water (100–1,000 fathoms deep). However, in shallow water a sonar beam that struck the bottom only very glancingly could propagate to a considerable distance, comparable with convergence zone performance.

The discovery of the importance of the vertical structure of the water led naturally to the invention of a means of measuring the way temperature varied with depth: the bathythermograph (BT), which became a vital ASW (and counter-ASW) tool during World War II.

Conditions in the Mediterranean, a vital Cold War operating area, were different enough from those in the deep oceans that they demanded different techniques. Although it has deep spots, the Mediterranean is typically much shallower than the Atlantic, nominally 1,500 rather than 2,500 fathoms deep in the major basins. In spring and summer, the surface layer is shallow or nonexistent. Around September surface cooling and mixing creates a deeper surface layer; in winter the water is nearly isothermal (i.e., is one layer). The sea is split between eastern and western basins by Sicily. Each of the two basins is further split by other land masses, with the western comprising the Balearic Basin west of Sardinia and

The layer and the deep channel are relatively close to the surface in the Mediterranean. The French navy became interested in dipping sonars below the layer. This DUBV 43 fish was on board the experimental destroyer *La Galissonière*. It worked with the destroyer's DUBV 23 bow sonar. The fish was typically towed at a depth of 300 m (984 ft) at speeds of 4 to 24 kts; it operated at about 5 kHz, like the hull sonar. DUBV 43 was developed in 1964–67; it could operate either independently or in conjunction with the hull sonar, the object being solid direct-path coverage at ranges between 10 and 15 km to support the Malafon ASW missile. The formal requirement for the VDS was issued in 1963–64. USNI

the Tyrrhenian Basin to the east. The eastern basin is split between the Ionian Basin in the west and the Levantine Basin to the east of Crete, with a subsurface ridge and current between them. The basins are essentially closed, their dimensions making them reverberation chambers at lower frequencies. They amplify shipping noise, so that at low frequencies, noise in the Balearic and Ionian basins is comparable to that just outside New York Harbor. In most seasons the "deep" sound channel is only 400 ft deep, so it can trap surface and near-surface sound. In the summer, the "deep" channel is so shallow (below about 250 ft) that it can be used by hull-mounted variable-depth sonars and by helicopter dipping sonars.[17]

There are further complications for sonar use in the Mediterranean. High temperatures there cause considerable evaporation, which cannot be replenished by runoff and water from major rivers. Cold water from the Atlantic therefore flows into the Mediterranean. Salty (due to evaporation) warm Mediterranean water flows out into the Atlantic, forming a layer there. The salty surface layer in the Mediterranean (due to evaporation) is heavier than the surrounding water, so it sinks, this layer refracting sonar signals.

World War II researchers do not seem to have envisaged yet another sound propagation path. Sound projected downward at too steep an angle to bend up to the convergence zone may bounce off the

The Royal Navy became interested in bottom-bounce propagation because it promised gapless submarine detection rather than the gapped transmission available from the U.S. SQS-26. The destroyer *Matapan*, shown here, was modified as a bottom-bounce testbed in a cooperative program partly financed by the U.S. Navy. In the 1970s bottom bounce was seen as the key "high-quality sonar" of the 1980s, and then of the 1990s. The idea was overtaken by towed arrays, which were far less demanding of the ship. The big array weighed about 100 tons. Its 26 m × 2.4 m (85.5 × 8 ft) planar surface was tilted down at an angle of 26 degrees; it consisted of 2,128 elements. This was smaller than the projected operational array, which would have been 150 ft long. The array could operate in surface duct, bottom-bounce, and convergence zone modes, active or passive. The active band, 2.5–3.5 kHz, was similar to that of SQS-26, but the transducer was much larger. The passive band was either 0.4–2.0 or 2.0–5.0 kHz. There was a single steerable transmitting beam. Maximum source level on the narrowest possible beam (1.2 × 1.3 deg) was 257 dB. Each of two independently steerable receiving beams consisted of 16 sub-beams. Range scales, which indicated expected performance, were 5,000, 10,000, 20,000, 40,000, and 80,000 yds. To provide sufficient transmitting energy, the ship had two big flywheel motor-generators. BAKER COLLECTION

bottom. Whether the bottom-bounce path is actually available depends on factors such as the reflectivity of the bottom, surface wind, and reverberation. The depression angle has to be computed. On the other hand, the distance the sonar signal has to travel does not depend much on target range, since it is mainly the path down and up from the bottom. A submarine might be detected at considerable range using bottom bounce. A related possibility was that a deeply submerged sonar looking up would be able to see, undistorted, within a cone, which might be quite wide near the surface if the sonar was deep enough. In the mid-1960s that nearly vertical path was called the reliable acoustic path (RAP).

NON-ACOUSTIC SENSORS

The operational World War II non-acoustic sensor was MAD. Any large metal body distorts the earth's magnetic field, and that distortion can be detected. However, the MAD technology available during World War II offered only very limited range, about 400 ft. MAD was therefore important only when no other technique was available. The Strait of Gibraltar seems to have been the main case in point; streams of water from the Atlantic and the Mediterranean meet there in such a way that sonar is ineffective. The U.S. Navy placed MAD detectors on board Catalina flying boats that patrolled the Strait. They were armed with backward-propelled "retro-bombs." The backward motion of the bombs canceled the forward motion of the airplane, so that the bomb fell in the position indicated by the MAD detector. The Gibraltar Catalinas were credited with participation in the destruction of three U-boats between 26 February and 5 May 1944, using their MADs for initial detection as well as localization. MAD was also installed on board blimps. It does not seem to have been very effective during the war, but improved versions became important postwar. MAD was often valued as the final proof that a submarine was present. After the Falklands Conflict it appeared that the British had attacked many false targets because their helicopters, their main ASW attackers, lacked MAD. An attempt to use MAD and retro-bombs on board one or more Avengers on board an escort carrier was apparently not successful.

MAD did not feature in the initial postwar HUK team aircraft, but the Grumman S2F-1 (S-2A) had ASQ-8, which was credited with a slant range of 1,000 ft. The redesigned S2F-3 (S-2D/E) had ASQ-10, which seems to have had a similar range. ASQ-81, introduced in 1970, in 1979 reportedly had about triple ASQ-10 range, using the new technology of optically pumped helium. In 1964 a British group visiting Texas Instruments, which was developing ASQ-81, thought that as carried by a helicopter it would have a range of 2,300 ft in straight and level flight.

Experiments showed that MAD was useful to localize a submarine, but not to search for one. In 1947 the Operational Development Force tried suspending the MAD bird from a blimp but tracked a submarine only once. At the time it seemed doubtful that MAD performance could be much improved using the current type of sensor. Development continued after the war, and MAD came to be seen as a means of classifying a contact as a submarine rather than as a means of detecting a submarine in the first place; it confirmed that an airplane was right over a submarine and, for example, should drop a weapon (films of U.S. ASW forces often feature the cry "MADman, MADman" indicating a MAD contact). The Naval Research Laboratory (NRL) developed a magnetometer that could be mounted in a buoy to indicate the position of a submarine. It was involved in a program called LORELI, in connection with the drone antisubmarine helicopter (DASH). LORELI involved vectoring an auxiliary vehicle carrying a sonar transducer to the vicinity of a datum for classification at short range. By the 1990s the U.S. Navy was mapping geologic (magnetic) noise spectra using aircraft, to support a program to deploy fixed electromagnetic barrier sensors. The sensors could be used to cancel magnetic noise and thus to improve MAD performance.

Underwater electric potential (UEP) was another wartime non-acoustic sensor. UEP detected the corrosion current generated between a ship's steel hull and her bronze propeller. UEP was considered, then and much later, to be a mine sensor, and late in World War II it was also considered to be a depth charge proximity sensor.

The difficulty of extending sonar range led to considerable postwar interest in non-acoustic phenomena. The most successful was the "sniffer," an automatic chemical analyzer that could detect submarine exhausts. This hot ionized gas formed a persistent airborne trail that could be detected and followed. Both the British and the Americans pursued this idea, the British proving more successful. The British Autolycus project apparently began in earnest in 1948; by 1953 the British considered the device effective. Air was drawn in continuously and sampled to measure the concentration of condensation nuclei by measuring the intensity of light scattered from a water drop formed by expansion in a chamber. When an airplane passed through the exhaust plume of a submarine, the concentration of nuclei would rise suddenly. Average detection range was 35 nm. The trail detector was said to complement radar and sonobuoys because its performance increased with wind and increasing sea state, the opposite of what happened to radar and buoys. The best search altitude was between 100 and 1,500 ft, depending on weather conditions. The U.S. 1959 Undersea Warfare Conference was told that the sniffer's light weight (under 45 lbs) made it a natural for all ASW aircraft. RAF Coastal Command used Autolycus Mk 3 in aircraft, including Nimrods; 1969 specifications for Nimrod included Autolycus Mk 3. The U.S.

Navy initially adopted Autolycus as AN/ASH-2, but the version widely deployed in the 1960s was AN/ASR-3. Other NATO countries also used Autolycus. Interest declined in the 1970s, and it was not carried, for example, by the U.S. carrier-based S-3A Viking, although it had featured in models of the previous S2F (S-2) Tracker from the S2F-3 (S-2C) onward. Presumably the problem was the increasing use of diesels by merchant ships. Sniffers also suffered from atmospheric contamination associated with coastal industrial areas.

The other near-term non-acoustic possibility was thermal wake detection. Because water temperature varies with depth, a submerged submarine should churn up a wake that a thermal detector can distinguish. Although the idea keeps recurring, thermal wake detection seems never to have been successful. For example, according to a post–Cold War account of the Be-6 maritime patrol aircraft by a former crewmember, this airplane carried an infrared (IR) submarine detector that should have been effective (the article cited the reasoning involved, and the figures used). The writer commented drily that although the device was entirely unable to find submarines, it did effectively distinguish land from sea. But there were easier ways to do that.

In connection with the SSBN Security Program established in 1968 (to make sure that SSBNs remained difficult to detect), the United States invested in testing schemes for wake detection, some of which the Soviets were already thought to be using. A submarine at depth churns deeper water to the surface, where its temperature differs from that of surface water. That creates a thermal scar or wake. Previously observed submarine IR scars were created by the vortex wake of the submarine sail. Submarine-induced internal waves did not produce detectable IR signatures. Even so, in the 1990s U.S. submarine R & D programs showed considerable interest in reducing submarine wakes (such as turbulent wakes) through better maneuvering control.

The SSBN Security Program conducted the first AC and DC signature measurements of a submerged U.S. submarine. It also made the first in-air measurements of a submarine ELFE (UEP) signature and conducted the first active magnetohydrodynamic experiment (presumably measurement of the magnetic field due to turbulence in the water, which is a conducting medium in the earth's or other magnetic field).

Yet another possibility was LIDAR (laser radar) employing blue-green lasers whose beams could penetrate water (a blue-green laser was later used for mine detection). The surface of the water scatters back much of the energy in a blue-green pulse, but it is possible to cancel out the scattered radiation by range-gating. About 1962 it was discovered that blue-green lasers could indeed penetrate sea water to a considerable depth. A primitive laser radar figured in advanced U.S. ASW development in the early 1960s; and in the late 1960s such a device, UDACS, manufactured by Kollsman, was publicly described. By that time it was fairly clear that, although the beams were good enough for one-way communication down to a submarine (albeit not down to great depths), there was not enough of a backscattered signal for useful detection or tracking. Even so, lasers remain attractive in really shallow water. The SSBN Security Program promoted means of enabling submarines to seek areas in which lasers would be less effective.

The Soviets experimented extensively with non-acoustic submarine detectors, apparently with no great success. Many of their nuclear attack submarines had a variety of sensors (mainly protrusions from their sails) that had to be non-acoustic. They were apparently part of the *Sistema Obnaruzheniya Kil'vaternogo Sleda* (SOKS) wake trace detection system. A credible Russian drawing of the sail of a Victor I (*K 147*) shows both the protrusions and a retractable boxlike structure rising from the top of the sail. The box may have been standard on Victor II–class (Project 671RT) submarines. It was apparently on board at least some Victor IIIs (Project 671RTM). In an Akula (Project 971) it is apparently on deck forward of the sail. There are also protrusions from the sail. In a Sierra (Project 945) these devices protrude from the starboard side of the sail about halfway down, and again the protrusions are evident on the fore side of the sail. The box structure is also evident atop the sail of the Beluga-class (Project 1710) experimental submarine. A model displayed in 2003 at St. Petersburg, apparently representing an Akula, showed a triplet of uncovered needles on deck and others under the hull. Apparently, this sensor system is not present on missile submarines (both strategic and cruise).

Presumably the SOKS system is designed to help the submarine track a submerged target by remaining within its wake. The sensors involved may be turbulence or temperature or chemical wake detectors. In 2000 it was reported that the sensors in front of the sail are nucleotide detectors. The Russian drawings seem to show similar needlelike objects protruding from the sail and contained within (actually protruding from) the retractable box. The open box is said to be an optical turbulence detector (based on changes in the conductance of light through water passing through it). A likely interpretation would be that the box and the needles offer independent indications that the submarine is in a wake, the two sets of needles confirming that the submarine is pointing along the path of concentration of whatever they are sampling. A major issue would be how long that wake would persist. The standard Soviet surface ship suite of the late 1960s included an MI-110R infrared wake detector (there was no NATO codename).

At the end of the Cold War the U.S. Navy and the Royal Navy gained access to much of the Soviet Union's non-acoustic research. Both navies subsequently deployed "environmental" sensors on board some submarines. They may have been wake detectors using Soviet-inspired technology, or they may have been intended to help guide submarine commanders into waters where wake detection would be more difficult. These devices have been less visible in recent years, perhaps reflecting experience showing that they were not particularly effective. Widespread Soviet deployment would not have proved otherwise; the Soviets sometimes deployed sensors that proved altogether useless.

3
LEGACY

THROUGH MOST OF THE COLD WAR, Western strategists assumed that a war against the Soviet navy would include a submarine campaign like the World War II Battle of the Atlantic, but with updated resources on both sides. The single most important fact of the battle was that, until well into 1944, submarines were surface ships that could dive to conceal themselves if they had to. Their sustained underwater speed was slow, so they tried to avoid submerging. Aircraft were thus vital ASW weapons. Initially U-boats could operate freely on the surface at night, but the introduction of aircraft radar changed that. The Germans restored their ability to operate freely in the open ocean in 1944 when they introduced snorkels that enabled submarines to run their engines while submerged. Existing airborne radars could not detect snorkel heads beyond a very limited distance. The expected more or less complete loss of air ASW became the most important fact of early postwar Western ASW. In the event, snorkels did not prove decisive, because by 1944 other limitations on U-boat operation overshadowed the virtual end of airborne detection.

GERMAN SUBMARINE STRATEGY

German submarine tactics were developed before the war by CAPT (later ADM) Karl Doenitz. During

Defender: the Flower-class corvette HMS *Lotus* in 1942. Her size limited both range and crew comfort. Before the war it had been assumed that a combination of limited ocean surveillance and distance from bases to patrol areas would keep U-boats in the waters around the British Isles. *Lotus* has a minesweeping wire reel (and davits) aft, reflecting British concerns with the other major German antishipping weapon, the mine. She also shows an initial adaptation to North Atlantic conditions: a built-up forecastle. The device visible on her bow supported a noisemaker, a defense against German acoustic mines. The lantern-like structure above her bridge houses the antenna of a Type 271 surface search radar, a defense against U-boats attacking on the surface at night. Not visible are her depth charge throwers (her stern depth charge tracks are visible). NHHC

World War I, the U-boat Doenitz commanded had been sunk during a 1918 attack on a convoy in the Mediterranean. He later wrote that Allied convoy strategy had been effective for three reasons. First, it concentrated ships in relatively small areas, making it far more difficult for U-boats to find targets. Second, a submarine could attack only a few targets in the brief period during which a convoy swept by. Most of the convoyed ships would therefore escape. Third, the convoy concentrated ASW ships around the targets, making it much more likely that an attacking submarine would be sunk. This last was true even though existing ASW sensors were ineffective, and escorts carried few depth charges. Relatively few U-boats were sunk in World War I convoy battles. Doenitz used his experience to formulate his new ideas before the Germans had built many U-boats; he assumed that in wartime he would have a large submarine fleet to prosecute the tonnage war he had in mind.

Doenitz envisaged a central U-boat commander, *Befehlshaber der Unterseebooten* (BdU), who would direct groups of U-boats (wolf packs) to attack convoys based on the plot of enemy shipping activity he maintained. His intelligence information would probably be imprecise, so BdU would form his boats into patrol lines across expected convoy tracks. Once the convoy was spotted, the entire patrol line could concentrate against it, the U-boats exploiting their high surface speed to approach beyond the convoy's lines of sight. Multiple U-boats attacking together would overcome the numbers problem.

In wartime, a typical U-boat patrol line would consist of up to 25 U-boats spread about 20 miles apart in a line abreast, athwart the estimated convoy track. Each U-boat would patrol at slow speed. When one spotted a convoy, she radioed BdU; she did not communicate directly with the other U-boats. BdU in turn vectored U-boats to concentrate for a wolf pack attack. The attack itself was not coordinated, although the U-boats might keep in touch with each other by radio. U-boats attacked at their own discretion, but BdU decided when to pull them all back. Doenitz envisaged night attacks, when the U-boats would be effectively invisible to the escorts. Surfaced U-boats were much faster than convoys, so they could attack, turn away to reload, and reattack. Operating on the surface, each U-boat commander could easily understand the tactical situation. For example, he could gauge the convoy's speed and course simply by keeping station on it. A submerged U-boat lost the advantage of speed and typically got only one chance to fire. To provide enough U-boats, Doenitz concentrated construction on a relatively small type. He issued orders to sink as many merchant ships as possible. U-boats were to try to sink a merchant ship with each torpedo by firing from close range, at most about 600 yds. The U-boat's radio link back to BdU (in wartime typically Doenitz or a deputy) made it possible for him to monitor which submarines still had enough fuel and enough torpedoes to execute an attack.

Doenitz needed some means of ocean surveillance to cue his submarines to attack, particularly in the open ocean. The two alternatives were air reconnaissance and radio intercepts (codebreaking). Doenitz relied mainly on intercepts, partly because the Luftwaffe, which came to control all German aircraft, usually brushed off the naval support mission (and had not trained its pilots to navigate over the featureless sea). In any case, the Germans never had aircraft with the range required to reach the mid-Atlantic.

Doenitz relied mainly the German naval codebreaking organization (*B-dienst*) and suffered when its access to British codes was blacked out. German and Italian naval codebreakers benefited from the extensive ship-to-ship communication within the British fleet during the Mediterranean crisis of 1936, when Britain and Italy nearly fought. The Germans were quite successful in 1939–40. That did not much affect the U-boat war, because Doenitz had too few U-boats to exploit what his naval codebreakers provided. German codebreaking did lead to early successes such as the sinking of the carrier HMS *Glorious* in the North Sea.

When German codebreaking failed late in 1940, Doenitz tried to fall back on air reconnaissance. At his behest, in January 1941 Hitler ordered a Luftwaffe

Midocean defender: the Canadian River-class frigate HMCS *Hyulloise* postwar. She and other River-class frigates were initially classed as "twin-screw corvettes." Their important improvements over the corvettes were higher speed, much better seakeeping, and far greater range, enabling them to cross the Atlantic with a convoy. At her masthead is the antenna of a U.S. SU surface-search radar, replacing the wartime Type 271 in its "lantern." Aft is a stub mast carrying an HF/DF antenna, which enabled her to exploit the German U-boat command and control system. Her Hedgehog is forward of her forward superstructure, which carries a twin 4 in. mount rather than the single mounting of wartime British Rivers. Aft she shows her depth charges: four side-throwers and two depth charge rails. USNI

unit, KG 40 (of long-range FW 200s), placed under Doenitz' operational control. Two new commands were set up in March 1941: Fliegerführer Atlantik and Fliegerführer Nord, of which the Atlantic command was far more important for U-boats. Some of its long-rang aircraft operated out of Norwegian bases, and the Norwegian effort was strengthened with a squadron (*Staffel*) of medium bombers (He 111s). However, aircraft–U-boat cooperation was consistently unsuccessful, as indicated in BdU's war diary, due to a combination of poor navigation and an inability to continuously shadow convoys found at the limits of aircraft range. I/KG 40 also lost an increasing number of aircraft in its other role as a maritime attack force. The aircraft were withdrawn to the Gibraltar–UK convoy route to home U-boats onto convoys there. In theory, on finding a convoy the airplane signaled BdU, who signaled U-boats to prepare to cooperate with it. The FW 200 had a beacon that transmitted not only its own position but also the range and bearing to the target convoy. U-boats could take bearings on the beacon's signals, enabling them to locate the convoy. The BdU organization would take cross-bearings on the U-boats, enabling it to locate the convoy. The airplane would transmit convoy course and speed. That in turn would enable BdU to deploy a scouting line across the convoy's path. These U-boats would not have to reveal themselves before attacking. This complex technique worked only occasionally, the examples cited in the communications intelligence headquarters (GCHQ) history being OG 60 and OG 71. Air shadowing could not really be continuous, and spotting was insufficient because convoys could change course between sightings. Too, British communications intelligence (COMINT) could often detect the formation of the lines of U-boats intended to intercept a convoy. This was aside from fighters sailing with a convoy on board catapult ships and escort carriers, which could destroy the shadowing FW 200.[1] To some extent the Germans tried to exploit British aircraft reports, but they found it difficult to coordinate friendly data with enemy data. Among other problems, German air navigation was often poor, and aircrews were unable to tell whether they spotted the same convoy on different sorties. Even so, these "snoopers" could sometimes lead packs of U-boats to effective attacks, and the question of how to deal with such aircraft became important after the war.

Initially the British paid insufficient attention to the security of their naval codes. Their World War I codebreaking success against the Germans had been widely publicized, and they therefore considered it unlikely that any major navy would risk using radio extensively in future. In fact, all navies, including the Royal Navy, used radio (hence codes and ciphers) extensively during the war, and codebreaking became a major source of intelligence amounting to a means of ocean surveillance. Doenitz' heavy reliance on codebreaking made the struggle for and against his form of ocean surveillance very much one of success and failure in coding and codebreaking. The struggle highlights the role of ocean surveillance in oceanic submarine warfare. It is not clear to what extent later Soviet successes, based heavily on assistance from the Walker spy ring, provided something like the codebreaking success Doenitz enjoyed, albeit intermittently, until mid-1943. At the least, the codebreaking struggle is a useful proxy for later attempts at ocean surveillance, highlighting their importance to both sides. After the war the British authors of a study of Germany's use of codebreaking pointed out that even noncurrent decodes were valuable to the Germans, because they revealed the convoy cycle and typical routes: "So vast a freighting undertaking could hardly be carried on in complete secrecy or without recourse to a certain stereotyped pattern."

Doenitz was well aware that communications to and from his U-boats could be intercepted. But he believed that the machine codes (Ultra) his submarines were using were unbreakable, and on several occasions he refused to accept evidence that they were not. Doenitz was aware of high-frequency direction-finding (HF/DF), but he was convinced that this technique would be limited to land sites, and hence would be too approximate to enable attacks on U-boats transmitting messages. He also hoped that limiting the duration of transmissions would offer security. Initially that was achieved by using compressed codes in which a few letters sufficed. By the end of the war the Germans were experimenting with transmitters (*Kurier*) that compressed a message into a short burst. The Allies did not yet have a shipboard intercept receiver for *Kurier*, but some of their land sites intercepted a few *Kurier* messages. Despite urgent efforts, in 1945 the Allies could not establish the direction of *Kurier* messages, either from ashore or from the sea. Solving the *Kurier* problem became an urgent postwar project.

Doenitz' concept of communication-based submarine warfare was almost the only choice available. The U.S. submarine offensive in the Pacific was similarly intelligence-based, and it was largely centrally directed, although the operation differed considerably in detail from German methods.[2] Similarly, British submarines in the Mediterranean relied heavily on codebreaking in their war against Italian shipping supplying Rommel in North Africa.

Doenitz and his U-boat commanders were aware that the British had a secret U-boat detector they called asdic (sonar), but no one knew what it was. Doenitz told his commanders that whatever it was, it would not work against a surfaced U-boat. Having had little success with active sonar themselves, as late as 1944 the Germans seem to have believed that the British were relying on hydrophones like theirs.[3]

The British seem not to have known about Doenitz' new concept of U-boat operation, although they did obtain good information about U-boat numbers and characteristics.[4] They assumed that U-boats would operate much as in 1918, the main difference being their own possession of asdic. In 1918, with Allied shipping crossing the open seas in well-protected convoys, the Germans operated in coastal areas, such as the approaches to ports, where shipping had to concentrate. Doenitz' analysis was not reflected in the German official history of the World War I U-boat war, which the British translated as its volumes were published. The last prewar volume mentioned some attempts to mount multi-boat attacks, but only on a very small scale and without resort to any sort of reconnaissance. German naval writers who did discuss the new ideas were not considered authoritative, and the Royal Navy seems not to have taken their work seriously. To the British, 1918 tactics meant that U-boats would attack singly, and the escorts would drive off the attacker. There were not enough escorts to pursue an attacker to destruction, however. Pursuit to destruction became an essential wartime Allied tactic.

Before the war the British built some long-range escorts (sloops), but when the crisis came in 1939 their emergency plan concentrated on escorts that would operate near the British Isles: trawlers and corvettes and inshore ASW craft (Fairmiles). It was impossible to build anything like enough ocean escorts. The British thought that the U-boat threat would be confined

Winning the war of numbers required mass-produced escorts. The U.S. Navy ordered a thousand destroyer escorts to ensure that enough would soon be available. With the crisis of the Battle of the Atlantic ended by late 1943, many of them were canceled, and others were completed as fast transports (APDs). Still others already completed as destroyer escorts were converted to fast transports. Ultimately 565 were completed as DEs and APDs. Industrial effort allotted to the destroyer escort program was redirected to amphibious ships. *Brackett* (DE 41), shown here off Puget Sound Navy Yard on 1 November 1943, was one of the original group designed to Admiralty requirements (originally for Lend-Lease). She therefore had a British-style open bridge, with a sonar hut in the windowless structure below it, the wheelhouse being the windowed enclosure below. The project seems to have been inspired by the British "Hunt"-class escort destroyers, with much greater endurance (6,000 nm at 12 kts) but reduced speed (23 to 24 kts). The all-antiaircraft armament did not match that of the British ships, but these ships had a much heavier ASW battery, including a Hedgehog abaft No. 1 gun. Once ships were being built, the United States was mobilizing, and there was insufficient gear-cutting capacity for the steam turbines envisaged. Instead, many destroyer escorts, like this one, were given diesel engines with half the designed horsepower (in this case, GM diesels driving generators, as in *Barnegat*-class seaplane tenders). Later destroyer escorts had longer hulls (306 rather than 289 ft long). Many were driven by steam turbines driving generators, their greater length (which reduced resistance) making up for their greater-than-designed displacement. The British received 32 short-hull and 46 long-hull destroyer escorts as their "Captain" class. Two short-hull destroyer escorts were transferred to China in 1945. All other surviving short-hull destroyer escorts were discarded at the end of the war. NHHC

USS *Whitehurst* was a turboelectric (TE) destroyer escort with much of the originally designed performance. Her funnel shows her characteristic pair of trunked uptakes, leading to two boiler rooms separated by an engine room. Note her triple torpedo tubes and her relatively light antiaircraft battery, including an obsolete quadruple 1.1 in. gun superfiring over her after 3 in. gun. She has the standard destroyer escort outfit of four throwers on each side and two tracks aft, as well as a Hedgehog firing over No. 1 3 in. gun. In the British-style bridge, the officers conning the ship can look down into the front of the sonar hut. This photograph is undated but was probably taken not long after the ship was completed, 19 November 1943. Further turbine ships were completed to modified designs incorporating 5 in. guns and a very different bridge structure. NHHC

to the eastern Atlantic because, lacking bases on the French and Norwegian coasts, any U-boat headed for the Atlantic had to expend much of its endurance en route from Germany around Scotland to the Atlantic (the shorter route through the Channel would doubtless be blocked).

The Royal Navy also had to reckon with other threats to vital shipping. The Spanish Civil War demonstrated the potential of air attacks on merchant ships. Sloops were initially conceived as antisubmarine ships, but those built from 1936 on had powerful antiaircraft batteries. The small Hunt-class destroyers (described as fast escorts; i.e., fast sloops) were given heavy antiaircraft batteries. During the run-up to World War II the Royal Navy began converting old V & W–class destroyers into air defense escorts.

Too, the British assumed that beyond coastal waters the main threat to trade would be surface raiders, because surface ships were far better adapted to finding targets in the open ocean. Thus, at the outset their convoys were protected by cruisers—often merchant ships converted into auxiliary cruisers. The few ASW ships available would be concentrated near the British Isles, where U-boats were expected to operate, leaving convoys largely unescorted for much of their journey. Convoys would pick up ASW escorts when they approached the waters around the British Isles (at the Rockall Bank)

Finally, there were mines. The British had faced a massive mine threat during World War I, and their interwar programs included adding numerous large minesweepers, which could also function as ASW escorts. They planned to mobilize trawlers as sweepers. In the 1930s the Royal Navy began building smaller sweepers to reduce pressure on builders who might instead build escorts.

ORGANIZATION FOR ASW DEVELOPMENT

During the interwar period the Royal Navy's Admiralty Research Laboratory (ARL) developed asdic (sonar) and electronic devices. The principal asdic research establishment was HMS *Osprey* at Portland. The U.S. Naval Research Laboratory (NRL) developed sonars independently of the British. The civilian Woods Hole Oceanographic Institution was important in that regard both during and after the war. In 1940 the British allowed U.S. naval representatives nearly free rein, and the U.S. Navy adopted two important British asdic (sonar) innovations: the sonar dome, which made it possible to use sonar at significantly higher speeds; and the chemical recorder. Both countries had important civilian government laboratories—the National Physical Laboratory (NPL) in Britain, and the Bureau of Standards in the United States. The United States also had the leading civilian R & D organization in the world: Bell Telephone Laboratories.

Both Allied countries created additional wartime organizations with important ASW applications. The Admiralty formed the Department of Miscellaneous Weapon Development (DMWD), which invented the wartime Hedgehog ASW weapon. The DMWD was allowed to operate more freely than other Admiralty offices, but it lacked links with the naval staff and other naval organizations, leaving its creations without the necessary postinvention support. This was particularly true of Hedgehog. The new Admiralty Operational Research group headed by P. M. S. Blackett was concerned mainly with tactics but also suggested some new devices, such as the sonobuoy.

The United States created the National Defense Research Committee (NDRC) to exploit university science departments and other civilian scientists; it was later subsumed into the larger Office of Scientific Research and Development (OSRD). Its naval arm operated despite resistance from NRL. A special subcommittee on submarine detection was created in October 1940. In January 1941 the subcommittee gave a critical report on U.S. Navy progress in that field, noting that training was inadequate and too little had been done to understand sound transmission in the sea. As a result, NDRC was assigned to work on ASW, a new C-4 (later NDRC Division 6) section being created for that purpose. NDRC created new sound laboratories. The Columbia University Division of War Research operated the U.S. Navy Underwater Sound Laboratory at New London, Harvard operated its own Harvard Undersea Sound Laboratory (HUSL), and the University of California Division of War Research took over the U.S. Navy Radio and Sound Laboratory at Point Loma, San Diego. At the end of the war New London became the U.S. Navy Underwater Sound Laboratory, and Point Loma became the Naval Electronics Laboratory. Torpedo work was concentrated in the new Ordnance Research Laboratory at Pennsylvania State College.

The wartime British government was impressed by the extent to which the United States had mobilized its scientific talent.[5] The British were also impressed with the way the Americans designed equipment with a view to putting it into production as quickly as possible. Immediately after the war, an American officer said that the British were very good at conceiving equipment, but the Americans beat them into production every time. The sheer size of the U.S. industrial base made it possible to tolerate duplication or triplication of effort. The Admiralty was surprised that the Americans often deliberately funded competing prototypes.

In 1943 the U.S. Navy created the Tenth Fleet specifically to run its ASW war. Related organizations were the ASW Operations Research Group (ASWORG) and the ASW Development Group (ASDevLant) in the Atlantic Fleet. After the war, ASWORG evolved into the Operational Evaluation Group (OEG), which was intended to test devices and tactics. The OEG merged with an AAW (anti-kamikaze) organization to form the Operational Development Force, an important factor in postwar ASW.

SURFACE ASW ATTACK TACTICS

In 1939 the only effective weapon for attacking a submerged U-boat was the depth charge, a cannister of explosives preset to explode at a particular depth (chosen from among several alternatives). There was no hope of aiming a depth charge precisely. Instead, a ship laid a pattern of charges around the estimated position of the U-boat, hoping that one would explode close enough to the U-boat to damage or sink it. The

The stern of USS *Seid* (DE 256) at Bethlehem Steel's Alameda yard on 22 May 1944 shows her heavy depth charge battery: eight side-throwers (K-guns) and two stern tracks. The battery was designed to lay down two patterns of depth charges that would sandwich a submarine. The patterns would extend 60 to 150 ft to each side of the ship, the stern tracks laying down charges along the centerline. None of the charges could be thrown ahead, so a successful attack required that the stern of the ship be maneuvered more or less directly over the submarine, a difficult feat. Nevertheless, depth charges accounted for most of the U-boats sunk by surface ships during World War II. According to a 1947 report, of 216 U-boats sunk by British surface ships, 155 were destroyed by depth charges, 45 by Hedgehogs, and 16 by Squids. NHHC

This 1 April 1946 exhibit showed both conventional cylindrical depth charges and streamlined fast-sinkers (at far left and far right), which could be given influence fuzes. Not visible here were the vanes on the tail of the fast-sinker, which were intended to spin it so that it sank at a more perpendicular angle, hence closer to the intended position despite increased depth. The object on the far right is a Mousetrap rocket-thrown charge, similar to a Hedgehog charge but suited to ships and craft that could not withstand the deck reaction of a Hedgehog. NHHC

sheer weight of a pattern of depth charges limited the number of patterns a ship could carry. Charges were rolled off a ship's stern or fired off to the side. Typically, charge depths were set in hopes of sandwiching the target submarine between two layers. Ships were fitted for salvoes as large as 14 or 19 charges, although by the end of the war somewhat smaller salvoes were standard. Further, the depth-charging ship had to be moving fast enough to escape the shock of these explosions. Depth charge explosions created enough noise and confusion to require a ship to search to regain contact with a submarine. Through the war it was discovered that U-boats were tougher than expected, so more charges had to be used.

A ship attacking with depth charges had to pass directly over the U-boat, by which time sonar contact had been lost; hence the importance of the chemical range recorder. Typically, an escort accelerated to close the gap to the U-boat as quickly as possible. The submarine commander could hear that acceleration and try to maneuver away. The escort tried to maneuver her stern (from which the charges were dropped) over the invisible target. During the war, the British developed a "creeping barrage" team tactic in which one ship tracked the target and coached the other in to drop the depth charges. Since the target was always being tracked, the attacker could run in at low speed, denying the target warning. The creeping barrage was intended particularly to deal with deep U-boats.

The prewar Royal Navy was aware of the limitations of depth-charging, but its attempt to develop a depth charge gun failed. Presumably the asdic beam was too broad to be used for aiming. The gun could

The destroyers *Epperson* (DDE 719) and *Sarsfield* (DD 837) drop depth charges during a 15 June 1959 exercise. The depth charges had to be set to go off long enough after the ships passed not to damage them. *Sarsfield* has also used her K-guns; the splashes of the depth charges are visible (they have not yet exploded). NHHC

not be fired rapidly enough to make up for that, and there was no means of measuring the depth of the U-boat target. The idea of an aimable ahead-thrown weapon was revived early in World War II.

The British pursued two alternatives, multiple depth charges and a contact-fuzed weapon. Depth charges had the important virtue that they could be lethal even if they did not hit the target directly. Even if they were not lethal, they might well damage the U-boat; at the least, a near hit would demoralize the crew. On the other hand, a contact-fuzed weapon could be quite small; a 30 lb charge, a tenth or twentieth of that carried by a depth charge, could break a submarine's pressure hull.[6] If the charge failed to hit its target, it would make no noise, and thus would not cause the firing ship to break contact with the submarine. The weapon had to project enough charges to have a reasonable chance of hitting the target with one of them. After trials in 1941, the British chose contact-fuzing. The spigot mortar they produced was named Hedgehog due to its resemblance to that spiny animal. Hedgehog was manually stabilized. It fired dead ahead, but tilting it could shift its aim point slightly (but would distort its pattern). Its on-mount

The starboard Hedgehog being loaded on an Atlantic Fleet destroyer, 1956. It took time for crews to learn to use this weapon effectively. Royal Navy Hedgehog attacks in the second half of 1943 were only 7.5 percent successful, but that rose to 23 percent in the first quarter of 1945.NHHC

A partly loaded Hedgehog (note the bare covered spigots). The spigots could be tilted from side to side to stabilize them, one of the operators doing that manually. USNI

The destroyer *Sarsfield* runs toward the two circular patterns of splashes created by her two Hedgehogs, 15 June 1950. The ASW destroyer *Epperson*, armed with Weapon A, is in the foreground. Note that there are no big waterspouts from explosions. Hedgehog projectiles exploded only on contact with a submarine (and then with nothing like the force of a depth charge). Initially crews were reluctant to fire Hedgehogs because they did not see spectacular waterspouts as they did with depth charges (and Squid). NHHC

operator fired its 24 charges on command from the bridge. The U.S. Navy adopted Hedgehog.[7]

Ahead-thrown weapons reduced the blind-time problem, but they could not eliminate it. A Hedgehog fired a circular or elliptical pattern of charges whose center was about 250 yds ahead of the attacker. Even if the attacker was tracking the submarine at the moment of firing, the submarine had some time to evade while the projectiles were in the air or were sinking toward it. However, blind time was reduced from the 45–90 seconds inherent in a single-ship depth charge attack to 15–20 seconds. Because the charges were contact-fuzed, there was no need to set a depth. In March 1942 Hedgehog charges were made more lethal by filling them with Torpex, an explosive 1.7 times more powerful than TNT on a volume-for-volume basis. Hedgehog was declared ready for production in September 1941.

Crews initially disliked Hedgehog because they saw no evidence that it was working effectively unless it hit the submarine. They much preferred depth charges. Although a pattern of depth charges had only a small probability of destroying a U-boat, the noise and water plumes convinced crews that they were doing something effective. It took time to overcome this unexpected problem; Hedgehog success rate did not reach the expected level until late in 1944. Hedgehog was important in the postwar U.S. Navy, but not in the postwar Royal Navy.

The British prewar asdic development organization, relocated to Fairlie in wartime, pursued its own ahead-firing weapon, which fired depth charges rather than contact-fuzed weapons. Even a near miss could damage or destroy a submarine. Moreover, as in the U.S. Navy, crews were more willing to fire weapons that provided audible evidence that they had exploded. Advocates of this type of weapon also claimed that using it required less skill than a hit-to-kill device such as Hedgehog.

Fairlie developed Mortar A (Parsnip) and Mortar B (Squid). Parsnip used two sets of ten barrels each slung from trunnions to produce a circular pattern much like that of Hedgehog; it was abandoned in August 1941. Mortar B, tested in May 1943, created a triangular pattern with 120 ft sides. The depth of explosion was set automatically using the new Type 147B asdic. Trials showed that Squid could be set to within about 20 ft of the depth of a submarine as long as the latter was below 100 ft. A postwar Royal Navy analysis showed that a depth charge pattern had about a 6 percent chance of killing a U-boat, compared with 20 percent for a Hedgehog attack and about 50 percent for a Squid attack. Squid was deployed on board corvettes and frigates of the late-war Loch and Castle classes, initially on board HMS *Hadleigh Castle* in September 1943. After the war, Squid survived in British naval service, equipping many destroyers and other ships. The British installed one on board the frigate USS *Asheville* to demonstrate its abilities, but the U.S. Navy did not adopt it, presumably because adopting a completely new weapon in 1944 would have been unduly disruptive. However, U.S. officers said after the war that they considered Squid by far the best ASW weapon they had seen during the war (the British took it home at the end of the war).

The wartime U.S. Navy developed an antisubmarine attack plotter (ASAP) as an attack aid. ASAP took information from the dead-reckoning analyzer, the ship's sonar, and the ship's gyro compass. The sonar automatically indicated the bearing of its beam, and range was taken from the range recorder. Own-ship course and speed were automatically input. Manual inputs included weapon range, dead time, and flight time. The operator projected the submarine's path and estimated how the ship had to be oriented to hit it.

Squid on HMS *Arunta*. Escorts generally had a pair of Squids to create a sandwich of explosions above and below a submarine. Like Hedgehog, Squid was fixed in train, although the point of impact could be adjusted slightly by tilting its barrels. Squid was considered the best World War II ASW weapon. Dual-Squid attacks in the second half of 1944 were 33.3 percent successful, but that rose to 62 percent in the first quarter of 1945. DR. JOSEF STRACZEK

Squid turned on its side for reloading on HMAS *Arunta*. DR. JOSEF STRACZEK

Squid was manually loaded, as here on HMAS *Arunta*; it was not intended for rapid fire. DR. JOSEF STRACZEK

Loch Ard displays the new priorities associated with Squid in a postwar photograph. In contrast to a River, she has her ahead-throwing weapons (two Squids) in the protected B position before her bridge, with her 4 in. gun in the less protected position occupied by a Hedgehog in a River. That is because her Squids were her main battery, whereas in a River the Hedgehog was added after the design had been framed (at the end of 1940, before Hedgehog had been adopted). She has no depth charges at all. The whip radio antennas were adopted after the war, as were the "candlestick" antennas on her yardarms. At her masthead she has the standard antenna of the British FH-4 HF/DF set. Although the antenna of her Type 277 radar could tilt for height-finding, in her case it was a surface search set, the successor to the Type 271 in its "lantern." The object right aft is presumably her towed torpedo decoy. *Loch Ard* and other Loch-class frigates were built from prefabricated sections to limit the time they spent on the slip vulnerable to German bombing. Much the same rationale led the Germans to prefabricate parts of Type XXI submarines. *Loch Ard* was transferred to South Africa on completion, retaining her British pennant number as part of the Commonwealth fleet. USNI

ASAP could also project the appropriate course for a depth charge attack (weapon range was set to zero). The Hedgehog firing key was typically adjacent to the ASAP. Course to steer was transmitted electrically to the helmsman or to the weapon control station. The display was small (250 yds/in.) and offered limited resolution. The range displayed was the slant range that the sonar measured, but it was difficult to translate slant into horizontal range, which was what was really wanted. Because the sonar automatically indicated the target's position, ASAP could be confused by reverberation. Much depended on the skill of the operator. Finally, the device was not suited to postwar trainable weapons such as torpedoes and Weapon A. In 1949 ASAP was being modified (as Mod 3) to work with the trainable Weapon A. About a thousand were made. ASAP was introduced in 1942; about 700 remained in stock in 1949. The postwar successor was the Attack Director, which was embedded in the Underwater Fire Control System.

HMS *Barrosa* steams through the pattern of water thrown up by the explosions of the three projectiles from her single Squid. USNI

ASW AIRCRAFT

In 1939 the Admiralty considered aircraft a secondary ASW measure. Shore-based aircraft were operated by the RAF, which was not particularly interested in ASW. The only British shore-based maritime reconnaissance aircraft were small Avro Ansons with limited range and payload. Coastal Command's longer-range flying boats were considered mainly as reconnaissance and antiship weapons.

During the war it became clear that long-range aircraft were a vital means of limiting U-boats' mobility—and, often, of sinking them. By 1943 they were sinking more U-boats than surface escorts, and it was clear that only air support could break up a big wolf pack attack. The RAF, which prioritized the strategic air offensive against Germany, resisted giving Coastal Command long-range bombers. It also much preferred offensive ASW operations to what it considered defensive direct convoy support. The main case in point was its insistence on the value of the Bay Offensive in the Bay of Biscay in 1942–43. The postwar U.S. view was that the Bay Offensive had certainly been an effective way of killing U-boats, but the Admiralty's position was that the point of ASW was protecting merchant ships. The best use of long-range aircraft was to operate them near convoys. Too, RAF Bomber Command resisted calls to bomb German U-boat shelters while they were under construction, calling them "panacea targets." As a consequence, Coastal Command had to work largely with obsolescent medium bombers, whose deficiencies left an area in the North Atlantic outside the umbrella of shore-based aircraft. The Germans soon discovered that they could operate freely in this gap. One reason the U-boats were so badly defeated in 1943 was that political leaders in both Britain and the United States understood how serious the air gap problem was and ordered very-long-range bombers diverted to solve it. The U.S. B-24 Liberator seems to have been the most important. RAF Bomber Command released some Halifax bombers but would not relinquish the desired Lancasters.

As for the United States, in 1939 the U.S. Navy was restricted to seaplanes (and carrier aircraft) under an agreement reached in 1933 by then CNO ADM William V. Pratt and then Army Chief of Staff GEN Douglas MacArthur. When the Navy was assigned to enforce a U.S. neutrality zone in the Atlantic, it learned that its big Catalina flying boats, which had been developed to fight a Pacific war, suffered badly in wintery weather. The Navy was authorized to buy Lockheed Hudson patrol bombers, which were also being bought for Coastal Command.

At this time Army air missions included coast defense, so the Army assigned bombers to ASW operations off the U.S. coast. This mission is significant

Closing the mid-Atlantic gap: a PB-4Y-1, the naval version of the long-range B-24 Liberator, shown in Atlantic camouflage in mid-1944. Not visible is the antenna of her underbelly centimetric radar, which was used to detect surfaced U-boats. Its housing replaced the usual belly turret. NARA

here because the Army ASW units welcomed the first sonobuoys, while the U.S. Navy initially resisted them. The inherent conflict between Army and Navy air ASW was not resolved until 1943, when Army Liberator bombers assigned to ASW were transferred to the Navy. The mission transfer survived the end of the war, leaving shore-based U.S. naval aircraft with a major Cold War ASW mission.

Escort carriers were the other application of naval air power to ASW. Their mobility enabled them to attack U-boats located by signals intelligence. Here aircraft line up on the deck of the carrier *Anzio*, 21 April 1945. NHHC

Aircraft could range widely to attack U-boats revealed by SIGINT. Here *U 848* (Type IXD) is attacked by long-range land-based ASW aircraft—Liberators from VB 107 and two B-25s—southwest of Ascension in the South Atlantic, 5 November 1943. She sank with all hands. The submarine's augmented antiaircraft battery (two twin 20 mm) did her no good; note that the crew are trying to take cover. The housing at left covers the antenna of a surface-search radar. The post forward of it carries a radar search receiver antenna, which can just be made out. Note the water thrown up by the depth bomb explosion in the distance. NHHC

The U.S. strategic bomber force was more willing than its British counterpart to answer the call to attack German U-boat bases, but unfortunately its bombers were limited to 2,000 lb bombs, which were inadequate for the task.

By 1943 both the U.S. Navy and the Royal Navy were operating escort carriers equipped with standard carrier attack aircraft that could deliver such ASW weapons as depth bombs and rockets. The two navies used their escort carriers very differently. The Royal Navy attached them to support groups intended to help convoy escorts deal with heavy U-boat attacks. The U.S. Navy, which was not responsible for transatlantic convoys, created hunter-killer (HUK) groups intended to attack U-boats revealed by intelligence—generally communications intelligence. Such operations horrified the British, who feared that the attacks would reveal the Allies' success in code-breaking. Typically a HUK task group consisted of an escort carrier and four destroyer escorts or comparable ships. This difference in practice was significant after the war.

The U.S. Navy alone used blimps—nonrigid airships—for ASW. The Royal Navy had used them in World War I, but they had disappeared from British service once the Royal Naval Air Service was absorbed into the Royal Air Force.

The main airborne sensor was radar, which could detect a surfaced U-boat, even at night. By 1940 Coastal Command aircraft were being fitted with the ASV Mk II, a metric-wave set using fixed antennas (the airplane had to turn toward the U-boat to detect it, and range was limited). Typically, the radar echo disappeared into sea clutter as the airplane neared the U-boat, so final aim had to be visual. The British devised the high-intensity Leigh light, which was widely installed on board land-based ASW aircraft from 1942 onward.

Once the Germans realized that the British were using airborne radar (they captured a Mk II set in Tunisia in the spring of 1942), they began work on a search receiver that would warn a U-boat that an airplane was approaching. U-boats were being attacked in the Bay of Biscay, and the Germans decided that radar was responsible. The French Metox firm developed a search receiver (the system was called Metox, and its antenna the Biscay Cross). Metox covered the range between 60 and 265 cm (with harmonics), and probably had a maximum detection range of about 6 nm, but it could not distinguish radar from nonradar signals, it had a poor antenna (which often broke down), it could not indicate bearing, and it could not detect short transmissions. It also produced powerful signals of its own. Even so, it halved attacks on U-boats in the Bay of Biscay.

Intercepted U-boat radio traffic frequently reported radar detection. The Allies were well aware of the reason why, so in October 1942 an urgent request went in for centimetric (S-band) sets because Metox would be unable to detect their radiation. A U.S. lab crash-built 14 sets (DMS-1000s), which went into British aircraft. By the end of 1942, the U.S. Navy APS-2 was entering service. The British S-band ASV III was in widespread service in February and March 1943, greatly improving the efficacy of Bay of Biscay air patrols through July 1943. About five U-boats were being sunk monthly. The Germans had previously been assured by their radar experts that really-short-wave (i.e., centimetric) radar was not feasible. Initially, then, they thought the Allies were relying on IR detection (U-boats reported attacks "without radar"). U-boats were covered in paint intended to absorb IR radiation. One U-boat commander did report "a new kind of radar," but the report seems to have had no impact. Thinking that the Allies' new radars were simply producing headphone tones outside the audible range, the Germans introduced the "magic eye" detector, a CRT. In the summer of 1943 the Germans became obsessed with the idea that Allied aircraft were detecting radiation from their search receivers (radio receivers often generate radiation to detect signals by heterodyning).

The Germans did not conclude that the Allies were using 10 cm aircraft radar until September 1943, even though they had captured one from a crashed bomber in March 1943. In October the first centimetric detector, Naxos (8–12 cm), was produced. In theory it could detect Allied radars at a range of 5–10 nm, but in service it did considerably worse. Its aerial could not be submerged, so it had to be set up each time the submarine surfaced. Too, until the spring of 1944 German U-boat commanders were reluctant to use their own radars for fear that Allied aircraft had search receivers.

When an Allied bomber equipped with 3 cm radar (X-band) crashed, the Germans found that they had to cover that band, too. They added an antenna and receiver to cover 2–4 cm, and these were integrated with Naxos to form a single unit called Tunis. By this time Naxos had been fitted with a new reflector antenna that considerably increased its range. Snorkels were fitted with a search receiver (Naxos) effective over the 8–12 cm band. Although many U-boats managed to dive as airplanes approached, many were surprised on the surface due to the inefficiency of Naxos. The Germans sent out *U 406* with their best radar expert on board, only to have him captured when the U-boat was sunk. Another radar experimental U-boat was sunk in April. As airborne radar contacts declined, the Allies developed countermeasures of their own. For example, an attenuator would reduce radar power so that it appeared to a U-boat that the airplane was turning away when it was actually approaching. Too, a U-boat that dived when it detected 10 cm airborne radar was effectively immobilized, which was valuable in itself. German attempts to develop radar camouflage at this time seem not to have been very successful.

After the war, Westerners assumed that the Soviets had captured German search receivers and were now aware of the characteristics of Allied air radars. The question was whether effective search receivers made airborne radars useless. By that time, too, it was assumed that the Soviets would put radars on board their submarines (German U-boat radar had been ineffective, but the Soviets had Western centimetric technology). Airborne search receivers therefore became important ASW sensors.

Sonobuoys in the bomb bay of a TBM Avenger bomber on USS *Guadalcanal*, 7 October 1944. These omnidirectional CRT-1 buoys weighed about 15 lbs; their resin-coated paper tubes were 5 in. in diameter and about 3 ft long, later standardized as "A-size." Buoys were powered by flashlight batteries, and their endurance could be increased by taping additional batteries to them. Hydrophones were suspended 24 ft below the buoy, which had a radio antenna transmitting on any one of six color-coded frequencies. The buoy picked up audible sound (initially in the range 300 Hz to 8 kHz, later 100 Hz to 10 kHz). The buoy was lowered into the sea by parachute. The capsule visible to the left, on top of the buoy, houses the parachute and antenna.
AUTHOR'S COLLECTION

Once the U-boat dived, the hunter airplane could drop sonobuoys around its supposed position to maintain contact. The idea seems to have originated with British physicist P. M. S. Blackett, who suggested that surveillance aircraft could drop such buoys in the track of a convoy to warn of approaching U-boats.[8] RCA began developing buoys in June 1941 under an NDRC contract. It tested a prototype on 12 September, but then the project was suspended due to other more urgent needs. The project was revived in 1942 as a blimp-droppable sensor, demonstrated in Long Island Sound on 7 March 1942. By this time it was the "bomb buoy," incorporating an FM transmitter and batteries giving it a four-hour life (later buoys had additional batteries strapped to them to prolong their life in the water). A modified version developed for faster aircraft entered production as AN/CRT-1. This was an omnidirectional buoy. In October 1942 BuShips ordered 100 receivers and 1,000 buoys. The Army Air Corps was much more enthusiastic; it ordered 6,410 buoys while the U.S. Navy initially ordered 1,800. By early 1943—about the time it took over the Army's coastal ASW role—the U.S. Navy was again interested. Sonobuoys were first used, by carrier-based Avengers, in October 1943. Work on a directional CRT-4 buoy began in February 1943. The CRT-4 was held in place by a sea anchor while wave action rotated its hydrophone. Tests began early in 1945. By the end of the war more than 160,000 omnidirectional and 7,000 directional buoys had been made. Surviving sonobuoys were used extensively in postwar tactical experiments.

Typically, an airplane sighting a diving U-boat laid a sonobuoy pattern to determine where the U-boat was heading, as a preliminary to attacking with depth charges or, later, with a homing torpedo. Sonobuoys were preset to operate at any of six different frequencies, which were denoted by color. The airplane typically laid five (with a sixth as spare) in a cross-shaped pattern with one in the center. Spacing from the center buoy varied. Against a submarine that dived after visual contact, U.S. fleet instructions (January 1945) called for 1,500 yds between buoys. Against a disappearing radar contact, instructions called for a four-mile square with another buoy in the center. Standard U.S. pattern on a suspicious surface indication had the buoys spaced 3,000 yds from the center buoy. RAF Coastal Command envisaged a pattern in which buoys were 2,500 yds apart. The operator was expected to estimate U-boat course based on comparative sound intensity from the buoys, which the operator could monitor in sequence.

When helicopters became available in 1942, advocates in the U.S. Coast Guard pressed to adapt them for ASW. The U.S. Navy decided in July of that year to obtain helicopters for evaluation, ordering a YR-4B from the Army later that year, and two more in March 1943; the first was delivered at the end of October 1943. By this time the Coast Guard (part of the Navy in wartime) was responsible for naval helicopter development. In February 1943 CNO ADM Ernest J. King directed that helicopters be evaluated for their potential to operate from convoy merchant ships. The British were also interested. Responsibility for U.S. tests was vested in the Coast Guard, which had assumed major ASW responsibilities when the war broke out. In May 1943 the Combined Board for the Evaluation of the Helicopter in ASW was formed, with representatives from the Navy (both OpNav and BuAer), the Coast Guard, and the British Admiralty Delegation. The Navy bought an Army R-4B under the designation HNS-1. Some board members saw the helicopter as an attack weapon carrying a depth charge and based on destroyers, which would direct them by radio. The Coast Guard was more interested in operating helicopters from merchant ships, arguing that the helicopter would be more valuable as a scout spotting submarines over the convoy's horizon. There was not much point in arming a helicopter with a single depth charge, because the chance of success was very small. The first sea trials, conducted on 7 May 1943 from the tanker *Bunker Hill*, showed that helicopters could operate under visibility conditions precluding normal air operations. Two ships were converted into helicopter platforms: the British SS *Daghestan* (taken over under Lend-Lease) and the Coast Guard cutter (ex-merchant ship) *Governor Cobb*. Further tests were conducted on board the transport *James Parker* cruising between New York and Norfolk. On 6 January 1944 *Daghestan* sailed in convoy from New York to Liverpool, but only 3 of the 16 days at sea were suitable for helicopter operations (*Governor Cobb* was never used operationally). Flight operations were successful, but the helicopter proved underpowered; in September 1944 it was downrated to training status. As a result, the FY 45 Navy program was cut from 210 to 38 helicopters. By that time work was underway to adapt the dipping sonar already used on board blimps to helicopters.

Initially U.S. carrier ASW aircraft operated largely in daylight, so they depended mainly on airborne radars. By 1944 U-boats were surfacing almost exclusively at night, so the radar aircraft were provided with searchlights. Aircraft often operated in pairs: a radar-equipped hunter coaching a killer into position. The killer often relied on sonobuoys to confirm the position of a submerged U-boat for an attack using depth charges or an acoustic torpedo, both the sonobuoys and the torpedo being carried in the airplane's bomb bay. Most aircraft on ASW duty dropped depth charges (depth bombs) at the spot where submarines disappeared as they dove.

A Mk 24 homing torpedo (Fido) in the bomb bay of a TBM Avenger on the carrier USS *Manila Bay*, 24 February 1944. The two objects in the background are depth bombs. The fins of the homing torpedo are enclosed in a box tail to stabilize it as it drops. NARA

As an alternative to depth charges the United States developed an air-dropped homing torpedo called "Fido" or Mine Mk 24. The mine designation was a cover for the weapon's true character. Mk 24 entered service in 1943. It had to be air-launched well away from friendly surface ships so as not to home on them. Special security measures governed its use. Surfaced submarines could be attacked only during darkness, when the chance of compromise was slight. A submarine making over 10 kts would probably escape, but surfaced submarines attacked at night would probably submerge and thus become vulnerable. A diving submarine could be attacked until 90

A Mk 24 is recovered by a torpedo retriever, 14 June 1950. At this time the wartime Mk 24 was the only available U.S. air-launched homing torpedo. NHHC

seconds after it was completely submerged. After that Fido was to be used in conjunction with sonobuoys. If a periscope was visible, the mine was more likely to be effective than a depth charge.

Fido was so secret that it was not mentioned in the postwar Confidential report on ASW prepared by the U.S. Operational Evaluation Group. After the war, it turned out that the Germans had received, but not believed, reports describing it.

ASW aircraft could drop Fido on a U-boat in the act of diving or on a submerged submarine whose presence was indicated by sonobuoys. Fido sank its first two U-boats on 14 May 1943 and went on to become the most successful Allied air-launched ASW weapon. Allied forces dropped about 340 Fidos in 264 attacks, of which 60 were on non-submarines. The remaining 204 sank 37 U-boats and seriously damaged another 18. U.S. forces had better training with the device, so they achieved somewhat better results: 142 attacks sank 31 U-boats and damaged another 15, for a total of 32 percent rather than the overall 27 percent. Although the Germans never understood what Fido was, U-boat commanders soon realized that the Allies had a very dangerous weapon and dived quickly on spotting an airplane. Many attacks were probably made after the U-boat got too far from the swirl that the airplane used as an aim point. Initially the Germans' tactics reduced Fido's effectiveness, but sonobuoys solved the problem. Many Mk 24s were lost while aircrew learned how to use them. Overall, Mk 24s sank U-boats 22 percent of the time, compared with 9.5 percent of U-boats attacked with depth charges. Tests at Key West showed that Mk 24s should have been making about twice as many kills. Analysis late in the war showed that carrier-based bombers were achieving about 75 percent of the theoretical figure, whereas land-based patrol aircraft were doing much worse. It was not clear why this was so, but Mk 24s may have been better maintained on board carriers.

SHIPBUILDING

The ASW war was affected by the rate of shipbuilding, both naval and merchant. Although Britain had by far the largest shipbuilding industry in the world in 1939, it had limited resources, and it took time to assemble the materials necessary to accelerate warship production to produce escorts. ASW escorts had not been built in large numbers before the war because they had little or no role in a peacetime navy intended to support national policy by impressing other governments. Thus, coastal escorts such as the Flower-class corvettes were not ordered until war was clearly imminent. Since foreign navies also had no interest in escorts, none were available in 1939 to be purchased or taken over by the Royal Navy. The British were able to take over some destroyers being built for other navies (mainly Brazilian) in 1939, but their numbers were limited.

Warship building competed with merchant ship building and repair, because both were urgently needed to make up for war losses. The British took over the large Norwegian merchant fleet when the Germans overran Norway in 1940, and they gained tonnage from other countries the Germans conquered. The United States entered the war with a considerable fleet of laid-up merchant ships built under World War I programs. Ultimately the British hoped to build large numbers of standard merchant ships in the United States and Canada as well as in the United Kingdom. This program produced the U.S. Liberty ship and its Canadian equivalent, the Fort. The design for maximum production limited the ships' size and performance, with consequences for the way convoys operated. Merchant shipbuilding did not exceed losses until the climactic period of the ASW war, between October 1942 and June 1943. New construction of about 1,026,000 gross tons per month was about twice the monthly losses. The total stock of merchant ships, which had fallen from about 40 million tons in 1939 to as little as 31 million, now grew to about 36.5 million gross tons.

At the outbreak of war, the British had about 220 antisubmarine craft fitted with asdic: 165 destroyers, 35 patrol craft, and 20 requisitioned trawlers. That compared with more than 3,000 ships and craft employed on ASW in 1918.

THE COURSE OF THE WAR

When the war broke out, the British assumed that convoys would be subject mainly to attack by submerged U-boats whose captains were unwilling to risk counterattack on the surface.[9] "Limiting lines of approach" to the convoy were determined by the convoy's speed. The U-boats had to attack the convoy from ahead. The faster the convoy, the narrower the lane in which a submarine had to be in order to attack. Conversely, the faster the submarine prior to attacking, the broader the area from which it could attack, and hence the larger the area escorts had to guard. Doenitz had relatively few U-boats at the time, and only about half of them were large enough to operate freely in the open ocean. On average, early in the war only six of them were on station at any one time, and they had to attack individually. Initially, too, German commanders were unwilling to accept Doenitz' argument that they should attack on the surface at night; they attacked submerged during the day, because they had to see their targets through their periscopes.

The situation was complicated by international agreements forbidding unrestricted (sink on sight) submarine warfare. It was widely understood that the United States had been propelled into World War I largely by Germany's adoption of unrestricted submarine warfare in February 1917. To the extent that America's entry had been fatal for Germany, it could be argued that the Germans would be unlikely to repeat their mistake. We now know that the Germans initially ordered their U-boats to honor the rules. They abandoned this restraint after one of their commanders sank the British liner *Athenia* (falsely claiming that she was armed) on 3 September 1939, immediately after the outbreak of war.

The British rightly doubted that the Germans would wait long before introducing unrestricted submarine warfare. They introduced convoys on the outbreak of war; their first one sailed on 6 September. Despite very limited escort resources, the convoys succeeded. Between September 1939 and June 1940 (the fall of France) about 2,500 ships per month were escorted. Of these, about 5 per month were sunk: 2.5 in escorted convoys, 1.5 in unescorted convoys, and 1 straggler. Unescorted (independent) merchant ships were sunk at about four times this rate.

The British also deployed Coastal Command aircraft, whose main value at the outset was probably that they drastically limited U-boat mobility by forcing all U-boats who spotted them to dive. They also helped discourage U-boats from operating near the British coast. The British were well aware of this value, and in November 1939 they began to use light unarmed aircraft called "Scarecrows" to patrol their coasts.[10] Some aircraft escorted convoys directly, holding down U-boats in daytime; others patrolled. In the peak month, June 1940, Coastal Command flew 9,500 hours, including about 6,700 on convoy escort. Each month aircraft spotted about 20 U-boats and attacked 12, damaging some of them.

During this period the Germans lost 21 U-boats, 15 of them to surface ships and 1 to coordinated action by 2 surface ships and an airplane. The Allies lost about 280,000 gross tons of merchant shipping per month (223,000 to enemy action), against about 88,000 tons of new construction. U-boats accounted for 48 percent of Allied merchant ship losses. Mines accounted for 26 percent. Much of that was probably due to the Germans' new magnetic mines laid by aircraft. By June 1940 the British had mastered this threat, but keeping ports and sea lanes open in the face of mining required very large numbers of minesweepers. Many of them were about the size of small escorts, and presumably consumed resources (including personnel) that might otherwise have gone into ASW.

The war on land influenced what could be done at sea. After the Germans conquered western Europe, they could base U-boats on the French and Norwegian coasts, much closer to their Atlantic hunting grounds. They could operate much farther west, so convoys required ASW escorts over greater distances. The limited number of available escorts weakened the protection they received.

Too, with the Germans on the French and Belgian coasts, the British faced a real threat of invasion. Destroyers that might have escorted convoys had to be concentrated in the Channel to meet it. The German conquest also brought the Luftwaffe within easy range of the Channel and the North Sea. The invasion threat did not abate until the Germans invaded Russia in June 1941. Even then it took some months for the British to decide that it could be discounted.

The German conquest of western Europe brought U-boats to the Atlantic coast, and thus greatly extended their endurance (and hence range) in the Atlantic. For example, those based on the French coast could proceed through the Bay of Biscay directly into the Atlantic instead of having to go all the way around Scotland as in 1939–40. Anyone contemplating World War III had to reckon with the possibility that the Soviets, too, would reach the Channel and base submarines there. With the advent of new submarines that were particularly difficult to counter, the idea of attacking them in their bases became more and more attractive—and more and more possible given nuclear weapons that could overcome the kind of protection visible here, in a protected U-boat pen at Brest still being extended when the port fell to the Allies (this photograph was taken on 7 September 1944). NHHC; ARMY SIGNAL CORPS PHOTO

Too, the British lost the services of the substantial French fleet. The prewar French fleet had depended on hydrophones for ASW, French research on active sonar having been abandoned after the end of World War I. After war broke out, the British began fitting French ASW craft with asdic. It is not clear how far this program went, but it was not disclosed to the Germans after France fell.

In July 1940 German submarines began to adopt night surface attack tactics. Successes by a few particularly aggressive U-boats inspired others to follow, and night surface attacks became the rule. In October 1940, for example, 71 percent of attacks were carried out at night. The Germans still had only a limited number of long-range U-boats, although new construction had been ordered in 1939.

Typically, a single U-boat that spotted a convoy shadowed it at maximum visibility distance on the bow or beam until nightfall, when it closed in and tried to gain a position broad on the bow. The British assumed that attacks would come from ahead (given the assumption of limiting lines of approach), so that was where the available escorts were. The attacking U-boat would try to pass astern of the escorts on its approach to the convoy. At Doenitz' urging that U-boat commanders fire at minimum range, some U-boats fired at ranges as short as 600 yds, although others fired full salvoes from much farther away.

The British were forced to open the spacing of ships in convoys to reduce the chance that a single salvo from a U-boat would hit more than one ship. Escorts had to be placed farther from the convoy, and tactics were developed to use star shells to search for surfaced U-boats at night. The Admiralty formed escorts into groups that would work together as far as possible, and thus coordinate their defense. The Admiralty also took over responsibility for routing ocean convoys, translating intelligence about U-boat location into quick rerouting for evasion.

Although the British had developed radar before World War II, compact radars suited to aircraft and escorts did not appear until 1940. While centimetric (10 cm) radar was still in development, radars operated at metric wavelengths, which limited their ability to deal with small surface targets.[11] Centimetric radar could form a much tighter beam that was much less subject to surface reflections, hence much better for surface search. The centimetric Type 271 was being tested by March 1941.

The advent of surface ship radar, which roughly coincided with the Germans' new night-attack tactics, made it possible to spot surfaced U-boats before they could attack. On 21 November 1940 the corvette *Rhododendron*, following a convoy as a rescue ship, spotted an object 1,500 yds away about an hour after a U-boat attack on a convoy. She approached the object, which dived, and laid down depth charges that killed the U-boat. This was the first successful counterattack against a night U-boat attack.

During this period the British developed effective shipboard high-frequency direction-finding (HF/DF) that enabled an escort to determine the bearing of a shadowing U-boat. Shipboard HF/DF could pick up the surface wave emitted by a U-boat at a range of about 30 nm, so a ship running down a bearing at maximum speed might well be able to destroy a shadowing U-boat.

The British also developed radio fingerprinting (RFP), which recorded details of transmission by photography. RFP made it possible to identify sources of HF signals by determining the characteristics of the transmitters and the "fists" of the radio operators. During World War II, RFP seems to have been associated with shore HF/DF stations, but it was provided experimentally to a frigate in a 1949 British exercise. The convoy's senior officer on the frigate could tell that one submarine that had been detected ahead of the convoy had been overrun and was now astern. In a large-scale exercise all 17 of the submarines could be distinguished. None of this would have been possible before RFP.

The British characterized the period between May 1940 and March 1941, when U-boats operated mainly in the eastern Atlantic, as the Battle of the Western Approaches (i.e., the western approaches to the UK). The British credited the improvement of their ASW forces with driving U-boats out of the Western Approaches. The Germans moved their U-boats farther west, partly in hopes of intercepting convoys before their ASW escorts could join them. The British dispersed convoys as far as possible. This was not evasive routing in response to known U-boat positions; that came later when the British were reading German messages.

Once the Germans were forced out of the relatively confined waters around the United Kingdom, they had to find convoys in the open ocean. Their principal means of doing so was radio intelligence. To put his prewar ideas into effect, Doenitz had to form a picture of convoy operations in the open ocean and communicate it to U-boats at sea. Initially the German codebreakers, the *B-dienst*, were able to provide that information. According to an early postwar British account based on German records, German naval codebreakers could read about half the messages in the British Naval Code and Cipher from September 1939 to the early fall of 1940, but the German fleet was not powerful enough to exploit this success. The first use of communications intelligence for attacking shipping seems to have been by *U 53*, which signaled on 17 November 1939 that she had contacted a convoy already reported by decoded messages. Doenitz wrote in his war diary: "If subsequent reports prove equally reliable and come in with equally good timing, they will be of the greatest value for the conduct of the war." He had previously been unable to exploit decodes because boats were not in the areas involved.

The Germans learned not only about specific convoys but also about the overall convoy system, enabling them both to continue breaking into British messages and also to estimate convoy positions when they failed to decode specific messages. Early in the war, when the British had few ASW escorts, convoys picked up their ASW escorts at sea for the run into British ports. Codebreaking revealed the pickup area: the Rockall Bank. Convoys also signaled their position, course, and speed and ocean escort while they were still in the western Atlantic. All of this gave the Germans an accurate picture of the convoy cycle and also of the time and place where convoys would approach Britain. Thus, the Germans registered the July 1940 change from approaching the UK from the southeast to the northwest.

The British changed codes and ciphers periodically. The first important wartime change came on 20 August 1940, when the basic books of the British Naval Code and Cipher were changed. The signaling procedure was changed on 1 October 1940, and on 20 January 1941 the British introduced fundamental changes in procedure (reciphering starting point indicators and introducing reciphering tables by area). After that, German sources of information dried up for a time. According to the British official history, after the last change the Germans' decoding success dropped to less than 10 percent of the previous level. The logs of the U-boat command show that Doenitz did not initially realize the magnitude of the loss, which to some extent the Germans were able to overcome by relying on their previous knowledge of the convoy cycle. U-boats themselves were ill-suited to reconnaissance; during the winter of 1940–41 Doenitz increasingly demanded air reconnaissance, which was not initially forthcoming, at least to the extent he wanted.

The British suspected that the Germans were reading their codes, but the Germans never realized that the British were reading theirs, even when the daily signal sent by Admiral Western Approaches giving his estimate of U-boat positions became unreasonably accurate. The sources of that intelligence were attributed to radar and shore D/F, and the Germans apparently never realized that most of it came from codebreaking. The British authors of a postwar study of the use of codebreaking in the Battle of the Atlantic wrote that the Germans "did not realize that their own operational U-boat signals and not the physical U-boats themselves were the source of that reflection of U-boat movements which was apparent in [German signals] intelligence. . . . [T]he Germans violated their own security regulations and transmitted their special intelligence to U-boats without disguising it." This latter practice gave away the Germans' ability to read Allied convoy codes during a March 1943 convoy battle. The Germans were ordering their U-boats to change course to deal with diversion orders sent out to the convoy in code.

For the British, initially the most important use of broken German codes was to support evasive routing of convoys.[12] The postwar GCHQ history of the Battle of the Atlantic describes convoys' increased survival in April–December 1941 as an Allied defensive victory due to evasive routing. That in turn was possible because there was so much more sea room in the open Atlantic.

Evasive routing was not always successful. It might be difficult to route several convoys clear of a large concentration of U-boats, and convoys might be spotted by German maritime reconnaissance aircraft (FW 200s). Once or twice, too, the escort of one convoy was strengthened to enable it to sail through a known danger area to draw off enemy forces from more valuable convoys that were not so strongly escorted. Later, such tactics had an offensive purpose, but in 1941 they were used only when it was not possible to route all convoys safely. Otherwise the offensive use of COMINT was limited to directing aircraft and surface vessels in independent ASW patrols.

The Germans suffered a codebreaking blackout in December 1940 and January 1941 as the result of changes in the British Naval Cipher. The Germans made no attacks at all against escorted convoys during that period, with overall convoy losses dropping dramatically even though the Germans had doubled the average number of U-boats in the Atlantic from 6 to 12. U-boats sank stragglers and unescorted ships, but overall they were far less successful. In March, U-boats began operating in groups of three to five, converging on whichever U-boat spotted a convoy. Frustrated by the failure of the *B-dienst* to find convoys, Doenitz went to Hitler to demand air reconnaissance support.

Convoy escorts were becoming effective enough to dissuade U-boat commanders from making the short-range attacks that had been so successful in the past. In March 1941, escorts put the top three U-boat commanders—Günther Prien, Otto Kretschmer, and Joachim Schepke—out of action. Schepke was killed during a joint attack on convoy HX 112 on 16–17 March 1941. Kretschmer was captured during the same battle. Later analysis showed how important such losses could be. The commander of a submarine, like a fighter pilot, determines how aggressively his ship will fight. In the case of fighter aircraft, it is often said that 5 percent of the pilots account for 90 or 95 percent of aircraft shot down. After World War I, for example, the overall U-boat commander, Andreas Michelsen, wrote that too many U-boat commanders were unwilling to take chances against convoy escorts. Michelsen advocated group attacks specifically because commanders would be unwilling to show what he called cowardice in front of other commanders.

As an indication of the Germans' decreased confidence, toward the end of August 1941 there were indications of long-range (browning) attacks on convoys—salvoes fired from outside the convoy in hopes of hitting at least one ship. U-boats also began to attack escorts, but by this time there were too many escorts for that to be very profitable. And the escorts were now counterattacking the wolf packs. The typical convoy escort of one destroyer and three or four corvettes was still insufficient to prevent a U-boat attack but was enough to shake off U-boats, particularly with air assistance. In mid-September the U.S. Navy announced that it would take over covering ships between the United States and Iceland. This alone considerably increased the available escort force, although the U.S. Navy was as yet quite inexperienced. The destroyer USS *Reuben James* was lost to a U-boat in October 1941 (she was helping escort convoy HX 156), and the destroyer USS *Kearny* was damaged during another convoy operation. German operations were also affected by orders to concentrate U-boats in the Mediterranean to support resistance to the British offensive in Libya. The British tried to close the Strait of Gibraltar to U-boats, and two U-boats were sunk trying to get through.

For a time the German command lost confidence in the *B-dienst*. In June 1941 nearly the whole operational strength of the U-boat force in the Atlantic, 25 boats, was engaged in reconnaissance—an unprofitable endeavor because a single U-boat could not see very far. Doenitz was unable to concentrate his force against any particular convoy. During the last week of July 1941, however, the *B-dienst* was again successful. The Germans managed a mass U-boat attack against convoy OG 69 bound for Gibraltar. At least nine ships were sunk. A few days later U-boats attacked a convoy located entirely by codebreaking. It helped that in September 1941 the British simplified their Naval Code and Cipher.

During the period between April and December 1941 Doenitz finally had enough U-boats at sea to form the wolf packs he had envisaged before the war. In August 1941 he had an average of about 36 U-boats at sea in the Atlantic. He moved them farther west in hopes of attacking convoys before their escorts could join, beyond aircraft range. He also moved U-boats south to the Azores and the area around Freetown in West Africa. To extend escorts farther west, the British occupied Iceland and stationed escorts there. These ships joined convoys when the England-based escorts had to be detached. This practice created a bottleneck that the Germans could and did exploit. Whether or not they were reading British convoy codes, they could expect to find convoys near Iceland. To some extent the concentration near Iceland could be traced to the limited endurance of corvettes. By late 1940 the British were seeking escorts with sufficient endurance to escort convoys all the way across the Atlantic. Results included the River-class frigates and converted destroyers in which boilers were replaced by fuel tanks.

By basing escorts in Newfoundland, the British were able to provide continuous escort across the Atlantic, although with fewer total escorts. By this time the Royal Canadian Navy was beginning to operate a substantial corvette force (Canada had about 35 ASW ships). The first escorts from St. John's, Newfoundland, sailed on 31 May 1941. Asdic-fitted minesweepers used closer to England added to the escort numbers.

When codebreaking alerted the British to the U-boats' new night surface-attack tactics, they stationed escorts down the sides of convoys, since U-boats attacking on the surface were not affected by limiting lines of approach. As radar-equipped escorts became available, they were stationed on each beam of the convoy. Early radars suffered from back echoes—that is, signals projecting to the rear as well as the front, bouncing off ships behind the radar. To avoid creating back echoes the escorts had to be stationed well outside the convoy, typically 4 nm out. Ships at the rear of the convoy became the most likely to come under attack.

The Coastal Command aircraft (Sunderlands and Hudsons) deployed to Iceland to extend convoy air cover westward proved lethal to U-boats.[13] There was now evidence that the Germans wanted to avoid Coastal Command. In September 1941 about 75 percent of sinkings were outside the range of its aircraft (about 400 nm from shore). In October, no ships were sunk within 400 nm of Coastal Command bases. Aircraft appearing even briefly near a convoy could break up an attack. Conversely, the Germans found that a concentrated attack could break up a convoy formation and render even a strong escort ineffective.

The ASW aircraft based in Newfoundland lacked the range to provide air cover in the mid-Atlantic air gap. The Royal Air Force retained its very-long-range aircraft (mainly Liberators) for Bomber Command. The U.S. Army Air Force did use some of its Liberators for ASW (before transferring them to the U.S. Navy), but its efforts seem not to have sufficed. Recognizing the advantage, the Germans concentrated U-boats in the gap.

Doenitz was particularly upset by a series of failures that began on 4 November 1941, despite *B-dienst* successes. His scouting lines were failing to find convoys, presumably because the British were ordering successful evasive action. This failure was the culmination of about nine months of failures to find convoys in the open Atlantic. By December, as U-boats were sent into the Mediterranean and to the Arctic, the U-boat war against Atlantic shipping had almost stopped. The convoys that could be found, on the Gibraltar-UK run, were too heavily defended.

During the period between April and December 1941, surface ships sank 13 of the 23 U-boats sunk. British submarines patrolling off the U-boats' French bases sank five U-boats in September and another in December. The Germans reacted by submerging when entering or leaving their bases, which slightly reduced their available patrol time. New U-boat tactics reduced U-boat loss rates, but U-boats were less productive than in the past. The limited British successes could not keep up with the expanded U-boat construction program, however; the Germans commissioned about 45 oceangoing submarines while losing only 18 of them. The effective German ocean U-boat fleet doubled. On the other hand, the rapid expansion combined with the loss of their best U-boat commanders left them with relatively inexperienced crews. Many U-boats were sunk on their first cruise. Although the rate of Allied merchant shipbuilding had increased to about 114,000 gross tons per month, it was much less than the increased rate of loss, about 456,000 gross tons per month. U-boats accounted for about 55 percent of the tonnage lost. As the number of U-boats increased, so did the number of antisubmarine craft; the number suitable for ocean escort increased from about 233 to about 375 at the end of this period. This number includes the 50 ex-U.S. destroyers transferred to the Royal Navy in September 1940.

The combination of convoy (including evasion) and air ASW proved effective. Even though the average number of U-boats at sea tripled to about 30 between April and December 1941, they sank 23 percent fewer ships. Shipping was still being sunk faster than it could be replaced, but losses were no longer rising very rapidly. The United States was preparing to mass-produce Liberty ships, which would more than overcome the existing rate of losses. The U.S. Lend-Lease plan also contemplated mass production of ocean escorts (destroyer escorts) to a British specification.

The United States entered the war early in December, after the attack on Pearl Harbor. Until that point the growing U.S. naval force had been concentrated in the Atlantic, but now the U.S. Navy moved more forces into the Pacific. The United States was also committed to assisting in escorting transatlantic convoys. U.S. forces were untrained and inexperienced in that regard. A shortage of escorts for the Western Hemisphere made it impossible to set up a coastal convoy system when the country entered the war. It did not help that coastal cities refused to adopt a blackout. Ships silhouetted against the onshore lights were easy targets for U-boats at night. They were particularly effective against the tankers that carried oil from the oilfields in Texas to the users on the East Coast. In January 1942 losses were worse than in any month of the previous period, and they continued to mount as more U-boats appeared off the coast. The U.S. Navy was finally able to mount East Coast convoys in mid-March 1942, but the U-boats moved to the unprotected Gulf of Mexico. Worldwide shipping losses reached 707,000 gross tons in June 1942. As it became possible to extend the U.S. convoy system, losses along the U.S. coast declined, despite the increased number of U-boats. Worldwide shipping losses declined to 472,000 gross tons in July 1942.

In January–September 1942 the Germans concentrated U-boats off the U.S. East Coast to exploit the very limited ASW measures the United States could mount. They were extremely successful; the Germans called this period the "Happy Time." British historians of *B-dienst* and its impact on the U-boat war consider it natural that a frustrated Doenitz would concentrate his U-boats in U.S. coastal waters to attack the soft target presented by unescorted merchant shipping. Since the voyage to and from the operational area was long and fuel was limited, there was no surplus of time or fuel for attacks on transatlantic convoys.

As defenses along the U.S. coast improved, the U-boats moved back into the mid-Atlantic gap between air coverage from the U.S. and British coasts. Atlantic convoys accounted for about 10 percent of convoyed shipping. Typically, about 14 ships were sunk per month out of 360 convoyed, more than half the total losses to convoyed shipping. This was more than five times as high as for other convoys. The British characterized the period between August 1942, when the U-boats withdrew from the U.S. coast, and May 1943 as the second battle of the convoy routes.

When the U.S. Navy began to assist in convoying merchant ships in the fall of 1941, a combined Anglo-U.S. Naval Cipher had to be introduced (October 1941). This Naval Cypher No. 3 carried the bulk of convoy traffic. The Germans recognized its similarity to the British naval ciphers that had been broken in 1939–40 and managed to break into it by February 1942. However, their success can be overemphasized. Postwar examination of German records revealed considerable time lags in decoding, although the lags were offset to some extent by the time it took a convoy to cross the Atlantic: a message read a week or ten days late might still be valuable. Many gaps could not be filled for weeks. Perhaps the main advantage the Germans gained was much better insight into the convoy cycle, which enabled them to deduce some convoy routes from other information, such as sightings.

Through the first half of 1942 Doenitz was unable to plan his campaign on the basis of solid information on British convoy operations. He was compelled to rely on U-boats not only to attack Atlantic convoys but also to find them in the first place. Between January and May 1942 there were 21 U-boat contacts with convoys. With one exception they were single U-boats on patrol or on passage; in half the cases no other U-boats were in position to join in the attack. The Germans concluded that they needed two large concentrations of U-boats, toward the eastern and western ends of the transatlantic convoy route, to have a reasonable chance of sighting convoys without access to COMINT or some other form of ocean surveillance. Even those wolf packs could often be evaded.[14] The Luftwaffe was concentrating on the Gibraltar-to-UK route, which, unlike the central Atlantic, was within its range. Convoy attacks therefore occurred much more often on the Gibraltar-to-UK route than on the transatlantic routes. The waters involved were narrower, with fewer possibilities of evasion, and were closer to the Biscay U-boat bases.

With the considerable improvement in German codebreaking and also the increased U.S. Navy ASW effort in coastal waters in mind, on 6 May 1942 Admiral Doenitz decided to revive his planned U-boat dispositions intended to intercept convoys. He sent eight U-boats, including three commanded by convoy warfare veterans, en route across the Atlantic. His greatest problem, he told Hitler, was finding Atlantic convoys; more U-boats would solve that. Anticipating the time when American coastal convoys would become effective, Doenitz said that he now had enough U-boats to probe for other weaknesses in Allied ASW defenses that would force the Allies to disperse their efforts.[15] By October 1942 U-boats were operating not only in the North Atlantic but also from the Gulf of St. Lawrence on the Canadian coast to the Caribbean Sea and the African coast.

Doenitz' main effort against Atlantic convoys began in August 1942. During June and July U-boats had begun to operate effectively against convoys. The percentage of ships sunk in convoys versus independently rose from 4 percent in April to 10 percent in May and June and then to 25 percent in July.[16] On 1 July, a month before the planned resumption of full-scale attacks on Atlantic convoys, 101 "frontline" U-boats were assigned to the Atlantic, making it possible to maintain an average of 69.6 at sea, including 21.7 in operational areas.[17] By 1 October there were 160 frontline boats, making it possible to keep an average of 105.4 at sea, of which 44.7 were in operational areas. With so many U-boats in the Atlantic, the Germans could form packs of as many as 20. A pack of this size badly mauled convoy ONS 154 as it left air cover. The wolf pack outnumbered the escorts by more than two to one.

"Tanker" U-boats were sent out to refuel the packs, extending the time U-boats could spend at sea. During August, one group of U-boats at a time covered the routes between England and the United States, but by September two groups operated at a time, one in the eastern Atlantic west of the British Isles, and another in the western Atlantic northeast of the Newfoundland Bank. The eastern group was intended to intercept westbound convoys and fight them across the Atlantic, while the western group was to attack eastbound convoys. The operations that followed could be distinguished from those of May through June by the sheer number of U-boats involved. For example, the fight against SC 94 (early August) eventually involved 17 U-boats. Global Allied shipping losses reached their highest point to date, but they declined in December as the U-boats were concentrated in the Atlantic approaches to the western Mediterranean.

To some extent Doenitz' attempt to concentrate his whole force in the North Atlantic was frustrated by forced diversions to the Arctic and the Mediterranean, the latter even well before the Allied invasion of North Africa.[18] The diversion to the north was part of the Germans' effort to interdict convoys to the Soviet Union. The first convoy to northern Russia sailed in August 1941. In October the Northern Waters U-boat group was formed at Hitler's direct order. The Italians asked for help (initially for the invasion of Greece) beginning in January 1941. The Germans tried to fend off requests for U-boats to operate against the British lines of communications in the Mediterranean, initially arguing that it might be better for the Italians to provide the submarines themselves by withdrawing them from the Atlantic (where the Germans felt they were useless). In August the Italians renewed their demands as the British built up their Eighth Army. At the beginning of September 1941 BdU was ordered to send six boats. Demands for more boats kept rising, particularly after the British mounted an offensive into Libya in November. BdU sent 4 U-boats at the beginning of the month, then another 11 a few days later. The naval war staff decided to maintain 10 boats in the eastern and 15 in the western Mediterranean, including the area west of the Strait. Together with the demands of the Arctic area, these deployments so drained the force available for the Atlantic that operations there stalled. The German Supreme Command tolerated this situation because the initial battle of the convoy routes had virtually been lost by this time.

In November 1942 the Allied invasion of North Africa—which *B-dienst* had not predicted—demanded a substantial diversion to the Mediterranean: 31 boats were immediately drawn off to attack the invading forces and their supplies. That left only nine U-boats in position to attack the North Atlantic convoy route; these boats were low on fuel, and some were damaged. U-boats in and near the Mediterranean had to operate in range of Allied ASW aircraft and suffered accordingly. In December Doenitz had only 20 U-boats in the North Atlantic. He would not have enough to fight major convoy actions for another two months. As for North Africa, Doenitz managed to reduce demands to a patrol line of 12 U-boats operating to the west of Gibraltar, far enough out to be safe from Allied land-based aircraft. They found nothing, and in December 1942 Doenitz was allowed to abandon this effort.

The Germans broke into the Allies' combined cipher at about the same time as Allied codebreaking failed. On 1 February 1942 the Germans introduced a four-wheel (rather than three-wheel) Enigma for U-boats, blacking out British access to U-boat orders. This new machine code was not broken until late in December 1942. Lacking intelligence on U-boat dispositions, the Allies found it much more difficult to route convoys evasively. It was still possible to estimate U-boat positions from other information, particularly shore HF/DF, but not nearly as precisely. Current traffic for the 24 hours from 1200 on 25 December to 1200 on 26 December became readable at about 2100 on the 26th, beginning a period of virtually current decryption through 2 January 1943.[19] Convoys were once again diverted between the end of December and early January 1943, and losses to U-boats dropped sharply. Current reading continued through the first three weeks of January, but for five days between 19 and 31 January nothing at all was read. Reading on other days was delayed. The loss of this intelligence contributed to successful U-boat attacks early in February. A loss of current reading had also contributed to a disastrous German attack on a tanker convoy in the Caribbean in January, helped by the Germans' insight into convoy operating patterns.

The Germans had their own problems. The Allies learned to omit key information (routing) from the convoy sailing telegrams. Without *B-dienst* support, in January 1943 Doenitz felt compelled to use his U-boats for reconnaissance. With his U-boats dispersed, he could not make massed attacks. He formed most of his boats into a long patrol line extending south from Greenland. This effort largely failed because the Allies were now reading much of the German radio traffic, although a contact by the southernmost U-boat of the line made it possible for Doenitz to mount a massed attack on convoy ON 166. Of 19 U-boats ordered to attack, 13 managed to engage.

By this time the North Atlantic fight against convoys was the only one left to the U-boats. Defenses in distant waters had been strengthened, and few Allied ships were sailing independently. When convoy attacks began to fail, Doenitz blamed it on the diversion of U-boats to tasks he considered secondary.

However, diversions continued after Doenitz became commander in chief of the German navy at the end of January 1943 (he retained personal command of the U-boats).

Twice as many ships sailed in convoys in March 1943 as in February, as the force that would invade France in 1944 was built up. Convoy defenses were strengthened accordingly, particularly air defenses. The U-boat fleet in the Atlantic reached a peak, with an average of 116 at sea (about 50 operating against convoys). German losses also mounted, 19 U-boats being sunk in February, against a monthly average of 20 completed. Also in January, Allied merchant ship completions finally exceeded sinkings. Aware that he had to win the "tonnage war," Doenitz mounted a maximum offensive in March 1943. The Germans were finally reading the Allied code concurrently, and *B-dienst* could provide current convoy positions, diversion signals, and also the daily list of British estimates of U-boat positions. Doenitz was surprised by the accuracy of the estimates but refused to believe that they might be the fruit of codebreaking.

The result of Doenitz' offensive was the devastating March 1943 convoy battle. U-boats now often greatly outnumbered escorts; the largest pack employed up to that time comprised 40 U-boats. A pack of this size attacked convoys HX 229 and SC 122, which had been routed close together. While out of range of protective aircraft, they lost 13 ships in a single night. Some American historians of the U-boat war consider that engagement on 20 March 1943 the high point of the U-boat offensive.

During the March 1943 battle, U.S. codebreakers realized that U-boats were being ordered to make radical course changes to deal with Allied diversion orders. They were obviously reading the convoy code. The British had resisted changing it; now they gave in.[20] Allied codebreakers in turn were reading Doenitz' messages attacking shirkers and defeatists. Doenitz was also telling his U-boats, incorrectly, that the Allies were using devices directed specifically against U-boat morale.

In March 1943 U-boats sank 600,000 tons, a figure exceeded only in June and November 1942. In the previous months, most of the sunken tonnage had been unprotected ships sailing independently. This time it was ships in convoy.

During this period about 70 percent of attempted U-boat attacks on convoys were frustrated by escorts. Each successful attack sank about two ships, but only in half of them did the U-boat escape detection. Since U-boats were operating largely on the surface, radar was responsible for more than half of U-boat detections. This was the first period of the war in which aircraft were the leading U-boat killers; they sank 45 percent of the total, and another 6 percent in combination with surface craft, a total of 86 U-boats. Ten of the U-boats were sunk by carrier-based aircraft. Surface ships accounted for 35 percent of U-boat sinkings, and submarines another 10 percent. Improved crew training showed in a rising percentage of surface attacks that at least damaged a U-boat: about 25 percent. Another 10 percent sank a U-boat.

Continuing losses of merchant ships made for high-level interest in ASW. An Allied ASW Conference was convened. In Britain a cabinet-level Battle of the Atlantic Committee was created. The Atlantic Convoy Conference held in Washington in March 1943 allocated 260 very-long-range aircraft (VLRs, mostly Liberators but also British Halifaxes; Lancasters were wanted but not supplied) to close the mid-Atlantic air gap. These aircraft had the necessary endurance of 2,000–2,500 nm. Escort carriers were entering service now, too, closing the gap by accompanying convoys all the way across the Atlantic. The U.S. Navy surrendered its responsibility for some transatlantic convoys to the Royal Navy and the Royal Canadian Navy and took on complete responsibility for all convoys between the United States and North Africa, for special oil convoys between Aruba and the UK, and for those between Aruba and Africa. On 20 May 1943 the U.S. Navy formed the Tenth Fleet, directly under CNO ADM Ernest J. King, to coordinate ASW operations and tactics. The Tenth Fleet sponsored the Atlantic Antisubmarine Development Detachment, which developed tactics and new weapons. Probably at least as important, by this time escort crews and aircrews had gained considerable experience.

The increasing numbers of ASW aircraft seem to have been a particular threat to the morale of German submariners because they could appear unexpectedly anywhere U-boats were operating. Aircraft made Doenitz' favored tactic of sweeping ahead of a convoy less and less practicable. Doenitz found it impossible to formulate a consistent policy to deal with the aircraft, although he pressed for development of radar detectors that would enable U-boats on the surface at night to dive before being detected and attacked. In May–July 1943 Doenitz apparently failed to realize that the mid-Atlantic air gap had been closed by a combination of land-based VLR aircraft and escort carriers.

By this time the large ASW building programs ordered from 1941 onward were beginning to produce numerous escorts as well as escort carriers. The Royal Navy assigned Home Fleet destroyers to support groups that could reinforce convoy escorts in areas of particular danger. During April 1943 five such groups were operating in the North Atlantic, and two of them included escort carriers.

In April 1943 merchant ship sinkings fell by about half. U-boat sinkings reached their highest point to date, and airplanes from an escort carrier (HMS *Biter*) found and attacked a U-boat, which was sunk by the destroyer HMS *Pathfinder*.

Despite the losses, German U-boat strength in the Atlantic peaked at about 120 in May, half of them in the northwest Atlantic. That month, about 40 U-boats attacked convoy ONS 5. The convoy consisted of about 30 ships and 4 stragglers and was covered by an average of 8 escorts. The U-boats began attacking shortly after midnight. They sank 12 ships on 5 May: half at night and half the next day. However, the escorts did much better the following night, frustrating 24 attacks and inflicting considerable losses and damage; at least 5 U-boats were sunk. The U-boats broke off the action the next day. The American history of the ASW war notes that U-boats never displayed the same determination in any later convoy battle.

Perhaps at least as important, in May Allied ASW forces sank 44 U-boats, nearly twice the previous high, of which 38 were in the Atlantic (13 in the northwest Atlantic). On 24 May Doenitz decided to withdraw U-boats from the North Atlantic until he could find some way of countering sonar and ASW aircraft. When he ordered his U-boats out of the North Atlantic, Doenitz told their commanders and crews that new technology that would reverse the situation was coming. The new technology amounted to a combination of radar countermeasures, more powerful antiaircraft defenses, and the *Zaunkoenig* homing torpedo for counterattacking escorts.

Doenitz dispersed the U-boats widely in hopes of finding targets nearer to enemy coasts. He also deployed tanker U-boats to keep his force operating as long as possible. Arranging fueling rendezvous required considerable messaging, which the Allies could now read. The U.S. Navy deployed its escort carriers against the tankers and sank nearly all of them, and often the U-boats waiting to fuel as well. There is some evidence that the British objected to this operation on the ground that it could reveal to the Germans that their code was being broken. But as before, the Germans refused to admit that their code might be insecure.

Because U-boats still had to transit the Bay of Biscay en route to and from operational areas, Coastal Command renewed its offensive there. Coastal Command caught numerous U-boats returning to base from the mid-Atlantic. Doenitz ordered his boats to fight on the surface when attacked, and eventually armed them with more powerful antiaircraft guns, the quadruple 20 mm. The tactic was ineffective; the Allies sank 11 U-boats in the bay in May, and 17 more in June and July. In August Doenitz ordered U-boats to transit the bay submerged and dispersed. The escort groups attacked U-boats trying to evade Coastal Command by moving close to the shore. The Germans countered with their new guided antiship missiles, their first victim being the sloop HMS *Egret* on 27 August 1943.

On 10 June 1943 Naval Cipher No. 3 was replaced by the Combined Anglo-U.S. Naval Cipher No. 5, which the *B-dienst* failed to break.[21] This event coincided roughly with the temporary withdrawal of U-boats from the central Atlantic, and hence may seem less important than it was. Although *B-dienst* never broke into Combined Cipher No. 5, in July it succeeded in reading the Naval Code, which had become unreadable in most areas on 1 March and in all areas on 20 April. Reading the Naval Code, the Merchant Ship Code, and traffic analysis enabled the *B-dienst* to track convoys. Using various radio sources, the *B-dienst* was able to track 44 of the 49 convoys that traveled between the United States and Britain between 25 May and 3 September 1943. After that time German performance declined sharply, as the *B-dienst* was reading fewer and fewer messages, and traffic analysis was becoming less effective. The *B-dienst* hoped to break into Naval Cipher No. 5 in October, but that never happened. There were further blows. On 1 December a change in the Naval Code deprived the Germans of that source of information, and messages in the Merchant Ship Code began to disguise positions. As of 1 May 1944 the only naval code the Germans could read was the Merchant Ship Code, and even then they could not read convoy positions after the first one to four days.

By the fall of 1943, the new technology Doenitz had promised his U-boats was available, and he resumed operations in the mid-Atlantic. The British characterized the period between about 1 September 1943 to the dispersal of the Preussen group on 22 March 1944 as the third battle of the convoy routes. The antiradar elements were the new Hagenuk radar search receiver and the Aphrodite radar decoy, a balloon carrying metal strips. The antiaircraft weapon was the quadruple 20 mm gun. The mid-Atlantic fight resumed on completion of the *Zaunkoenig* program. Thus equipped, the 21 U-boats of the Leuthen group were sent out in September 1943 to form a patrol line against westbound convoys. Seventeen of these boats carried four *Zaunkoenig* homing torpedoes, two in the bow and two in the stern. Some had special "B" parties on board to intercept and exploit enemy communications, especially convoy R/T.[22] U-boats were ordered to dive to periscope depth if they met an escort and fire a homing torpedo at it. Orders were now to destroy the escorts first, then attack the convoy. U-boats operated in pairs: one would entice the escorts away, and perhaps even sink them; the other would attack the merchant ships. Hagenuk was to be used to keep contact with escorts using radars. Unfortunately for the Germans, Hagenuk covered the 120–80 cm band, not the 10 cm band escorts had been using since 1941. U-boats were to use Aphrodite during night attacks.

Using *B-dienst* information, the Leuthen group contacted convoy ONS 202 on 30 September 1943. Within 24 hours it claimed 7 destroyers sunk and 3 probably sunk, plus 2 merchant ships certainly sunk and 1 probable. On the second day, with visibility poor, ONS 202 merged with the slower convoy ONS 18. By the third day, aircraft reappeared and harassed the U-boats. BdU ordered the U-boats to break off on the fourth day because the convoy was too close to Newfoundland. By now the Germans claimed 12 destroyers sunk and 3 probably sunk, and 9 merchant ships sunk and 2 more torpedoed. Actual losses were 4 escorts and 7 merchant ships, with 3 U-boats sunk and 6 severely damaged. Doenitz announced a great victory for the new technology. He attributed the failure to sink more than nine (actually seven) merchant ships to the unusually heavy escort. His order that U-boats remain on the surface had enabled them to maintain contact. He concluded that *Zaunkoenig* was a great success but wanted it to arm at a greater distance so that a U-boat could remain surfaced after firing without risking that the torpedo would home on it. He issued orders that all outgoing U-boats should have it by 3 October.

Allied codebreaking had already revealed the existence and nature of the anti-escort homing torpedo. By October escorts were using the Foxer towed noisemaker to divert it. The Allies had already developed Foxer due to an erroneous interpretation in spring 1942 of the nature of the first German pattern-running torpedo, FAT, whose existence had been revealed by codebreaking.

Regular reading of German communications began late in September, causing the failure of two later German patrol lines. BdU continued to assume that his code was secure, attributing failure to enemy air reconnaissance, which he hoped his new search receiver would overcome. He ordered U-boats to limit their radio transmissions and to quit using beacons (which had made it possible for a U-boat in contact with a convoy to home others onto it). He initiated an elaborate off-frequency radio technique to frustrate Allied D/F (but it was betrayed by codebreaking). When the *B-dienst* broke a British message setting up a straggler route in connection with convoy ON 204, Doenitz interpreted it as clear evidence of evasive action. His possible explanations were Allied air reconnaissance and treachery, in the form of information provided by the Italians after they surrendered. Some of the Italians were aware that the Germans had broken Allied ciphers, and he thought that the British were deliberately putting deceptive information out in this form.[23]

Doenitz managed to steer a U-boat group into contact with a convoy in October. Its unsuccessful battle with heavily armed bombers demonstrated the fallacy of his policy of remaining on the surface to fight. The U-boats sank a single merchant ship but lost 6 of their own number out of an initial strength of 14. By the end of October, 15 U-boats had been lost, of which Doenitz estimated that two-thirds had been sunk by aircraft. He decided that the 20 mm guns were not enough and hoped that 37 mm guns would solve the problem. For the moment, U-boats should dive as soon as aircraft approached. U-boat patrol lines would have to submerge during the day. Instead of operating in the paths of aircraft based in Iceland and Greenland, the next U-boat line was ordered to form off Newfoundland. It missed the expected convoy. Doenitz also changed battle tactics. U-boat groups would form into short lines of four or five boats each, with a single scouting U-boat placed ahead of each gap in the overall group. Traffic from Canada to Britain was now organized in two lanes, and U-boats properly placed should be able to cover both. Allied codebreaking revealed the locations of the scouts, and the U-boats consistently failed to find convoys. The loss of the U-boat tankers made it impossible to sustain combat on the far side of the Atlantic. U-boats were diverted to the eastern Atlantic and to the convoy route between Gibraltar and the UK.

By this time escorts were so powerful that convoys could survive without loss even when diversion failed. Doenitz did receive additional air support, including He 177s armed with guided missiles, but they seem to have done little good.

Doenitz pushed hard for Luftwaffe cooperation for reconnaissance, but what he received proved insufficient. He had to fall back on U-boats to find convoys. His orders for the end of 1943 and the beginning of 1944 show that he had no idea of current convoy routes. Thinking that convoys were scattered over the Atlantic, he set up long patrol lines in hopes of finding them. The British, aware of those boats' position and extent, routed convoys around the ends of the lines. Overall, during the seven months of the "third convoy battle" Doenitz managed to maintain an average of 60 U-boats across Atlantic convoy routes. He consistently overestimated their successes, which vanished almost completely after the initial battles of the Leuthen group. Doenitz lost 50 U-boats and 2 U-tankers during this period. He attributed the losses to enemy radar and too few German aircraft. The aircraft were, moreover, too vulnerable to carrier-based fighters.

Once again Doenitz hoped to resume the offensive with new technology—in this case snorkels, with improved torpedoes and the Type XXI U-boat soon to follow.

THE SNORKEL

In June 1944 the Germans finally introduced an effective counter to ASW aircraft: the snorkel ("snort" in British parlance, Schnorchel in German). The snorkel was a breathing tube that enabled the U-boat to continue to run her engines while at periscope depth (they exhausted either through an accompanying pipe or into the water just below the surface). The snorkels deployed at this time had only one-ninth the radar cross-section of a surfaced U-boat, and the Germans hoped to reduce that even further by using radar camouflage. Doenitz was confident that the snorkel made possible "total underwater warfare," the U-boat never having to surface at all.

The Allies learned of the snorkel in decrypted messages in January 1944 referring to a boat in the Baltic that was presumably testing it. The message indicated only that the snorkel was a mechanical device associated with the submarine's diesel engine. It is not clear how quickly the snorkel was understood for what it was. From February 1944 onward, U-boats were asked to report their experiences with the snorkel; some of these reports were intercepted and decoded. A general U-boat order broadcast on 31 May stated that several U-boats had used snorkels successfully to cruise entirely submerged for up to several days, that batteries could be charged using snorkels four to five hours daily (some U-boats filled up with smoke while using them), and that radar detection range of a snorkeling boat was considerably shorter than that at which a surfaced boat could be detected. U-boats were told to interrupt snorkeling about every 20 minutes for a self-protection hydrophone sweep. The Germans considered the snorkel so important that in August 1944 they made its installation the most important task to be accomplished when U-boats were in port.

The snorkel of *U 889*, folded down. The fitting to connect with the induction pipe alongside the conning tower is visible. NHHC

The surrendered *U 889* shows the pipe connecting her snorkel to her induction on the left. The snorkel is down. With the advent of snorkels, Allied ASW aircraft lost much of their efficacy, since a snorkel had only a small fraction of the radar cross-section of a surfaced U-boat. Observers on board aircraft could occasionally spot the exhaust smoke of a snorkel, but their radars generally could not detect a snorkel at a useful range. The Royal Navy and the U.S. Navy modified submarines with mock snorkels to familiarize aircrews with the appearance of snorkel heads. NHHC

By this time the Allies knew a lot more, from radio traffic, from prisoners of war, and from captured documents. NID UC Report 458 (March 1944) described German experiments with extensible underwater intakes and exhausts that would enable U-boats to proceed at periscope depth on their main engines. NID UC 452 of 18 March 1944 had stated that the Germans had learned that their new weapons and equipment, such as sonar decoys, had failed and had "probably for some time recognized the need for a radical change." That would most likely include increased speed (surface and submerged), increased submerged endurance (to reduce the danger from aircraft and radar while proceeding to operational areas "and to allow batteries to be charged submerged"), and increased diving depth.[24] Prisoners of war stated that a great deal of snorkel equipment was visible at U-boat fitting-out docks, indicating that snorkel installation was a large program. Prisoners also said that by mid-August 1944 a third of Type VIIC and nearly all Type IXs had been fitted with them. Type XXI and XXIII would also be fitted, and their snorkels would "possibly [be] raised in the same manner as the periscope." This was correct; the earlier snorkels were carried folded down on deck and raised for use. By this time it was known that *U 1230*, which had brought German saboteurs to Maine, had spent 40 of her 54-day voyage from Norway submerged. She normally snorkeled twice every 24 hours, each time for 2.5 to 3 hours.

It seems that initially many German U-boat commanders did not want to use their snorkels and remained on the surface instead. After six of them were sunk by aircraft between 7 and 10 June, the survivors began to use their snorkels.[25] Between June and 2 November 1944, 34 U-boats were equipped with snorkels for transfer to Norway, 4 being lost en route.

Snorkels almost completely disrupted air ASW, because aircraft could see snorkels and their exhaust plumes only at short range, and by that point U-boats had time to reach a safe depth. The British maintained air patrols, however, to keep U-boats submerged. That shifted the burden of ASW to escorts, which could detect snorkel heads using their radars, but only at a much closer range than they could spot a surfaced U-boat. The inshore escorts needed to distinguish bottomed U-boats from wrecks, so known wrecks were mapped; some were marked with pinging buoys. Several short-range sightings of snorkels by escorts demonstrated that U-boats snorkeling at night were unable to detect escorts proceeding at slow speed without operating noisemakers. The snorkels were not fitted with radar detectors. Patrols of small craft were organized to prevent U-boats from snorkeling in quiet areas and to report incidents. Small deep trap minefields of contact mines were laid in suitable areas and near convoy routes to destroy U-boats proceeding below snorkel depth and to catch U-boats bottoming. The British tried placing obstructions to foul snorkels but had no luck.

The British felt compelled to convoy all merchant ships in coastal waters, not least to make it possible for a hunting group to be nearby in case of attacks. U-boats seeking targets would find coastal traffic routes and lie in wait for passing convoys. A cautious U-boat lying quietly near a wreck and or other submarine-like asdic target could avoid detection if it did not reveal itself by attacking.

Although U-boats could fire homing torpedoes without using their periscope, U-boat commanders clearly preferred to see their target, so they tended to attack in daylight or in bright moonlight. None of the 75 merchant ships torpedoed in inshore waters between 1 June 1944 and March 1945 was hit when light conditions completely prevented using a periscope. The Germans were increasingly using homing torpedoes against merchant ships, but slow (quiet) speed and noisemakers would defeat them.

The Germans hoped that by using snorkels they could maintain a close blockade of the British Isles. Snorkels enabled U-boats to operate in British waters denied them since 1941. U-boats laid mines as well as awaiting coastal traffic. Further, in coastal waters they enjoyed a constant stream of information, including numerous *B-dienst* reports.[26] The first five U-boats operating in coastal waters were unexpectedly successful. Then the British reinforced the early very weak ASW patrols. Of seven U-boats dispatched in August 1944, only one survived. BdU took comfort in a claim that his force had tied down a larger Allied force. He withdrew the U-boats temporarily from the Channel, resuming the offensive in December in support of the German attack in the Ardennes. Seven merchant ships and two escorts were sunk in attacks between 18 and 29 December. In January 1945, *U 245* operated in the Thames estuary and used the new *Kurier* "squash" transmitter to communicate with BdU. In April she returned in company with *U 2322*, a Type XXIII coastal U-boat, and sank two ships (the smaller submarine sank one). U-boats also operated in other Allied coastal areas where targets were likely to be found, such as the approaches to Gibraltar; the sea off Reykjavik, Iceland; the Nova Scotia–Newfoundland area; and the U.S. East Coast from Cape Hatteras north.

Against the advantages offered by snorkels, a snorkeling U-boat had to proceed at low speed. Her commander could not take in the tactical situation at a glance, as he could from a surfaced U-boat. Communication was far more difficult, although the Germans found that by using low frequency they could transmit to a U-boat at periscope depth. However, the U-boat had to put an antenna through the surface to send messages by HF. Doenitz demanded full situation reports as U-boats completed patrols, but commanders operating in narrow coastal waters did not want to risk detection. Navigation while snorkeling was difficult; U-boats had to rely on dead reckoning and radio beacons. Prolonged submersion made the inside of the U-boat damper, damaging electronics. Pack operations were impossible because it was impossible to coordinate U-boats invisible to each other.

Without effective ocean surveillance, the Germans were compelled to operate, as in World War I, in focal areas where dense shipping could be expected. Most hunted in the Channel and in British coastal waters, in some cases subject to specially laid ASW minefields. At least one U-boat reached the North American coast and operated in the Gulf of St. Lawrence.

U-boats operating in coastal areas often bottomed, waiting for the sound of an approaching convoy. Asdic was relatively ineffective in such situations, to the point that at the end of the war the British were developing a bottom-profiling asdic (Type 162, Cockchafer) specifically to identify bottomed objects.

Allied ASW groups quickly learned how to deal with the new German tactics. Hunters sank 19 U-boats in February 1945, including 12 in the waters around England; compared with a total of 6 in January. Some U-boats were also sunk en route to their operational areas. The high rate of kills continued in March; in April, 36 submarines were sunk, including 31 in the Atlantic. The April kills included four (of a group of six) known to be headed for the U.S. East Coast; there was some fear that they were armed with missiles.[27] They were intercepted by a barrier patrol (the first in U.S. practice) consisting of escorts attached to two U.S. carrier task groups. Each of the escort carriers worked with 12 to 14 destroyer escorts to create a 120-mile barrier. The escort carrier with its own close screen operated 35 miles beyond the ends of the barrier. An air patrol was maintained 65 nm out beyond the barrier. The whole barrier was thus about 250 nm long, and it was moved in response to intelligence information. The barrier was effective despite bad weather and rough seas.

During May, U-boats tried to escape from European waters to Norway, and about ten were sunk by aircraft in Danish waters. Others may have been sunk in the heavy bombing of German ports in the Baltic. U-boats continued attacking until Germany surrendered on 8 May 1945.

Some late-war air attacks showed that ASW aircraft could still sometimes be effective against snorkeling U-boats.[28] On 30 March 1945 a Coastal Command Liberator (86 Squadron) with the new X-band radar detected a snorkel and dropped a standard five-sonobuoy pattern over a three-mile-square area. One of the buoys heard characteristic U-boat sounds. Another snorkel detection localized the U-boat, and the Liberator dropped two Mk 24 torpedoes. After six minutes the sonobuoy operator heard a long reverberating sound that was later identified as the death of *U 905*. Two days later another British Liberator (120 Squadron) detected a snorkel and dropped two homing torpedoes that killed another U-boat. On 25 April a U.S. Navy Liberator of VP-103 sank *U 1107* with another two Mk 24s.

By this time the advance of the Allied armies was significantly affecting the battle at sea. When the Russians overran East Prussia, the Germans lost access to their working-up bases. In Silesia the Russians seriously dislocated the Type XXI program; Danzig was supposed to be building almost a third of these boats.

Between 1 September 1944, when U-boat anti-invasion operations ended and the inshore offensive was considered to have begun, and 31 March 1945, 54 merchant ships were torpedoed and 42 sunk. In addition, 25 escorts were torpedoed and 18 sunk. The British estimated that 37 U-boats had been sunk, 24 by A/S vessels, 9 by aircraft, 2 by mine, and 2 by other causes.

4

NEW KINDS OF SUBMARINE

THE BRITISH DOUBTED that the Germans would abandon U-boat warfare after their defeat in the Atlantic in 1943. A panel convened in February 1944 to consider the implications of likely German countermeasures concluded that the Germans could respond in three possible ways: deeper diving, increased submerged speed, and increased submerged endurance.[1] The British were already aware of Germans' interest in closed-cycle propulsion as a means of increasing submerged endurance.[2]

The deeper the U-boat, the more difficult it was to attack. Depth charges and other projectiles took longer to reach it, giving the U-boat more time to maneuver away. Although the fast-sinking (45 ft/sec) Squid would remain effective against a deep U-boat, Hedgehog projectiles would take three times as long

The future of submarines, 1945: *U 3008* at Wilhelmshaven in June 1945. She would be ceded to the United States a few months later, providing the U.S. Navy with its initial insight into the capabilities of the new submarines. NHHC

Initially, British naval intelligence thought that the Germans' response to the ASW defeat in the Atlantic would be to dive deeper, so they developed depth-determining sonar, mainly the Type 147 "sword." The Royal Navy gave 50 of them to the U.S. Navy for installation on board destroyer escorts. These ships could be recognized by the enlarged sonar space built out ahead of the bridge. USS *George* (DE 697) is shown off Mare Island Naval Shipyard, 27 August 1947. She had been rearmed with two 5 in./38 guns in place of her previous single 3 in./50s, and with three twin 40 mm guns aft, as part of a late-war anti-kamikaze upgrade. Long- and short-hull ships received the additional sonar. NHHC

to sink to 1,300 ft and would be less effective against the predicted thicker pressure hull. Depth charges would take longer to sink and would be quite inaccurate. Depth-determining asdics would need further modification and refinement.

The Germans had already modified their standard Type VIIC U-boat for slightly greater diving depth—120 rather than 100 m. They typically sought a collapse depth twice the test depth; in this case the calculated collapse depth was 250 m (820 ft). When survival was at stake, U-boat commanders sometimes took their boats close to collapse depth.

British intelligence initially pointed to greater diving depth as the Germans' primary response to ASW; the Royal Navy's ASW experts thought that the Germans were aiming at 400 m (1,300 ft). To get there they would either thicken pressure hull plating to 1.5 in. or adopt a special steel. Constructing the thicker hull would be no problem, but it would increase overall displacement by about 20 percent. There were no reports of the Germans having special steel. The British thought it unlikely that the Germans would redesign their torpedoes and torpedo tubes, hence a U-boat evading at great depth would be unable to counterattack. By early April 1944 the British thought the Germans were aiming at diving depths of 1,500 or 2,000 ft. The Allies' existing ASW weapons and sensors were not very effective at depths much greater than 1,200 ft and would need considerable modification. All the way to the end of the war the Allies continued to believe that the Germans planned to greatly increase diving depth, despite a lack of evidence.

There was no conclusive evidence of an attempt to increase underwater speed either, but the panel considered it "a logical development, and must be expected before long as it is known that experiments with closed-cycle propulsion are proceeding." Greater submerged speed would reduce risk from aircraft and surface escorts while the U-boat was attacking or disengaging afterward, and would improve her chances of escaping if surprised on the surface. A fast U-boat could approach submerged at "comparatively high" speed to gain a favorable attacking position; she would not be subject to limiting lines of approach because she could attack from any bearing, not just from ahead. She could attack upwind in rough weather. It would no longer be safe to reinforce the van of a convoy at the expense of the rear. Convoys (and the fleet) would need more escorts. High underwater speed would throw off the accuracy of ASW attacks and would make it easier for the U-boat to escape hunters. It would be particularly useful for a U-boat trying to escape in bad weather, which would hamper the hunters. The U-boat would also be able to submerge farther away from targets, complicating the task of escorting aircraft. On the other hand, the U-boat would probably have to slow down to attack, to avoid being deafened by her own noise, and hence would be unable to judge where attacking ASW vessels were. The fast U-boat would likely use her battery for bursts of high speed.

High speed would make U-boat attacks using homing torpedoes more effective. Escorts and hunters would have to proceed at so high a speed that they would have to tow noisemakers, accepting a considerable loss of asdic efficiency. Ships might have to use expendable noisemakers, which were currently being tested. There was some hope that asdic could be made to operate at higher speed, but that refinement was unlikely to be available in the near future, and it would benefit only a few escorts.

A faster U-boat would be nearly immune to depth charge attack because she could hear the run-up and move to evade. Attacks by ahead-thrown weapons, normally conducted at slower speed, would be less detectable by the U-boat, but they were still subject to blind time; the faster the U-boat, the better the chance that she could evade. Probably all attacks would have to be team attacks, one ship tracking the U-boat and conning the attacker. The best near-term countermeasure seemed to be faster-sinking projectiles (including depth charges). The best longer-term countermeasure would be homing torpedoes launched by surface ships and aircraft.

Available data and the calculations of Director of Naval Construction (DNC) suggested that if the Germans sought higher submerged speed, they would

Unfinished Type XXI U-boats at the Deschimag yard at Bremen in May 1945 show their slender, fish-shaped hulls, in contrast to the fat hulls with prominent ballast tanks of earlier submarines. The boat at center-right seems not yet to have all six of her bow tubes, whereas the boat next to her has the tubes but not her bow. She has the pressure-proof structure inside her conning tower, but not the fairing over it (which the boat to the left has). NHHC

Bows of Type XXI submarines, captured on the slip at Bremen in 1945, show the big *Balkon* (Balcony) sonar dome underneath them. Note that the bow in the foreground has not yet been attached to the rest of the submarine hull. *Balkon* was copied by both the U.S. Navy and the Soviet navy. In contrast to previous practice, the XXIs had no stern tubes. NHHC

aim at 12–13 kts (there was some speculation that underwater speed might rise to 15–20 kts). Assumed underwater endurance was 7 hours at 12.5 kts, 12.5 hours at 8 kts, and 20–25 hours at 3 kts.

The Allies considered it unlikely that the Germans would try to combine greater diving depth with closed-cycle propulsion, as that would require a much larger submarine, perhaps 40 percent larger. Longer underwater endurance seems mainly to have meant either increased battery capacity or closed-cycle power (it took time to understand that snorkels would solve this problem).

THE BRITISH RESPONSE

At the end of the war the British were envisaging a two-stage future submarine threat. Initially they would face 15 kt submarines that could dive to 1,000 ft; after 1950 there would be 25 kt submarines capable of diving to 1,500 ft. The first prediction tallied with the supposed capabilities of the new German Type XXI (though the assumed depth was much greater than it actually was). The second was associated with the German Type XXVI, but this kind of performance did not actually materialize until nuclear submarines appeared in 1955. Initial trials with the fast target HMS *Seraph* suggested that when moving at high speed she could be tracked passively by her HE (hydrophone effect, meaning her sounds). When she slowed, existing asdics seemed to be good enough.[3] It seemed that maintenance of asdic contact was, if anything, easier against fast U-boats than with ordinary U-boats. If an attack failed, the U-boat could be held down to exhaustion. Development of a scanning asdic was thus not urgent. Immediate postwar development concentrated on counters to submarines that might have become operational had the war lasted longer. Development of entirely new equipment stopped in favor of basic research.

The initial British response to faster U-boats seems to have been to develop new weapons, but not necessarily new sonars. In 1944 the Director of the Admiralty Anti-Submarine Warfare Division (DA/SW) wrote that the tactical limitations of current fixed-range ahead-thrown weapons would make it very difficult for a hunter to reach a firing position against a fast submarine. The U-boat would be changing bearing so rapidly (perhaps 1 deg/sec) at close firing range that the escort would be unable to pivot quickly enough to hit it. A future weapon would need both considerably greater range and wider bearing arcs; it would have to be laid and trained like a gun. Control techniques would guide a ship to a good firing position and then place projectiles accurately, independent of the ship's position. The mounting would have to fire accurately at targets between 200 and 600 yds (or more) and at bearings from ahead to 90 degrees or more on either side. It would have to fire repeated salvoes at a rate of four- or six-projectile salvoes per minute. Squid projectiles seemed adequate for that.

At a 6 April 1945 meeting, a formal staff target was defined. Range was set at 200 to 1,500 yds, with all-round training subject only to constraints due to ship structure. The weapon had to be effective down to 2,000 ft, and the associated detection device had to be capable of dealing with a 25 kt U-boat. A ship should accommodate at least 30 salvoes. Rate of fire should be about that of the existing Squid. An additional requirement added later was that projectiles should enter the water at the same angle at all ranges,

Limbo was the much more successful British equivalent to Weapon A. The aim point was swept by tilting the mounting and the three barrels. Range was adjusted by letting variable amounts of gas out of the breeches of the three tubes. A British frigate shows a typical installation in a well deck aft. Ports in the sides of the well were used for loading through the muzzle, enabling Limbo to maintain a high rate of fire. NARA

greatly simplifying fire control. This weapon seems to have begun as Mortar D. By 1947 it had acquired its later name, Limbo. Mortar C, which never materialized, was presumably a replacement for Hedgehog.

Limbo's range could be varied between 400 and 1,000 yds by opening gas vents. Its three barrels were aimed by a combination of roll and pitch (also used to stabilize the mounting), with spacing between rounds achieved by short time delays between firing the barrels. Tilting the mounting and the barrels independently made it possible to cover targets over all 360 degrees of bearing. As with Squid, each round weighed 300 lbs, including 207 lbs of explosive. Estimated lethal range was 5 yds. A typical installation was 2 mortars (3 barrels each) plus a magazine holding enough bombs for 17 salvoes. Limbo was associated with the postwar Type 170 (four-square) attack (target tracking) sonar.

It later turned out that Limbo projectiles were often inaccurately set for depth; the pattern might be 100 ft or more above or below target depth. Modern submarines could change depth very rapidly. A British O-class submarine, for example, could change its

Limbo reloading on HMAS *Derwent*. In addition to the Royal Navy, Limbo was used by the Indian navy, the Royal Australian Navy, the Royal Canadian Navy, and the Royal New Zealand Navy.
DR. JOSEF STRACZEK

depth by 300 ft in 30 seconds. In 1966, therefore, the Royal Navy raised a requirement for a proximity fuze, eventually buying a magnetic fuze that had been developed for the Royal Canadian Navy.[4] At this time the British were reducing to one Limbo mounting per ship, hoping this would be acceptable provided that depth errors could be overcome by using a proximity fuze. In the 1970s they began to eliminate Limbo altogether, planning to rely on lightweight homing torpedoes in shipboard tubes as a replacement urgent-attack weapon—a decision they would rue in the Falklands.

In line with British policy, attention initially concentrated on improving the existing destroyer/frigate asdic.[5] An improved depth-determining adjunct to Type 144 had increased power, its range increasing from about 1,000 yds to 1,400 yds to deal with deeper submarines. The associated depth recorder was modified to increase recordable depth to 1,500 ft and to make provision for errors in depth due to temperature gradients in the sea and also to evading action by the submarine at depth. A new range recorder introduced new plotting scales: up to 3,000 (rather than 2,500) yds for search and up to 1,500 yds (rather than 1,000) for attack. The use of a single scale for attack made it possible to completely redesign the associated predictor. The predictor on the bearing recorder was modified to allow for depths down to 1,500 ft. A new ship's component mechanism automatically removed the effect of own-ship speed across from the bearing recorder plot (previously this had been done manually). That increased freedom of maneuver; the correct gun bearing (for the ahead-thrown weapon) could be predicted, and the sound beam automatically trained to compensate for all movement of own-ship and target. The asdic operator could step his beam across an apparently static target. These improvements were incorporated in the follow-on Type 164, which became standard in the postwar and refitted destroyers.

Ships had to be faster to deal with faster submarines; a new A/S 26 asdic dome enabled operation at 30 kts (100 in. long, aspect ratio 4:1). It was designed to house two Type 144 oscillators, one for the current Type 144Q and one for a Type 144 without a bearing recorder, with a depth-determining (Q) oscillator atop the after (lower) main oscillator. This combination, Type 160X, was developed into the postwar standard Type 164 searchlight sonar.[6]

After the war, destroyers were fitted with Squid, which replaced their depth charges aft. The ships already fitted with Squid were too slow to attack the expected fast submarines.

The most difficult problem revealed by the coastal U-boat campaign waged at the end of the war was that U-boats could bottom in shallow water, where their asdic echoes became less distinct. Not surprisingly, the first new postwar sonar was Type 162 (Cockchafer), a bottom-profiling set that could distinguish a U-boat from, say, a wreck. Until it appeared, ASW ships relied on a "dual purpose echo sounder" to produce a bottom profile that would show the U-boat if the hunter ran over her lengthwise.

By 1947 it was clear that the existing searchlight sonar could not cope with the expected next-generation submarine. A submarine running at over 15 kts would run through the area swept by the asdic so quickly that she might well avoid detection altogether; each sweep over 80 degrees to each side of the ship took too long. If the asdic did catch the submarine, it might be at too short a range. World War II average detection range was only 1,300 yds, even when the rated range of the asdic was 2,500. Work was therefore proceeding on the British scanning sonar, but it seems to have ceased in 1948 after an initial trial.

For target location and tracking, the step-across technique used in the past would be too slow, so the British became interested in applying split beams. Target bearing could be obtained on each pulse. The same technique could be applied to target depth. Initial tests were encouraging. The only problem was in measuring shallow depths, due to interference caused by reflection off the surface. By 1947 the split-beam sonar (Type 170) was associated with the new Limbo mortar.

THE U.S. PROGRAM

The U.S. Navy devised a long-term ASW upgrade program to meet the threat posed by the new submarines.[7] This program resulted in the new weapons and sensors of the immediate postwar period. It first addressed the weaknesses of existing equipment: short detection range, inaccurate sonar bearings, no depth determination, problems due to limited ship maneuverability (particularly depth charge attack, which demanded accurate conning, but also Hedgehog attack), and the lack of direction equipment. Of these, only inherently short sonar range was not being addressed. Other than depth charge attacks, ahead-thrown attacks were badly affected by adverse weather, as the roll of a ship would throw off aim. Squid could place depth charges more accurately than could the usual depth charge equipment, but it suffered from large control errors.

If ahead-thrown weapons could be made trainable, with variable range, they would be far more effective per salvo than existing fixed-train (and range) weapons. They could deliver many more attacks within a given period of time, and they would remain effective as the firing ship maneuvered.

The new program was laid out in a report written in 1944 and revised in the spring of 1945. The report summarized what was currently available and what could be available on an interim (1946) basis and ultimately (1948). Although depth charges were falling out of favor because they were less effective than ahead-thrown weapons and homing torpedoes, a new Mk 14 fast-sinking depth charge with an acoustic Doppler firing mechanism and the usual hydrostatic mechanism that could fire down to a depth of 1,500 ft was under development. There was also a magnetic pistol for the standard Mk 8 depth charge. Alternative-influence pistols were being investigated. Development of the Mk 35 ASW homing torpedo was ordered in October 1944 (see appendix C).

BuOrd had considered adopting Squid as the ultimate U.S. ASW weapon (U.S. naval officers who tried it liked it). But a much more effective weapon could be developed without much added complication by giving Squid train and variable range (using variable elevation). Faster submarines had to be countered by more flexible weapons.

BuOrd thought the time was right to develop an effective fire control system for the ahead-thrown

The U.S. Navy developed Weapon A (or Able or Alfa, depending on the phonetic alphabet of the time) as a counter to the new fast submarines. It was an automatically reloaded rocket gun that fired a fast-sinking projectile. The project was considered so vital that Weapon A was ordered into production after the war without having been operationally evaluated. This one, in No. 2 position, is on board USS *Waller*, a *Fletcher*-class destroyer converted for ASW as a DDE. The ship is firing her port Hedgehog. USNI

weapons. The Attack Director Mk 4 was the result. Its gun fire control systems were all regenerative. The target's estimated course and speed (including depression, in the case of a submarine) were fed into the system, which calculated (generated) target position based on them. Comparison between generated target position and measured target position was fed back into the system to correct assumed target course and speed; when they matched, the solution was verified, and weapons could be aimed and fired. Sonar data fed into Mk 4 would be specially smoothed (filtered), and the sonar projector would be stabilized.

Four preproduction QHB scanning sonars were undergoing operational tests. Scanning was expected to increase effective sonar range by making search more effective, offsetting to some extent current short contact ranges. A laboratory version of a stabilized tilting-beam sonar was being tested on board the destroyer escort *Foss*; it was intended to support initial tests of Attack Director Mk 4 "in the very near future," which would mean soon after April 1945.

Current plans included installing modernized sonars and depth-determining sonars (QDA, equivalent to the British Type 147 sword) on 500 ships of the destroyer, destroyer escort, frigate (PF), and 220 ft minesweeper classes. Target depth could then be determined within 45 degrees of the ship's bow. A prototype vertical scanning sonar (the preferred depth-determining option) was being tested in the sonar trials ship *Cynthera*; another (designed as part of an integrated sonar) was expected to be ready in June or July 1945.

In the "1946" scenario, weapons would be much the same as in 1945. Ships would have Attack Director Mk 4 Mod 1 (without depression rate or thermal gradient correction) as well as fire control equipment to improve the effectiveness of Hedgehog and depth charge attacks. Ships would have the current searchlight sonars, but they might also have horizontal scanning sonars, "providing certain unknowns are established." They would have tilting sword-type (QDA) depth-determining sonars. Horizontal-scan sonars might be modified to feed their data to attack plotters.

Current programs would reach fruition in 1948. BuOrd was working to improve ahead-thrown weapons. Its Project I would increase the sinking rate and the explosive charge of Hedgehog projectiles.

Weapon A was to have been the standard postwar U.S. Navy ahead-thrown weapon, arming a wide range of new destroyers as well as destroyer escorts. The ASW cruiser *Norfolk*, actually a large destroyer intended to lead ASW groups, is shown on 19 September 1957. She had two more Weapon A turrets aft, as well as four fixed torpedo tubes for homing weapons (Mk 35 torpedoes). The muzzles of two of them can be seen on the side of the after deck house, one almost hidden by a ship's boat between the two superstructure elements. The twin 3 in./70 guns were considered the best available weapons. Note the radar dish, feeding an on-mount director, on No. 2 gun mounting, and the Mk 56 director atop the forward superstructure. A planned second ship was also authorized in FY 48, but she was deferred in 1949 (presumably due to the budget crunch) and then canceled in 1951. Another four *Mitscher*-class large destroyers projected for the FY 49 program were not authorized. NHHC

Project II would develop a new trainable ahead-thrown weapon (Weapon A) to give it a variable range of 400–800 yds, a charge of about 250 lbs, and a firing arc within about 140 degrees on either side of the bow. The weapon would be elevatable and stabilized. Sinking rate would be at least 40 ft/sec. Project III was Weapon B, a smaller equivalent of Weapon A (50 lb charge) to fit smaller ships. For both, rate of fire should be as rapid as possible.

To gain maximum benefit from the proposed combination of sonars and more sophisticated fire control equipment, the two bureaus (BuShips and BuOrd) recommended rearranging the equipment. Ships currently carried their sonar racks and attack plotters in a sonar hut that was typically adjacent to the bridge. In "1946" the Attack Director might be located there. The CIC would have remote indicators for the sonars and for the Attack Director. Ultimately, however, control of an antisubmarine attack should be centralized in a new station that would serve much as the plotting room currently served for gunfire control in a major combatant. For destroyers the new space would be adjacent to, or part of, the plotting room. The bridge and the CIC would have only necessary indicators and firing safety devices. The two bureau chiefs recommended that by 1 January 1946 ships should be provided to test this arrangement; by 1947 they should be available as a unit to demonstrate effective surface ship ASW.

This report laid out much of the U.S. surface ASW development of the five postwar years, particularly the concentration of equipment and the advent of trainable ahead-thrown weapons. Weapon B never materialized, but Weapon A (Able or Alfa, as the phonetic alphabet changed) became a standard postwar ASW weapon.

THE NEW U-BOAT: TYPE XXI

Type XXI, the "electro-U-boat," was the first entirely new U-boat. It was capable of higher underwater speed, and its underwater endurance was greatly increased thanks to its much larger battery. A Type XXI could make 15 kts submerged for an hour. More important, it could sustain a high enough submerged speed to remain beneath a convoy while firing torpedoes up into the ships above, using the noise of the convoy to avoid detection. It also dived somewhat deeper. Design test depth was increased from the 120 m of Type VIIC/41 to 135 m (443 ft), which would normally equate to a crush depth of 270 m (886 ft; i.e., the usual safety factor of 2.0).[8] The Type XXI's designed safety factor was 2.5, giving a crush depth of 1,100 ft for a test depth of 440 ft.[9] In 1947 the British calculated the Type XXI's test depth to their standards (four-sevenths of crush depth) as 450 ft. Their structural model of a Type XXI collapsed at 780 ft, but the British continued to believe that the Type XXI was a very deep diver. In fact, due to design weaknesses a Type XXI could not exceed 180 m (590 ft).

The U.S. Navy received two Type XXIs as part of its share of surrendered German U-boats. *U 3008*, her conning tower drastically streamlined, is shown off Portsmouth Naval Shipyard, 30 August 1946. The British were also supposed to receive two Type XXIs, but a battery explosion convinced them that refitting Type XXIs sufficiently to make them suitable for experimental service would be too expensive. They transferred the better of their two to France (*U 2518*, which became *Roland Morillot*). NHHC

The Allies gained insight into what the Germans planned to do from decrypted messages from the Japanese naval attaché in Berlin.[10] In a 31 December 1943 message, read in June 1944, the attaché described two new types of U-boat based on information from one of his subordinates who had been allowed to view plans and drawings and had glimpsed parts of one of the craft. The following month the attaché saw plans at the Deschimag yard. The chief feature of the new U-boats would be high underwater speed rather than greatly increased diving depth. The attaché thought that the new U-boats were already under construction, the first to be completed around May 1944 and in service about August or September. In August he provided speeds and underwater ranges of both Type XXI and the coastal Type XXIII.

As early as October 1943 the British Naval Intelligence Division asked DNC to investigate the possibility of converting an existing British submarine into a fast underwater target in order to develop tactics to counter fast U-boats. DNC estimated that a fully streamlined British submarine could make 13 kts for about 20 minutes. Late in May 1944 the British received evidence of the German Type XXI program in the form of reports of prefabrication of a new type of U-boat displacing about 1,000 tons with fine hull lines. They began to test tactics using Attack Teachers, but they really needed a fast target submarine.

Given his earlier studies, DNC was ready with a conversion proposal when he attended a 6 June 1944 meeting to discuss the threat presented by Type XXI.[11] HMS *Seraph*, an existing S-class submarine, was hurriedly modified for tactical trials to be undertaken in September–October 1944. She was given higher-powered motors and new propellers (for 13 percent additional underwater power) and was drastically streamlined, with reduced planes and a cut-down bridge structure (for 55 percent reduced drag). As converted, she could maintain 12 kts for 45 minutes.

Early trials showed that although the echo from a streamlined submarine was not much reduced broadside-on, it was considerably reduced end-on. However, when moving at high speed the Doppler-shifted echoes the submarine generated enabled operators to distinguish the weak echo against background noise and other interference. Using Doppler, an operator could hear echoes even when the asdic recorder could not distinguish them. Against that, the submarine's underwater wake interfered with asdic operation over a 20 degree sector dead astern. The *Seraph* tests also revealed that she had a faster "silent" (noncavitating) speed because at lower speeds she needed less power and was turning her screws more slowly.[12]

On learning of the characteristics of Type XXI, the British quickly converted HMS *Seraph* into a fast target in order to develop applicable tactics and train ASW personnel. Additional submarines were converted after the war, on the theory that the Soviets would soon build equivalents to Type XXI. The British used fast-running motors with inherently noisy propellers in the conversion, and were thus convinced that passive sonar would be a viable means of detecting the new submarines. That was hardly true; the Germans paid considerable attention to silencing their propellers. This was a much less expensive program than its U.S. equivalent, which was to convert existing submarines into equivalents to Type XXI under the Guppy program. HMS *Seraph* is shown on 30 March 1953. BAKER COLLECTION

Initial conclusions from the trials proposed that convoy ships should slow (for self-silencing) and conduct an all-round hydrophone search every five minutes. Ships with well-trained plotting parties could find a fast submarine by exchanging passive bearings from such searches. By this time the Germans were using homing anti-escort torpedoes, and escorts were towing noisemakers, which could interfere with their sonars. Escorts might have to intersperse bursts of high speed with slow-speed listening.[13]

Tactical exercises against *Seraph* were initially disappointing, but the escorts learned from them, and by the end of the trials they managed to hold contact on *Seraph* (which was making 9 kts nearly 40 percent of the time) for two hours, making five attacks. Hydrophone effect was used to hold contact while *Seraph* was running at high speed, and the group of escorts waited for her to slow down before attacking—as she had to do eventually because her endurance was limited at high speed. It appeared that a fast U-boat could elude ships altogether only in rough weather. Trials also showed that the automatic-stepping Type 144 asdic with bearing recorder and helmsman's indictor enjoyed a tremendous advantage over earlier versions.

The naval establishment suspected (correctly) that a Type XXI could drastically reduce her running noise by going deep. It also turned out that the Germans had effectively silenced the Type XXI's propellers. The A/S establishment at Fairlie pointed out in March 1945 that because of their limited high-speed endurance, the new U-boats typically would not be proceeding at high speed when attacked, and need not be attacked if they were running at high speed. Further, high speed would be a considerable aid to classification because it was evidence that the asdic ping indicated a real target. Unless it had been fitted with gear to heterodyne asdic pings, the U-boat would be acoustically deaf at speed. Fairlie concluded that its small margin of speed did not make the new fast U-boat a very serious problem. Typically, escorts would be attacking U-boats running at normal speeds. Attacks on U-boats at high speed would not necessarily be less accurate.

No one could gauge how noisy a Type XXI was until one was available for tests. The British were unable to run one through their sound range, but the U.S. Navy took over two Type XXIs and tested them thoroughly. Initial assumptions that a Type XXI would be quite loud at 12 kts proved incorrect; tests showed that she was quieter at 12 kts than the best U.S. fleet submarines were at 6 kts. Up to 10 kts the Type XXI was extremely quiet; above that speed there was some cavitation, and passive sonar range improved.

The great question was how quickly the Germans could build Type XXIs and make them operational. The Germans continually postponed the date. On 26 July 1944 the Japanese ambassador to Germany informed his government that the U-boat's "effect will first be felt from about September, and by November or December it is said that greater results will be effected by its use in great quantity." Doenitz was later quoted as saying that he wanted to introduce the new submarines in quantity, meaning that they would not appear until the winter of 1944–45. On 3 December 1944 the Japanese attaché reported that nothing was likely to happen until the following March. That month he reported that boats would be ready in quantity in May or June. Even that was overoptimistic.

Delays in the program could be attributed partly to Allied bombing. In hopes of completing a large Type XXI fleet, the Germans had dispersed the production of prefabricated sections, which would be brought together and assembled only at a few selected shipbuilding centers. Prefabrication imposed a high degree of quality control to ensure that the pieces fit together, and the final assembly yards had to do extra work to make that possible. Further, the program required that details of the Type XXI be given to many shipyards, some of which fell to Allied armies well before Germany collapsed.[14]

Plans to build Type XXIs dated to May–June 1943; the Deschimag firm began work on a sketch design on 1 July.[15] Deschimag lacked sufficient capacity, so a central drawing office was established, with a goal of finishing all drawings by December. The program seems to have been delayed because the high command (OKM) did not want to prioritize a new kind of submarine after the U-boat fleet had been defeated in mid-1943. The XXI's production was assigned to Albert Speer, who made excessive promises. Among other things, he promised to deliver 4 boats in April 1944, 10 or 11 in May, and reach a peak production rate of 33 per month from August 1944 onward. Production began before the drawings were ready. In some cases, prefabricated sections arrived at the building (assembly) yards only 35–50 percent complete. The designs turned out to be fatally flawed, and the yards had to modify them. Endless modifications delayed the program by about a year. The first Type XXI was completed by May 1944, but its teething troubles were never resolved, and it was never operational. The first submarine to embody all the required modifications, *U 2511*, was completed nearly a year later. Air attacks added considerably to fitting-out efforts, the workload per boat increasing from a planned 84,000 man-hours to 110,000. Assembly effort on the slip rose from 65,000 to 90,000 man-hours. Alterations to make the first boats combat-ready averaged 85,000 man-hours.

Construction never lived up to the plans. According to a 1945 British report based on German documents and testimony, as of April 1945 the shipyards had delivered 115 Type XXI submarines rather than the 287 promised by that time, with planned production rising to 32 per month in October 1944. In fact, the best month for deliveries was December 1944, when 29 were delivered; in January 1945 that fell to 14, with 8 in February, 7 in March, and none in April 1945.

Could Type XXIs have overwhelmed convoys using the wolf-pack tactics that had been so successful in the mid-Atlantic gap in 1942–43? Maintaining contact while a pack concentrated had been difficult even in 1942–43. For example, it was difficult for a shadowing U-boat to stay in contact when forced to dive while a convoy executed an emergency turn. When U-boats could operate freely on the surface, about a third of contacts led to attacks, and about a third of attacks produced hits. U-boats in a pack depended on cueing from the U-boat command ashore, so navigational errors could easily frustrate attacks. Existing torpedoes, moreover, limited firing positions.

Postwar interrogation of U-boat officers, including the first three assigned to Type XXIs, established that the Germans had not thought through most of the questions that had to be resolved to use them effectively. The Germans expected to penetrate convoy screens at a silent speed of about 5 kts, the submarine being bow-on to limit her echoing area (for asdic). Once through the screen, the U-boat would take up station under the convoy and fire torpedoes up at the ships. For the moment, at least, it seemed that the Germans underestimated the possibility that the convoy would execute an emergency turn and expose the U-boat. Overall, it seemed that the Germans placed far too much faith in further technological advances. For example, they envisaged pack attacks but grossly underestimated the coordination problem.

Once the British were aware of the new U-boats, they took Doenitz' building program very seriously. In mid-1944 they expected that 15 would be operational by the end of 1944, and perhaps 95 by April 1945.[16] A 31 January 1945 paper issued by the British Combined Chiefs of Staff predicted that about 30 Type XXIs and 15 Type XXIIIs could be ready for operations during February—sufficient numbers to start a campaign in strength. The First Sea Lord predicted that the main U-boat offensive in strength could begin to take shape about mid-February. He went so far as to suggest that the situation was coming to resemble that of 1918. If U-boat construction and training continued at the present rate and the necessary diesel oil (10,000–12,000 tons per month) was available, the Germans might be able to maintain about 70 U-boats on patrol, increasing to more than 90, compared with about 60 during the height of the U-boat campaign in the spring of 1943. If the German submariners were able to maintain their morale, they might be able to sink about 70, rising to 90, ships per month compared to 60 per month in the spring of 1943. The Naval Staff estimated that the Germans employed a total of 120 operational boats during the period 1 October–31 December 1944: an average of 6 or 7 (rising to 10) in British inshore waters, 20 on Russian convoy routes, and about 10 on patrol elsewhere. During this period 13 merchant ships and 10 U-boats were sunk.

The Naval Staff assumed that the U-boats' relative lack of success in the target-rich British coastal waters was due to inexperience and "lack of offensive spirit." They speculated that German commanders were reluctant to risk their lives for a war so obviously lost, but also that they had been encouraged by the recent German ground offensive (in the Bulge); of 26 successful attacks, 12 had been delivered since the start of the offensive. Over the last fortnight (to 6 January 1945) merchant ship sinkings had increased without any commensurate increase in U-boat sinkings.

At this time the Naval Staff estimated that by mid-February the Germans would have delivered 40 Type XXIs and would add 15 each month after that. They would also have completed 20 Type XXIIIs and would complete 8 each succeeding month. The Germans had 180 older-type submarines and were expected to add 10 each month; but production of this type had ended.

Prisoners of war had told the British that the Type XXI offensive was to begin no later than mid-March, and the Naval Staff had no explanation for why the new boats were not already in service. Possibly the yards were experiencing technical problems, but there was no evidence of that. It was therefore assumed that the first wave of U-boats would leave the Baltic in mid-January and reach patrol areas around the United Kingdom by early February. The Type XXIs would operate in packs against Atlantic convoys, with an estimated one-third of the fleet on patrol at a time. Their higher speed would make it possible for them to spend more of their time on patrol. A combination of snorting and underwater cruising on battery would enable a Type XXI to average 8 kts, and thus to follow a convoy without ever surfacing. The new burst transmitter (*Kurier*) would make any warning of a convoy attack unlikely and would seriously restrict offensive activity by the escorts. The British tracking organization would be hampered by lack of fixes, and diversionary routing would become less effective. On the other hand, it appeared that U-boats would still have to run on the surface to intercept convoys and hence would be subject to air attack.

Fortunately for the Allies, nothing like this number got to sea, although reportedly as many as 140 were complete when Germany surrendered. The first XXI sent on a war cruise, *U 2511*, was outbound the day Germany surrendered.

As for coastal waters, due to their small size, only a fifth of the complementary coastal Type XXIIIs could be maintained on patrol. Only four of them made war cruises.

By the fall of 1944 the Western Allies were looking toward the end of the war and considering the use they might make of captured German U-boats. In October 1944, for example, the senior British submarine officer (FOSM) wanted to maintain two operational Type XXIs and two Type XXIIIs (one to commission and one to cannibalize). Immediately after Germany surrendered, during the summer of 1945, British crews manned *U 2502* and the Type XXIII submarine *U 2326* for trials. *U 2502*, the oldest of the type, suffered from serious defects, and was discarded in favor of *U 3017*. These submarines were available because the Admiralty was acting as caretaker of surrendered U-boats. Most Type XXIs were still in German ports at the end of the war, never having become fully operational, but 12 were in British hands. Under an Anglo-American agreement, the two navies were to share all tactical test data.

The British planned to run a Type XXI over a sound range and also to subject her to thermal detection tests. The latter suggests an early attempt to restore the ability of maritime patrol aircraft to find submarines in the open ocean. Through the summer of 1945 the Type XXIs in British hands began to experience real problems. An August 1945 battery explosion disabled *U 3017*. A Court of Inquiry decided that this had probably occurred because the British were using the Germans' operating practices rather than their own. Main electrical cables run in the bilges under the battery seem to have been poorly insulated; a German engineer recalled four battery explosions. Extensive shipyard work seemed necessary to make *U 3017* safe enough for trials, and she was instead reduced to care and maintenance.[17] Later, when the U.S. Navy was running two Type XXIs experimentally, a U.S. officer remarked that these submarines had been designed for a wartime operational life of about one year; by 1947 the two were running down, and the U.S. tactical trials unit (OpDevFor) was looking forward to replacing them with converted U.S. submarines (Guppies). The shortage of spares and even batteries in postwar Germany complicated experiments.[18]

The British soon realized that their converted S-class submarines could not match Type XXI's performance, and thus could not provide full experience of high-speed submarine operation. They and the U.S. Navy badly wanted a really fast submarine that would enable them to explore the problems of submarine control at high speed. The British approach was to strip the submarine *Scotsman*, remove her diesels, and fill her with batteries. *Scotsman* is shown as completed in 1948.
BAKER COLLECTION

Seraph was a good enough model to give the British a sense of what it would be like to fight a Type XXI, but they were well aware that they needed something faster. They modified another S-class submarine, *Scotsman*, much more extensively. She attained 16 kts submerged but, like *Seraph*, lacked sufficient power to recharge her battery.

TYPE XXI TACTICS

A group of British naval officers led by CAPT G. H. Roberts, RN, visited Germany in the summer of 1945 to interview U-boat officers; they were particularly interested in tactics planned for Type XXI. Captain Roberts was the senior British ASW tactician, having headed the Western Approaches Tactical Unit (WATU). He and his colleagues were surprised to learn that the Germans did not have a coherent operational concept for Type XXI.[19] Battle instructions for Type XXI were based on trials using single units.

A summary of Doenitz' Type XXI instructions was published in the final issue of the *Admiralty Anti-Submarine Report* for September–December 1945, CB 04050/45(7).[20] XXI commanders were urged to proceed submerged in areas subject to heavy enemy air patrols, surfacing to charge only if the sea state precluded snorkeling. Appended data showed how far the submarine could run using various combinations of snorkeling and surface running. For example, a boat that spent only 3 hours on the surface, otherwise snorkeling or crawling, could make good 114 nm in a day. In an area with weak air patrols, she might spend as much as 11 hours per day surfaced, and she could make good about 180 nm.

When discovered by aircraft, the U-boat should alter course radically and make off at maximum submerged speed, subject to the state of her battery. She should put as much distance as possible between herself and the datum where she had been found by the airplane. That would be particularly important if the airplane had called in a surface group.

Although Type XXI had no stern tubes, her six bow tubes could be reloaded quickly: a second salvo could be fired after 5 minutes, and a third after 20 minutes. XXI commanders were told that the XXI's "fighting power is adequate to destroy powerful surface forces, but it has not sufficient strength to fight its way on the surface and it must, therefore, sneak up on its enemy unobserved." Most important, the submarine had to remain unobserved in the operational area at all costs; "this you can do with the Type XXI." U-boats had to operate in the open sea because areas near enemy ports were too well patrolled. "When convoys are scattered the lack of aerial reconnaissance and the small visual range of the U-boat make the finding of the enemy the most difficult problem; the problem is the expanse of the Atlantic. The countermeasures, the quickness of their appearance, the number and excellence of available aircraft and ships and of their equipment mean that for the enemy there is no such thing as the expanse of the ocean. For him it is merely a pond."

Doenitz seems to have imagined that better German aircraft operating in numbers far from their bases might solve the problem of contacting convoys in the open ocean. However, no such aircraft existed in 1944–45, and cooperation with the Luftwaffe was always problematic.

Much of the document was written to reassure U-boat commanders that Type XXI offered revolutionary improvements, and hence they should be willing to go back to sea to face the Allies' excellent countermeasures. Skeptical commanders were assured that "each Type XXI represents a more serious threat to the enemy than any earlier U-boat. Each boat has to be intensively attacked and, therefore, causes a much greater splitting up of enemy defense forces. The struggle for mere existence in the operational area will be much easier than in comparison with present efforts."

Central control would assign a large patrol area to each U-boat. If submerged, the submarine would rely on her hydrophones. The boat should surface if the contact was lost and proceed along the last bearing to conserve battery power. She should submerge only if aircraft approached. On hearing enemy radio telephone traffic (i.e., short-range tactical radio) she would run along the indicated bearing at maximum speed to place herself ahead of the convoy. If enemy aircraft were sighted, the U-boat would remain submerged and operate along the enemy's presumed mean course. When within visual range, the U-boat would come to periscope depth. By night, the U-boat would rely on her hydrophones, using her radar search receiver to warn of enemy aircraft. "By skillful use of [search receiver], *Balkon* [passive sonar array], hydrophones, and MF/DF, surface attacks will frequently be possible." When submerged at night, the U-boat would carry out a blind attack using her hydrophones and her *Nibelung* (active sonar). "In general after judging the weather and hydrophone listening conditions, the CO must decide whether he is going to approach surfaced or submerged, crawling or noisy."

Doenitz did include instructions for Type XXIs operating in a coastal area, but that role was generally to be reserved for the small Type XXIIIs.

Doenitz envisaged a variety of dispositions to locate convoys. One possibility was to station many boats in a checkerboard pattern around a chokepoint. Boats would not report contact, but any boat in the direct path of a convoy would attack it. An alternative was to dispose boats across the bottleneck as scouts working with one or more groups outside, submerged. The scouts were to report to the operational control center via either conventional radio or *Kurier*. Control would set up the attack. If the convoy changed course after it emerged from the chokepoint, however, the ordered attack might fail. Another possibility was a patrol line placed athwart a convoy route, Type XXI being more survivable than in the past because it could patrol submerged. However, convoy information would be unreliable, so only a few of the boats in any patrol line would make contact, and it would no longer be possible for U-boats to get ahead of a convoy thanks to their superior sustained speed. Yet another possibility would be for boats to be led to a convoy by aircraft beacons. Doenitz admitted that constant convoy reconnaissance by aircraft, which he had briefly enjoyed in 1943, might not be available, but if it was, patrol lines could be more concentrated. On receipt of an aircraft report, boats would move off singly ahead of the convoy or form small patrol lines of two to five boats to increase the chance of finding the convoy. All this would be done submerged, contact with the convoy being obtained entirely from hydrophone bearings. Boats would report, if possible, by *Kurier* before attacking. "If an attack is not possible the boat's most important task is sending the report." If necessary, it should surface to radio. Control would contact boats using low-frequency radio, which they could receive at periscope depth.

In laying out attack procedures, Doenitz pointed out that Type XXI was "not a submersible, but a submarine proper." If an enemy was found, the submarine should run in submerged, even at a "reckless" cost in battery charge. "Even if you fire with a battery capacity of only 20 percent of the total, that is sufficient to last you through a long period of depth-charging." "As you are unable to tell whether you will have a second chance of attaining an attacking position, you must hit with your first attack. . . . [I]n spite of long-range and LUT torpedoes, you must get as close as possible." Although sonar (*Balkon* and *Nibelung*) made it possible to discover and attack the enemy from a greater depth, "your eyes provide the best view. If you can carry out your attack as a surfaced attack by night or a periscope attack by day, try one of these methods first. You can always change to an attack with your technical eyes later."

Tactical possibilities included an unobserved attack from outside the close escort of the convoy, the U-boat taking a firing position 3–5 nm away. The U-boat should fire from dead ahead to get the most out of a LUT pattern-runner. The U-boat should slip through the outer escort, crawling (at minimum speed) to avoid enemy hydrophones. The boat should attack using her periscope (*Nibelung* would not always work at extreme range). Immediately after firing, the U-boat should reload to attack again. Long-range shots would have only a limited chance of success because firing data would be inaccurate, the pattern-runner would have few effective legs, and the enemy's evasive action would be effective. The situation would improve once the expected long-range torpedo (12,000 m at 40 kts) entered service.

Another possibility was to attack unobserved from inside the escort screen. Compared with earlier

U-boats, Type XXI was a larger asdic target, but she was very quiet when crawling at five knots. Another option was to attack from average depth using *Nibelung* and not showing a periscope, although the attack should begin with a survey of the situation by periscope. The U-boat would then approach at 50 m or deeper (choosing the best depth to avoid detection) using mainly her hydrophones but with slight use of her *Nibelung* active sonar. She should fire from a depth of 30–50 m when in the best position (right ahead). Outside the screen, *Nibelung* was to be used sparingly and at minimum power; that did not apply once the U-boat was inside the screen.

Another alternative was a forced attack, meaning forcing through the convoy screen at high speed (12–15 kts). The U-boat should aim to dive unobserved under the outer screen to delay countermeasures during the long run between the outer and inner screens. She should approach the inner screen at the "lowest high speed" necessary to reach a firing position. As soon as the U-boat was spotted, she should break through the screen at high speed, disregarding pursuit or depth-charging.

"If possible, fire on your approach run. If this is not possible, get under the convoy and attack from below." Doenitz devoted a separate paragraph to this "shark attack." Doenitz pointed out that a convoy's merchant ships offered some protection to a U-boat hiding beneath the convoy. The U-boat should attack from ahead or astern, at a depth of 30–50 m (about 100–164 ft). The boat's high underwater speed would enable her to stay under the convoy and make several attacks. There would be little danger from sinking ships.

After the last attack, the U-boat should go deep and crawl, then turn about and move off laterally astern. She could use her *Nibelung* and hydrophones to avoid "after sweepers." They would be running at high speed and rescuing survivors, hence the chance of detection would be slight. Compared with a periscope, *Nibelung* was more difficult to discover, and there was no risk of the boat being rammed. Against that, the boat would see less during the approach and would not know whether she had been seen by the enemy. Type XXI could change position quickly, and could turn quickly than earlier submarines.[21] If discovered, she could quickly reach a safe depth. She could attack in heavy weather, which would limit enemy countermeasures.

If the U-boat was seen, battery charge permitting, she should make for the convoy at high speed, diving under the escorts and carrying out a forced attack, using *Nibelung* either from the flank or from inside the convoy. She should proceed with the convoy, reload, and fire again. If battery capacity did not permit, the U-boat should make for the convoy at high speed (on the surface), dive to a considerable depth, and get ahead by crawling under the convoy.

Overall, Type XXI sonars offered both all-round listening and the ability to find the direction of sounds they detected. They also had radar (*Hohentwiel*).

Doenitz seems to have been unaware of the Allies' ahead-thrown weapons and homing torpedoes, so he devoted a section of the instructions to an analysis of counters to depth-charging. Type XXI, he wrote, could reach greater depths (135 m if a safety factor of 2.5 was accepted), had high crawling speed, high underwater speed (making it possible to handle greater alterations in trim and in weight), larger battery capacity (to make hold-down tactics practically impossible), the ability to pump out more water (the boat had two bilge pumps), and much better hydrophones to make better sense of escort movements. Boats were to go deep (to 300 m or more) and go off at maximum crawling speed to gain space (maximum asdic range was given as 5,000 m). If forced to go at higher speed, the boat should carry out wide evasive maneuvers and alter course again when slowing to crawl speed. The boat was to seek a good counter-attacking position to fire Gnat. "The days of sitting impotently in the cellar are past and gone," he said confidently.

Doenitz listed Type XXI endurance at various speeds, emphasizing how much shorter its passage even to distant patrol areas in the South Atlantic and Indian Ocean would be. A Type IXC would require 50 days to reach Panama, but a Type XXI would need only 28.

The Germans planned to use Type XXI to revive anti-convoy warfare in the open Atlantic, even though the battle instructions issued to commanders were based on tests by single units. As yet they had no detailed concept for operating packs of Type XXIs, particularly if the packs were to operate submerged. They thought that the new *Kurier* burst transmitter would defeat Allied HF/DF. On the other hand, they admitted that because Type XXI was more complicated than earlier U-boats, the submarine would probably would not stand up as well to depth-charging.

Experienced German U-boat officers emphasized the value of the sustained medium submerged speed of 10–11 kts, which left sufficient battery capacity for the submarine to evade after attacking. The boat would penetrate the screen silently, at about five knots; even at eight or nine knots she was no noisier than a Type VII at three or four. The British were aware that U-boats might try to shelter under a convoy and had found that a sudden change of convoy course could uncover the intruder—which might not detect the change. Captain Roberts asked a captured U-boat commander how much he could deduce of the convoy's movements while beneath it but did not want to hint at the new evasion technique. One of the British interrogators concluded that the British were about three months ahead in anticipating the impact of Type XXI, and that they tended to give the Germans more credit than their tactical ability warranted.

The British officers concluded that as yet the Germans had no coherent concept of Type XXI operations, simply believing that the boat's new technology would solve operational problems. Looking back in 1948, the director of the Torpedo and Anti-Submarine Division (of the Admiralty) wrote that the increased torpedo range was the most important new development because it would enable a submarine to attack before escorts could detect it. Too, a fast submarine could fire from any position around a convoy. In the past, given their limited underwater speed and short-range torpedoes, submarines had had to fire from ahead. The new submarine could, moreover, shadow a convoy while submerged, at a speed of about 10 kts and could expect to attack any target moving slower than 25 kts from any position other than broad on the beam, including from astern. A submarine could stay under a convoy while firing all her torpedoes, and she could stay deep for days to avoid detection. The combination of longer range and faster reloading would have made up for the difficulty of using wolf packs of submerged submarines, which could not easily communicate among themselves except by underwater sound (in 1948 the British considered the detectability of such communication overrated).[22]

There was some question as to whether the Germans had overengineered Type XXI. Some American officers who examined the captured Type XXIs thought that they were far sturdier than their U.S. counterparts, clearly reflecting the lessons of the Battle of the Atlantic. The chief U.S. submarine designer, however, wrote in 1946 that Type XXI was too fragile and too complex to maintain. By that time the U.S. Navy had learned just how difficult it was to maintain its two Type XXIs, partly because spares were scarce in devastated postwar Germany. By 1947 the U.S. Navy was modifying existing submarines to achieve the sustained high underwater speed associated with Type XXI. It was also designing its own fast battery submarines, justifying them as a way of exploring the potential of the new submarine technology so that countermeasures could be developed.

DISTRIBUTING TYPE XXIS POSTWAR

In July 1945 the Potsdam Conference participants decided that most captured U-boats would be scuttled. The rest would be shared equally between the three Allied countries.[23] Ultimately this meant that each would receive ten U-boats, presumably from those in British hands. The Tripartite Naval Commission did not make initial allocations until October 1945. By that time two Type XXIs had been covertly transferred to the U.S. Navy (in August): U 2513 and U 3008; the Navy cannibalized U 2506 for spares. The U.S. Navy did not want any of the Type XXIII coastal submarines. The commission allocated two Type XXIs to the Royal Navy: U 2518 and U 3017. The British had already tested a coastal Type XXIII, presumably of special interest before the end of the war because it seemed likely that the Japanese would have comparable submarines.

The Soviets received four Type XXIs plus one coastal Type XXIII and five conventional U-boats with snorkels: four Type VIICs (*U 1057*, *U 1058*, *U 1064*, and *U 1305*) and one Type IXC (*U 1231*). All of these submarines remained operational after the war, whereas the Royal Navy and the U.S. Navy soon disposed of their ex-German conventional submarines. The four ex-German Type XXIs (*U 3515* [*N 27*, later *B 27* and then *BSh 28*], *U 2529* [*N 28*, later *B 28*], *U 3035* [*N 29*, later *B 29*], and *U 3041* [*N 30*, later *B 30*]) served in the Baltic Fleet until 1955. *B 27* survived until 1972 as a pier trainer; the others were broken up in 1958–59.

The situation was complicated by the Soviets' capture of Danzig, a center for Type XXI assembly, on 30 March 1945. The Soviets captured 20 incomplete Type XXIs: *TS 5* (ex-*U 3538*, renamed *R 1* in 1947), *TS 6* (ex-*U 3539*, renamed *R 2* in 1947), *TS 7* (ex-*U 3539*, renamed *R 3*), *TS 8* (ex-*U 3541*, renamed *R 4*), *TS 9* (ex-*U 3542*, renamed *R 5*), *TS 10* (ex-*U 3543*, renamed *R 6*), *TS 11* (ex-U 3544, renamed *R 7*), *TS 12* (ex-*U 3545*, renamed *R 8*), *TS 13* (ex-*U 3546*), *TS 15* (ex-*U 3547*), *TS 17* (ex-*U 3548*), *TS 18* (ex-*U 3549*), *TS 19* (ex-*U 3550*), *TS 32* (ex-*U 3551*), *TS 33* (ex-*U 3551*), *TS 33* (ex-*U 3552*), *TS 34* (ex-*U 3553*), *TS 35* (ex-*U 3554*), *TS 36* (ex-*U 3556*), and *TS 38* (ex-*U 3557*). All were destroyed or scrapped in 1947–48. The Soviets also captured parts for another 14 submarines and exploited their technology.

These submarines were the subject of postwar Western speculation. At the end of the war U.S. and British members of the Tripartite Naval Commission pressed the Soviets to honor their agreements to dispose of the submarines but received somewhat ambiguous replies. Moreover, they were aware of only 8 Type XXIs captured incomplete, rather than the 20 the Soviets actually took. Five of the 20 (*U 3535*–*U 3539*) were on the slip nearly ready for launching, and they were launched on 15 July 1945.[24]

Initially the Soviets planned to complete the incomplete submarines under Project 614. That was not done, probably because key components were not available. According to an official post–Cold War Russian article on the origin of the Whiskey class, the Soviets disliked the Type XXI's power plant. They considered its drive shaft arrangement too complex, they rejected its short-life engine, and they considered its insulation grossly inadequate for high latitudes. Type XXI also did not meet their standards for watertight subdivision and had inadequate reserve buoyancy (hence was difficult to handle on the surface and was insufficiently seaworthy). The emphasis on surface qualities and reserve buoyancy seems to have lasted well into the nuclear era, helping kill a proposal for a single-hull submarine about 1960.

The Soviets scuttled the five launched boats in deep water on 7–8 August 1947. Three less advanced hulls on the slip were scrapped in February 1948. By April 1947 the incomplete sections of other submarines had been broken up. The scuttling and breaking up were all kept secret, partly because the mere existence of incomplete Type XXIs beyond the allocated quartet was proof that the Soviets had violated the Potsdam agreement; it was also proof that the postwar Soviet navy, already embroiled in the Cold War, was not as formidable as Westerners imagined. For their part, Westerners apparently accepted that the Soviets had captured only eight incomplete Type XXIs in Danzig, and these nonexistent submarines continued to feature in Western intelligence accounts for nearly a decade. Thus, in September 1954 the British Joint Intelligence Committee (JIC) claimed that "about eight Type XXIs, 80 percent complete, together with thirty other types, had been captured and removed illegally from Germany by the USSR. Most of these are believed still to be lying out of commission in various ports."

The Potsdam allocation did not include the prototype Walter submarines obtained by Britain and the United States, which became significant in postwar British thinking.

The Western Allies seem to have been particularly determined to limit experiments with the new U-boat technologies. The French were prevented from completing the Type XXIs they initially seized

The two Type XXIs were nearly impossible to maintain given their complexity (some said overengineering), limited workmanship, and the shortage of spares in postwar Germany. The U.S. Navy began its own conversion program and produced operational submarines under the Guppy (Greater Underwater Propulsive Power) program. Boats were given high-capacity batteries to drive their motors at greater speed (the motors were already powerful enough, having been used for surface propulsion). Hulls were streamlined, as shown here, and periscopes and snorkels were enclosed in a fairing. *Jallao* is shown as a Guppy IIA converted under the FY 52 program (completed in 1954). One of four diesel engines was removed to allow relocation of auxiliary machinery away from her bow sonar. The large object visible above her sail is her air-search radar antenna. Guppies were designed to facilitate later conversion to ASW submarines (SSKs). USNI

in their zone of Germany, although about a year later the British loaned them (and later transferred) a Type XXI. The Swedes were forced to relinquish *U 3505*, a Type XXI scuttled by her crew outside Göteborg on 8 May 1945 (her design influenced later Swedish submarine designs).

U.S. and British efforts to modify existing submarines for high underwater speed and endurance showed that Type XXI was anything but extraordinary.[25] The initial pair of U.S. Guppy conversions were rated at 19 kts for a half hour, at 16 kts for 1 hour (Type XXI: 15 kts), and at 8 kts for 8 hours (Type XXI: 8 kts for 10 hours). Snorkel speed was 10 kts rather than the 9–12 of a Type XXI. The Guppies had greater range (10,000 nm at 10 kts vs. 11,000 nm at 8 kts), carried more torpedoes (22 vs. 20), and could dive deeper (515 vs. 430 ft, based on a British standard taking test depth as four-sevenths of collapse depth). They were larger, too: 1,988 versus 1,625 tons. In 1947 the British estimated that their Guppy-like conversion of a T-class submarine would make 16 kts for a half hour, 14 for 1 hour, and 8 kts for 7 hours; it would dive to 350 ft; and would be armed with 12 torpedoes. The success of the Guppy project added a new fear: that the Soviets would likewise convert some existing submarines. They might simply install German-type snorkels, or they might go further.

One perhaps unexpected consequence of high sustained underwater speed was that the motor-generators involved had high battery-charging capacity. A fast battery submarine did not have to spend nearly as long charging her batteries when snorkeling (which came to be called "indiscretion time"), although she would be most vulnerable then; high underwater speed was thus associated with a low indiscretion rate.

THE NEXT STEP: TYPE XXVI

In 1945 the Germans were developing submarines with much higher sustained underwater speed than Type XXI, mainly Type XXVI. As the British had predicted in 1944, the new submarine employed closed-cycle propulsion. Until early in 1945 it was not known that high speed was being achieved using high-test hydrogen peroxide (HTP), which had been an important theme in U-boat development as early as 1940.[26] Since March 1943 "intelligence reports of varying reliability" had mentioned a "somewhat alarming" fast U-boat then under development called a "Walter boat" after the designer, a Dr. Walter of Kiel. Her speed might be up to 30 kts submerged, and fuel was thought to be HTP, stowed in leather bags, with "machinery compartments in the stern isolated from the rest of the vessel." Only a few of the boats had been ordered, but "some of them were dangerously near completion." A U.S. Navy report dated 20 December 1943 quoted NID 401 (29 November) on new types of U-boats, including the statement that "a new method of underwater propulsion giving much higher submerged speeds has also been mentioned in connection with these small U-boats," referring to reports that the Germans had already based, or were about to base, small U-boats on the Channel ports.

The new submarines were easily confused with other German projects. For example, during the run-up to D-Day there were claims that the Germans were building a very fast submersible motor torpedo boat called a "W-boat."

In the 1930s Dr. Helmut Walter had conceived a U-boat driven by a closed-cycle engine, the submarine carrying both oxidant and fuel. He chose HTP as his oxidant (oxygen carrier) after trying several alternatives. Initially he imagined a closed-cycle diesel fed by HTP decomposed by a catalyst, carbon dioxide, and steam in the exhaust being mechanically removed. He soon discovered that most of the heat of decomposition of the hydrogen peroxide could be used in a

The future, as understood (and feared) in 1945: a damaged experimental Walter U-boat on the slip at Germania, Kiel. She would have had a single propeller, as later submarines did. The Germans had done their best to demolish *U 794* (Type XVIIG) before Kiel fell. Her single propeller was missing. The planned production Walter boat was Type XXVI. NHHC

reciprocating engine, so he turned to a turbine. He first tested his trials boat, *V 80*, in April 1940. By that time he was working with HTP in 80–85 percent concentration rather than the initial 30–40 percent. *V 80* used a simple cycle, HTP being decomposed by a catalyst to yield steam that ran a 2,000 SHP turbine. Because he was interested mainly in high underwater speed, Walter adopted a novel hull form, which he tested in a wind tunnel, a considerable innovation. The boat displaced 80 tons and achieved 26 kts on trials, a figure not exceeded by later German Walter boats.[27]

Walter envisaged a more powerful plant in which HTP was both an oxidant and a direct source of steam, generated when it broke down under the influence of a catalyst. Fuel would be injected to add energy to the steam generated from the HTP; the result was often described as a gas turbine. Further development was slow, with Walter blaming German naval authorities who disliked anything so radical. In 1942 he argued that submarines should be 50 percent faster than their targets, and that future convoys would make 16 kts. That made the speed of his ideal U-boat 24 kts. A cautious German navy wanted to try a smaller boat first, with an estimated speed of about 19–20 kts. At this time the German navy showed little interest in high underwater speed, presumably because conventional U-boats were proving so effective. It was willing to order a 300-tonner from Blohm & Voss, which appeared in 1943 as Type XVIIB. Project drawings were sent to the builder in the spring of 1942. The Germania yard also wanted to participate in the program (Type XVIIG). Each yard received a two-boat contract. The first two, one from each builder (*U 792* from Blohm & Voss and *U 794* from Germania), were delivered for trials in November 1943. The other two of this series, *U 793* and *U 795*, arrived in May 1944. All were officially accepted by December 1944. Trials showed that the Germania boats were unsatisfactory, their hull form costing about 2 kts. The Blohm & Voss boats were acceptable, and initial plans called for building another 196 of them. The remaining three Type XVIIs passed down the Kiel Canal a few hours before 30 Assault Unit, the British intelligence seizure force, entered the plant. It is not clear that they were really operational, having spent only a very short time under HTP power. Walter also claimed that he had invented the snorkel and had presented it personally to Doenitz.

The Type XVIIB's plant was quite complex, which helps explain why it took so long to become operational. HTP was fed into a chamber where the catalyst converted it into hot (800–900°F) steam and oxygen at a pressure of about 520 lb/in^2. That was fed into a combustion chamber in which diesel fuel was burned, superheating the steam to a temperature of about 3,500–4,000°F. Fresh water was injected to reduce the mixture of steam and CO_2 to a temperature suitable for a turbine. The resulting mixture of CO_2, steam, and some oxygen was run through a dirt separator before going to the turbine. The system was considered so dangerous that initial tests of the combustion system were conducted from behind concrete shields. CO_2 had to be compressed to be pumped overboard against sea pressure, so the deeper the submarine, the less power was produced. Estimated output at 100 m (328 ft) depth was 80 percent of that at periscope depth. Some CO_2 would inevitably be released into the engine room through the turbine glands. To avoid the danger of poisoning the engine room personnel, the engine room had to be isolated completely from the rest of the boat while the turbine was running. In theory, once the turbine had run, the engine room could not be reentered until the boat surfaced and was thoroughly ventilated. Walter called this cycle the Direct Process. He was working on an alternative Indirect Process when the war ended. Instead of injecting water into the stream of hot steam, in this version the hot gas heated steam, which drove the turbine. There would be no need for a CO_2 compressor. Walter claimed that his engine could drive a Type XVII at 20 kts for 180 nm (9 hours) or at 14 kts for 230 nm (16.4 hours), in each case at 25 m (82 ft) depth.

Initially Walter convinced Admiral Doenitz to move toward an open-ocean HTP submarine, Type XVIII, a 1,750-tonner powered by two 6,000 SHP turbines and two 1,750 BHP diesels, with a maximum submerged speed of 25 kts. The Allies' ASW successes in mid-1943 convinced Doenitz that he needed something new in the near term, however. Type XVIII was a step too far away; development was too slow. Its design incorporated a lower pressure hull mainly to carry the dangerous HTP stored in rubber bags that would be surrounded by sea water. Walter suggested that a satisfactory conventional submarine could be built using the Type XVIII hull. It would have much greater submerged endurance, due to more batteries replacing the underslung HTP. Streamlining adopted for the HTP design would enable the new type of U-boat to run submerged at much higher speed. She was given much more powerful electric motors. The result was Type XXI.

Doenitz ordered yards to concentrate on "electro"-U-boats—Type XXI and the coastal Type XXIII. Plans for Type XVII were cut back to 24 hulls (12 from each yard), then to 12 at Blohm & Voss, and then to only 6. The first three (*U 1405–U 1407*), the only ones built, were delivered in February–April 1945. They had only one turbine (2,500 SHP) due to delays in delivering turbine blades. These boats were credited with a maximum speed of 20 kts on turbine power. Later boats would have had two turbines. At the beginning of May 1945, the crews scuttled all three boats.

Meanwhile, work on a big HTP submarine continued. A new Type XXVI was designed, powered by a single Walter turbine (initially 6,000 SHP, but later 7,500 SHP). At 1,030 tons, she was to have made 24 kts on 6,000 SHP. The design was described as a compromise between Types XVIIB and XVIII. Initially, XXVI was conceived as an enlarged Type XVII with half the power plant of Type XVIII and a different torpedo armament. According to Walter, the first two trial boats were to have been delivered in the summer of 1945, followed by 6 and then 12 per month; the test rig was to have been ready in June 1945. This was quite optimistic; the smaller HTP boats that had been completed had proved less than reliable.

In 1945 the British were told that Type XXVI could sustain 23 kts for 6 hours, after which all its HTP would have been expended. For the rest of its patrol it would make do with a diesel-battery system roughly comparable to that in a Type VII.

The early Walter boats were seen from the air but not understood as the radical innovations they were. As an example of misleading information spread at the time, a U.S. Navy report dated 20 March 1944 on new U-boat developments combined an accurate description of the new snorkels with paragraphs on Walter boats that described them as submersible MTBs (E-boats) employing some new method of underwater propulsion to achieve a submerged speed between 18 and 30 kts, using some new method of propulsion. A U.S. Navy report wrongly identified a new type of U-boat as Type VIIC42, mentioning that both closed-cycle turbines and double battery capacity had been reported (presumably the Germans had mentioned Type VIIC42 in communications, and it was not yet understood that the boat was merely a slightly deeper-diving version of Type VIIC). Size was likely to be that of Type VII; diving depth 400 m (1,300 ft); and submerged speed would be higher than that of Type VII. An intelligence report noted, "It is not established that the Type VIIC42 has closed-cycle propulsion. If closed-cycle engines are fitted, they might be designed for economy of weight, not necessarily for higher speeds." This description mixed characteristics of Type XXI and the semi-experimental Walter Type XVIIB.

Reports of German submarines with unusually high underwater speed were largely discounted at the time as exaggerations meant to improve German morale, but they turned out to be close to the truth. The Western Allies seem not to have understood the character of the new U-boat until the Japanese naval attaché in Berlin described the new kind of submarine in considerable detail in his reports home.[28] As a result, the Walter plant in Kiel became a prime intelligence target for 30 Assault Unit. The first positive evidence of the Walter program was found when British troops entered the Blohm & Voss yard at Hamburg and found two Walter boats lying on the jetty: *U 1409* and *U 1410*. Their engines had been removed, and they had been severely damaged by a bomb that fell between them. After some unwillingness to discuss their machinery, Rudolf Blohm and a technical director revealed the nature of the engine and the performance figures for Type XVII. Soon afterward another unit of 30 Assault Unit entered Kiel ahead of the first British army units and made the Walterwerke its headquarters. Most of the plant was undamaged. Walter, who was living nearby, was "secured." The stern of a Type XVII submarine was found, complete except that essential parts of the combustion system were missing, having been destroyed. Initially Walter was unwilling to divulge details. He became much more talkative after the British Naval Officer in Charge at Kiel issued an order that all oaths of secrecy to the Reich had been rendered null by the unconditional surrender. Doenitz sent a German captain to tell Walter not to withhold anything. Although papers had been burned, fortunately all the drawings had been microfilmed, and the films were found. Walter told his interrogators that from the first his objective had been a fast U-boat, his many other developments being offshoots. The Walterwerke was ordered to reassemble the Type XVII stern and to complete a Type XVIII test unit; it was also ordered to complete a large number of HTP torpedoes.

The British were fascinated by the HTP submarine, to the point that when British troops overran Hamburg they asked Blohm & Voss, which built Walter's submarines, to build them a Type XXVI. The embarrassed Germans told them the necessary equipment and machinery did not actually exist. The British and the Americans agreed not to mention HTP or Walter to the Soviets. Because none of the Walter prototypes remained afloat, they saw no need to include them in the list of surrendered U-boats.[29]

The Admiralty overrated the abilities of the Walter submarine, fascinated by its high underwater speed and, perhaps, its technological novelty. For example, ADM Max Horton, the wartime Western Approaches commander, hence the most senior British ASW officer, said that the future U-boat virtually need never surface, combining as it did high underwater speed with great submerged endurance—although at the cost of high noise levels. He ignored the limited high-speed endurance of a Type XXVI.

The effort to keep the information from the Soviets proved pointless. The Soviets also captured considerable Walter material, even the test version of the Type XXVI engine (26X; the British had the 18X test engine intended for the Type XVIII's plant). They considered building Type XXVI submarines (their Project 616) using the plans they had seized but were forced to abandon the project because they lacked key components. However, they were able to build a trials submarine, *S 99* (Project 617), which they completed in December 1955. Unlike the two experimental HTP submarines the Royal Navy completed, *S 99* was armed, with six bow torpedo tubes. She was rated at 20 kts for 6 hours. On 17 May 1956 she suffered an internal explosion while at 80 m depth. Although she managed to surface, she was laid up at Leningrad and never operated again. The published Russian account states that the Russians had only the one captured Walter turbine, and hence could not repair her. Project 617 was unknown to the West, but Western naval attachés did see *S 99* moored in Leningrad (presumably after the explosion) and speculated as to her purpose. She was unofficially nicknamed "the Whale."

Also unknown to the West was a Soviet program to develop a closed-cycle submarine, which became operational in the form of 30 Project 615 ("Quebec" type) coastal submarines.[30]

The Germans had developed no coherent doctrine for the new submarines, even less than for Type XXI. Attacks on fast submarines would be complicated not so much by their sheer speed as by the effect of their wake and loud propellers (at high speed). It was already known from the *Seraph* trials that the wake would interfere with pinging. Really loud propellers would mask pings, to the extent that they might not show on the asdic recorders and the operator might be unable to hear pings on his headphones. It might be particularly difficult to hit a submarine crossing the attacker's path. The speed itself might be a problem as well, as existing sonar domes were designed for use at speeds up to 25 kts. In the summer of 1945, a new dome for operation at 25–30 kts was being developed as a high priority.

The British raised *U 1407* and took her back to the United Kingdom as the basis for an HTP development project. She was placed in service as HMS *Meteorite*. They installed Professor Walter at Vickers in Barrow-in-Furness, which became the center of their postwar HTP program. *U 1406* was raised and given to the Americans. She was never placed in service, but her power plant was set up at the U.S. Navy engineering experimental station in Annapolis. The third boat, *U 1405*, was not raised, perhaps to preclude her being allocated to the Soviets.

The U.S. Navy view was that Walter had showed that the next submarine generation would be powered by closed-cycle engines, but not necessarily his design. It therefore set up a program to develop a wide variety of such plants, ultimately to produce 15,000 SHP, the goal being a submerged speed of 25 kts. Possibilities included closed-cycle diesels (which had had German adherents) and a pressure-fired boiler. As a longer-turn proposition, this program included a nuclear submarine. By 1948 it was clear that the companies contracted to develop the closed-cycle engines saw little commercial future in them and were thus reluctant to assign their best engineers. The only exception was the nuclear program. CAPT Hyman Rickover told the 1948 Submarine Officers' Conference that his program, in contrast to the others, enjoyed enthusiastic commercial support; companies wanted to build the necessary laboratories themselves because they could see a great future for nuclear power.

For both navies, Type XXVI was a terrifying prospect because she came much closer to the "true submarine," a vessel that would no longer have to surface or even show a snorkel or a periscope. She could be armed with the long-range torpedoes the Germans had been testing at the end of the war, and most analysts seem to have accepted that the type might employ a new fire control system based on the low-frequency passive arrays the Germans had, it seems, perfected.

TORPEDOES AND FIRE CONTROL

The new submarines themselves were only part of the new underwater challenge. Westerners assumed that the Soviets had captured all of the German naval technology and would soon copy it. In 1945 the Germans were developing new long-range torpedoes that a submarine could fire from outside sonar range.[31] They were also developing a sonar-only fire control system for Type XXI that would enable a submarine to fire from beneath a convoy, whose noise might shield her. During the war the Germans fired slightly over 10,000 naval torpedoes, of which 7,000 were G7e electric torpedoes (electric to avoid making a visible wake) and 2,300 were G7a steam torpedoes. Pattern-running torpedoes accounted for 5 percent of the total, and homing torpedoes another 6 percent. A German summary of torpedoes fired in action up to January 1945 included 640 acoustic torpedoes.

Early in the war Doenitz emphasized the need to make every torpedo count, because the standard Type VIIC U-boat carried only 12. U-boat commanders were therefore encouraged to fire no farther away than about 600 yds. By 1943 U-boats were compelled to fire at 2,000 to 3,000 yds, and hits by single torpedoes were unlikely. To make longer-range fire profitable, the Germans introduced the FAT pattern-runner torpedo. The FAT was still aimed, but if it missed, it turned back and got a second chance. The pattern was such that a U-boat had to fire from roughly abeam of a convoy. FAT required considerable torpedo range, and it could be converted only from a steam torpedo (most U-boats used electric torpedoes because they had no detectable wake). FAT II could be made to circle at the end of its initial run as an anti-escort measure. FAT II guidance was applied not only to the standard steam torpedo (G7a) but also to a new longer-range (8,000 yds) electric torpedo (T-IIIa) that appeared in 1943.

A further development, LUT, made it possible to set a pattern at an angle to the initial torpedo course, so a convoy could be attacked from any bearing. The pattern itself was far more flexible. The British found LUT formidable, but fortunately it did not become available until late in the war; only about 70 were fired.

The Germans had begun experiments on homing torpedoes at least as early as 1936. By 1940 they were concentrating on electric torpedoes, both because they left no visible track and because they were inherently quieter. The prewar object was to attack fast warships, so the desired speed was at least 30 kts; that in turn made for unacceptable self-noise. The targets were changed to merchant ships, for which a torpedo speed of 20 kts would have sufficed. A small initial batch of about 30 of these *Falke* or T IV torpedoes was issued in 1943. By that time the important need was for an anti-escort torpedo: T V (*Zaunkoenig*). T V and T IV were the only acoustic torpedoes the Germans used operationally. The British called T V "Gnat" (German Naval Acoustic Torpedo). Gnat was a passive homer that used a pair of hydrophones in its nose. Comparison of sound intensity received by the two hydrophones indicated how far off the torpedo was from a path directly to the target. The hydrophones were arranged to avoid hearing sound (torpedo self-noise) from aft. T V could attack ships proceeding at 10–18 kts. Below 10 kts, target sound was insufficient; above 18 the torpedo could not home due to its self-noise. Torpedo speed was 24 kts. In action, T V's hit success was about 9 percent versus an overall figure of 20 percent for all German naval torpedoes. In 1952 the British official report on German torpedoes pointed to the sheer complexity of the homing torpedo, which would greatly complicate maintenance.

Late in the war the Germans deployed the wire-guided *Spinne* (T X) electric torpedo. The T X was conceived mainly for shore batteries, but wire guidance was of considerable interest after the war for submarines and ASW ships. *Spinne* could be ordered to surface so that the operator could see its position and guide it. The 5,000 m wire was on a fixed coil either on board the firing ship or at a shore site. About 200 were made, and some firing positions were built on the Belgian coast. There was also a wire-guided acoustic homing torpedo, *Lerche* (the human controller heard what the torpedo heard). At the end of the war *Lerche* had not yet been tested on board a U-boat; the British thought that its use would require considerable operator training.

For really long range at high speed a new kind of torpedo engine was needed. The main German project, still experimental in 1945, was the hydrogen peroxide torpedo, using high test peroxide. HTP already had an important role in submarine propulsion. In 1940 the Germans adopted a steam turbine for their experimental HTP torpedo. Twelve different HTP torpedoes were under development at the end of the war, all of which were basically similar. Speed was 45 kts. In 1945 the closest to production was *Steinbarsch*, rated at 8,000 m range at 45 kts. The Germans estimated that the more advanced *Steinwal* would run 20,000 m at 45 kts, but its developers still had to overcome major problems. When Germany collapsed, a large number of *Steinbarsch* torpedoes had been made, but no trials had been run. It became apparent that the Germans were willing to issue the torpedoes untested.

The Germans had also worked on a closed-cycle oxygen torpedo. Many of the Germans preferred oxygen to HTP but admitted that HTP technology was so much further advanced that it was a natural choice in wartime. The oxygen project was shut down in 1943.

Postwar British trials showed that none of the HTP torpedoes was mature, although they did make their rated performance. All left nearly no wake, but their depth-keeping was poor. "A very high degree of care" was necessary in handling HTP, as "small mistakes can have very dangerous consequences." That proved true when an HTP torpedo exploded on board the submarine *Sidon* and blew off her bow in June 1955.

Homing technology made it worthwhile to provide torpedoes with much greater range. The wartime T V had a range of 5,700 m at 24 kts; in 1945 the Germans were working on T XI, which had a new battery and was expected to achieve 9,000 m. T XI had not yet been perfected; the extra weight of the battery reduced its stability, and the British suspected that its depth-keeping would not have been acceptable. T XI offered variable arming range (300, 600, 1,100, and 1,900 m rather than the 400 m of T V). It could also hit targets moving at higher speeds. At the lower end of the scale, reduction of self-noise made it possible for the torpedo to attack ships as slow as

8 kts. The homing mechanism was changed to reduce vulnerability to a towed noisemaker.[32]

Late in 1942 the Germans decided to develop an active homing torpedo (*Geier*). Among its homing modes was something similar to wake-following, the Germans having discovered that a ship's wake could be a strong sonar target. *Geier* superseded two other wake-homing projects, *Ibis* and *Ackermann*.

The important development in torpedo control was the ability to fire based entirely on acoustic data—that is, without exposing a periscope. Type IX U-boats (1938) introduced a semiautomatic dead reckoner (only in Type IXB), which in theory made underwater sonar-only shots possible. It was not used for such control until 1944, when it was revived for the Type XXI, which had the same fire control system as Type IXB.[33] In his orders to Type XXI commanders, Doenitz made it clear that shots based on periscope data were preferable, but he mentioned all-acoustic shots, perhaps using acoustic torpedoes to make up for inaccuracy in sonar bearings (this was not explicitly stated in references to all-acoustic shots). The Germans became interested in the "shark attack" toward the end of the war—firing from below a convoy or other formation entirely on the basis of acoustic data. T XI was therefore designed for firing at depths as great as 50 m (164 ft).[34]

COMMUNICATIONS

The system of central control for all U-boats required not only that all of them be able to receive orders from the controller, but also that all of them report back; the controller needed to know where each U-boat was and its state (damaged, low on fuel, low on torpedoes, etc.). To reach submarines, Doenitz relied on a combination of HF and VLF radio. HF offered reliable communication with a good data rate, but it could not reach a submerged submarine. To do that, the Germans used VLF, which a submarine could receive at periscope depth. For tactical coordination, for example to home wolf-pack submarines on one that was in contact with a convoy, the Germans relied on medium-frequency wireless, whose important virtue seems to have been its relatively short range.

VLF became increasingly important after the Germans adopted snorkels, and U-boats could operate entirely submerged. The Germans set up several VLF transmitters, but by the 1945 they had only two—one at Nauen and one they called Goliath; another, at Kremplemor, was brought into service at the end of the war. The Soviets captured Goliath late in the war, disassembled it, and took it home. Afterward the British in particular anxiously watched for the Soviets to reactivate it, because that would indicate that they were preparing for a German-style centrally controlled submarine war. Goliath was never reactivated; the Soviets apparently relied on HF, for which the submarine had to put an antenna above the surface. Compared with VLF, HF made for a much higher message data transmission rate.

Because U-boats had to use HF to contact U-boat control ashore, Doenitz and his staff were always looking for ways the submarines could communicate without endangering themselves. At least in theory, shore stations could measure the direction of signals from U-boats (DF-ing them), and a net of shore direction-finding stations could locate a transmitting U-boat, at least approximately. Doenitz tried several methods of ensuring security.

The first approach was to vary the transmitting frequency. The Allied shore stations did not normally roam the HF spectrum looking for signals. They listened on the frequencies the Germans typically used. The Germans tried to defeat the shore stations by using "off-frequency" transmissions at ordered differences (typically a few kHz) from the usual ones. By 1943 they wanted to be able to frequently vary the difference in frequency. The submarine command sent out a table of off-frequency differences for the U-boats to use, but the strategy failed when the British got hold of the table in about October 1943.

Other approaches exploited the time delay inherent in current forms of precision DF. An operator used a rotating goniometer, an adjunct to the main antenna that imposed a relatively narrow receiving pattern. The operator turned the goniometer back and forth looking for a null (which was much easier to sense than a maximum). The null was 180 degrees from the direction of the signal. Location required cross-fixes, so DF stations formed a network that was alerted by flash signals when one station picked up a signal. The need to coordinate multiple stations added more time to the fix. The shorter the signal, the smaller the chance that one station, let alone a whole net, would manage to make enough DF "cuts" to localize the emitter. Doenitz never realized that ships could also DF signals. In their case, cross-fixes were not needed. Because they were picking up only the surface wave, they could simply run down the line of bearing they found, knowing that the U-boat was within 30 nm.[35]

Doenitz first tried to thwart DF by reducing message length, ultimately substituting single letters for words. But sometimes U-boats had to send lengthy messages.

The ultimate countermeasure to HF/DF was the *Kurier* "squash transmitter," which drastically compressed messages.[36] The idea of message compression seems to have been proposed about 1943. Communications trials of a device called *Zusma* (a compression of the German word for "signaling apparatus") were held between Admiral Northern Waters and Gruppe Nord on 1–9 1943; further tests were made farther south during the summer of 1943. The British found that these experiments were connected with some form of "off-frequency" transmission. The receiver was a 12 × 12 square whose elements could be illuminated, the whole then being photographed so that it could be read. Much of the jargon later used with *Kurier* appeared for the first time, including a reference to time settings. The trials were abandoned because the equipment involved was too delicate, but the idea survived. An improved *Zusma* was ready early in 1944, and it was later fitted in a U-boat for trials that summer.

The next step was after *Zusma* was *Kurier*. The first reference to *Kurier* in special intelligence (codebreaking) was a 16 June 1944 order for an engineer to go to Gdynia to instruct *U 878* about U-*Kurier*. A signal referring to a further trial with U-boats of the 7th and 15th Flotillas and the St. Nazaire Dockyard was intercepted on 22 June, but it is not clear that the trials were ever undertaken, given the invasion of France. A captured document referred to an order concerning "Short Signal Procedure *Kurier*," but unfortunately all that was obtained was the index entry. It appears that the first operational trial was conducted 4–7 August 1944 using six U-boats from the Baltic. The equipment had to be adjusted afterward, so work was suspended until 21 August. This time 12 U-boats were involved. On 19 August, for the first time, a *Kurier* signal from a U-boat was successfully received ashore. *U 978* successfully sent four messages on 22–23 August, but the transmitter was still not considered reliable enough to adopt. Experiments continued, and in November control received messages fairly successfully. Tests through December included transmission from the dipole antenna surrounding a snorkel head. Trials had to be stopped on 2 February 1945 when the special receiving equipment was evacuated from Bernau due to the Allied advance ashore.

By the end of March two U-boats were ready for trials with an improved transmitter: *U 2511*, the first Type XXI to leave the Baltic; and *U 878* operating from St. Nazaire. It does not appear that the planned trials were carried out.

U-boats were to carry only the transmitter (*Geber*). The heart of the device was a disc carrying iron studs, which could be pushed in. Morse symbols for the signal were set by pulling out studs, one for a dot and two for a dash. The disc was scanned by an arm that could detect the pattern of studs. Both changed the current passing through the arm; the changing current was amplified and used to modulate the U-boat wireless transmitter. *Geber* had a switchboard that could turn off the disc and enable the U-boat wireless to be keyed conventionally (i.e., manually). Transmission was limited to one revolution of the disc, a maximum of 436 msec. Each pulse was 2 msec long, with a gap of 2 msec between pulses. The 25 fixed contacts formed the call-up or triggering portion of the message. It was followed by a 16 msec gap, and then 304 msec for the message, enough for 7 of the longest Morse letters. The receiver used a CRT to display the pattern of dots, which was photographed to be read.

The system operated on four basic frequencies, each with tuning to off-frequencies. *Kurier* would always use an off-frequency to avoid interception. The frequency deviation involved changed each day, the amount involved being provided by radio signal. The deviations were later encrypted. Once that happened, the British were unable to determine the desired deviations. Normal methods of interception were inadequate, but a station at Scarborough managed to photograph a series of *Kurier* transmissions on 7 and 8 December 1944, before the Germans encrypted the frequency deviations they were using. The *Kurier* problem became an important theme of postwar Allied ASW, although it was largely unpublicized because developments in communication interception were closely guarded.

5

FACING THE POSTWAR NIGHTMARE

THE TWO MAJOR WESTERN NAVIES ended World War II well equipped to deal with the conventional U-boats they had fought the open-ocean Battle of the Atlantic. They faced a new kind of threat, however, in the form of large numbers of Type XXIs, and possibly later HTP submarines. In effect, the Western antisubmarine effort of the period between the end of the war and 1950 was largely what would have been done had the war somehow continued and had the Germans fielded the new submarines.

In June 1946 U.S. Chief of Naval Operations FADM Chester Nimitz designated antisubmarine warfare the U.S. Navy's first priority, on a level with the nuclear tests conducted in Bikini (and, presumably, the effort to adopt naval nuclear weapons). A coordinator of undersea warfare was designated. His office called a high-level (Top Secret) ASW conference and then a series of annual Secret-level conferences.[1]

Unfortunately, the resources readily available to the Navy in wartime had vanished. The expert personnel who had made Allied ASW so effective were mostly gone as well. The commander of the Fleet Sonar School said that the connection between training and effectiveness was more marked in sonar than in other work due to increasing complexity and to the need for men with a special ability to recognize sounds. Some frontline ships (e.g., in the western Pacific) absorbed disproportionate numbers. Late in 1947, for example, the Pacific fleet destroyer command (DesPac) had only 72 percent of required qualified personnel—30 percent of whom would soon leave as their enlistments expired. DesPac ships were considered "satisfactory" on average, but below wartime standards. Only about

The nightmare: massive Soviet construction of submarines equivalent to the German Type XXI. This is the sail of a Whiskey (Project 613), the subject of the largest peacetime submarine-building program in history. The object emerging diagonally from the sail is her snorkel exhaust, which would normally be underwater. Next to it is a radio direction-finding loop. The object emerging from the circular hole in the top of the sail is a radar warner (NATO Stop Light, after its appearance). The prominent fairing at the after end of the sail seems to cover the ammunition-passing hatch installed to serve a 3.9 in. gun, which was soon removed (the first units also had antiaircraft machine guns forward of the sail, no trace of which remains here). Whiskeys were produced in large numbers: 215 were built for the Soviet navy, and another 21 were assembled in China from Soviet-supplied components. Remembering how German Type VIICs, which were considerably smaller, had roamed the Atlantic, Allied navies saw the Whiskey program as an attempt to build the capacity to fight a new Battle of the Atlantic. They assumed that the Soviets would also modernize their considerable fleet of earlier submarines. Initially, the Soviets seem to have seen the Whiskeys as a means of coast defense, reserving the ocean mission to the larger Zulus (Project 611). USNI

a third of the required 605 officer graduates of sonar schools were available, of which about 78 percent were qualified unit commanders and commanding officers, hence would not supervise sonar and underwater batteries. Many junior officers had left. Designated officers were being required to take the DesPac correspondence course on board ship. Sound men found echo recognition very difficult. Men who had had one or two cruises with older equipment were familiar with it and asked for it.

DesPac ships performed well in Hawaiian waters, largely due to excellent sonar conditions; sonar teams consistently achieved ranges of about 3,000 yds. Radar and sonobuoy performance, however, was not up to standards; in exercises most air contacts were made visually. Night radar interpretation was often defective. Sonobuoys averaged one defective buoy per pattern. A submarine had an excellent chance of escaping if two buoys in a pattern were defective. Rough seas increased the percentage of inoperative buoys. Shortages became acute. A Pacific Fleet air commander told the 1947 conference that he had to think twice before dropping a buoy because sonobuoys had to be recovered after use, a time-consuming and expensive endeavor.

On 1 June 1947 the Atlantic destroyer force had only 39 percent of its authorized sonar strength; it fell to 33 percent on 1 July and was expected to fall to 28 percent on 1 December. Sonar strength included sailors designated for sonar, but many of them were barely qualified.[2] There were not enough seaman trainees with the necessary aptitude and intelligence ratings for sonar training.

The 1948 increase in defense spending somewhat eased the situation. The Navy was able to attract more personnel, who got better training. With more money available, the Pacific Fleet was able to double training of sonar men in 1949. Good personnel could make up for obsolescent equipment. Opposed by Guppies, a unit at Key West rated 100 percent ready detected 12 of 13 attempted penetrations of a convoy screen. More commonly readiness was about 30 percent, and only about 25–30 percent of penetrations were detected. Intense training could certainly improve matters. Screening vessels rated 40 percent ready in February–March were about 60 percent ready in the fourth quarter of a year. Unfortunately, this improvement did not last, because in 1949 the ASW budget was drastically cut.[3]

The fleet still had the fruits of wartime mobilization, with which it would have to fight if war broke out. At least in the near term there would be few new weapons and sensors, so the Operational Development Force (OpDevFor) sought to develop new tactics that would make existing ships, sensors, and weapons more effective. Production moved much more slowly than in World War II, not only because much less money was available, but because the wartime days of 21-shift weeks were gone; new production would come from 1 shift per week. The British military was particularly badly starved because rebuilding the export trade was essential; at the end of the war the country was nearly bankrupt.

Nearly all the scientists who had created new sensors and weapons in wartime returned to civilian life.[4] In an attempt to keep the link between the Navy and the civilian scientific community alive, after the 1946 Undersea Warfare Conference the new Office of Naval Research (ONR) enlisted the National Academy of Sciences, which created the nine-member Committee on Undersea Warfare (CUW). The new committee included Dr. C. O. D. Iselin of Woods Hole Oceanographic Institution, who had been very active during the war, Professor Morse of MIT, and an engineer from Bell Telephone Laboratory, at that time the leading U.S. industrial research center. Veterans of Division VI (ASW) of NDRC feared that the United States would repeat the situation at the outbreak of World War II, when it took so long for civilian scientists to learn enough about ASW to be able to help the Navy. CUW therefore held conferences on ASW developments, but it did not sponsor research.

The wartime alliance between the United States and the United Kingdom had been dissolved, but close naval relations survived. At the June 1946 high-level U.S. ASW conference, the chief of the Bureau of Ordnance said that he had recently returned from the United Kingdom and had talked with both CNO Nimitz and the Secretary of the Navy. He considered cooperation with the British essential, as the British were particularly strong in underwater warfare. British representatives attended all the later Secret-level Undersea Warfare Conferences, describing their own work and reporting back on U.S. work. British delegations were also welcomed at U.S. research facilities, as is evident from their surviving reports. The U.S. Navy sought permission to share its new weapons and sensors with the Royal Navy. In 1949 its view the British were far more sophisticated tactically. Their tactics made up for their less sophisticated equipment, so that the two navies were at about the same level in overall ASW effectiveness.[5]

THE NEW MISSILE THREAT

In 1945 naval strategists thought it possible that future submarines might be armed with land-attack missiles. German researchers had sketched a submersible barge containing a V-2 missile that a U-boat could tow. The barge had not been built, but there was no reason to think that it was impracticable. It is not clear whether the Germans were interested in launching the air-breathing V-1s from a U-boat, but that too was clearly possible. The U.S. Navy followed up on the idea after the war, successfully firing a U.S. version of the German V-1 cruise missile from the submarine Cusk on 7 March 1947.[6]

Views on how seriously submarine-borne missiles should be taken varied over time. At the 1947 Undersea Warfare Conference that fall, LCDR Edward L. Beach, later a prominent naval author, said that "the problem of early detection of a submerged submarine approaching the coast of this country for the purpose of releasing a guided missile has assumed major proportions." Although it might be several years before such an attack became possible, the problem was important because the missiles might carry nuclear warheads. It was vital to develop some means of detecting and destroying such submarines well offshore. An informal Naval War College study found no feasible defense against such submarines. Navy planners were also working on the problem.

The Navy's planning group told the next year's Undersea Warfare Conference that the Soviets were showing considerable interest in both of the long-range German missiles—the air-breathing V-1 and the ballistic V-2—although as yet there was no information about a submarine program. A large submarine could carry four to six V-1 cruise missiles but probably only one V-2. Adaptations for missiles would make the submarine more vulnerable, since hangars for cruise missiles would reduce underwater speed and lengthen the time to crash-dive. Cruise missiles would have to be assembled while the vessel

A new nightmare: missile attack from the sea. USS *Cusk* launches a Loon (a U.S. version of the German V-1) off Point Mugu. Missile attack from the sea impressed the U.S. Navy far more than it did the Royal Navy, which feared Soviet land-based missiles if the Soviet army advanced into West Germany. That missile threat justified maintenance of a British army on the Rhine, whose cost affected the Royal Navy. NHHC

was on the surface, and the submarine would then have to position herself to fire them. U.S. intelligence thought the submarine would have to remain on the surface to guide the missile, although control might be passed to another submarine. Missile range would be short, compelling the submarine to approach the coast to fire. Given such limitations and the limited accuracy of a V-1, the psychological effect of a land attack would be less than that of attacks on shipping. In effect the planners were dismissing Beach's main point: that a few submarine-launched nuclear missiles could have an enormous impact. The planners considered the V-2 or ballistic missile threat remote. It later turned out that the barges were impractical. U.S. and Soviet attempts to make them work failed.

Above all, Americans viewed submarine-launched missiles as giving the Soviets a way of directly attacking the United States, a possibility otherwise largely denied them. The British, of course, already faced the possibility of land-based missiles. At this time Britain's decision to defend Germany against a possible Soviet assault was justified by the need to prevent the Soviets from getting to within V-2 range of England.

It is not clear why U.S. planners dismissed the missile submarine threat at just the time their own Navy was developing it in the form of the Regulus missile and dedicated missile-firing submarines (plus dedicated guidance submarines to extend missile range). However, U.S. ASW documents of this period discuss only the defense of shipping. Only in 1954, when Regulus was close to becoming operational, did the British naval attaché in Washington warn the First Sea Lord that the Americans might abandon defense of ocean shipping in favor of homeland defense against missile submarines.

ESTIMATING THE SITUATION

The Soviets had built the world's largest submarine fleet before the war, and it seemed obvious that they would soon adopt the wartime German U-boat technology they had captured. In 1946 the U.S. Operational Evaluation Group estimated the impact of the new submarines based on tactical parameters: the probabilities of making contact, attacking, and killing ships.[7] The exchange rate would be about five times the worst experienced during World War II. Something like 100 merchant ships would be sunk for every submarine sunk, instead of the half of a merchant ship per submarine of 1945. Assuming the same number of contacts on ships by each submarine per unit time as in 1944–45, the new submarines could close or attack twice as often. Half again as many ships would be sunk in each attack. Compared with the situation in 1944–45, OEG assumed that 75 percent of contacts would be attacked. On the ASW side, only one-sixth as many contacts would be sunk as in World War II. This arbitrary estimate assumed that aircraft would make almost no contacts because they would find it difficult to spot snorkels, and submarines would never operate on the surface. OEG concluded that it would be much more important for escorts to kill submarines than to stay in a convoy screen. This rather pessimistic picture of ASW justified an intensive search for alternative strategies, such as attacks at source (i.e., at submarine bases).

Using long-range pattern-running torpedoes (developed by the Germans but not perfected by 1945), submarines could attack from outside the sonar range of convoy escorts. Because it would be more difficult to find and kill submarines during the approach, attack, and disengagement phases, convoys needed more screening ships (in a 7:3 ratio). At a minimum, using World War II tactics but also better airborne radars, the Allies would lose 38 merchant ships for every submarine sunk in convoy warfare. Among important factors was inherently short sonar range, the estimated limit (in 1948) being 5,000 yds.[8] OEG admitted that the low-frequency sonars used by the Germans were a promising way to increase detection range.

OEG revisited the convoy problem in 1949. Better tactics would increase the number of contacts by ASW forces per unit time from 0.17 to 0.3. All of the attacking submarines might be sunk rather than only 13 percent of them. The expected exchange rate might fall to 12–20, still very bad but not so horrible as estimated in 1946. In presenting these figures at the 1949 Undersea Warfare Conference, OEG pointed out that the calculated exchange rate might be misleadingly pessimistic. For example, if the submarines were cautious, or if there were too few of them, the fact that each might sink 100 merchant ships before being sunk would have little bearing on the outcome.

OEG admitted in 1949 that it had been far too pessimistic in 1946 in evaluating surface ship attacks on submarines. It had ignored the possibilities offered by repeated attacks. OpDevFor experiments showed that apparently small improvements, such as scanning sonar, significantly improved the chance that contact could be held, and successive attacks delivered, until the submarine was destroyed or its batteries exhausted. A well-trained ship had a 35 percent chance of destroying a new-type submarine, which compared well with the wartime figure of 38 percent. If full advantage was taken of the new developments, the proportion of successful attacks might nearly double. Near-term improvements would include larger depth charge patterns employing faster-sinking depth charges. Soon, new weapons such as Weapon A and homing torpedoes might further improve matters. The new APS-20 airborne early warning (AEW) radar promised to detect snorkels at something like the range wartime radars had achieved against surfaced U-boats, but only in sea state 2 and below. However, it seemed unlikely that submarines would show their snorkels for as much time as their predecessors had spent on the surface. The advent of the new radar therefore did not fully restore the wartime advantage enjoyed by ASW aircraft.

The OEG exchange rate of 12–20 did not take account of other medium- or long-term possibilities: (1) further antitorpedo measures; (2) other passive measures, such as improved formations, better routing, and zigzagging; (3) greatly improved weapons and fire control; (4) improvements in sonar; and (5) increased effectiveness in screening as the result of substantially increased forces. Nor had OEG taken into account measures to keep enemy submarines from reaching their patrol areas, such as attacks on bases, mine blockades, and blockades by ASW submarines. However, OEG also had not taken into account the medium-term threats of long-range pattern runners or the advent of true submarines. The near-term threat was thus smaller than it had seemed in 1946, but it was still uncomfortably large.

Type XXI was already so severe a threat, and so undetectable except when attacking, that the feared future "true" submarine seemed unlikely to make matters much worse. Any of these submarines using long-range torpedoes was apt to announce its presence only as a ship sank. The best means of protecting merchant ships might be to equip them with torpedo-killing weapons—if these could be developed. At the least, a convoy should be escorted by enough ships to keep a viable escort with the convoy when several ships had been detached to run down an attacking submarine. OEG (and others) asked whether convoy screens could be expanded using expendable or invulnerable pickets such as helicopters using dipping sonars or airships towing sonars.

The only available wide-area antisubmarine sensor was HF/DF. To fight an open-ocean war, submarines still had to communicate with their central command, so they would still use HF radio in wartime. The U.S. Navy kept its Atlantic HF/DF net in commission and sponsored research to solve the problem the Soviets would present if they adopted *Kurier*, as seemed inevitable.[9] U.S. destroyers converted into specialized ASW escorts carried wartime-type HF/DF antennas.

Many in the U.S. Navy concluded that, given limited escort numbers, hunter-killer groups might be more effective than convoy escorts.[10] Even if no long-range sensor could cue the groups, they could search designated areas and, for example, sanitize convoy routes. The first Undersea Warfare Conference (1946) recommended creation of two experimental hunter-killer groups, one built around an escort carrier to fight the immediate Type XXI threat, and one built around a light fleet carrier to fight the future threat of Type XXVI. Both would be operated by OpDevFor, and both would employ existing or near-term weapons, sensors, and fire controls. Destroyers and destroy escorts were modified specifically for these groups.

The possibility of attacking submarines at source was an important justification for building the big (abortive) carrier *United States*. It was designed specifically to launch heavy bombers, which at the time seemed to be the only aircraft that could deliver the currently massive atomic bombs. Such bombs were the only practicable means of effective attack at source. There was historical justification for attacking at source. In 1945, attacks on the German U-boat industry destroyed 41 of the 153 U-boats sunk that year. Until then it had been impossible to convince the British and U.S. strategic air forces to concentrate on U-boat targets, a subject of bitter contention between the U.S. and Royal navies and their respective air forces.[11] A variation on attack at source was aerial mining of areas near enemy submarine bases (in the 1950s the U.S. Navy's Bureau of Ordnance developed a new series of antisubmarine mines). The British became interested in other forms of attack at source as well, including the use of a midget submarine to plant a large nuclear bomb in Leningrad harbor.[12]

Another alternative, which the U.S. Navy developed, was deploying submarines as barriers to destroy enemy submarines in transit between their bases and their patrol areas. This was part of a larger effort to develop submarines as ASW weapons. Submarines sank enemy submarines in both world wars, but only once did an Allied submarine sink a submerged U-boat: on 9 February 1945, HMS *Venturer* sank the snorkeling *U 864*. The U.S. Navy planned to adapt wartime German long-range low-frequency sonars and a new generation of homing torpedoes, which would initially be combined in low-cost mass production ASW submarines (SSKs). Although the SSKs were not very successful, the idea of higher-performance submarines in barriers became a staple of the U.S. ASW strategy, particularly after nuclear submarines were available in numbers. The British also became interested in submarine ASW, but they did not envisage building specialized submarine-killing submarines on anything like the scale considered by the U.S. Navy.

When he attended the U.S. 1947 Undersea Warfare Conference, CDR R. G. C. Haines (of the British Admiralty Delegation in Washington) cautioned that his country could not wait for U.S.-type offensive ASW to wipe out the enemy's submarines; Britain would starve while it waited. The British therefore concentrated on defending 10 kt, 80–100 ship trade convoys. They viewed the new submarines' ability to get under a convoy and fire upward, sheltered by the noise of the merchant ships, as the worst tactical problem. Hunting the submarine while dodging through the columns of the convoy would be a difficult task.

The British tested the tactics they planned to use by gaming. Initially they assumed that the enemy submarine commanders were extremely brave, determined, and skilled—men who would press their attacks regardless of opposition. Certainly the Germans had had such men—whose very determination had led them to take fatal chances. Then the British tried a submarine commander who was skillful but cautious, and hence willing to make several attempts to secure an undetected approach. Another submarine commander in the game was brave but rather obtuse. He had also heard that U-boats that went under the convoy were not likely to live to tell the tale (if only, the British speaker said, there was some prospect of this being true!).

The British devised the "Cauliflower" tactic to get a submarine out from under a convoy. The convoy turned together through 90 degrees, with the escorts sweeping along its leading edge and past the stern. When it was tried with an abbreviated convoy (ships at the four corners, none inside), the submarine was revealed without realizing that the convoy had altered course. Throughout the Cold War, however, the problem of a submarine hiding under a convoy was never considered solved.

The British found that they had to place escorts farther from the convoy to give them maneuvering room; they might place some faster ships (pouncers) inside this extended screen. They also tried sonobuoy barriers on each bow of the convoy to create a wide sanitized lane. Any submarine not right ahead of the convoy would have to cross that hedge to get into a good attacking position. In exercises, submarines had been caught in this hedge. The gamers concluded: "We should like to have a group disposed about 20 nm ahead of the convoy, well placed to follow up these contacts, but we shall lucky if we have enough escorts for this purpose." At least in theory, aircraft could lay and monitor a lane of sonobuoys across the Atlantic. That operation would cost as much as two merchant ships, but it might be worthwhile.

Full-scale tactical trials by the Royal Navy's Londonderry Flotilla showed that successful hunting demanded an unprecedented level of communication among the ships involved; an escort group's senior officer needed almost as many VHF lines as a fleet flagship, and the lines had to be manned by the A/S teams themselves for quick action.

The British also found that an enemy submarine force would face limitations of its own. The force would need air reconnaissance unless it could rely on codebreaking, which seems to have been considered unlikely at this time. Reconnaissance information would have to be communicated. Without precision long-range navigation, there would probably have to be a radio link between airplane and submarine—and the submarine would have to put an antenna through the surface to establish it. Much the same would be the case with a HF link to submarine control at home. The Soviets apparently never fielded the alternative: low-data-rate LF radio that a submarine could receive entirely submerged.

A submarine approaching a convoy would have to use her periscope and/or her radar—and would therefore have to run shallow. It also seemed likely that a submarine trying to close in on a convoy would snorkel to save the battery until in contact with escorts or with the convoy itself.[13]

For the British, the great lesson of the Battle of the Atlantic was that they needed numerous escorts. These might be unaffordable in peacetime, but designs could and should be drawn for emergency construction. The British retained as many war-built escorts as they could, and they planned to convert wartime destroyers into escorts. By 1947 design work was proceeding, and the British were beginning to upgrade fleet destroyers with Squid, the best wartime ahead-thrown weapon.

By 1949 British exercises were showing considerable success against a submarine forced to seek out a target. Hunting a submarine trying to avoid detection would be considerably more difficult. Exercise experience indicated that the most profitable way to use ASW aircraft was to make it difficult for a submarine to gain contact with a convoy, by having aircraft search beyond the range at which a submarine could gain contact with any part of the convoy. Such aircraft could provide the convoy with early warning on which to base evasion.

Asked to evaluate the capabilities and limitations of all available ASW and submarine equipment, OEG broadened the problem to include all measures affecting the security of overseas transportation, including whether all merchant ships should be convoyed and how efficiently they could be turned around at ports.[14] The most striking initial conclusion was that in the immediate future, the threat of the enemy mining overseas terminals was far more serious than the submarine threat. It also seemed that the air threat to the overseas terminals had been overlooked. Later it was shown that reducing time in port by a third (from the wartime 30 days or more) would gain shipping capacity equal to half of all tonnage (14 million tons) sunk by submarines during World War II. Port facilities at home were the greatest bottleneck. Additional modern, efficient berths could be built at a cost approximately equal to that of building a ship. The extent to which port improvement was undertaken is not clear. The OEG analysis probably contributed heavily to the emphasis on mine countermeasures in NATO naval planning and the considerable U.S. support for construction of new mine countermeasures craft.

A major U.S. exercise conducted in November 1948 gives a sense of the pre–Korean War fleet at its ASW peak before the 1949 budget cuts.[15] The operating area was north and south of a rhumb line 710 nm long from a point just east of the Virginia Capes to a point south of Newfoundland. A striking force (TF 28) and a logistics force (TF 29) were required to pass along this line within 71 hours; they were unrestricted as to speed or evasive action. Each had to pass within a specified distance of a fixed point at each end. Two HUK groups and land-based ASW aircraft operated independently over the whole area, but no patrol aircraft were assigned to support surface forces directly. The opposing force was 14 submarines, 5 of them Guppies and 1 a non-Guppy snorkeler. They were not given any specifics of the movement of the target forces, and they were precluded from concentrating closer than 50 nm to the start and finish points.

The Atlantic HF/DF net broadcast all the information it obtained. It made 184 contacts as submarines used their radios to pass along contact information, revealing their positions early in the exercise. The HF/DF net uncovered the whole line of submarines attempting to intercept the surface forces. That enabled the strike force to avoid attack. On three known submarine positions, the HF/DF fix was accurate to within 10 nm. Trailers and concentrations of submarines were found; it was difficult to see how any submarines were able to close for attacks. Cued by HF/DF, HUK forces got 10 contacts, made 6 attacks, and were credited with sinking 4 submarines.

The logistics support force made 22 contacts, conducted 6 attacks, and was credited with 2 submarines. The striking force made 13 contacts, made 6 attacks, and was credited with 2 submarines. Land-based patrol planes made 24 contacts, attacked 17 of them, and was credited with 14 submarines. Of the total of 22 submarines counted as out of action, 8 were confirmed by submarine reports.

Carrier aircraft were initially grounded by weather, but they were able to fly after the afternoon of the third day. Once airborne they made it impossible for the submarines to trail and track both target forces. Because land-based aircraft could fly despite the weather, they caught many submarines on the surface. The lesson was that land-based all-weather aircraft should be assigned to direct support of surface forces when carrier aircraft could not fly. When carrier aircraft could fly, the land-based ASW aircraft could be released for offensive scouting.

The submarine force was credited with 125 contacts. It made 12 attacks and was credited with 17 ships out of action, including the carrier *Siboney* in the HUK group covering TF 29; the carriers *Mindoro* and *Wright* and five destroyers in another HUK group (TF 24); and the fleet carrier *Midway*, the cruisers *Albany* and *Roanoke*, and a destroyer in TF 28.

Submarines were sometimes able to penetrate screens undetected: two against the striking force, two against the logistics force, and one on each of the HUK groups. On the other hand, at least one submarine attempting to penetrate a screen was detected, and its attack nullified by the screens of both the strike force and the logistics force. The exercise seemed to show that surface screens were relatively ineffective. The submariners drew the lesson that they should counterattack destroyers trying to attack them. Many saw counterattack as perhaps the soundest submarine evasive maneuver. The five destroyers put "out of action" reflected this point of view. A commentator wrote that should the enemy adopt such tactics, they "would make an interesting war, to say the least."

Another exercise simulated a force retiring after a strike against enemy bases in northern waters. One Guppy got under the target formation (logistics force) and stayed there for 38 minutes, making 6 separate attacks. An attack on the supporting HUK group was detected and the submarine put out of action by its screen, but she managed to sink two of the three destroyers in the process. Another exercise that simulated a strike on an enemy base, opposed by land-based aircraft and submarines, was hampered by bad weather. The strike force managed to detect submarines four times, in each case attacking; it put one submarine out of action. The 2 submarines in the exercise, however, made 53 contacts resulting in 3 attacks. They claimed the carriers *Midway* and *Philippine Sea*, the cruisers *Newport News* and *Roanoke*, and 3 destroyers. Submarines penetrated screens undetected three times, making excellent attacks. The absence of support by land-based air was strongly felt; one submarine managed to trail the attacking force for almost a whole day. Some improvement in screen sonar performance was noted, but there were at least 19 undetected screen penetrations; all of the surface forces engaged were penetrated.

The main lesson was that land-based air was vitally important for its ability to restrict submarine communications and movements. All carriers should have airplanes carrying APS-20 radars (nominally for airborne early warning, but increasingly appreciated for snorkel detection) so that they could mount effective ASW patrols.

THE ESTIMATED THREAT

The Soviets' security made it impossible to gauge the level of the submarine threat, which was easy for Westerners to exaggerate. The wartime Soviet submarine fleet had accomplished almost nothing, even though some of its craft were technically quite adequate. The Germans were favorably impressed by Soviet submarines and equipment they captured. They rated Soviet diesels highly. The Soviets had built 200 submarines under their 1935 Five-Year Plan, and they had built a submarine yard at Gorkiy on the Volga River specifically to mass-produce submarines. Known building yards were at Leningrad (capacity 35 boats, as understood in 1946), Gorkiy (capacity unknown), and Molotovsk (later Severomorsk, capacity unknown). Knowledge of yards like Gorkiy had come largely from wartime German aerial reconnaissance; the Germans had never reached the Soviet Far East, and so had no information on the Far Eastern yard at Komsomolsk-on-Amur.

It now seems that although the Soviets adapted the seized German material for their own use, they were unwilling to learn from German submariners. That was not obvious at the time. For example, in June 1946 a U.S. Navy intelligence officer told the Navy's first high-level Undersea Warfare Conference that in 1945 the Soviets had wanted the four Type XXIs allocated to them delivered with their crews, presumably to act as instructors.[16]

In East Germany and Poland the Soviets seized Type XXI yards at Danzig, Elbing, Gdynia, and Konigsberg. Reportedly the Elbing and Gdynia yards were being moved with all their equipment to Kronstadt near Leningrad. Reparations claimed by the Soviets included the facilities of the Deschimag submarine yard at Bremen, which had not yet built any Type XXIs. The Type XXI program had emphasized prefabrication away from building yards; 16 inland plants in East Germany and Poland had been involved. Vendor plants had also fallen into Russian hands, the only exception being those that manufactured diesels. The Russians successfully recruited German workmen and engineers (plus some crewmen, although on a smaller scale). It was believed that they had acquired at least 74 Type XXI sections at Elbing, and they had probably seized many more in various stages of completion. ONI thought that in 1947 the Soviets would have 8 Type XXIs. In 1948 they would have 10–20, rising to 30–50 in 1949, and to 100–300 in 1950. The ONI spokesman considered these figures conservative.

In 1948 the British Joint Intelligence Committee pointed out that although the Soviets apparently had not yet fitted any of their own submarines with snorkels, such equipment was present on board German submarines they had taken over, so "it can be expected that the Russians will convert their present fleet and incorporate it in all ships now building and in future construction."[17] The British credited the Soviets with 230 submarines.[18] In addition to the 10 U-boats they had been allocated, they believed that the Russians had acquired at least 10 additional seaworthy U-boats and had taken as many as 40 unfinished U-boats to shipyards in the Baltic. The JIC went on to say: "The development of Walter type high speed submarines is being energetically pursued, but even if this is successful the completion of a large number is not expected before 1957. Nevertheless, it is expected that a type of submarine with comparatively high underwater speed (Type XXI or an improvement on it) will have been built in large numbers by that date."

A June 1949 estimate gave current Soviet submarine strength as 11 Type XXIs, 5 ex-German oceangoing slow type, 79 large slow type, 81 medium slow type, 131 coastal type, and 70 midget.[19] The net increase in operational ships by the end of 1950 was expected to be 7 Type XXI, 5 large slow type, and 9 medium slow type. It was considered unlikely that the Russians would complete successful sea trials of a redesigned German Type XXVI before the end of 1951; significant numbers were not expected before the end of 1954. It seemed obvious that the Soviets would upgrade their existing submarines, much as the Americans and the British were doing, by streamlining them and adding more batteries. That was expected to begin in 1950. British intelligence, which was sometimes shared with the Americans, was probably particularly important at this time because U.S. intelligence resources had not yet been built up. The JIC papers are a proxy for U.S. naval intelligence estimates, which are not available.

In 1955 the British (and presumably also the Americans) assumed that two-thirds of oceangoing and one-third of coastal submarines had been fitted with snorkels. It was known that existing Soviet submarines were being refitted on a large scale, but not what improvements were being made. It now appears that no existing Soviet submarines received snorkels, although some design work may have been done. Aside from the difficulty of mass conversions, the reason may be that the Soviets had selected the two-stroke Kolomna 37/D diesel for their postwar submarines. The four-stroke engines the Germans used placed different demands on a snorkel. The U.S. Navy also used two-stroke diesels, and it had to develop its own snorkel. Soviet snorkel development was apparently protracted; early W-class submarines lacked snorkels. The Soviets did use four-stroke diesels for some of their existing submarines, but presumably they were unwilling to develop two different kinds of snorkel, one of which would be used only on obsolescent submarines.

It was not clear when the Soviets might begin building new types of submarines. In 1948 ONI estimated that Soviet submarine construction capacity was 50–100 per year. Work on an improved Type XXI had begun, and production would likely be on an assembly-line basis.[20] ONI thought that some new submarines had been built, though that was not the case. In 1949 ONI credited the Soviets with 277 submarines, down from 335 the previous year. ONI credited the Soviets with 11 Type XXIs, although it had firm information only on the 4 transferred under the 1945 Allied agreement. It was reevaluating the existence of the others, "concerning which our information is vague." The Soviets were credited with 63 modern ocean patrol submarines, 64 medium-range submarines, and 103 coastal units. There were also 36 obsolescent submarines, 14–20 years old (after 20 years ONI dropped them). In ONI's view it was feasible to install snorkels and to streamline some but not all of the Soviet submarines. However, it seemed that the equivalent of full Guppy conversion (more batteries and more powerful motors) could not be accomplished without installing entirely new propulsion, which the age of the hulls would preclude.

It appeared that the Soviets had nearly completed restoring their prewar submarine-building infrastructure, which the Germans had largely destroyed. The Soviets had the strategic materials they needed, and they had seized enough German workers to make up for any gaps in their own skills. Their future submarine program would probably be limited only by state planners' decision as to its importance.

For 1955, ONI projected a total of 400 Soviet submarines, including 50 ocean patrol, 50 medium (600 tons), 100 coastal (300 tons), and 200 obsolescent. The current fleet, basically World War II types, demanded training to make up for the loss of World War II expertise, with some new development based on existing sensors and weapons.

ONI pointed out that the Soviets' naval orientation was defensive, and they might not be planning to wage a new Battle of the Atlantic. If they did, however, during the early months of a war, a Soviet submarine offensive might cost the Allies about 30 merchant ships (200,000 tons) per month.[21]

The West first received solid information on what the Soviets were doing in 1950, when they began repatriating Germans they had recruited (or seized) in 1945–46. It became clear that early postwar Soviet submarine construction had been limited to about 50 coastal M-class submarines; the open-ocean threat was still in the future. At least some longer-range submarines were undergoing large refits.

What follows is based on British papers, but they almost certainly reflect a consensus with U.S. analysis, which is not currently available. In 1950, both the British JIC and the U.S. Navy estimated that the threat posed by new Soviet submarines would probably become evident about 1954 and would grow rapidly after that.[22] By 1954 about 35 percent of Soviet oceangoing (and, presumably, medium) submarines would likely have high submerged speed.

In 1951, based on reports from German returnees, the JIC dramatically scaled down its estimate of Soviet submarine building: "We estimate that only 50 submarines a year are now being produced. These are almost entirely coastal types and replacements for old boats. The Soviets had nearly completed rebuilding their submarine construction industry, but at present very few new larger submarines are being built."[23] "It is probable that only slight progress has been made by a German research team in the Soviet Union for the development of the 'Walter' type propulsion, but it is thought that Soviet submarine research is proceeding independently." The same report credited the Soviets with a maximum building capacity of 150 submarines per year. In 1954, 35 percent of the Soviet oceangoing underwater fleet would be capable of 20 kts submerged. A summary of intelligence for SHAEF mentioned ominous evidence that Soviet submarines had undertaken long voyages for training.[24]

A 1 September 1951 periodic British intelligence summary seems to have been the first to mention that new submarine construction would probably concentrate on medium submarines, which had not been built in the Soviet Union since the war.[25] A 27 February 1952 report on Soviet R & D noted that "there is reason to believe that it will be a medium-size vessel with fairly high underwater speed, armed with at least six torpedo tubes, besides being fitted with snort and designed to carry mines."[26] This was a reasonable description of the first postwar Soviet mass-production submarine, Project 613 (NATO nickname Whiskey). It was estimated that 4 of the new submarines (described as Type XXI equivalents) were available in February 1952, and there would be 80 in 1955. In fact, the prototype (*S 80*) was completed on 2 December 1951, and the next (*S 64*) on 4 November 1952. It appears that 151 had been completed by the end of 1955. The March 1953 report on Soviet R & D stated that the Soviets were now building large and medium ocean patrol submarines, with less being known about the large ones. By 1954 agreed nicknames had been applied to both new types: W (Whiskey) for the medium ones and Z (Zulu, Project 611) for the large ones.[27]

In 1954 Whiskey was believed to be about 240 ft long, 1,600 tons submerged, with a maximum surface speed of 17 kts and an endurance of 13,000 nm. Zulu was described as very large, apparently 320 ft long, with a submerged displacement of about 3,000 tons; it was believed to have a maximum surface speed of 20 kts and an endurance of 20,000 nm. Both types were armed mainly with torpedoes but also had deck guns. It was believed that both had maximum underwater speeds of 16 kts for 1 hour. The present construction rate was believed to be 18 very large and 22 large submarines each year, an overestimate for the Z class but a serious underestimate for the W class.

For all their supposed lack of interest in Soviet submarine-launched missiles, the British reported on that possibility. In 1954 they saw no evidence of interest in or progress in shipboard missiles, although the Soviets had the capacity to produce a V-1 type missile (i.e., a cruise missile) carrying a one-ton warhead.[28] Should the Soviets decide to use submarines to launch V-1s, they could convert long-range submarines (the two new types or the old long-range K class) to carry two each. It was estimated that they could convert 8–10 submarines at any given time. If they could make 10 conversions in a 10 week period, they could produce 50 missile submarines in a year—enough to mount a large-scale assault. The British expected to receive warning of any such program within a month of its inception. The Soviets would need 16 months to mount a major attack.

After the new submarines had been at sea for a protracted period, they could be observed in action.[29] The Whiskey was credited with only about 12 kts submerged at the one hour rate, compared to an earlier estimate of 16 kts (endurance 20 nm). Displacement was now estimated to be 1,100 tons surfaced and 1,600 tons submerged. Estimated armament was 6 bow and 2 stern torpedo tubes, with an estimated load of 18 torpedoes. Gun armament was one 3.9-in. and possibly others. The estimates were fairly accurate. A Whiskey actually displaced 1,050/1,340 tons. Rated underwater speed was actually 13.1 kts. Submerged endurance was 166 hours, 335 nm at 2 kts. Surfaced endurance was 8,580 nm at 10 kts. Whiskeys had only 4 bow tubes (no stern tubes) and carried 12 torpedoes. Zulu was now thought to displace 1,480/2,095 tons, with a surfaced speed of 20 kts (endurance 5,000 nm at 20 kts) and a maximum submerged speed of 16 kts. Estimated armament included 6 bow and 2 stern tubes (24 torpedoes). Zulu actually displaced 1,900/2,350 tons. She was credited with 18 kts surfaced and 16 submerged, with an endurance of 20,000 nm surfaced and 9,500 nm at 8 kts snorkeling. Armament was 22 torpedoes. Both types initially had gun armament of a twin 57 mm and a twin 25 mm mounting.

The last of the 1950s trio of Soviet submarines was the successor to the earlier coastal type, Project 615 (NATO Q or Quebec class). The first was completed on 30 May 1953, but NATO seems not to have known about it until sometime in 1955 and took even longer to realize that it was not conventionally powered.

In addition to the big submarines, the Soviets were reportedly building midgets, having captured 50 German *Seehunds* at the end of the war. By 1955 they were thought to have their own designs in production.[30] Details were not forthcoming, but the Soviets certainly built midgets later, and presumably Western intelligence did not observe craft that did not go into the open sea.

We now know that Stalin was far more interested in building a seagoing surface fleet than Westerners believed in the 1940s. He therefore limited his immediate (1945–55) submarine program to 40 large submarines, 204 medium submarines, and 123 small submarines. By 1955 the Soviets had completed 6 large, 151 medium, and 66 small submarines, the latter of prewar design except for the prototype Q. After Stalin died, the submarine program received considerably greater emphasis, as already noted. Successors to the initial Whiskey and Zulu classes were designed: the medium Romeo (Project 633) and the large Foxtrot (Project 641). Initial plans called for construction of 560 Romeos, but the program was curtailed with the success of the Soviet nuclear propulsion program, and also as a result of Khrushchev's radical policy shift.

The U.S. and Royal navies were uncomfortably aware of the potential offered by the much-longer-range torpedoes the Germans were testing in 1945: they could be fired from beyond sonar range, so the first indication that a submarine was present might be a torpedo explosion. The first solid information on Soviet torpedo projects came from returning Germans, probably in 1954.[31] They said that the Soviets were pursuing both HTP and semi-closed-cycle oxygen torpedo designs to gain long range.[32] The first fruit of the long-range program was 53-57, a 53 cm (21 in.) torpedo adopted in 1957. It used a German-type HTP power plant.

SONARS

The new much-longer-range torpedoes demanded new developments in sonar. Sonar range had to be extended to the point where a submarine could not fire effectively from beyond it. By 1947 the U.S. Navy wanted to double the average World War II performance, which was a first-contact range of 1,400 yds. An officer said the ASW problem would be 80 percent solved if sonar range could be doubled from 1,500 to 3,000 yds. The U.S. Navy had decided to replace existing searchlight sonars with QHB scanning sonars. Initially QHB was expected to outperform the searchlights as a search sonar, but by 1949 it was clear that it was no better than a searchlight in that role. Under operational conditions the average contact range of QHB was only about 2,000 yds. Its limitations may have been due to operators missing the brief spots of light indicating targets on their display. This problem surfaced again and again in the 1950s, ultimately leading both the U.S. Navy and the Royal Navy to adopt computer-aided detection. Until then effective sonar range was generally considerably shorter than the predicted range.

QHB excelled, however, in its ability to maintain contact on a fast submarine. In controlled tests, QHB maintained contact at up to 30 kts, compared with 24 kts with the wartime standard QGB searchlight.[33]

It maintained contact with a fast evading submarine: 5 hours with a U.S. fleet submarine and 2.5 hours with a Type XXI. In each case the sonar was still holding the submarine when the exercise ended. By 1948 QHB was in production and was being installed on board destroyers. Development of a companion SQG-1 attack sonar began in 1946. As of mid-1948 an experimental model was expected to be available for operational test and evaluation by July 1950.

The obvious way to extend sonar range was to lower the frequency it used, because the sea absorbs less of a lower-frequency sound. A sonar operating at lower frequency needed a proportionately larger transducer, which set an ultimate limit on hull sonars. By 1948 the Naval Research Laboratory (NRL) was building an 8 kHz searchlight sonar to test the theory. In August 1949 it measured propagation losses at 7.8 and 16 kHz over areas of the North Atlantic between Long Island, Bermuda, and Cape Hatteras. Water conditions were poor for sonar operation, with an isothermal layer about 100 ft deep over a sharp negative gradient. Sonar projectors (searchlights) were suspended from the destroyer escort *Maloy* (DE 791). A string of 6 receivers at depths ranging from 15 to 400 ft was suspended from the ASW ship *PCE(R) 849*. The performance of the deep hydrophones was astoundingly good. Calculations indicated that any range up to 70,000 yds could be attained if pings were powerful enough. NRL estimated that it could reach 20,000 yds with a 5 kHz sonar, and 70,000 yds (the convergence zone) with a 2.5 kHz sonar. The chair of the 1949 Undersea Warfare Conference, ADM Charles Momsen, called NRL's new data the most significant ASW development of the past year.[34] The immediate result was SQS-4, in effect a lower-frequency (8–12 kHz) version of the 22 kHz QHB (see appendix A for details). It became the standard U.S. Navy sonar of the 1950s.

Another way to improve sonar performance was to reduce ship self-noise. The 1948 Undersea Warfare Conference was told that reducing a ship's noise by 6 dB (a factor of 4) would increase sonar range at 20 kHz by at least 50 percent. The range at which a submarine could detect the ship would be reduced by 20 percent, and the activation range of passively homing torpedoes would be reduced by 35 percent. The U.S. and Royal navies developed bubble screens to quiet propellers. The U.S. Prairie-Masker emitted air bubbles around the machinery spaces and other bubbles from the propeller. The bubble screen around the machinery was intended to block the water path for machinery noise to enter the ship's sonar. The British Nightshirt emitted bubbles from the propeller. Both navies also worked on quieter propellers. The U.S. Navy became interested in pump-jets, which were initially installed on the destroyer *Witek* in 1959. A different form of pump-jet became an important means of silencing submarines in the 1960s and beyond.

The U.S. Navy carried out its early postwar sonar experiments off New London and Key West. It just happened that in both places there was a shallow (30 ft deep) surface layer that trapped sonar pings. The false perception that this was a more general phenomenon led the U.S. Navy to try shipboard and helicopter dipping sonars, because they could be lowered under this layer. It was soon discovered that the surface layer in the open sea was generally much deeper, but interest in dipping sonars survived.

There was also interest in passive sonar, particularly for submarines that could operate silently and preferred not to give themselves away by pinging. Passive sonars had no visual displays; operators depended entirely on their ears as receivers. The second U.S. Navy Undersea Warfare Conference (December 1947) recommended development of some visual means of displaying the output of a passive sonar. Probably based on a recommendation by Bell Telephone Labs, it suggested using a spectrogram or a "fine structure" presentation of frequency, time, and intensity. The spectrogram would show the intensity of received sound at different frequencies. This recommendation turned out to be very important.

FIRE CONTROL

Ships fighting fast submarines would have to maneuver much more radically than their World War II forebears. They needed to keep track of the situation in geographic rather than own-ship terms. Otherwise, their picture of the maneuvers executed by the submarine would be confusing at best. The World War II A/S Attack Plotter already automatically displayed a geographic picture of an engagement. The U.S. Navy's synchros transmitted data quickly between weapons and sensors.[35] In accord with the 1945 BuOrd-BuShips program, the postwar Navy developed regenerative underwater fire control systems: Mk 101 for submarines and Mk 102 for surface ships. They presented the tactical situation in geographic terms. Mk 102 was built around the Mk 4 Attack Director conceived in 1944–45 to replace the wartime A/S Attack Plotter.

The inputs (estimated target course and speed) for Mks 101 and 102 were produced by analyzers that fitted sequences of on-target range and bearing to either a straight line or a curve, deducing from that the target's course and speed. Both generated future target position for comparison with what the ship's sonar actually heard. The human operator of the wartime A/S Attack Plotter had had to estimate target speed. In Mk 102, the Mk 4 Attack Director accepted analyzer input and pointed the ship's sonar at the estimated target position. The operator could evaluate the solution using a paper plotter. The full Mk 102 could eventually produce a solution in 20 seconds.

Mk 102 was considerably more massive than earlier types (about five tons), and it was expensive ($400,000–$500,000, compared with $35,000 for the Attack Director and $2,000 for the Attack Plotter). It required three or four operators compared to only one for the Attack Plotter or the Attack Director, but because it was so automated, its operators would require much less training—a very important advantage in a fleet badly limited in qualified personnel. Estimated accuracy was 0.3 kts for a solution time of 45 seconds, and 0.6 kts for a 20 second solution. If the automatic solution seemed inadequate, the system could accept manual inputs. The Attack Director itself was credited with a maximum range of 2,000 yds, compatible with most of the existing weapons. Claimed Attack Director accuracy was about 0.7 kts, based on the current scanning (QHB) sonar, which was credited with an accuracy of about 20 yds in range and 2 degrees in bearing (which might not be achievable at sea).

In 1947, with the Attack Director still in the design stage, CNO considered it so important that he ordered installation on board all active ASW ships. A production contract for the first 80 was let in May 1947. Twelve prototype Mk 4 Attack Directors were installed on board the converted *Gearing*-class destroyers of DesDiv 81 and on board the other OpDevFor destroyers. Of the 80 on order in 1948, 60 were to be completed as Mod 2, capable of computing data for fixed Hedgehogs and depth charges, to fit destroyers not scheduled for trainable Hedgehogs. It transmitted target course and speed. Mod 3 included a solution for the trainable Hedgehog and had space reserved for components needed to compute gun train and elevation orders for the future Weapon A. As of early 1950, the first full Mk 102, incorporating the Attack Director, was expected in May, with tests expected beginning in September.

The existing A/S Attack Plotters were retained; they were even considered valuable adjuncts to Mk 4. The Operational Development Force developed some simple improvements, including lengthening the own-ship predictor line to allow for attack speeds greater than 18 kts. Modified Attack Plotters entered service early in 1949.

While waiting for Mk 102 to appear, OpDevFor combined existing equipment so that it could test new weapons and tactics. The resulting Mk 100 was installed on board four ships. The first two were for the two prototype DDEs, *Epperson* (DDE 719) and *Robert A. Owens* (DDE 827). Both had provision for Mk 35 torpedoes, but at this time only *Epperson* had provision for a trainable Hedgehog. The two DDKs (*Basilone* [DDK 824] and *Carpenter* [DDK 825]) had modified Mk 100s. Because components were not yet available, there were two versions. Type A was built around an Attack Director Mk 4 Mod 3, a depth computer, and a modified submarine torpedo data computer (as its position keeper). It also incorporated a paper plotter as an evaluation tool. Type B analyzer (target course and speed deducer) and position-keeper functions were provided by a modified

PT boat device. The required ballistic computers for Weapon A were not immediately available. Version A was slow because the Attack Director was slow, and its torpedo data computer required manual inputs and was also slow. Version B had a direct connection between its angle solver and its position keeper, but it too was slow. The Mk 100 designation made it the first of a series of underwater fire control systems. Although there was no production order, Mk 100 was valued as a prototype interim system for use in an emergency.

In accord with the wartime plan, the new underwater fire control systems were treated like surface fire control systems, with spaces devoted to them. New belowdecks Underwater Battery Plots would contain computing, plotting, communications, and controls needed to solve the underwater fire control problem and to pass orders to the underwater battery, just as the plotting rooms of a surface ship controlled her guns. Data would be fed to the plot from an adjacent sonar control room containing the sonar operator's console. Tactical information would be fed automatically to the CO Tactical Plot (COTP), a weatherproof station immediately accessible to the pilothouse and to the open bridge in which the CO could find a display showing all the information needed to fight the ship.

Ships would fight in groups, so by 1948 the U.S. Navy envisaged an automatic data link connecting the integrated display systems of the ships. The system would transmit target range and bearing as observed by each ship, the ship receiving the data combining it with the relative positions of the transmitting ships, so that the receiving ship could reach a firing solution without having to use her own sonar. This would permit attacks analogous to the creeping barrages of World War II, with the target submarine not knowing that she was under attack until weapons arrived. The link would have been a fully analog system, using synchro signals transmitted by radio. In 1948 an evaluation version was expected at the end of FY 49, but it is not clear that it was ever completed. Within a few years the U.S. Navy would be interested in digital means of data transmission.

The British and the Canadians were also interested in maintaining a valid ASW picture. The British developed the analog Cambria tactical table (display) as an offshoot of the analog Comprehensive Display System (CDS) developed for fleet air defense. It became the Automatic Surface Plot, the prototype entering service in 1953. A planned data link to connect ASW ships' displays was never adopted. The Canadian equivalent, the NC-2 tactical table, was adopted by the U.S. Navy, also without any intership link.

LT James Belyea of the Royal Canadian Navy seems to have been the first to argue that the tactical system required to deal with fast submarines should be digital, including a ship-to-ship digital link. Belyea drew in the Canadian arm of the British Ferranti Company, which was interested in digital computers. In 1950 Ferranti succeeded in transmitting digital radar data via a radio link. In 1951 the Canadian government let a contract for the digital system, DATAR (Digital Automatic Tracking and Remoting). Ultimately the RCN failed to adopt DATAR, but before that the digital link had been demonstrated, first ashore and then between two Canadian minesweepers and a shore station. It became Link 11, the standard U.S. Cold War tactical link.

NEW SHIPBOARD WEAPONS

In 1945 Weapon A (or Alfa or Able) was the planned future U.S. Navy ASW weapon, featuring in the designs of new ships (the ASW cruiser *Norfolk*, the *Mitscher*-class destroyers, and planned conversions of destroyers). As of the summer of 1948, pilot production of 200 firing mechanisms was underway to make final development evaluation possible. It appeared that Weapon A could replace one quintuple set of torpedo tubes on board a general-purpose destroyer. It was conceived as a gun that could rapidly fire rockets to a range considerably beyond the reach of Hedgehog: initially 760 yds (compared with 188 for Hedgehog) and then, with a new rocket, to 975 yds. Weapon A was considered so important that it was ordered into production without having been evaluated.

While Weapon A was being developed, the U.S. Navy's Operational Development Force needed a trainable ahead-thrown weapon to test tactics against submarines that might cross an escort's path at high speed. In 1947 it placed a Hedgehog on a quadruple Bofors antiaircraft mounting as Projector Mk 14 (the earlier Hedgehogs were Projectors Mks 10 and 11). It was power-trained, remotely controlled from the Underwater Battery Plot. Given the high priority accorded ASW, CNO and BuOrd expedited the project, and the new weapon was available within a year. Nine were made. The first ships equipped with it were the destroyer escort *Jack W. Wilke* (DE 800) and the destroyer *Sarsfield* (DD 837), in that order. In each case the Hedgehog replaced a main battery gun mounting. In *Wilke*, it initially replaced No. 1 3-in. gun, and then the superfiring No. 2. This makeshift proved successful, so a more sophisticated Mk 15 followed in 1948. It added remote control of spigot elevation to better compensate for ship's roll. By 1949 it had been accepted for production, still prior to evaluation. In some cases, Mk 15 was installed pending availability of Weapon A, but some ships received it instead.[36] OpDevFor also determined that it was essential to attack the submarine as quickly as possible and at maximum range, so they recommended immediate efforts to increase the range of ahead-thrown weapons to at least 300 yds.[37]

The British already had Squid, which the U.S. Navy had tested in 1945. It armed the first postwar U.S. destroyer escort, USS *Dealey* (her sisters all had Weapon A, which reloaded faster than Limbo). Faster reloading was said to be necessary because given high relative speeds, a ship might not have time to get off more than a few rounds. Other countries developed their own ahead-firing weapons.

The U.S. Navy already had fast-sinking ("teardrop") depth charges. OpDevFor, charged with getting the most out of existing weapons, considered improving depth charge attacks. Existing rule-of-thumb tactics were not good enough: depth-fuzed charges would probably be no more than 1.3 percent effective against Type XXIs at 250 ft in a single-pattern attack. Analysis of why that was the case began with the attack error of an attack—the distance between the center of the exploding barrage and the submarine. Adopting proximity fuzes alone would make a pattern five times as effective. Faster-sinking charges (the original 21 ft/sec increased to 38 ft/sec in teardrops) would reduce blind time. The new sinking

Facing the nightmare: the U.S. Navy became interested in ways to quickly grow its ASW fleet. One idea was to convert *Fletcher*-class destroyers, all of which were in reserve, into fast escorts (DDEs) by removing three of their guns and installing ahead-thrown weapons and new sonars. Initially, plans called for converting ships and then returning them to reserve. USS *Nicholas* is shown off Mare Island Navy Yard, 2 March 1951, with an interim battery consisting of a trainable Hedgehog (rather than Weapon A) and her original forward torpedo tubes rather than the planned quartet of fixed tubes. Note her HF/DF mast. Wartime-type H/DF was still valued because the Soviets had not yet adopted anything like the German *Kurier* "squash" transmitter. NHHC

USS *Cony* shows the planned Weapon A (in No. 2 position) as well as one of two fixed Hedgehogs, fruits of a later plan to upgrade the entire destroyer force. The muzzles of a pair of fixed long torpedo tubes are visible just above her deck, in the deck house section abaft her after funnel. She shows early postwar ESM antennas on the stub mast around her after funnel, but the HF/DF mast originally installed is gone. Note that she retained a depth charge track right aft. USNI

The Operational Development Force sought ways to keep the large fleet left over from World War II effective. With development of Weapon A, OpDevFor experimented with trainable Hedgehogs, initially on twin 40 mm gun mountings. USS *Jack W. Wilke* is shown in the 1950s or later, with a trainable Mk 15 Hedgehog in No. 2 position. Earlier she had had a Hedgehog in No. 1 position, and the gun in No. 2. Note that she has an expanded sonar space, reinforced (as in other destroyer escorts) against the sea. NHHC

rate was just short of that achieved by Weapon A and Squid. Runs on an attack teacher showed that current attack errors against Type XXIs were smaller than wartime errors against slow submarines. Theoretical analysis justified using a 19-charge pattern against a Type XXI, the charges spaced 160 ft apart (the pattern extended over 848 ft). To avoid increasing topweight, the number of standby charges on each side of a ship would be reduced. Replacing the usual TNT with more powerful HBX would increase the lethal radius of each charge from 14.3 to 18.3 ft, and it could be further increased by changing the influence range of the depth charge proximity fuze. Normally it was set to 35 ft, but it could be set instead at 50 ft, which it was argued would have a valuable psychological effect on submarine crews. It might be increasingly important against deeper submarines. OpDevFor cited recent Woods Hole data. For an additional boost, the weight of explosive could be increased from 195 to 250 lbs.

Proximity-fuzed charges, improved attack procedures, faster-sinking charges, revised patterns, and more lethal ammunition brought lethality against a submarine at 250 ft to about 50 percent, and to about 25 percent at 500 ft. Adding greater fuze range, larger charges (for part of the pattern), and the extra lethality predicted at depth brought the probability of

Depth charges were still valued, particularly given analyses by the Operational Development Force. Off Korea on 27 July 1951, *Renshaw* (DDE 499) shows a row of depth-charge throwers behind her bulwark. She also shows the muzzles of two fixed torpedo tubes just above her deck and forward of the sponson of her port twin 3 in./50 antiaircraft gun. NHHC

damage at 500 ft above 70 percent. Even 40 percent probability was 31 times the previous figure.

An important caveat was that no full-scale experiments had been conducted against Type XXI submarines to confirm the estimates of damage probability. Those had to wait for the U.S. Navy to convert enough submarines to Guppies to allow retirement of the two Type XXIs so they could be subjected to underwater explosions.

The U.S. Navy also had ASW homing torpedoes. By 1947 the highest priority in U.S. torpedo development was being given to ASW, though other types of torpedoes were being developed (see appendix C).

Given the potential range advantage of enemy torpedoes, both the U.S. Navy and the Royal Navy were vitally interested in torpedo countermeasures. The sheer variety of torpedoes made the problem very difficult. They ranged from noisy unguided torpedoes such as the wartime U.S. Mk 14 to quiet electric homing torpedoes such as the U.S. Mk 28 and the German T-5. Pattern-running torpedoes had to be defeated. The U.S. Project General Type 2 used towed explosive streamers with active fuzes to protect merchant ships by destroying torpedoes that crossed the line of explosives.[38] An early version had been used during World War II. British visitors thought this technique might protect merchant ships but would embarrass a maneuvering warship.

Early estimates of active protection for escorts were depressing: the weight of antitorpedo weapons needed for passage across the Atlantic might exceed the weight of ASW weapons an escort could carry. Soon after (if not during) World War II the British began work on Ruler, a mortar-firing proximity-fuzed bomb, to destroy an incoming torpedo.[39] Once they realized that Ruler would be too heavy for an escort, they planned to limit it to cruisers and larger ships. A Ruler salvo would be triggered by a torpedo detection sonar (a torpedo approaching from astern would be a problem). After Ruler failed, the British turned to antitorpedo torpedoes (Camrose). The question was whether nearby explosions of such weapons would defeat a torpedo designed to be fired in a salvo, and hence subject to nearby torpedo explosions. The British seem to have abandoned these projects about 1955. The Type 184 asdic, eventually an important ASW sensor, was conceived as a torpedo detector, presumably initially to control antitorpedo weapons and, if necessary, to warn a major unit in time to evade.

The Allies had used decoys to deal with homing torpedoes. Work continued after the war. Conversions of light aircraft carriers for ASW operations included QHB as a torpedo evasion sonar and also as a means of triggering Mk 31 rocket launchers firing NAE "beacons" (decoys). Tests showed that QHB could detect noisy torpedoes at 1,500 yds and sometimes even at 2,500 yds, but it missed quiet homing torpedoes, and it heard nothing in the baffles aft. A possible alternative was the German passive array sonar (GHG) captured on the cruiser *Prinz Eugen* because it proved capable of hearing quiet torpedoes. This application was not pursued, but GHG became the basis of a U.S. submarine sonar. The rocket-thrown decoys were abandoned because they were unacceptably dangerous: their launcher had to train and fire automatically whenever a torpedo was detected—that is, without warning.

That left towed decoys, which are still the standard U.S. torpedo countermeasure, though some navies use thrown decoys. Postwar decoys had volume controls that made it possible to use them while pinging. In 1948 the U.S. Navy tested Fanfare (T Mk 6) and determined that the noisemaker could be switched off for no more than 10 seconds at a time and still leave the escort with about 95 percent protection and all search capability. This device became the standard U.S. torpedo countermeasure of the 1950s and 1960s. The British Type 182 was analogous.

ASW SHIPS

Both of the largest Western navies were acutely aware that ASW demanded large numbers of escorts. During NATO negotiations in 1950, the U.S. Navy argued that a through convoy escort should have a 50 percent probability of detecting a submarine attempting to penetrate the screen of a 60-ship convoy. Exercises showed that it would take 4 escorts to deal with 1 submarine, so the through escort was set at 8 escorts (with another 4 for particularly dangerous areas). The Royal Navy argued for a through escort of 2 destroyers and 4 frigates, but envisaged support groups in the approaches to the UK. When the NATO regional force plans were added together, the result was clearly unaffordable. Instead, the NATO standing group adopted the results of a U.S. global force study. It included two 6-carrier task forces (one Atlantic, one Mediterranean), each of which would require 40 destroyers. Remaining forces included 17 light and escort carriers, 297 destroyers, 512 destroyer escorts, 114 submarines, and 877 minesweepers. This was much the force envisaged in the April 1950 NATO medium-term plan: 12 fleet carriers (the 2 task forces), 19 escort carriers (cores of HUK groups), 920 destroyers and larger escorts, and 107 submarines.[40]

Other estimates reached similar conclusions. In the 1949 Dropshot study, the U.S. Joint Chiefs of Staff estimated the forces needed to fight a 1957 war (i.e., in the "year of maximum danger"). On average, 8 Atlantic and 8 Pacific convoys would sail each month, and 2 convoys would be at sea in the Mediterranean at the same time. A typical Atlantic escort group would consist of a light carrier (CVL) screened by 4 destroyer escorts, supported by an ASW surface attack unit of 4 ASW destroyers (DDEs). Facing greater air opposition, a Mediterranean convoy would be screened by 2 light carriers with only 2 destroyer escorts and a 2-destroyer surface action unit. There were not enough light fleet carriers, so 6 escort carriers were needed. In addition, the U.S. Navy planned to form 9 HUK groups within 6 months of the outbreak of war: 3 in the North Atlantic, 3 in the Mediterranean, 1 in the South Atlantic, and 2 in the Far East, each consisting of a carrier, an ASW flagship, and 12 destroyers. These mobile HUK groups would be backed by 8 fixed-area patrol/HUK groups.

Against such figures, both the United States and the United Kingdom still retained very large numbers of war-built ships. In 1950 the U.S. Navy had 351 destroyers: 52 *Benson-Livermore* class (about 1,630 tons), 150 *Fletcher* class, 53 *Sumner* class, and 96 *Gearing* class (including 5 ships suspended incomplete). These figures include destroyers converted for ASW and as radar pickets, but not minelayers and minesweepers. In addition, there were slower destroyer escorts of six types: the nominally 24 kt steam turbine classes (TE [55 *Buckley* class], TEV [20 *Rudderow* class], and WGT [77 *John C. Butler* class]) and two types of slower (21 kt) diesel ships (FMR [85 *Edsall* class] and 9 DETs), a total of 246. Many DETs had already been transferred, and more were earmarked for transfer to postwar Allies. A program to convert destroyer escorts into radar pickets for continental defense was about to begin. Seven TEs that had been converted into radar pickets in 1945 to support the expected invasion of Japan were not reactivated.

The Royal Navy had mounted a massive wartime building program of its own, but it had also lost more destroyers. After scrappings and transfers, the Royal Navy retained 119 destroyers, with another 12 on order (4 "Weapons" and 8 *Darings*). There were also fast frigates (Hunt class), but they lacked endurance and were not modernized. All other British frigates were relatively slow, capable of about 18–19 kts: 28 sloops, 37 River class, 25 Loch class, 21 Bay class, and 24 Castle class (corvettes), a total of 135.

The Royal Canadian Navy, in 1945 the third largest (by numbers) in the world, struck its River-class frigates but did not sell most of them. It took them back into service, so that in 1950 it had 23 River-class frigates, 2 of which became weather ships. The others were modernized beginning in 1951. Canada also had six destroyers.

Together, the U.S. Navy, the Royal Navy, and the Royal Canadian Navy had a total of 878 destroyers and large escorts, a more than adequate number against the existing, mainly obsolescent, Soviet submarine force. The other NATO navies could contribute 44 more ships, so that total NATO naval strength barely exceeded the required figure.[41] This was, of course, something of an illusion, because the United States had to split its naval strength between the Atlantic and the Pacific. Presumably the NATO figure took into account the assignment of destroyers as carrier escorts (in an ASW role if attack at source were taken into account) and of frigates to HUK groups.

There was little hope either in 1946 or in 1950 that governments would be able to repeat the mass escort production of World War II.

Many of the escorts were relatively slow, and in 1950 nearly all of them were still equipped with World War II sonars and weapons. That was why the U.S. Navy's Operational Development Force devoted considerable attention to developing new methods of using existing equipment, such as depth charges. When some participants at the 1948 Undersea Warfare Conference described this as a waste of time, OpDevForce's commander replied, "If this thing ever pops in your face, you are going to find 90 percent of your ASW ships are going to start where they were at the end of the war."[42]

The expected Soviet Type XXIs were a particular problem, because in theory an ASW escort should be 50 percent faster than the submarines it hunted—an anticipated 27 kts to deal with the new generation of fast submarines. In fact, the Whiskey turned out to have a maximum speed of only 13 kts, and the ocean-going Zulu made only 16 kts. That cut the required speeds to hunt such submarines to 20 and 24 kts; a speed that many surviving wartime escorts could achieve.

To obtain fast escorts without new mass production, both the principal navies became interested in converting fleet destroyers. Initially the U.S. Navy envisaged two different kinds of converted destroyers: "killers" and "escorts." Killers (DDKs) would have light antiaircraft armament and the most advanced ASW armament available. They would have a CIC on the first platform deck with an A/S plot below it. The positions of other ships of the killer group and the target would be displayed automatically and three-dimensionally. Escorts (DDEs) would have sonars, a light ASW armament, and good antiaircraft armament. By 1949 the two types had merged under the DDE designation.

The 1946 U.S. Navy Undersea Warfare Conference called for formation of two experimental hunter-killer groups, one to develop tactics against medium-speed deep-diving submarines (Type XXI) and one to develop tactics against fast submarines (Type XXVI). The medium-speed HUK group would be built around an escort carrier (*Mindoro* was chosen), two *Gearing*-class destroyers, and five fast destroyer escorts. Numbers of destroyers and destroyer escorts were reversed in the spring of 1947, and modifications begun on 1 June, scheduled for completion by October.[43] The fast HUK group would be built around a converted light carrier (*Saipan* was chosen) and five *Gearing*-class destroyers, which were to be converted in West Coast yards. Initially they were to have been completed in 1949, but by late 1947 that had been advanced to 1948. Conversions would depend on the performance of the first group of destroyers, which formed Destroyer Division (DesDiv) 81. OpDevForce was already converting some ships to experimental ASW units, beginning with the destroyer escort *Jack W. Wilke* (DE 800) and soon followed by the *Gearing*-class destroyer *Sarsfield*. By late 1947 both had been fitted with the latest available ASW fire control system and the Mk 14 projector (the unstabilized trainable Hedgehog). *Sarsfield* was considered the prototype of the first five converted *Gearing*s, but she was never redesignated, as they were, as a DDE. A second TE-type destroyer escort, *Foss* (DE 59), also assigned to OpDevForce, was given the new sonar but apparently not a trainable Hedgehog.

By late 1947 seven escort carriers had been ordered converted to form the basis of wartime HUK groups. All were of the largest wartime type, the specially built *Commencement Bay* class. *Mindoro* had already been converted; *Badoeng Strait* was in a yard, scheduled to complete in November; *Palau* was scheduled for completion in December 1947; and *Sicily* was scheduled for June 1948. *Siboney*, *Bairoko*, and *Rendova* had recently been designated for conversion. Three would be employed in peacetime for ASW, the others operating as Marine Corps carriers, available in an emergency as ASW ships. The two latest light carriers, *Saipan* and *Wright*, were to be converted for ASW. As of late 1947, this work was scheduled to be completed by July 1949. In 1948 this program was extended to two more light carriers, *Cabot* and *Bataan*. All of these ships were

Escort carriers built late in World War II were adapted for postwar ASW as the cores of hunter-killer groups. *Bairoko* is shown on 13 January 1953, with an all-Avenger air group. The third airplane back on the port side has an APS-20 surface search radar, considered an effective means of detecting snorkels. USNI

given antitorpedo systems employing QHB sonars and Mk 31 decoy launchers, HF/DF facilities, the latest radars, sonobuoy stowage, and facilities to stow and maintain air ASW weapons such as homing torpedoes.

By 1949 fleet operators were complaining that the converted escort carriers were inadequate. The Atlantic HUK commander argued that to prosecute ASW on a 24-hour basis an ASW carrier needed a large and relatively stable flight deck offering flexibility and safety, which the small flight decks of the escort and light carriers could not provide. A larger carrier would impose fewer limitations on future ASW aircraft. The ship needed radio and radar sufficient both to support operations away from the carrier and for safe and rapid recovery. Living accommodations had to provide acceptable rest for all personnel. The ship needed enough speed and endurance to search over a wide area. The ship also needed protection against air attack, meaning both fighters and a sufficient antiaircraft battery. Instead of small escort carriers, *Essex*-class fleet carriers were needed; one or two should be assigned exclusively to HUK operations. The Pacific Fleet ASW carrier commander could operate continuously in average to marginal weather conditions, but he was hampered by the limited equipment on board his escort carriers, limited instrumentation of his aircraft, and cuts to training funds. Not surprisingly, he expected his force to be effective against World War II–type submarines but only moderately effective against the fast new submarines. In a November 1948 exercise, an aircraft from the carrier laid a large sonobuoy barrier (thought to be the U.S. Navy's first), in a square 30 nm on a side, with nine buoys on each side. Two escort carriers operated inside the square. All submarines trying to penetrate the square were detected, and they never attacked the carriers. He tried single-line barriers in May and October 1949. In contrast to the Atlantic Fleet, he tried operations controlled by aircraft with APS-20 early warning radars, which could detect snorkels in fairly calm water.

The exercises showed that the escort carriers were too slow, even when the ships they were covering slowed to 13 kts. They could barely keep up because they had to keep turning into the wind to launch and recover aircraft. Each time, the gap between them and the ships they were protecting widened; on some occasions the escort carriers were 80 nm behind the forces they were covering. It took as much as an hour for the carrier to recover her ASW patrol and launch the next one; during that interval the task group or convoy had no ASW cover at all. Breaking the air patrol into three sections, as had been proposed, would make the problem worse, because recoveries and launches would be more frequent.

During 1948 exercises in bad weather in the North Pacific, escort carriers were so unstable compared with fleet carriers that they operated for a quarter less of the daytime. They could not even try night operations. Anything they could do, a fleet carrier could do better. The Pacific Fleet commander asked whether it might not be better to give standard fleet carrier air groups one all-weather attack squadron fully qualified and equipped for ASW. This idea may have been behind the development of ASW-capable versions of the standard attack bomber, the Douglas Skyraider.

The high priority assigned ASW justified inclusion of a large killer ship (the cruiser CLK *Norfolk*) in the FY 48 program. *Norfolk* was a large ship with specially silenced hull and machinery, and what was expected to be excellent seakeeping. She would be a wartime flagship for an ASW force, heavily armed with four Weapon A launchers and eight fixed tubes for ASW torpedoes. The FY 48 program also included four new *Mitscher*-class destroyers, with another four planned for FY 49; final characteristics were approved on 26 July 1947. They were not primarily ASW ships, but they were to be armed with two Weapon A and four fixed torpedo tubes for ASW. Four planned FY 49 ships were not built; nor was a second *Norfolk* also included in the FY 49 program.

Four *Gearing*-class destroyers suspended at the end of the war were completed, two as prototype escorts and two as prototype killers (DDKs). The

The first step toward countering the new fast submarines was to form hunter-killer groups around small carriers supported by modified *Gearing*-class destroyers. Initially the groups were to be led by massively upgraded DDKs working with more austere DDEs, but in 1950 the two series merged. The DDKs and some of the DDEs were *Gearing*s left incomplete in 1945. *Carpenter* is shown after her original pair of twin 3 in./50s were replaced (in 1957) by the originally planned pair of twin 3 in./70s in enclosed mountings. She had a Weapon A superfiring over each twin 3 in./70. As completed she had a trainable Hedgehog in a bandstand forward of her forward Weapon A, superfiring over her forward 3 in./50 mounting. Eliminating the 5 in./38s of the original design made it possible to replace the big Mk 37 director with a small Mk 56, freeing considerable space in the bridge structure for the desired sonar and ASW command-and-control spaces. Note that she lacks the HF/DF mast of *Fletcher*-class DDEs. She had a single sister, *Robert A. Owens*. NHHC

***Basilone* (DDE 824) was a prototype austere *Gearing* conversion. She is shown soon after completion off Boston Navy Yard, 21 July 1949, her No. 2 twin 5 in./38 mount having been replaced by Weapon A. Note the trainable Hedgehog in her waist; when this photograph was taken the platform on the other side had not yet been filled. She has two twin 3 in./50 antiaircraft guns, part of a fleetwide antiaircraft upgrade program planned in 1945 to deal with kamikazes. She had one sister, *Epperson*. Their two trainable Hedgehogs were later replaced by fixed ones on the 01 level forward.** NHHC

Other *Gearing*-class DDEs were converted from ships that had been completed as destroyers, and hence had somewhat limited capacity for major modification. They were intended for the two initial HUK groups: DDE 818–820, 847, and 871 in 1947, and DDE 764, 765, and 858–861 under the FY 50 program. *Norris* is shown in July 1952, with a trainable Hedgehog in No. 2 position and her original torpedo tubes between her funnels. She has an HF/DF mast. Note her K-guns and depth charge tracks aft. In addition, *Sarsfield* and *Witek* were converted for service with the Operational Development Force, not redesignated. Both had their No. 2 gun mounts replaced by a trainable Hedgehog; later each also had a pair of fixed Hedgehogs on the 01 level abeam the bridge. At this time the *Gearings* were the most capable U.S. destroyers. Converting many of them for ASW reflected the priority then assigned to that role. Three *Gearings* delivered incomplete at the end of the war were considered for DDE conversion and even tentatively redesignated, but they were never completed. NHHC

DDKs would have three Weapon A launchers (two forward and one aft), four fixed torpedo tubes, and two twin 3 in./70s, at that time the most advanced U.S. shipboard antiaircraft weapon. Their all-antiaircraft-gun battery could be controlled by the Mk 56 gun fire control system, whose belowdecks element was considerably more compact than that of the Mk 37 system on board ships retaining 5 in./38 guns. That reduction made possible a better arrangement of the CO Tactical Plot. The escorts (DDEs) would have one Weapon A forward and two stabilized trainable Hedgehogs (port and starboard, near the forward funnel), as well as four fixed torpedo tubes. These ships were in addition to *Gearings* converted specifically for service in the two experimental HUK groups.

Gearings were too valuable to be converted en masse to specialized ASW ships, but the 150 laid-up *Fletchers*—all in reserve—were suitable. Initially, 12 would be converted, to be completed between September 1949 and February 1950, 9 under the FY 48 program and another 3 under FY 49. Another 6 were converted under the FY 50 program (a planned seventh, *Charles Ausburne*, was not converted). ASW armament would be one Weapon A (replacing No. 2 5 in. gun), one stabilized Hedgehog (Mk 15), and four fixed torpedo tubes. Gun armament would be two (rather than five, as built) single 5 in./38 guns and two twin 3 in./50 antiaircraft weapons (replacing the No. 3 and No. 4 5 in. guns). At least ten ships initially had trainable Hedgehogs rather than Weapon A in the No. 2 position, but all were ultimately fitted with Weapon A. At least one ship (*Cony*) initially retained her forward trainable torpedo tubes.

There would never be enough fully converted destroyers, so in 1949 OpDevFor was asked what could be done for general-purpose destroyers. Studies showed that two fixed Hedgehogs could be installed forward at the cost of four 20 mm guns, which was considered a reasonable reduction in the ships' overall antiaircraft firepower. It was not clear whether sonar could be improved. This study seems to have been the basis for the later (1950) suggestion by VADM F. S. Low that fleet destroyers all be given two Hedgehogs and a QHB sonar.

On the other hand, ASW improvement to destroyer escorts would cost important antiaircraft firepower. BuAer argued in 1948 that the ships needed to be able defend against air threats to overseas shipping terminals. As of late 1949 BuShips was studying a destroyer escort upgrade. Later the two-Hedgehog upgrade was included in the destroyer escorts' class improvement plans, but few ships were upgraded. Late in 1947 12 destroyer escort ASW modernizations were contemplated, ultimately resulting in the SCB 63 and 63A programs a few years later.

By this time the Royal Navy was already looking toward construction of new frigates. The problem was combining the desired high speed with the required transoceanic endurance, typically given as 4,500 nm at 12 or 15 kts. This was not a significant problem for the U.S. Navy, which already had efficient high-temperature high-pressure steam plants. The British were beginning to build equivalent plants.[44] Until they were ready, however, the only available long-haul frigate plant was diesels, which could not give anything like the desired speed. Moreover, the British envisaged a standard hull for several different frigate roles: ASW, convoy air defense, and convoy aircraft direction (of shore-based aircraft). Their initial postwar frigates were designed for air defense and aircraft direction, with a common diesel-powered hull. For higher speed, their new ASW frigate (Type 12, the *Whitby* class) had a longer hull, essentially the earlier one with a hydrodynamic extension. The main ASW armament was double Limbo aft, supported by torpedo tubes.

With their experience in the Battle of the Atlantic in mind, in 1947 the British saw a wartime requirement for 180 frigates. The large number of standard destroyers that had survived World War II could be turned into fast frigates. The British conceived a high-quality conversion (Type 15) that would turn them into near equivalents to the new frigates,

The British solution to the Type XXI problem was the *Whitby*-class (Type 12) fast frigate. This is HMS *Torquay*. Construction had to wait for development of a new steam plant powerful enough to drive the ship at the desired 27 kts but economical enough for a transatlantic convoy run. The latter was equated to an endurance of 4,500 nm at 12 kts, to allow for ships running down contacts and returning to the convoy. The twin 4.5 in. gun was to have been replaced by a twin 3 in./70, a caliber under development by both Britain and the United States. The gun mountings were entirely different, but by using the same gun the British hoped to ensure sufficient ammunition supply in wartime. In the Royal Navy, only the *Tiger*-class cruisers were armed with this weapon, but it was adopted successfully by the Royal Canadian Navy. The ship's twin Limbos are hidden in a well aft. Initially these ships (and the other new frigates) were to have had the abortive British scanning sonar (Type 172) in addition to the Type 170 target tracker associated with Limbo. Since (as of 1949) Type 172 was not expected to be available before 1955, ships were given an interim outfit of Type 170 and Type 164 (a modified version of the wartime Type 144 searchlight). Type 172 never materialized, being replaced by the more massive Type 177. By the time this photo was taken, the Royal Navy had abandoned its plan to mount torpedo tubes in these ships. USNI

HMS *Torquay* shows her twin Limbos, in their well aft. Her deck shape forward can just be made out, the hull apparently pinched in forward of the 4.5 in. gun mounting. This shape reflects the origin of the class, which was conceived as part of a standardized postwar frigate program. When the others were designed, the British had no power plant suitable for a 27 kt frigate, so the other ships (Types 41 and 61) were powered by new standard Admiralty diesels. Their hulls were adapted to 23 kt maximum speed. To achieve the desired 27 kts, the Type 12s were given a "hydrodynamic projection" forward (a longer hull would meet less resistance). USNI

Because Type 12 was expensive, the Royal Navy sought a less expensive ASW escort: Type 14. On 31 May 1949 the Ship Design Policy Committee directed DTSD (responsible for collating ship Staff Requirements) to investigate the concept of "minimum requirements for A/S frigates." Initially DNC proposed using diesel machinery, but the resulting ship would have been too slow (23 kts vs. 27 for Type 12), so instead she was powered by one set (rather than two) of the new turbines developed for Type 12, though it would have to use a different type of boiler. Speed was slightly less than that of Type 12, and endurance also slightly less. A formal Staff Requirement was issued in December 1949. Compared with Type 12, the new frigate was armed only for self-defense against aircraft (Bofors guns vs. 4.5 in.) and carried fewer ASW torpedoes; she retained the double Limbo of Type 12. The new second-rate frigate would have the same asdics as Type 12. As the "Utility Frigate," Type 14 was proposed as the NATO second-rate frigate. HMS *Pellew* shows her twin Limbo main battery. Like Type 12, she was designed to carry homing torpedoes, but few ships of this class ever had them. Type 14s were not well liked because the big ASW war for which they had been conceived never happened. They were useless for most other naval tasks, although they were used for fishery protection. USNI

Some British war-built destroyers were rebuilt as near-equivalents to Type 12, but without their range. HMS *Urchin* shows her twin Limbo aft. Her sister HMS *Wizard* leads her. This design (like that of Type 12) was heavily influenced by the 1946 Bikini nuclear test performed on a fleet of surplus and captured ships. Their bridges were as low as possible, the helmsman placed belowdecks to protect him. The curved bridge face, not visible here, was intended to deflect water thrown up by a nuclear blast. Initial conversions had to make do with double Squid rather than double Limbo (*Rocket* and *Relentless*, the first of the class, apparently had initial production Limbo mountings): Group II (*Virago*, *Verulam*, *Volage*, and *Roebuck*) and Group III (*Rapid*, *Wakeful*, and *Whirlwind*). These ships all had Type 164B and Type 147F asdics (plus the Type 174 torpedo warner). Group II would be fitted with Type 170B to control Limbo. Group III would also get Type 176. Ships would ultimately be fitted with the Type 177 asdic expected about 1957 (except for Group II, described as structurally not prepared for 177). In 1949 the British contemplated 27 full (Type 15) conversions, with one prototype limited conversion in the 1950–51 program to be carried out by Rosyth Royal Dockyard. They actually converted 23 ships. USNI

though with less endurance. Since they needed numbers, the British conceived a less expensive second-rate frigate. The new-build version was Type 14 (*Blackwood* class). The second-rate destroyer conversion was Type 16.[45] Eventually there were 23 Type 15s and 10 Type 16s. The first Type 12, HMS *Whitby*, was ordered under the last pre–Korean War program (1949–50). The first two Type 14s were included in the 1951–52 program.

In April 1949 the Royal Navy held an exhibition at the Royal Naval War College, Greenwich, to display what it hoped it would have once recovery was complete: Project Trident.[46] In 1957 the Royal Navy would have 127 oceangoing ships intended primarily for ASW: 4 A/S destroyers (Weapon class), 12 new A/S frigates (Type 12), 68 old frigates, and 59 converted destroyers. The Weapon class, the new frigates, and 27 of the converted destroyers (Type 15) would all have double Limbo, the most effective British ahead-thrown weapon, as well as 12 ASW torpedoes. The other 32 converted destroyers (presumably Type 16) would have double Squid and 12 ASW torpedoes. The

HMS *Terpsichore* was a more austere Type 16 destroyer conversion. She is shown with her revised bridge, her Squids shrouded in canvas. Her sister HMS *Tenacious* was chosen in 1949 as the prototype for British Type 16 (limited conversion, i.e., second-rate) destroyer conversions to frigates. This was a more austere equivalent of the U.S. *Fletcher*-class DDE conversion. The limited conversion would apply to remaining M- through Z-class destroyers, but the text of the original Staff Requirement indicates that conversion of the Hunt class was also contemplated. The gun armament was intended only for self-defense and to engage a surfaced submarine, amounting to one twin 4 in. in B position and one twin and five single Bofors. A/S armament would amount to a twin Squid aft (20 salvoes) and a quadruple torpedo tube for A/S torpedoes. The asdic fit would be the standard wartime combination of Types 144Q2, 162, and the depth-determining Type 147F (later the postwar Type 166 searchlight set would be installed). The Staff Requirement was first drafted in mid-1949, the final version being issued on 1 April 1950. Preliminary drawings and a legend were submitted for board approval in June 1950. Sea opinion was skeptical as to whether Type 16 was worth what it cost. Officers felt that it would be better either to limit conversion to substituting Squid for one or more of the ship's guns or to complicate it by fitting Limbo and its associated Type 170 (four-square) sonar. The possibility of fitting single Limbo rather than double Squid was investigated in February 1950, and double Squid was accepted because the Limbo/170 combination would not be available before 1953. The redesigned bridge (with attached operations room) was designed in 1953 and fitted to *Tenacious* (as shown), *Terpsichore*, and *Tumult*. In February 1954 the Admiralty Board decided to end the Type 16 program and also not to begin a series of improved limited conversions (Type 18). *Troubridge* and *Zest*, both of which had been scheduled for "improved limited" conversion, instead received the full Type 15 conversion. By this time the cost of a limited conversion was nearly as great as that of a Type 15. It had already been decided that a proposal to arm the last Type 16s with U.S.-supplied twin 3 in./50 dual-purpose guns would be dropped. This left the Royal Navy with ten Type 16s. C. C. WRIGHT COLLECTION

HMS *Orwell* was another Type 16 retaining the original destroyer bridge. Her ASW weapon was a dual Squid right aft, with reloads in the small deckhouse visible there (a total of 124 rounds). She retained her after quadruple torpedo tubes for possible use for ASW torpedoes, as well as her minelaying fittings (she could accommodate 40 mines). Gun armament was reduced to the twin 4 in. Mk 19 in B position and one twin Bofors Mk 5 aft, with simple (STD) directors. Due to lack of ammunition stowage space and limited stability, she could not be fitted with the projected pair of single 40 mm guns amidships. Pakistan received three O-class destroyers, two of which were converted to Type 16s, financed by U.S. MDAP aid. The Royal Navy also fitted the eight Ca-class destroyers with twin Squids aft, the forward torpedo tubes being removed as weight compensation. Unlike Type 16s, they retained three of their original four 4.5 in. guns. About 1948 the British tried twin Squid against several fast submarines: the converted *Selene* (12 kts), *Scotsman* (16 kts), and the U.S. Guppies *Amberjack* (off Key West) and *Dogfish* (in UK waters). Kill rates were 25.7 percent against *Selene*, 13 percent against *Scotsman*, 15.4 percent against *Amberjack*, and 32.3 percent against *Dogfish* (it is not clear why). In each case, the submarine started with a fully charged battery, which would be unlikely in wartime. Data were reported to the 1949 Undersea Warfare Conference. C. C. WRIGHT COLLECTION

World War II–built frigates (Castle and Loch classes) would have their original single or double Squid, but they would also have 6 ASW torpedoes and 16 depth charges. Other ships would have some ASW capability: each of 58 destroyers would have a single Squid and 6 ASW torpedoes (in place of antiship torpedoes). Of 95 existing antiaircraft frigates (49 Hunts, 21 Black Swans, and 25 Bays), the Black Swans would be modernized with double Squids and the Bays with single Squids. Eleven new air defense frigates (Type 41) and 12 new aircraft direction frigates (Type 61)

The Canadian *St. Laurent* class was equivalent to Type 12. They had the same power plant and the same twin Limbo ASW weapon. They were designed with wartime experience in mind, their foredecks curved at the side to throw off water so that it could not freeze into ice. The deck cover right aft has been rolled back to show the Limbos. When rolled out, it protected them from ice. Unlike the Royal Navy, the RCN was not very concerned with resisting a nuclear blast, so the superstructure was considerably higher. Ships were completed with twin U.S.-type 3 in./50 antiaircraft guns fore and aft, but the forward mounting was replaced by a British-type enclosed 3 in./70, and the after mounting was enclosed. They had a U.S.-type SQS-10 scanning sonar and a U.S.-type SPS-12 air search radar. Procurement of the first three was approved by the Canadian Cabinet Defence Committee on 26 October 1948, before NATO had been formed. HMCS *St. Laurent* is shown as completed. The Canadians built seven of this type and seven of the modified *Restigouche* class but did not pursue plans for a Canadian equivalent to the British Type 14 (*Vancouver* class). That class was canceled in 1955 in favor of building four modified *Restigouche*s (*Mackenzie* class) followed by two *Annapolis* class adapted to operate large helicopters. USNI

would each have a single Squid. Note the absence of Type 14s.

In the mid-Atlantic, the minimum surface escort of a 60-ship convoy facing intermediate (B) submarines (i.e., Type XXI equivalents) was judged to be 8 in "normal" areas, 12 in focal areas or known areas of enemy submarine patrols, and 4 where submarine attack was only a possibility. Thus an 8 kt convoy bound from Britain to Halifax would require 8 escorts as it left, to be reinforced by 4 more (and a carrier) in especially dangerous areas. As the convoy approached the North American coast, 4 escorts could be detached. On this basis the British estimated that 11 escort groups (66 escorts) were needed in the North Atlantic, 4 in the Norwegian Sea (Methil–Bergen route: 24 ships), and 7 in the central Atlantic and to protect Gulf port tankers (42 escorts)—a total of 132 ships, close to the number the Royal Navy hoped to have. However, in addition, three patrol groups would be needed in the North Atlantic (each with one carrier and at least four escorts). Three carriers and about 130 escorts would provide a level of protection to British Atlantic shipping comparable to that of World War II. This figure made no allowance for casualties or for other escort requirements or training.

The ships needed for convoys were separate from those needed to protect coastal shipping not only against submarines but also against air attack and attack by fast boats like the wartime German E-boats. The latter would be a serious threat if the Russians managed to reach the Channel. There would have to be a large minesweeping force as well. Unfortunately, the advent of more sensitive magnetic mines late in the war had rendered most existing minesweepers obsolete. The estimate for 1957 was 66 ocean sweepers, 103 coastal sweepers, 57 inshore sweepers (including 19 new ones), and 92 minesweeping trawlers. During the war, the ocean sweepers sometimes functioned as escorts, but that does not seem to have been envisaged.

A view of *St. Laurent* from aft shows the Limbo pit and its cover. USNI

HMCS *Algonquin* was a Canadian equivalent to Type 15. She and a sister were rebuilt at Esquimault in 1952–56. Four of five Australian Q-class destroyer were also rebuilt to Type 15 standard (the fifth, *Quality*, was scrapped in 1958 due to hull corrosion while in reserve). Another eight Canadian destroyers were modified to something more than the British standard for their fleet destroyers; they had two Squids aft and an all dual-purpose 4 in. main battery. USNI

The Royal Canadian Navy rebuilt nearly all of its surviving World War II River-class frigates, 21 of them. HMCS *Buckingham* shows the built-up hull aft that surrounded a well containing her two Squids; its cover can be seen just abaft her twin 40 mm gun. Note the masthead HF/DF antenna, below which is a U.S.-type SU surface search radar and below that a Canadian Sperry Mk 2 precision surface search antenna. *Buckingham* later conducted the first RCN helicopter trials. USNI

Note that the number of escorts needed was based not on the scale of the submarine threat but on the number of merchant ships that had to be protected. The capability (hence the cost) of each escort was set by the capability of individual submarines, not by their numbers—a terrible problem as submarine capability increased. In 1949 it was estimated that the Soviets would have 30 large fast submarines and 20 large slow ones (conventional submarines with snorkels) on patrol in operational areas at any one time, with 50 and 30, respectively, in transit. Submarine numbers would determine the scale of shipping losses.

The U.S. Navy encouraged the European NATO navies to build their own modern escorts. The E50 class was the first postwar French escort, built to a French design but largely financed by the United States. *Le Brestois* (financed as DE 1017) is shown. She was armed with a unique French ASW weapon, the K2 torpedo, which had been conceived during the war by French naval constructor Louis Kahn, then in exile in the United Kingdom; later he became the chief French constructor. He offered it as an alternative to Squid. Each triple torpedo tube carried three K2s, which were fired together on parallel courses and timed to explode. The K2 was very fast (50 kts), with a short range (1,000 m at a depth of 300 m, or 1,500 m at a depth of 100 m [325 ft]). It was driven by alcohol burning in oxygen. The warhead carried 275 kg (610 lbs) of explosive. The structures behind the tubes each held three reloads. The guns were twin Bofors 57 mm. The subsequent E52 class also had the short triple tubes, as did the *Surcouf*-class destroyers. K2 was replaced by the L4 homing torpedo of the 1960s. USNI

The French soon supplemented their short torpedo tubes with more conventional ahead-throwing weapons. The E52 class frigate *Le Gascon* shows a Bofors sextuple 375 mm ASW rocket launcher in front of her forward twin 57 mm gun. The launcher was developed for the Royal Netherlands Navy in a four-barrel version, the six-barrel type shown here being developed specially for the French navy. The four-barrel version was used by many other navies, including those of Belgium, Egypt, Germany, Portugal, Spain, and Sweden. The French wanted ships suitable for operations in the French Union, with troop-carrying and shore bombardment capability, so in later ships they used a four-barrel 305 mm mortar that could fire shore bombardment rounds as well as ASW projectiles. Note that the short torpedo tubes all loaded via their muzzles, the reloads being located behind the tubes. Note also the British-style open bridge. *Le Gascon* is shown at the 1957 Jamestown Naval Review. Like *Le Brestois*, she was financed by the United States under the MDAP program. USNI

SUBMARINES

ASW might include submarines that could attack Soviet submarines between their bases and their patrol areas. A Submarine Force speaker at the 1948 Undersea Warfare Conference recalled the old saying that it takes a thief to catch a thief. The senior submariner, ADM James R. Fife Jr., assigned his Panama submarines (SubRon 6) to investigate the potential of submarines hunting enemy submarines. The group's representative said in 1948 that no other ASW field was as promising. Submariners were only beginning to detect snorkeling noise; as techniques became more sophisticated, any snorkeling submarine would find it difficult to penetrate an area patrolled by other submarines. By this time there was a formal program to develop specialized ASW submarines called SSKs, which would be backed by converted fleet submarines. SSKs would ambush enemy submarines, using their passive sonar to locate them and homing torpedoes to destroy them. The SSKs were distinguished from hunter-killers, which would detect and then run down enemy submarines. With the shortage of resources, the British could not envisage submerged picket lines and instead embraced the submarine hunter-killer role.

Early in 1949 CNO decided that these operations as well as studies showed that the SSK concept was sound enough to warrant more concentrated effort, particularly to find out how successful SSK operations were likely to be. A submarine division in each fleet was to specialize in SSK studies, which were known as Project Kayo (as in knockout, K.O.): SubDevGru 2 in the Atlantic and SubDiv 11 in the Pacific. SubDevGru 2 comprised the submarines *Tusk*, *Cochino*, *Toro*, and *Corsair*, of which *Cochino* was fitted with a *Balcon* array from the decommissioned Type XXI submarine *U 3008*. *Cochino* was sent north of Norway to test the array's capabilities in waters that would be blocked by U.S. submarine pickets in wartime. Unfortunately, she was lost due to a battery fire on 26 August 1949; she was replaced by *Halfbeak*.

SSKs were conceived for wartime mass production, to operate in multiple patrol lines to overcome their relative immobility and to increase the probability of detection. Production of the original small SSK was abandoned in favor of converted fleet submarines, the Guppies being designed so that they could later be turned into SSKs as a wartime mobilization measure.

From the outset, sonar performed better on submarines than on surface ships. By 1948 the Panama submariners and their active sonars were detecting other submarines at 8,000 yds—unheard of for surface ships. They could often close snorkeling submarines to attack undetected. Submarine sonars could hear a Guppy start her engines. A carefully silenced fleet submarine balancing on a density layer could detect a snorkeling Guppy 25,000 yds away. Much depended on choosing the optimum listening depth. The submariners using unmodified World War II fleet

The U.S. ASW submarine (SSK) *K 3* shows her big, passive BQR-4 bow sonar, adapted from the German GHG. Plans to mass-produce such submarines foundered. USNI

submarines investigated anchoring while submerged to maintain silence.

In 1949 the international situation was considered so desperate that SubDevGru 2 was ordered to seek short-term lessons applicable to a war that might break out as early as 1950. The group devoted much of its time to solving the problem of bearings-only fire control, without any range measurement, using only existing equipment. The SSK concept required submarines to form underwater picket lines, but it was soon clear that submarines in the Arctic, where skies were often overcast, would find it difficult to achieve the required navigational accuracy. Electronic navigation (Loran) coverage was inadequate and jammable. In 1949 submarines could not keep station 5,000 yds apart, so instead the front of the barrier would be covered in depth. Each SSK was expected to listen submerged more than 11 hours of every 12, snorkeling for slow propulsion and a battery charge during the other hour. While snorkeling she would continuously change course and keep a torpedo detection watch.

A 1949 exercise employed three SSKs limited to surface and snorkel operation. The first phase lasted four hours, but the third extended for three days. At least 35 percent of submarines passing through the SSK area were detected. SSKs were able to attain favorable firing positions on at least half of them. With so few runs, these results could not be considered statistically sound, but they did show that SSKs could significantly reduce the effectiveness of the enemy submarine campaign. Moreover, the results were attained even though the group could not devote all of its time to Kayo.

SubDevGru 2 learned to rely entirely on passive sonar so that targets were not warned, using a variety of plots for what was later called target motion analysis (TMA). It turned out that sea state, not quieting, was the limiting factor in hearing and stalking submarines. The German *Balcon* (GHG), which extended vertically as well as horizontally, outperformed the wartime U.S. JT linear array because it did not pick up noise from the surface down to turbulence around the submarine superstructure. *Balcon* was adopted as a standard U.S. passive submarine sonar (BQR-2). Another possibility was to suspend a hydrophone far below the submarine;

this idea was tried successfully in *Tusk* and then, in modified form, in *Corsair*.

Given the near-term emphasis, SubDevGru 2 limited itself to existing torpedoes: the straight-running Mk 14 and Mk 18 and the passive-homing Mk 27 and Mk 28. The target submarine could evade if it heard the torpedo soon enough, which seemed possible with the receiving element of a scanning sonar or the German GHG, usable even on the surface or snorkeling. Evasion became impossible at 1,500 to 2,000 yds or less. SubDevGru 2 wanted faster straight-runners, a spread in depth, and quieter torpedoes.

AIRCRAFT

World War II HUK groups succeeded largely because their aircraft could detect submarines at considerable distances, using radar and following up on HF/DF detection. The advent of snorkels precluded long-range radar detection, and in 1948 the U.S. Navy's OEG said that against a Type XXI skillfully operated, "the HUK group is reduced to the status of a police force."[47] That is, a HUK group could deny the submarine free use of the surface and could endanger a submarine caught while it attacked, but otherwise was unlikely to find it. By way of contrast, a modern HUK group could generally be sure of destroying a conventional submarine picked up by their HF/DF as much as 50 nm away. The great postwar question was whether new airborne sensors could reverse this situation. The U.S. Navy came to think that snorkel detection radar could be effective, based on experience with the big APS-20, which had been developed for airborne early warning and, to a lesser extent, surface search from the air. The modified wartime APS-15A, which still equipped land-based patrol aircraft converted for ASW, could detect a snorkel only at 2.5–3.5 nm up to sea state 1, and at only 1 nm in sea state 2 or 3. The original wartime APS-15 could detect a surfaced submarine no farther away than 10 nm. APS-20, however, had a 60–70 percent chance of determining whether a snorkel was present at 26 nm in sea states up to 2. In 1949 it was credited with the ability to detect a snorkel at 25 nm in sea state 3; it could detect a surfaced submarine 110 nm way. Doubling peak power (APS-20B) increased snorkel detection range to 30 nm and surfaced submarine detection range to 135 nm.

The U.S. Navy had been buying small numbers of APS-20 radars to detect low-flying enemy aircraft. A few carrier-based radar Avengers and nine converted B-17 Flying Fortresses carried the radar. Initially the APS-20 airplane sent its radar picture down to its parent ship via a specialized data link, the airborne radar functioning in effect as another ship radar. The huge APS-20 crowded out the other electronics on board carrier airplanes, so the radar airplane was paired with an attacker in a hunter-killer team, initially a TBM-3W carrying the radar and a TBM-3S attacking. At first the ship's CIC directed the attack, but that worked only if the aircraft were above the carrier's horizon. An air-to-air link was developed, initially with a reliable 100 nm line-of-sight range. The ideal was a "single-package" airplane combining the radar and attack roles. To make that possible for carrier aircraft, BuAer sponsored the development of more compact snorkel-detection radars. Their poorer

Sub-hunter: a TBM-3W Avenger with its big radome housing the antenna of an APS-20 radar. It was intended to detect snorkels and to direct TBM-3S "killers" against submarines. The airplane was initially conceived as a way of providing carriers with effective long-range warning of low-flying attackers; this "Cadillac" program was accelerated with the appearance of kamikazes. Only after the war was the radar's improved snorkel detection capacity discovered. This postwar photo (note the insignia) shows the vertical tail-top antenna enabling the airplane to transmit its radar picture. USNI

performance was acceptable because their presence would force a submarine to stop snorkeling and go deep. At the least, submarines would be discouraged from showing anything more than a snorkel. That would considerably reduce their mobility and their ability to understand the situation on the surface, and hence to attack surface ships. Depressingly, it turned out that snorkels could be detected by eye as far as by any radar other than APS-20A.

The radar hunter worked with a killer TBM-3S converted from a war-built TBM-3E. It was armed with a homing torpedo (Mk 34, or a Mk 24 plus two depth charges) and eight 5 in. rockets underwing. It retained its two 0.50-caliber wing guns. Typically, it carried the radar "bomb" of a short-range APS-4 under the starboard wing and a searchlight under the port wing, the APS-4 functioning as an attack radar. It also carried 14 SSQ-2 omnidirectional sonobuoys. There was also an intercept receiver, initially APR-5A (to be replaced by APR-9); an APA-70A direction-finder was added later. The TBM-3S2 version (1951) had a radar relay receiver (ARR-27) enabling it to cooperate with the hunter airplane, and a newer AVQ-2 searchlight. This TBM-3S lacks the tail fin radar relay antenna, and it does not show the usual underwing stores. In 1951, about 58 TBM-3Ss were in fleet use, and plans called for converting 225 mothballed TBM-3Es to TBM-3S2s (another 166 would be converted into TBM-3W2s). At that time 1,141 TBM-3Es were in storage, suitable for conversion. The TBM-3S was expected to end its allowed lifetime in December 1953. TBMs could operate from the *Casablanca*-class escort carriers in reserve, but none of the other ASW airplanes could. The Skyraider and Guardian could operate from the larger *Commencement Bay* class only with difficulty. It was, however, impossible to put the Avenger back into production because tooling was destroyed after World War II. Avenger hunters and killers were transferred to NATO navies, including the Royal Navy, the RCN, the French navy, and the Royal Netherlands Navy; they also went to the Japanese Maritime Self-Defense Force. Deliveries to the U.S. Navy began late in 1947. USNI

By 1950 BuAer had a more compact snorkel detection radar; it was produced in two versions, the forward-looking APS-31 and the all-around-looking APS-33. APS-31 was a stabilized version of APS-30, the planned wartime successor to APS-15. It occupied the wing nacelle radar on board the AF-2S Guardian. In 1949 APS-31 was credited with an average snorkel detection range of only 5 nm (32 nm on a surfaced submarine). APS-33 was the same radar with a much larger antenna. It was credited with a range of 10 nm on a snorkel and 48 nm on a surfaced submarine. In 1949 the OEG said that APS-33 was only about half as effective as the big APS-20 mounted in a modified B-17 (PB-1W) but was nearly as effective as the smaller version typically carried by a single-engine carrier airplane (at that time a TBM). To the extent that it was acceptable, APS-33 made it possible to build a single-package carrier ASW airplane, Grumman's S2F Tracker; the antenna was in a retractable ventral "dustbin." It was superseded in late-production S2F-1s by APS-38, which embodied detail improvements, better displays, and much more elaborate signal processing. Airships modified for ASW

There were parallel conversions of Skyraiders for fast carriers. This AD-4W has not yet been assigned to a squadron. The vertical antenna on the tail was for the data link to a ship or to a companion killer airplane. AD-4Ws were transferred to the Royal Navy as the Skyraider AEW-1. Douglas dated this photograph 15 March 1951. DOUGLAS VIA USNI

had APS-33. In December 1947 work began on the first radar designed specifically for snorkel detection, with optimum maximum range and minimum range performance, intended to achieve a sweep width of 50 nm (detection range 25 nm): APS-44. When it came into service in 1949 it was expected to outperform APS-20: 45 nm on a snorkel and 180 nm on a surfaced submarine, operating against snorkels in sea state 3 (rather than 2). APS-44A, which covered a 170 degree rather than a 360-degree arc, equipped early versions of the P5M Marlin.

The British did not invest in anything comparable. Presentations by RAF Coastal Command officers at U.S. Undersea Warfare Conferences in 1947–49 suggest that Coastal Command did not grasp the revolutionary impact of Type XXI. Among their other misapprehensions, they thought that submarines could be more or less immobilized by being kept down, not understanding that Type XXI was faster submerged than surfaced. They thought that snorkeling was so inherently disadvantageous that submarines would surface whenever possible. The limits of submarine battery capacity could be exploited. Snorting speed would be no more than 8 kts. A submerged submarine would find it difficult to grasp the tactical situation, and communications would be quite limited. It was unlikely that submerged submarines would be able to concentrate exactly on the convoy track. Submarines still had to come within limited lines of approach to attack. Given the mistaken assumption that new submarines would have limited submerged speed, it followed that submarines outside lines 35 degrees on the bow of a 10 kt convoy would have to surface to gain an attacking position. Submarines more than 10 nm off that track would be unable to run the convoy down while submerged. Coastal Command argued that air searches on the bows of the convoy out to 50 nm might prevent submarines from concentrating, and that patrols around the convoy could beat off shadowers. Unfortunately, such tactics required large numbers of aircraft, which did not exist. Moreover, the British did not yet have an air-launched homing torpedo. Coastal Command admitted that the best it could do was to maintain contact with a submarine (using sonobuoys) until surface ships arrived to sink it.

Both navies, lacking long-range airborne detectors, experimented with sonobuoys, using reconditioned wartime passive buoys, which had relatively short acoustic ranges. The buoys could be used to create barriers that could detect submarines crossing them. Alternatively, a flaming datum (a sinking ship) could become the center of a sonobuoy search, as an airplane could reach it much more quickly than a surface ship could.

In the summer of 1948 British forces exercised with the U.S. Guppy submarine *Trumpetfish*. In one exercise, the ships' hydrophones indicated that a submarine was in the area, but sonar conditions were so bad that they were unable to get close enough to attack. The surface commander ordered an aircraft barrier of 8 buoys on a circle of 5 nm radius centered on the surface force; the ships had sonobuoy receivers. One of them was able to track the submarine and to get close enough to attack. The British developed other sonobuoy patterns for such tactical cooperation. In another exercise, the submarine's periscope was sighted, but sonobuoys dropped where she had been seen did not hear her. The surface force commander then ordered a barrier (4 buoys, 6,000 yds apart) dropped across the suspected submarine track, 15 nm ahead of the ships. The submarine was tracked (by air) for an hour and a quarter, by which time ships were in sonar contact. The average range of initial detection by the buoys was 10,000 yds.

Aircraft could detect submarines' radar emissions, and often that would offer the longest-range warning. Radar detectors may have been the most important long-range aircraft sensor, but in 1949 they were only beginning to be installed on board ASW aircraft.

In 1947 the U.S. Navy began to adapt land-based patrol aircraft for ASW: the existing Privateer (PB4Y) and the Mariner (PBM-5), and also the new Neptune (P2V). The most important modifications to all these aircraft were sonobuoy capacity, searchlights, and electronic support measures (ESM). Only the Neptune could accommodate the big APS-20 radar. The ASW program expanded considerably after the United

States began to mobilize in 1948. Twenty-five patrol (VP) squadrons were designated as ASW units. About 1 January 1949 all patrol plane squadrons other than those with special purposes were ordered to specialize in ASW. This made a considerable change in the Atlantic, where previously only 3 of 16 VP squadrons had been assigned primarily to ASW.

Due to previous funding cuts, the equipment situation was grim.[48] Only two Mariner squadrons and one Privateer squadron had sonobuoy receivers. Patrol planes had APS-15 radars, with virtually no capability to detect snorkels. None had MAD gear or the searchlights needed for night attacks, and none had ECM gear. Two out of four Neptune squadrons, flying the latest patrol plane, had neither radar nor sonobuoy receivers. By the fall of 1948, almost all the Atlantic VP squadrons were equipped to some extent for ASW, but most of their equipment was obsolescent or obsolete. The most modern radar was APS-33 in Neptunes. All but one of the Neptune squadrons had "begged and stolen" sonobuoy receivers. Each Mariner squadron had four or five aircraft configured for ASW (PBM-5S). Given the recently ordered cuts in squadrons, aircraft could be reallocated to the remaining squadrons to fill them out with ASW aircraft.

By this time helicopters were considered very promising: they could dip their sonars some distance away from their own force, yet were invulnerable to submarine torpedoes. However, by 1947 it seemed that too many would be needed to cover a moving force. Since helicopters were also badly wanted for other naval purposes (e.g., gunnery spotting and amphibious assault), a reduction in their ASW priority did not affect their development. That in turn meant that helicopters were available when again wanted for ASW. By 1949 a sonar dunked from a helicopter achieved twice the predicted range. However, no existing helicopter offered a sufficient combination of endurance and weight capacity. Once that problem had been solved, the U.S. Navy had its first operational ASW helicopters in 1953, using the AQS-4 sonar. It was carried by Sikorsky HO4S helicopters originally developed to spot gunfire (hence the O in the designation). Operational range was 1,900 yds.

Blimps were a feature of postwar U.S. ASW, having proven effective during World War II. The angled tail surfaces identify this blimp as one of the postwar N-series (ZP2N). Blimps could tow sonars, but they could not hover as helicopters could. On the other hand, they had long endurance and could fly at low speeds. The towed sonars, a postwar development, were called ATERE (Airship Towed Echo-Ranging Equipment). USNI

The U.S. Navy also had blimps for ASW.[49] In 1948 ASW was made the primary blimp mission, and the number of blimp squadrons was increased from two to three. The blimp squadrons were fairly well equipped with MAD and sonobuoy receivers but had obsolescent radars (APS-2D) and no ECM gear. The towed blimp sonar, ATERE (Airship Towed Echo Ranging Equipment, AQS-2), employed two back-to-back QGB transducers. Blimps were also used later as part of the offshore barrier system.

6

THE CRISIS AND ITS SOLUTION

IN 1949 THE ASW SITUATION in the United States suddenly worsened. Not only did defense funding plummet, but that year CNO declared fleet air defense a priority on the same level as ASW. A senior participant at that year's Undersea Warfare Conference said, "The honeymoon is over." Active destroyer forces in both fleets were deeply cut, as were ASW modernization plans. Ultimately all destroyers were to be given scanning sonars, but for the moment that applied to only 16 of 42 Pacific Fleet destroyers. In 1949 the Atlantic Fleet operated four HUK groups built around three escort carriers (*Mindoro*, *Sicily*, and *Siboney*) and the light carrier *Wright*, each with eight destroyers. *Palau* was being added but would take time to get up to full efficiency. Under the budget-driven reorganization, *Siboney* was going into reserve and *Sicily* was going to the Pacific. New equipment was available in only token quantities. For example, each escort carrier had only two Avengers with APS-20 radars, and hence could not operate around the clock.

Only two Pacific carrier squadrons had ASW as their primary purpose, and only one of them had ASW-configured aircraft. There was only one carrier airborne early warning (APS-20) squadron for all purposes. The only land-based ASW aircraft that could be considered effective were Neptunes: two squadrons of P2V-2s and two of P2V-3s, a total of 36 aircraft, of which the P2V-3 was considered the most

Reality: the Soviets were unable to modernize their existing submarine fleet while setting up mass production of new craft. This gave Western navies breathing space to develop entirely new ASW techniques and weapons rather than rushing deployment of existing ones. The old L-class submarine in the foreground has no snorkel and has not been streamlined for greater underwater speed. She contrasts with the high-performance Zulu (Project 611) submarine in the background. This photograph was taken late in 1957, by which time the Soviets had a large fleet of modern submarines. NHHC

suitable ASW aircraft in the Pacific Fleet. Only two of 45 Privateers and 9 of 26 seaplanes were "S" (ASW) configured. Budget cuts limited back-fitting, so the only improvement foreseen was the replacement of 18 Privateers with P2V-4 Neptunes (single package aircraft). With too few patrol planes, teaming P2V-3W Neptunes (with APS-20s) with attack aircraft was not considered an overall operational improvement.

Less visible but probably more important, personnel and readiness were cut deeply. For example, flight instruction in advanced training of patrol plane students was halved. The Atlantic Fleet air force considered the shortage of electronics personnel almost critical. The Pacific Fleet air force reported that pilot training had been cut to 16 hours per month, when 50 was considered essential. It was proving impossible to provide second and third patrol plane pilots with any appreciable training during advanced and intertype exercises, and it became impossible to improve the standard of inexperienced personnel. Patrol squadrons were cut from 18 to 11 aircraft. To cut spending, the Pacific Fleet was ordered to drop highly qualified Reserve and temporary commissioned officers. Late in 1949 the Pacific Fleet air commander warned that the cuts would reduce operational readiness by the end of 1950.

Even as personnel were being cut, the need for experts grew. The overall ASW personnel situation was unstable. At the 1949 Undersea Warfare Symposium, the commander of the Fleet Sonar School at Key West pointed to the substantial training needed to produce an operator who could reliably distinguish a real submarine from false echoes. More and more such operators were needed. In the past an attack team had consisted of 2 operators supervised by the officer of the deck; now it was 10 or 12. Maintenance was also now far more complex. Past systems with 27 vacuum tubes were giving way to systems with 8,000 tubes.

THE LOW REPORT

VADM F. S. Low was asked to evaluate and report on the ASW situation, presumably in view of the budget crisis. In April 1950 he pointed out that the ASW emergency was several years away. As yet the Soviets had only 82 long-range submarines, of which only 13 seemed to have snorkels: 8 of a new "B" type (presumably the Whiskey class) and 5 ex-German submarines.[1] To meet the current threat, Low supported the upgrade to existing destroyers proposed by the Operational Development Force: two Hedgehogs on the 01 level (at a slight cost in antiaircraft firepower) and scanning sonars in place of wartime searchlights. The future, however, was much bleaker; the Soviets were reportedly building at a rate of 20 submarines per year, and that might accelerate. The weapons and sensors currently being developed would be inadequate to face this new threat. The limited funds available should be concentrated on developing entirely new weapons and sensors to deal with it.

Low pointed to the potential represented by the destroyer escorts in the reserve fleet, which had previously been discounted as too slow. He proposed ending the *Fletcher*-class DDE conversions and spending the money instead on *Albacore*, a fast underwater target that could simulate future submarines. This

In April 1950 VADM F. S. Low proposed stopping the urgent *Fletcher* DDE program and diverting funding for a nineteenth conversion to the experimental high-performance submarine *Albacore*. He suggested a much less ambitious fleet modernization program: installation of two Hedgehogs on the 01 level accompanied by a scanning sonar and, in *Fletchers*, a larger rudder. The radar picket destroyer *Higbee* is shown in 1955, her port Hedgehog visible on her 01 level forward, abreast the fore part of her bridge. NHHC

The *Fletcher*-class destroyer *Hopewell* shows the two visible upgrades of the 1950s: the Low Report Hedgehogs and the twin 3 in./50 antiaircraft gun, the latter of which required that one of her 5 in./38 gun be landed. *Hopewell* has landed her depth charge throwers, but she retains a depth charge track aft. The scanning sonar and the enlarged rudder are not visible. NHHC

was done. He also pressed for development of ASW helicopters with dipping sonars, an idea the contemporary Royal Navy also advocated.

Overall, Low considered attack at source, including extensive mining and bombing of submarine bases, the best counter to the Soviet submarine threat. Convoy operations were lower on his list of priorities but above HUK operations, which would be ineffective in the absence of effective HF/DF and codebreaking. Presumably Low assumed that the Soviets would soon have their own version of *Kurier* (that threat materialized only in 1960).

Low encouraged work on helicopter and blimp ASW. As of the spring of 1950 BuAer was buying Sikorsky S-55s (HO4S-1) as test vehicles, the planned sonar being the AQS-1. Ten helicopters were on order, the first to be delivered in July 1950, the last three in January 1951. Meanwhile BuAer had already run a design competition for an ASW helicopter, Bell Aircraft's twin-rotor HSL-1 having been selected. Its dimensions were set so that it could be stowed belowdecks on board the most numerous class of escort carriers, the small *Casablancas*. Bell had not yet built anything this complex, and BuAer began looking at alternatives.

Twin Hedgehogs were included in the destroyer escort Class Improvement Plan, although only ten ships received them, including the Operational Development Force ship *Foss* (DE 59). The others were DE 145, 148, 164, 220, 246, 252, 634, 699, and 797. USS *Whitehurst* served as a Naval Reserve training ship. She retained a few K-guns aft and a depth charge track (covered in black canvas). Light antiaircraft guns were eliminated to reduce the maintenance load. She retained her World War II air search radar but had a modern surface search set and postwar UHF radio antennas above it. USNI

The Royal Navy equivalent to the Low Report upgrade predated it: large fleet destroyers were given single Squids in place of their heavy wartime depth-charge batteries. The Squids fired forward over the ship's bridge structures, initial shots by HMS *Battleaxe* (with her Squids in X position) showing that the projectiles would clear the ship's bridge and mast. HMS *Agincourt* is shown about 1950 (note the three *King George V*-class battleships in the background, which would soon be laid up). A similar arrangement was adopted for the *Daring* class. BAKER COLLECTION

The OpDevFor air component (VX-1) ran preliminary tests in which a helicopter that detected a submerged noisemaker by sonar vectored another helicopter to attack it. The necessary lightweight torpedo was improvised from a Mk 30 torpedo body and the motor of a self-propelled decoy (NAD-10). It was a passive homer with a range of about 2 nm at 12 kts, capable of operating down to 500 ft.[2] It is not clear whether the experiments with the modified Mk 30 influenced the requirement set for a lightweight helicopter-borne torpedo.

BASIC RESEARCH: PROJECT HARTWELL

The Committee on Undersea Warfare (CUW) proposed the research study that set the agenda for the 1950s: Project Hartwell. In a 23 January 1950 letter to ACNO for Undersea Warfare RADM C. B. Momsen, the committee proposed a long-range ASW program that would bring together a group of "outstanding personnel providing a high degree of technical, scientific, and operational competence under the most skilled and vigorous leadership"—the sort of civilian scientists who had been so helpful during World War II. Momsen sent a memo to CNO ADM Forrest Sherman a month later, on 26 February 1950. At a conference in New York the next day Momsen and RADM T. A. Solberg (chief of the Office of Naval Research) met with Dr. M. J. Kelly, the director of research at Bell Telephone Labs. Also present were Dr. J. B. Fisk (who assisted Kelly at Bell Labs) and Dr. J. A. Stratton, the provost of MIT. Stratton said that MIT would cooperate with the Navy if the Navy considered advanced submarines a critical national problem. A contract was let to MIT, and the project was named Hartwell. The project director was Dr. J. R. Zacharias. A preliminary meeting was held in Washington on 27 April during the American Physical Society meeting, which many of the scientists the committee hoped to recruit were likely to attend. Twenty-two scientists and seven Navy representatives attended the meeting where Admiral Solberg explained the project. All the personnel were cleared Top Secret. Briefings began on 5 June and included one by Low on his recent report. The study concluded on 30 September 1950.

There was a sense of urgency in the proceedings. The previous August the Soviets had ended America's monopoly on atomic bombs, which had been considered the deterrent to Soviet aggression. The scientists assumed that in a European war the great prize would be control of the North Atlantic, since that would decide whether a Western army fighting an advancing Soviet army could be resupplied. Surely the Soviets would consider disruption of sea transport so important that they would mount a maximum effort to do so, particularly since investments in submarines and mine warfare would not seriously affect their economy. This was much the argument that Marshal Zhukov would make in 1953.

The scientists chose a two-year horizon, limiting recommendations to what they thought could be achieved by then.[3] Looking ahead, they estimated that in September 1952 the Soviets would have several hundred Type XXI equivalents, plus at least 200–300 submarines with snorkels. These submarines would be armed with long-range pattern-running and homing torpedoes. The Soviets would have enough influence mines for a major aerial mining campaign against European ports, and enough submarines to carry out small-scale mining on both sides of the Atlantic. It had to be assumed that the Soviets would have a substantial number of nuclear weapons. Even a minimum Soviet effort at disrupting SLOCs would have serious consequences.

Many of the Hartwell participants had worked on radar during World War II, so naturally they were interested in snorkel-detecting radar. They became very discouraged. At best, radar could detect a snorkel at 10–15 nm, but a submarine with a simple warning receiver could detect a radar at horizon range and dive. As a countermeasure, the Hartwell scientists emphasized the possibility of radar "floodlighting"—saturating an area with radar to the point that no submarine commander could afford to dive whenever the boat's radar warning receiver detected radar emissions. It was particularly important that snorkel detection radar not be easily distinguishable from other airborne radars. Thus, the Hartwell scientists rejected attempts to gain effectiveness by, for example, using dual frequencies. Their report included a chart showing how commercial aircraft with weather radars would add to the density of radar emissions a submarine in the Atlantic would experience.

The Hartwell report cited a historical example. In 1943 the British tried to create a barrier strip somewhat wider than the distance a U-boat could travel in one dive (60 nm). In a preliminary study, the Coastal Command's operational research arm estimated that a submarine that heard a radar every 15 minutes would find it impossible to dive on each alarm. During the actual "very successful" operation, it turned out that even a much lower alarm rate would keep U-boats on the surface and ignoring their radar detectors. The Hartwell scientists estimated that 0.25 alarms per hour would suffice to keep submarines from responding to radar alarms. The extent to which the early warning aircraft of the Atlantic and Pacific barrier forces were seen as contributors to radar floodlighting is unclear.

At this time BuAer was pressing development of single-package ASW aircraft to supersede the existing HUK teams. The Hartwell scientists disagreed, arguing that a single package would be too costly and would sacrifice too much in radar performance. They may have had in mind the decision to equip the new Grumman S2F Tracker with a radar much inferior to the APS-20 carried by previous ASW search aircraft.

Sonar

Most of the Hartwell scientists were unfamiliar with sonar when the project began, but they gained considerable respect for it during their study. They thought that existing sonars were being underutilized because operators had difficulty using the information. Data were received very slowly due to the low velocity of sound, and appeared only fleetingly on a PPI, whose bright flash quickly caused eye fatigue. The time between scans was so long that little scan-to-scan comparison was possible. The chemical recorder produced a permanent record, but it was difficult to interpret. Not all the data produced by the sonar were used to evaluate a target. Most sonar operators worked in hot, cramped, noisy conditions. The Navy did not seem to understand that the utility of an ASW ship depended on the performance of the operators as well as their equipment.

Worse, the factors that influenced an operator's ability to detect a target amid noise were poorly understood, so any attempts to improve the "operator factor" could only be guesses. Contact was often lost because an operator could not evaluate data at the moment a signal appeared on his screen. The Hartwell scientists insisted that operators needed some means of storing the data that would enable them to see how the signal was behaving over several scans. An operator given a promising ping should have tools for further analysis of it, such as examining the target for Doppler shifting. In any case, low sound velocity restricted the rate at which the sonar could pulse, hence the rate at which it could collect information. Perhaps the information rate could be increased by interleaving pulses at different frequencies; having two transmitters would double the information rate.

This issue was not really addressed until the 1960s, when free-play exercises showed that U.S. (and British) sonars failed to live up to their theoretical performance, both in average range and in probability of detection. The solution, computer-aided detection, was not available until the late 1960s.

Identification (classification) of the sonar signal was a crucial element of ASW. It had been solved in the past by relying on the operator to evaluate echo quality and target Doppler. Much depended on how the echo sounded to the operator, who needed perceptual abilities much like those required of a good musician. That implied specialized selection and training programs. But even with wartime manpower it would be difficult to find enough people with the requisite abilities. It was therefore essential to find some means of target identification that did not depend on specialized operator abilities.

Doppler was a promising possibility because it showed directly that a sonar target was moving. Exploiting Doppler would not require as much operator training and aptitude. Doppler would also make it easier to hear weak echoes in the presence of stationary targets or reverberation. A good operator could detect and isolate weak echoes with hearing alone, but that ability rapidly decreased as ping duration (length) was reduced. Hartwell recommended devising some means of visually presenting the frequency of a strong echo.

Changes in echo "quality" allowed the operator to detect the classic submarine evasive maneuver of producing a wake knuckle in the water. Experienced operators referred to the "mushiness" of echoes off knuckles. The operator needed a good musical ear, but some means of measuring or indicating the characteristics of an echo would be helpful in that regard.

The Hartwell scientists noticed that the scanning method that existing scanning sonars used sacrificed range. The suggested solution, preformed beams (their term), did not appear for some years. The existing QHB used (in effect) very short pulses due to scanning loss, hence could not exploit Doppler. Hartwell imagined a sonar consisting of a ring of searchlight sonars, their outputs phased so that they were equivalent to beams pointing in the desired direction. The searchlights would be connected to form multiple receiving beams simultaneously. Because multiple searchlights would be connected to form beams, the net gain of the combination would be several times that of any single searchlight. The next step, the one actually taken, was to separate reception (using preformed beams) from pulsing. This kind of sonar could use a very short pulse, and hence could gain considerably in range accuracy. At this time the British were working on the first true preformed-beam sonar, their Type 177.

Much lower frequencies, which promised much greater ranges, had not been well explored. Admiral Solberg raised the possibility of extremely long-range sonar operating at low frequency, below 1 kHz. The group saw particular potential in passive sonars operating at 100–1,000 Hz, most likely at 100–200 Hz. By this time it was clear that an ASW submarine equipped with a large passive low-frequency sonar could detect a snorkeling submarine at 100 nm or more.

In common with the ASW conferences, the Hartwell study suggested further work on non-acoustic methods of submarine detection, which if successful would restore the efficacy of airborne ASW. Infrared (IR) wake detection, sonic detection of bubbles in a submarine wake, and trail detection by detecting diesel exhaust were among the possibilities. Hartwell also suggested investigating the disturbance of the surface by a submerged submarine, which might involve displacement of surface water by cooler water from below. This effect would be most pronounced when there were strong thermal gradients between the submarine and the surface. Early experiments using ships showed that the background of irregular temperature differences in the ocean generally obscured this effect, but airborne IR was promising because it scanned an area of the ocean rather than simply a line. Even if sweep rates were too small for detection, IR might be useful as a means of rejecting false targets, an increasing problem as detection range increased. Bubble trails were created by cavitating propellers. Recent developments had considerably improved the sensitivity of bubble detection, and it had been discovered that there were surprisingly few bubbles in the open sea. Bubble-filled water from a submarine wake might rise to the surface. The study report also suggested further work on magnetic anomaly detection (MAD) using paramagnetic resonance, which looked promising on paper. Only exhaust trail detection and MAD proved useful enough to pursue.

The search for non-acoustic solutions to the submarine detection problem continued, but with little success. In 1959 the U.S. Navy was pursuing IR wake detection; it appeared that a sensitive IR system (capable of detecting a temperature difference of 0.001° C) could "see" a wake left by a submarine several hours before.[4] The mechanism of wake formation was not fully understood, but the possibility was so promising that an operational device for airships (AAR-23) was being built for evaluation. This program died, but the possibility of recognizing submarine wakes by the difference in temperature of water churned up from below the surface continues to excite interest.

NIDAR was a radar phenomenon sometimes observed above a submerged submarine. It was discovered by accident, and it was never entirely clear what caused it. Eventually the program died. In 1959 BuAer was about to install a high-resolution 8 mm side-looking radar in an airplane to determine its capability to "see" submerged wakes. This possibility had been suggested by photographic recordings of earlier flight tests. It too went nowhere.

CHLOE was an attempt to detect an anomaly caused by a submerged submarine, using receivers in fast high-flying aircraft. It was thought that power frequency radiation (CHLOE Type P) might be detectable; underwater electric potential (UEP) was also considered interesting (it was used in mines).

None of these approaches seems to have worked well enough to be adopted.

Convoys

The Hartwell group was told that for several years after war broke out, convoys would be composed mainly of 8 kt Liberty ships. The scientists argued that this was false economy because Liberty ships would be easy targets for homing and pattern-running torpedoes as well as for aircraft with nuclear weapons. Faster (19 kt) ships, in or out of convoy, would have a far greater chance of survival against both submarines and aircraft. Reducing the time a convoy spent at sea would reduce the overall number of escorts needed. Too, analysis showed that increasing convoy speed would greatly reduce the effectiveness of pattern-running torpedoes. The faster the convoy, the faster the torpedoes would have to be, but their range was inversely proportional to the square of their speed. In wartime, cargoes were so valuable that even a small decrease in vulnerability would be worth its cost. Hartwell argued that fast cargo ships equipped with ASW helicopters might be able to operate independently.

The Maritime Commission was soon building much the sort of cargo ship the study suggested (the Mariner class), though never in numbers approaching those Hartwell advocated (150 per year vs. a total of 35 Mariners). A mark of the Mariners' connection to the Hartwell study is the helicopter pad built atop the after deckhouse.

Other measures to improve overseas transport would be to extend and modernize port facilities at both ends of the shipping routes and to promote mine countermeasures. These ideas had already been raised by OEG the previous year.

Weapons

The Hartwell scientists were well aware of one fruit of the 1948 U.S. mobilization: a coming era of nuclear plenty. The United States already had enough fissionable material for large-scale tactical use, but much more was needed for smaller-scale weapons. The scientists decried the deep secrecy currently surrounding nuclear weapons. Such weapons either were

The fruit of nuclear plenty: a Lulu nuclear depth bomb shown on board a standard HSS-1 Seabat ASW helicopter. Proposed by the Hartwell group, the predecessor Betty (Bomb Mk 90) weighed between 2,467 and 2,514 lbs, of which 1,243 lbs was its pressure-resistant casing. Including its tail assembly, it was 10 ft 5 in. long, with a diameter of 31.3 in. The casing carried a parachute that braked the bomb's descent. The warhead was the AEC's W-7. In 1955, development of a much lighter bomb (boosted implosion) was approved as Lulu, shown here. It weighed half as much as Betty, with only a quarter of its volume (1,200 lbs, length 92.4 in., diameter 18 in.). The same W-34 warhead was used in the ASTOR nuclear torpedo. Lulu was withdrawn from service in 1971. A lighter Little Lulu was canceled, but the successor B57 was still smaller. At the end of 1957, the assistant secretary of defense asked the Atomic Energy Commission to cooperate with the Navy in a study of a lightweight air-dropped depth charge. The Air Force was also interested in a lightweight low-yield tactical bomb, the Navy being named as primary user. The resulting B57 bomb was 9 ft 10 in. long, diameter 14.75 in., and weighed 500 lbs. NHHC

already or soon would be quite inexpensive. Tactical nuclear weapons deliverable by carrier aircraft should be developed to knock out enemy port facilities. The Navy was already pressing for lighter nuclear weapons, which would soon appear. The smaller weapons could be delivered both by heavy carrier bombers and by the lighter attack aircraft that could fly from all the operational fleet carriers.

Since the nuclear technology already existed, the Hartwell scientists could consider nuclear depth bombs as alternatives to homing torpedoes. They argued that the cost of submarine detection was so high that weapons must be nearly certain of killing the submarines once located. That ruled out existing weapons such as Weapon A. Too, since surface ships were inherently noisy, they had to be able to attack submarines at such a distance that the submarines could not tell that they were under attack. Homing torpedoes or nuclear depth bombs should spend most of their time in the air. It was therefore vital to improve long-range detection and positive identification of submarine targets. This recommendation led to the development of rocket-boosted depth charges and torpedoes: ASROC. However, Hartwell's assumption that ships could engage submarines at 20,000 yds proved overoptimistic; as deployed, ASROC was a 10,000 yd weapon.

The Hartwell group hoped that nuclear depth bombs could have a lethal range as great as a mile. Since no one knew what that would require in terms of nuclear yield (explosive power), realistic underwater tests were necessary. The key test for underwater nuclear weapons was Wigwam (14 May 1955), in which a 30 kT device was detonated at a depth of 2,000 ft. The target array consisted of scale models of submarine hulls.

The first fruit of this proposal was the Mk 90 bomb ("Betty"), which used the same W-7 warhead as some land-based tactical missiles. The Mk 90, with a yield of 28 kT, was initially designed for use by fighter-bombers.[5] Production began in June 1955, and about 240 were made. Betty weighed between 2,467 and 2,514 lbs, of which 1,243 lbs was a heavy pressure-resistant casing. It was 10 ft 5 in. long and diameter was 31.3 in. The bomb bays of the Grumman Tracker (S2F) and the Martin Marlin (P5M) had to be bulged to accommodate it. Mk 101 ("Lulu") was a lighter-weight, lower-yield (9 kT) replacement (1,200 lbs, 92.4 in. long, diameter 18 in.). Lulu occupied a quarter of the space of a Betty, and its lower yield made delivery safer (presumably particularly by helicopters). Lulu did not have the heavy casing of the Betty depth charge. About 2,000 were made between August 1958 and late 1962. The last was retired in 1971. Mk 101 had the same W-34 nuclear warhead as the Mk 45 ASW torpedo (ASTOR). The successor was a version of the B-57 bomb (Mk 57) with a variable yield (0.5, 2, and 10 kT).

The Hartwell group pressed for helicopter development. This HO4S-3 (Sikorsky S-55) was the first operational U.S. ASW helicopter. It is shown in Canadian service, carrying a Mk 43 lightweight torpedo and dipping an AQS-4 sonar. USNI

The Hartwell group shared the general Navy view that ASW helicopters were attractive because they offered a high search rate but were invulnerable to torpedoes. They seemed to be the only way to screen a convoy against Soviet submarines armed with torpedoes that could strike from beyond sonar range. Helicopters seemed to be the single most promising ASW weapon system—if they could be armed. Hartwell argued for development of an all-weather helicopter with improved sonar, secure communications, and station-keeping navigation.

Hartwell pressed for a lightweight torpedo with sufficient performance to catch and kill a new-generation submarine: Mk 44. Here one is launched from its standard triple torpedo tube. It is recognizable by the ring around its control surfaces.
USNI

The group pressed for development of a torpedo light enough to arm a helicopter. The Navy was already developing the Mk 43 torpedo. It does not figure in the notebook of U.S. naval systems issued to members of the Hartwell committee, so presumably it remained an unfunded requirement until the outbreak of war in Korea released money. Mk 43 did not meet the requirements the group set. Speed when closing should be optimized for the fastest target speed, which would probably be 20 kts. Hartwell therefore recommended a 30 kt torpedo, far beyond what Mk 43 could do. Running range should be three times the sensing range. Since that was now about 1,300 yds, running range should be 4,000 yds. To minimize the torpedo's weight, there should be no allowance for reattack. The torpedo should weigh less than 500 lbs, carrying a 50–75 lb warhead. This torpedo became the Mk 44, the lightweight torpedo of the 1960s.

The lightweight torpedo was also needed for fixed-wing ASW aircraft. The currently planned Mk 41 was so massive that only one could be carried at a time, but the single-package ASW airplane needed more than one shot. A rocket could carry a 500 lb torpedo out to 20,000 yds. That would give a surface ship great flexibility, for example to engage a submarine detected by a helicopter. A surface ship so armed could exploit questionable contacts without leaving its position in a screen. With the long-range rocket weapon, a surface ship could outrange a submarine—assuming it could detect the submarine in time. Without such weapons, the surface ship would have to rely on homing torpedoes, which would take a considerable time to reach a distant contact—half an hour to reach 20,000 yds, even at 40 kts. These ideas presaged the RAT and ASROC rocket-thrown weapons, although it is not clear whether there was a direct connection.

Torpedo homing technology was essentially that available at the end of World War II. All U.S. passive acoustic torpedoes used much the same transducer patterns and electronic functions as the wartime Mk 24; they even used the same frequency. The two major active torpedoes (Mk 35 and Mk 37) showed only minor improvements. No studies had been made of the tactical use of wire-guided torpedoes. Very little work had been done on wake-following, although Mk 35 had some wake-following features. The ultimate limit to torpedo speed was self-noise, for example 27 kts in a Mk 35. Relatively little effort had gone into studies of self-noise and ways to reduce it. At its supersonic homing frequency, Mk 35 was one of the noisiest torpedoes ever built. Little data had been collected to make it possible to design a quieter torpedo, yet reducing self-noise should be a major design factor rather than an afterthought. For example, there was no data on the noise output of unconventional propulsion systems such as pump-jets and ducted propellers. Surely noise should be a major part of the effort of a single laboratory such as the Ordnance Research Lab at Penn State.

Mk 35 was also too expensive, and it lacked collision-course control. To overcome enemy evasion, the torpedo had to be 1.5 to 2 times the maximum target speed, with a range in the water about 3 times enemy active sonar range. It should have passive as well as active homing capability. All of this ruled out Mk 35—and it presaged the future Mk 48.

The Hartwell report strongly supported ASW mine warfare but pointed out that existing U.S. mines were designed to attack 5,000-ton merchant ships. The Naval Ordnance Laboratory at White Oak was already working on next-generation mines optimized for ASW, but it would be four years before they were in stock. The Kola Inlet, the major known Soviet submarine base with direct access to the Atlantic, presented particular problems because it was so deep (50–150 fathoms), and hence unsuitable for bottom mines. Magnetic mining was particularly difficult because the mines had to be sufficiently insensitive to protect them from detonation by the intense magnetic storms found in the north. Something more exotic might be needed, but its nature was not suggested.

The Beginning of SOSUS

The most spectacular outcome of the Hartwell study was the realization that submarines could be detected passively at very long ranges using the deep sound channel that Woods Hole had first demonstrated in 1944.[6] The key was that noise from a submarine had strong (and stable) peaks at particular frequencies: signatures. The Hartwell study emphasized the need for research into transmission in the deep sea, and also into the key issue of whether sounds so transmitted would be distorted.

It turned out that the deep channel would transmit sounds undistorted. A distant receiver could distinguish a submarine signature from surrounding noise, and thus a distant submarine could be detected and tracked. Concentrating on low frequencies would reduce attenuation in the ocean and make it easier to detect submarines at very long ranges. Sound generated at relatively shallow depths would radiate in all directions, including into the deep channel, which would carry it over great distances. Hunt came close to describing the future SOSUS system. At the fifth CUW undersea symposium, 15–16 May 1950, prior to the beginning of the Hartwell study, physicist Frederick Hunt put these things together and announced: "Have heart boys, we're back in business again."[7]

Bell Telephone Laboratories was using a spectrograph to display a snapshot (a spectrogram) of the frequency components of speech, and this technology

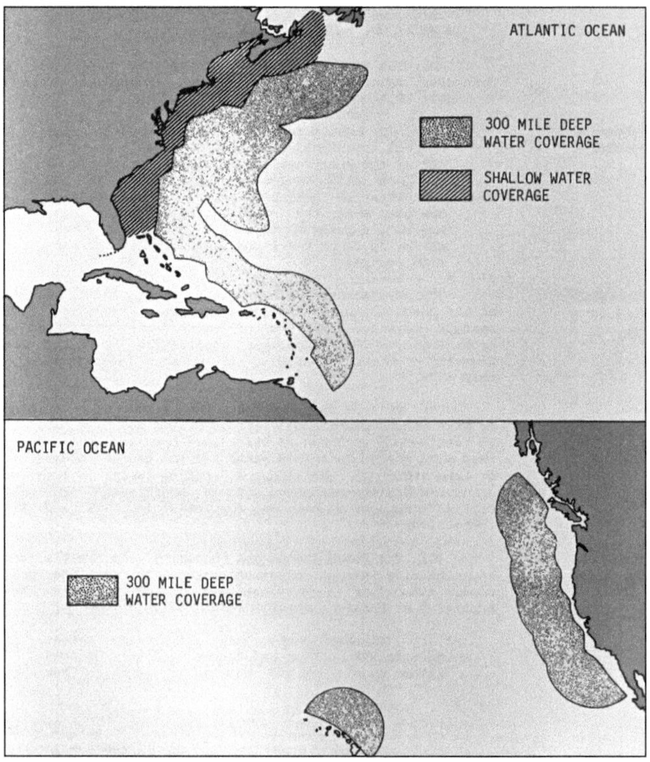

The beginning of SOSUS: coverage in 1958. At this time SOSUS was seen as a means of detecting submarines within missile range of the U.S. coast, and as a means of supporting other forms of coastal ASW. AUTHOR'S COLLECTION, FROM A DECLASSIFIED NHHC DOCUMENT

was applied to submarine sound analysis.[8] Up to this time "submarine sound" generally meant the sound of water or other liquids flowing over the submarine or through its piping. Such sounds varied relatively slowly and smoothly with frequency, and hence were called broadband. Early experiments using the Bell Labs spectrograph showed that in addition to this broadband—and generally not recognizable—sound, submarines produced sharp peaks associated with things like rotating machinery. These peaks, later called tonals or lines, became key to ASW from the mid-1950s on. On a spectrogram they were narrow vertical lines that emerged from the humps of broadband sound. In addition to the lines or tonals generated directly by submarine machinery, the submarine generated harmonics, tonals with basic frequencies multiplied by whole numbers. Sometimes the harmonics could be detected more readily than the basic tonals. Soon the cover name LOFAR was assigned to the frequency-analysis technique.

There was still a lot to be learned. It was already evident that sounds would penetrate the shallow thermal layers to reach the deep channel, along which they would propagate with little loss. But little was known about transmission in the open sea. Did submarine signatures propagate undistorted? Could sounds propagating along the deep channel be detected by hydrophones near the surface? Would sound received by one hydrophone be coherent with sound arriving at another 100 or 1,000 ft away, so that beams could be formed? How badly would multipath (sound going along multiple paths from source to receiver, potentially causing interference) distort signatures as received? The Hartwell group already knew that the peaks in a submarine signature would be very narrow (in frequency). But how narrow should the filters in a future system be? How much would be lost to self-noise in cables connecting hydrophones to the shore? Overall, the Hartwell group saw enormous promise in the deep channel and in the emerging knowledge of recognizable submarine acoustic signatures.

Once the LOFAR technique was translated into the sound surveillance system of underwater arrays (SOSUS), ASW aircraft were restored to much of their presnorkel prominence, at least in areas where the system was effective. SOSUS arrays created fans of beams. A submarine would typically be detected simultaneously by several arrays. She could be located by the crossing of the beams, which defined a SOSUS probability area (SSPA) that an airplane could search. SOSUS could also vector submarines and surface ships. A submarine could be detected in the open sea by her sound whether submerged or surfaced, whatever she was doing, in the open sea. Initially it was assumed that SOSUS range on a snorkeling submarine would be about 100 nm, but quite soon it was clear that SOSUS worked on an oceanic scale.

In 1950 all of this was still speculation, so the Hartwell report underplayed it. However, probably with the possibility of long-range submarine detection in mind, Hartwell advocated creating a U.S. East Coast barrier against enemy submarines and long-range aircraft, including an ASW barrier along the Nova Scotia–Bermuda–Puerto Rico line. The system would drastically reduce requirements for coastal escort and coastal and harbor defense, and would eliminate the requirement for Identification Friend or Foe (IFF) on the eastern seaboard. This barrier would be combined with a submarine barrier across the GIUK Gap.

The Hartwell group emphasized the value of low-frequency arrays on board submarines, which could make practical a picket line between Iceland and Norway blocking the route between Soviet home waters and the Atlantic. The longer the submarine array, the more effective it would be. Since sea noise was often a limiting factor in listening, the greater directivity of a long array would extend effective range. A linear array of low-frequency hydrophones might be mounted along the hull of a submarine or surface ship; a submarine would have the advantage of being able to take the array to greater depths.

SOSUS (called Project Caesar) became the most important ASW project of the 1950s. Western Electric, which ran Bell Labs, received a contract to develop the low-frequency spectral analysis technique (LOFAR) in 1950. The acronym LOFAR was spelled out in various ways to conceal its meaning. It was sometimes rendered as Low-Frequency Analyzing Recorder, but it was often given other meanings, such as Low Frequency Fixing and Ranging. The technique was also codenamed Jezebel, a term later applied to sonobuoys associated with LOFAR.

As applied by the Navy, the Bell Labs device had an output like that of the existing chemical recorder: a stylus crossing the paper indicated the content of a signal at various frequencies by darkening the paper in proportion to how strongly they were present in the signal. The content of the signal might not be obvious from any one pass of the stylus, but as the paper moved, the stylus passed over it again and again. The operator looking down the paper could see dark vertical lines indicating prominent frequencies—that is, indicating the frequency spectrum of a signal (or its absence). This approach was later called "visual integration," meaning a visual means of adding up the frequency spectra seen at different times. The idea was that the background noise in the sea was essentially random, while man-made objects made consistent, stable sounds. Identifying a particular spectrum with a particular type of submarine required that the SOSUS system learn the details of the spectra of various submarines. The output on chemically treated paper was called a Lofargram and usually referred to simply as a "gram." The term survived when the paper was replaced by an image on an electronic display. At least initially, in SOSUS, LOFAR normally covered the range between 5 and 150 Hz (SOSUS could cover the range up to 280 Hz if necessary).

In 1951 three codenames were assigned to the project: Jezebel, for LOFAR; Caesar, for procurement and installation; and Michael, for operation (oceanographic analysis). In 1952 the system was given the classified name SOSUS.

A pilot underwater array was built in shallow water off Sandy Hook, New Jersey.[9] By May 1951 it had demonstrated detection ranges of 40–50 nm, By January 1952 the prototype deep-water array had been installed off Eleuthera in the Caribbean at a depth of 240 fathoms. It employed a 1,000 ft linear array of 40 hydrophones wired together to produce a single steerable beam. A continuously snorkeling submarine was detected 9 out of 10 times at ranges beyond 200 nm. Tests begun in April 1952 were so successful that in June 1952 CNO ordered BuShips to install six stations off the East Coast. This prototype SOSUS array was first tested during ASDEVEX 1-54 between 26 April and 7 June 1954. The Eleuthera array and a second array off Bermuda worked with the ASW submarine *K 1* to provide data to a control center at Norfolk. The system vectored sonobuoy aircraft to the point at which a submarine had been detected. It proved successful 90 percent of the time.

In June 1952 BuShips was told to procure six stations; CNO suggested they should include Sable Island, Cape Hatteras, Bermuda (two arrays), and Culebra in Puerto Rico. About that time two LOFAR units were sent to the United Kingdom to analyze submarine signatures for the British. In September 1952 CNO expanded the East Coast system to nine stations and changed some locations. Three more (Barbuda, Barbados, and Newfoundland) were added in January 1954. In May, ten more stations were planned, six of them on the West Coast. Ramey (in Puerto Rico) was the first to go online, in November 1954. When CNO asked for a Pacific network in 1952, he called the system SOSUS for the first time.

SOSUS array acoustic data was fed into land stations called Naval Facilities (NAVFACs). The NAVFAC interlaced the output of the hydrophones to form beams 2–5 degrees wide, whose content was fed into banks of LOFAR analyzers. Personnel in the NAVFACs examined the resulting gram paper (called "smoky paper") to decide whether submarines were present. Data from multiple NAVFACs were collected at evaluation centers in the Atlantic (Norfolk) and the Pacific (Treasure Island), where targets could be triangulated. For security, the SOSUS system was publicly described as an oceanographic observation system, a preferable alternative to oceanographic research ships.

By 1956 plans called for 12 SOSUS arrays in the Atlantic and 7 in the Pacific (West Coast and Hawai'i); soon a shallow-water complement (Colossus) was being developed to protect the SOSUS cables returning to land. Ramey, Puerto Rico, became the first operational Atlantic SOSUS array in September 1954. At this time the estimated completion date of the Atlantic SOSUS network was December 1956; the Pacific network was to be completed in March 1958. From the first, the system included Canadian arrays and NAVFACs.

Three more SOSUS arrays had been added by 1959, at which time the whole system was expected to become operational by late 1960. An additional deep-water station northwest of Bermuda was completed in 1962. An array off Adak in the Aleutians became operational in November 1962. In 1963 a station in Norway became operational.[10] In 1966 a new system was installed at Keflavík, using three 16-element arrays (Keflavík was later credited with four arrays).[11] In 1968 new systems were installed at Midway and Guam. A NAVFAC was established at Barbers Point, Hawai‘i, in 1970; and at Brawdy, Wales, in 1974. There may have been an array in the Atlantic off Gibraltar.[12] Japan seems to have operated SOSUS arrays. Rumors of other arrays are probably mistaken.[13]

The initial planned SOSUS mission was early warning against Soviet antiship submarines, to cover the coastal area out to 100 nm. The threat of submarine minelaying was serious in those waters but vanished beyond the 100-fathom curve. Thus, the initial promise of detection out to 200 nm was adequate. However, by the time SOSUS was approaching operational status, the much greater perceived threat was Soviet missile submarines. SOSUS was now expected to cover the much larger area in which they might be expected to operate. By 1960 SOSUS was expected to locate Soviet submarines at their missile launch ranges, then assumed to be no more than 1,000 nm. Denial of unobserved operation inside this range was considered a valuable way of denying the Soviets the use of shorter-range missiles, which at that time were considered a severe threat. Planners admitted that location beyond 1,000 nm might be impractical, but apparently that was not the case.

Although SOSUS detections were generally subject to 15–35 nm errors, the system was given a tactical mission; initially that was mainly directing aircraft, but later submarines were added as well. Maritime patrol aircraft and, later, carrier ASW aircraft were given LOFAR readouts that could be matched to what the SOSUS system heard, and it was expected that U.S. submarines would have LOFAR by 1960.

The distances between arrays and NAVFACs were set by signal losses in the cables involved, so initially the NAVFACs had to be within a few hundred miles of the offshore arrays. By 1965, cable lengths in the Atlantic varied between 27 (Ramey) and 815 nm (Keflavík). Coaxial cable, used for the first time for the Adak array in the Aleutians, could carry signals as far as 1,000 miles (in the Adak case, it was 600 miles). By the early 1970s, advances in underwater cable design initially associated with transatlantic telephones made it possible to increase that distance considerably.[14]

The initial design requirement, soon exceeded, was to maintain a 50 percent probability of detecting a continuously snorkeling submarine out to a range of 250 nm from an array.[15] In 1954 tests, reliable SOSUS range on snorkeling submarines was about 300–400 nm, but some targets were detected at 600 nm. Such performance made SOSUS a coastal protection system, much as the Hartwell scientists had envisaged. By January 1956, the first eight Atlantic SOSUS stations were reporting 3,400 monthly contacts, 64 of them classified as probable non-U.S. submarines. Early on, the main limitations of the system were the lack of sufficient data on Soviet submarine signatures and the similarity of submarine signatures to those of some surface craft, particularly fishing boats.

In 1960, the Pacific SOSUS system reported 46 two-station fixes and 8 single-array fixes on submarines when operating against 8 Guppy submarines simulating missile submarines attacking Strategic Air Command (SAC) bases. Of the two-station fixes, eight were misinterpreted and were grossly in error. Of the good ones, half were in error by 12 miles or less, and 80 percent were in error by less than 30 nm.[16]

A complementary Colossus shallow-water system was created to cover the shallower water inshore of the deep area surveilled by SOSUS arrays. It employed pairs of lines of upward-looking active sonars buoyed about 15 ft up from the seabed, presumably acting as acoustic fences, the sonars based on fathometer technology. Two detections in sequence would indicate the presence of a submarine as well as its course and speed. There was also a 15–30 kHz passive capability against cavitating submarines. In 1960 Colossus I was described as a means of surveillance and detection in shallow water and in tactically important restricted passages. As of the spring of 1960, a small system employing 22 units (active and passive) was scheduled for installation that summer. Colossus later became the inshore complement to the SOSUS system.[17]

SOSUS and LOFAR were kept extremely secret because both were vulnerable to Soviet submarine silencing. Knowledge of these programs was therefore limited to the United States and Britain. Although they were fully informed of the evolving SOSUS program, the British did not adopt the new technology.[18] They were designing their *Porpoise*-class submarines, which had sound-mounted engines. They argued that within a few years the Soviets would adopt similar measures, and thus could evade long-range detection. British claims for their own silencing were vindicated in 1959 when HMS *Porpoise* could not be detected reliably beyond about 30 nm. Fortunately, the Soviets made no comparable efforts to silence their own submarines.

Initial U.S. efforts to bring the British into the SOSUS system ended in 1963. After that the British were denied access to the library of lofargrams the system was building. In 1967 the U.S. Navy again sought British involvement, this time in setting up a SOSUS array on the far side of the Atlantic. Among the advantages claimed by British advocates of this step (which was later taken) was access to the U.S. lofargram library. At the same time the British adopted other U.S. LOFAR-based systems, such as spectrum analyzers to be used by submarines.

The early SOSUS program seems to have inspired the British to undertake their own long-range passive Corsair (asdic Type 191) system.[19] Submarines could be designed to frustrate LOFAR. Too, at sufficiently low frequencies multiple arrays might offer much better bearing accuracies than the steerable beam then envisaged for SOSUS. At this time the best range the Americans had achieved had been about 400 nm, and it was not clear how often such performance was being achieved. Corsair focused on the broadband noise a submarine produced. The British thought that this noise could be distinguished from background noise because it was localized. British radio astronomers had demonstrated just the sort of localization needed, using interferometry to correlate signals detected by several separated antennas. By 1954 the British were undertaking laboratory tests, which confirmed expected system performance. The first full-scale tests seem to have been conducted in July 1956, against the test submarine *Scotsman*, which had an auxiliary diesel mounted on a raft that could be tied down rigidly to the hull. She was detected at 10 nm when the engine was rigidly mounted, but not at 7 nm when it was flexibly mounted.

Although early trials were considered successful (albeit at ranges much shorter than those achieved by SOSUS), the project was canceled as part of the 1957 defense cuts. It seems arguable that the policy shift toward a fleet oriented mainly to fight limited wars far from Britain made any kind of fixed ASW sensing system around the British Isles less attractive.

Although the French had no idea of what the “Anglo-Saxons” were doing, they deeply resented being excluded. Perhaps as a deliberate poke, they called their experimental ASW destroyer *La Galissonniére* (T 56) a “LOFAR destroyer,” meaning only that it was a test ship for their first relatively low-frequency hull and variable-depth sonars (DUBV 23/43).

HF/DF

HF/DF was the only other potential long-range submarine locator the Hartwell project identified. The Hartwell scientists acknowledged the continued value of shore HF/DF. The question was how soon the Soviets would field *Kurier*, and thus how soon the West would need totally new HF/DF technology. Hartwell expected the Soviets to use flash transmissions lasting about a second, approximately the minimum reaction time of a human operator (*Kurier* transmissions lasted 0.47 second). It seemed unlikely that the Soviets would field anything better for some years. Capacity would be limited by the nonlinear characteristics of the transmission path, so that it might take a second to transmit the equivalent of a 10-word telegram (but a more optimum approach could send 50 words in half a second). Hartwell was told that the interim system could be in place within two or three years, but it actually took longer; perhaps priorities were reduced after the end of the Korean War. According to the Canadian account, when the control station detected a signal, it flashed the frequency and call sign to other stations so that they could take bearings. The rather different Boresight recording procedure was highly compartmentalized.

The Hartwell group recommended against expending effort on anything beyond flash, both because it was unlikely to be adopted and because countering it would demand effort that could be better used elsewhere. An interim direction-finding network was being built. Hartwell recommended that it be modified to permit some interception of flash signals. The Hartwell scientists thought the net could be built mainly from existing equipment. The interim system comprised three HF/DF nets, one in the Atlantic and two in the Pacific.[20] The GRD-6 system covered the whole frequency range in four bands. A spinning goniometer fed a CRT direction display. The operator registered signal direction by turning a cursor to line up with the pattern on the CRT. The display was a propeller-like pattern pointing to the emitter and also directly away from it, because the array was symmetrical. The choice of which (the "sense") was determined using an additional omnidirectional (monopole) "sense" antenna.

Like earlier forms of HF/DF, GRD-6 tuned to a frequency after picking up a signal. Effort was concentrated on the order circuits used to alert DF stations in a net when one station detected a signal. Once alerted, stations had to set their tuners to the appropriate frequency before taking bearings. This technique worked for most signals 30 seconds long, but only rarely for those 15 seconds or less. Compared with earlier nets, the interim system reduced time to determine a bearing to 5 or 10 seconds after the control station discovered a transmission. Hartwell wanted the system modified to guard (listen to) multiple frequencies simultaneously. They argued that if the tuning controls of all the stations in a net were automatically adjustable from the control center, the entire net could rapidly be tuned to a signal. They estimated that the current net could detect half of all conventional transmissions but no flashes (*Kurier*-type bursts). The interim net would intercept 90 percent of conventional and 10 percent of flash transmissions. An ultimate system using information storage would detect 90 percent of conventional transmissions and 50 percent of flashes.

The Hartwell group recommended installing panoramic (wideband) receivers at some stations in a net. The receivers tuned continuously over a set band, so they spent very little time at any one frequency, and hence could not record messages. Associated with them would be specialized means of storing and presenting data, since the flash transmissions would be too quick for operators to see indications of them. Once enemy frequencies had been identified, receivers could be set to monitor them. It was deemed unlikely that the enemy used a new frequency for each transmission. At a specific distance, time of day, and time of year, HF propagation limited the useful band to 3–4 MHz; channel width was determined by message content. Compressing a message increased the range of frequencies it required. Probably 4–80 channels would be available at any one time and place. The smaller number was suggested by submariners and the larger by communication engineers. Hartwell estimated that 20 would be available to a fairly large submarine operation such as a 6-month campaign in the North Atlantic. The number could be reduced by jamming or by filling some channels with U.S. signals. Like earlier researchers, Hartwell doubted that previous HF/DF techniques would work quickly enough to handle flash signals. Instead the scientists argued for the inverse Loran technique first suggested in 1945. Comparing the times at which different stations heard the same signal would make

A Wullenweber (circularly disposed antenna array), part of the solution to the pressing *Kurier* problem. Because the array was wide open to signals from all directions, it had a high probability of detecting any particular short signal, as long as its receiver was tuned to the appropriate frequency. The Hartwell group pressed for this technology while proposing a partial interim solution. USNI

it possible to determine where the signal had originated. Data showed that this technique would fix transmitters with an average error of 10 nm, versus an average of 50–100 nm using DF methods.

All of this was speculative. As of 1950 no Soviet submarine radio transmissions had been identified.

Two alternative approaches to flash signals were then being pressed. The German Wullenweber (a wartime codename) circularly disposed antenna array was already being developed at the University of Illinois and at NRL.[21] Much as in a scanning sonar, a Wullenweber was scanned by a rapidly spinning commutator which formed beams in sequence. It was estimated that a high-speed dual goniometer (direction-finding commutator) would be effective against a transmission lasting half a second or more. The alternative was to rely on an intelligence estimate of the approximate probable enemy frequency. A fixed antenna would feed multiple receivers set at several frequencies. The latter approach seems to have worked, to the extent that by 1960 (before Wullenwebers) flashes were being intercepted regularly, presumably by an upgraded version of GRD-6.[22] The initial solution to *Kurier* was Boresight, a project that involved a subsystem installed at GRD-6 sites in 1960–61. Work seems to have begun in 1958, but it is not clear whether it was a solution to the long-studied *Kurier* problem or a response to Soviet communication changes. Bullseye sites received their own subsystems. The project may have been known as Clarinet Bullseye.

At that time the most active U.S. net, WestPac, was handling about 200–250 “flashes” per day and could fix about 40 percent of targets within about 100 nm. The chapter of the redacted NSA history that mentions the *Kurier* problem footnotes a history of Pacific HF/DF prior to the advent of Bullseye—that is, using GRD-6. Strategists feared that the Soviets would go well beyond *Kurier*, using compression techniques to create much shorter signals, as short as 1 millisecond (0.001 second). Reportedly the Navy found it possible to record incoming signals; flashes could be recognized retrospectively. Given accurate information as to the flashes’ timing, inverse Loran could be used to locate their source. It may also have been possible to associate flash signals with particular goniometer bearings. The technique was reportedly called retrospective direction-finding.

The next step was Wullenweber antennas coupled with new approaches to handling ultra-short signals. The U.S. Navy version of this circularly disposed antenna array was FRD-10.[23] Fourteen sites were built as part of the Bullseye program; another two were later built in Canada. The array consisted of two rings of antennas on 776 ft and 847 ft circles, backed by reflector screens, atop a reflecting plane, with a hut in the center housing a goniometer. The sheer size of the FRD-10 made its beams much narrower than those of the earlier GRD-6; bearings were said to be up to four times as accurate as those of the earlier system, better than within 0.5 degree. Gain was four times greater, and the FRD-10 could better filter out interfering signals and noise. A similar antenna was built for the U.S. Army and Air Force as FLR-9. Wullenwebers were sometimes called "elephant cages."

Nominal range was set by two signal bounces off the ionosphere, typically about 3,200 nm, although HF signals could be monitored at much greater distances. Like SOSUS, Bullseye had a cover story, in its case participation in a search-and-rescue net and support for research into basic problems of ship-to-shore communication.

HF/DF was limited in that stations had no ability to analyze the signals they intercepted; they simply reported all signals at the designated frequencies and with possibly relevant bearings. That made for a high false alarm rate and low credibility, as well as significant time delay. As a consequence, officers in tactical command generally disregarded information from Bullseye.

The situation seems to have changed in the 1970s. The U.S. Navy deployed a shipboard system called Classic Outboard beginning about 1973.[24] It was linked to the land-based net, which by that time had been automated (to reduce time delays) under a program called Classic Bulldog. The Royal Navy participated; six of its Type 22 frigates were fitted with Outboard so that two (to take cross-fixes) could operate with each ASW carrier.

The Bullseye system was shut down in the 1990s, after the end of the Cold War reduced the need to track Russian submarines (and, presumably, after the Russians shifted to satellite communications).

The Soviets also captured Wullenweber technology. About 1957 they deployed 20 or 30 Krug (Wullenweber) arrays.

The Outcome

Hartwell produced a long list of specific recommendations. In 1951 a member told the Undersea Warfare Conference that "when we started Project Hartwell we regarded the country as one regards a man with an inoperable cancer: doomed. When we finished, we felt that our earlier diagnosis had been wrong. Actually, the patient had rickets, halitosis, and a bad hangover, but nothing that occupational therapy could not cure."

The one essential ingredient for recovery was money. As the Hartwell study was underway, North Korean forces invaded South Korea. The U.S. government (and others in the West) saw the invasion as a first step toward a new world war. The United States mobilized, not just to fight the war in Korea but also to fight the expected wider war. Had mobilization not occurred, the Hartwell recommendations for further research would probably have been implemented, but they would not have led quickly (if at all) to large-scale production of new sensors and weapons. Moreover, the outbreak of war in Korea turned NATO from a political to a military alliance, greatly increasing the urgency of building new weapons and sensors. It took nearly two years for the post of SACLANT (Supreme Allied Commander Atlantic) to be filled, due to friction between the United States and the United Kingdom over the demarcation (and authority) of SACLANT versus CinCCHAN (Commander in Chief Channel), the latter becoming a British post. Once a SACLANT (double-hatted with the U.S. Atlantic Fleet commander) was appointed in 1952, he quickly drew up a war plan and called for a considerable increase in the NATO escort force.

7

THE MOBILIZATION DECADE

The Fifties

THE KOREAN WAR ENDED THE DAYS of austere U.S. Navy budgets. The United States and the United Kingdom mobilized not only to fight in Korea, but for the larger war with the Soviet bloc that they feared might soon come. The fiscal momentum generated by the outbreak of the Korean War lasted through the 1950s. The new ASW technology of the 1950s was largely what would have been required to deal with the threat of the new kind of diesel submarine the Germans first fielded in 1945. However, by the late 1950s the U.S. Navy was increasingly oriented toward neutralizing a new submarine threat: Soviet missile submarines capable of attacking the U.S. mainland. In effect, the success of the ASW campaign conducted during the Cuban Missile Crisis of 1962 tested (and validated) this development.

For the U.S. Navy, the most visible fruit of mobilization was activation of many of the ships laid up after World War II and after the 1949 cuts. On 1 January 1950 the Navy List showed 351 destroyers, 142 of which were active (plus 10 training ships in partial commission). Of 30 converted to mine

Mobilization more than reversed earlier fleet reductions. The *Dealey* class was put into production; in effect it was a U.S. counterpart of the British Type 12 frigate. USS *Cromwell*, the second ship of the class (and the first to have the planned Weapon A) is shown as completed except that triple lightweight torpedo tubes have replaced her original heavy depth-charge battery (three throwers on each side aft and the usual two tracks). For a time *Cromwell* had triple tubes but no Weapon A. The paired guns are standard 3 in./50s. Long torpedo tubes were never fitted. Thirteen ships of this class were built under the FY 51–55 programs, which might be compared to six Type 12 and another nine improved versions (*Rothesay* class) built under contemporary British programs—which, however, did not include the numerous larger U.S. ships and early nuclear submarines of the time. USNI

vessels, 4 destroyer minelayers were active, and 12 (of 20) destroyer minesweepers. Of 246 destroyer escorts, 10 were active. In 1950–51, 9 fleet carriers (including the rebuilt *Oriskany*), 2 light fleet carriers, 8 escort carriers, all of the inactive *Fletcher*-class destroyers, 78 destroyer escorts, and 38 submarines were activated. The activations more than doubled U.S. active fleet strength in most categories. Another 12 destroyer escorts went to the Coast Guard, which operated them in full naval configuration. The U.S. FY 53 naval operating plan showed 2,518 ships in commission, including 394 combatants, plus another 1,646 in reserve. By this time many of the existing *Essex*-class fleet carriers were being modernized, leaving seven unmodernized ships that could be converted for ASW under a new CVS designation. Later, other *Essex*-class carriers with hydraulic catapults were designated ASW ships.

By August 1952, 244 destroyers were active. Seventy-eight destroyer escorts had been activated, not counting the 12 transferred to the Coast Guard. Eighteen more had been removed from the sale list. Fleet expansion lasted beyond the Korean War. Thus, as of 1 January 1958 the Navy had 245 active destroyers out of a total of 368; 9 active ships were assigned to the Naval Reserve. In reserve were 68 *Fletcher*s and 60 *Benson-Gleaves* class (1,620–1,630-ton type). Thirty-five of 230 destroyer escorts were active, plus 36 recently converted radar pickets. By this time all ten destroyer minelayers had been laid up, and all of the destroyer minesweepers had been reclassified as destroyers (and were inactive).

Quick activation for the Korean War was possible because the equipment on board the ships, particularly for communication, was not yet obsolete. As Admiral Low had just argued, the current submarine threat was the 1943 threat, before snorkels and fast submarines, and ships with 1943 sonars and weapons were adequate for the time being.

The mobilization did not reverse the Low Report's verdict on the *Fletcher*-class DDEs; they were too valuable in their role as destroyers. With Pacific Coast yards extremely busy, many *Fletcher*s were initially activated in the same configuration in which they had been laid up. As time allowed, they were refitted with the Low Report's ASW improvements: the visible pair of Hedgehogs and the invisible QHB (or SQS-10) scanning sonar. The *Fletcher*s had been criticized as being much less maneuverable than their twin-rudder successors, the *Sumner* and *Gearing* classes, so they were given larger rudders. Under an antiaircraft upgrade program, ships received automatic twin and single 3 in./50 guns. These upgrades were made in stages. There were at least two attempts to develop minimum characteristics for a *Fletcher*-class ASW upgrade, SCB 7A (SCB 7 was the original DDE) and SCB 76. The latter was intended to support a prototype conversion of a reserve ship. It died, presumably because reactivated *Fletcher*s were being refitted to reflect the Low Report's suggested characteristics.

The shorter-legged 1,620–1,630-ton destroyers were not activated, and they were not upgraded, but their class improvement programs included the twin Hedgehogs and the sonar. An ASW conversion of two of them, called a corvette (DDC, SCB 130), was proposed for the FY 56 budget but not pursued. Characteristics approved 9 December 1954 envisaged replacing one boiler room with additional fuel tankage, as had been done in wartime to many British and Canadian destroyers.

Tweedy in Pensacola Bay, 4 December 1958. Her SCB 63 refit left her with four fixed Hedgehogs atop her rebuilt bridge. Massing them this way enabled her to fire two two-Hedgehog salvos in quick succession without reloading. She retained six depth charge throwers and two stern tracks. NHHC

USS *Vammen* after her SCB 63A modernization, which included a rebuilt bridge and two Hedgehogs in No. 2 position. She retained her depth charges, though she had only six throwers. Note the absence of an HF/DF antenna. NHHC

Prototype ASW upgrades of existing destroyer escorts were included in the FY 51 program under the SCB 63 and SCB 63A projects approved immediately after the outbreak of war in Korea, on 26 July 1950, characteristics having been recommended for approval by the Ship Characteristics Board (SCB) on 14 June. SCB 63 (DE 532 *Tweedy*, a WGT) was a full upgrade intended to include desired features of new escorts. SCB 63A (*Petersen* [DE 152, FMR type], *Lewis* [DE 535, WGT], and *Vammen* [DE 644, TE type]) was a more austere mobilization conversion. As of 26 July 1950, the SCB wanted one SCB 63 and two 63As in the FY 51 program. Conversion would make it possible to evaluate some features planned for new destroyer escorts. Tentative characteristics for SCB 63A showed conversion priorities: in order, one slow FMR, one fast TE with 3 in. guns, and one fast WGT.

In SCB 63 the DE superstructure would be modified to incorporate a consolidated control space with a combined CIC and underwater battery plot, a sonar control room, a captain's cabin, a radio and RCM room, and a conning station, all on the same level, except that the pilothouse and forward walkway would be raised to ensure a clear line of sight to the foot of the jack staff over the forward gun. Ships would have a scanning sonar and an attack sonar, with space and weight reserved for a variable-depth sonar or long-range sonar. The desired ASW armament was four fixed Hedgehogs on the 02 level and a depth charge battery suited to 19-charge patterns: 14 side-throwers and 2 stern depth charge tracks. Ideally there would be two ready-service patterns for each Hedgehog. ASW fire control would comprise an A/S Attack Plotter (Mk 1 or better) and an Attack Director Mk 5. For self-defense ships would have a Mk 3 beacon (i.e., active decoy) launching system.

In place of the elaborate consolidated control space of SCB 63, 63A called only for provision of an adequate underwater battery plot at bridge level rather than complete reconstruction of the bridge. In place of the four fixed Hedgehogs, these ships would be armed with two trainable Mk 15 depth charges. CIC instruments would be rearranged to reflect experience with two test ships: *Robinson* (EDE 220) and *Wilke* (EDE 800). In place of the beacon launcher, ships would tow torpedo countermeasures.

The class improvement programs for destroyer escorts envisaged further ASW upgrades. Ships with 5 in. guns had twin power-trained 40 mm guns superfiring over their forward mounts. Class improvement programs for these ships envisaged replacement of the 40 mm guns by a trainable Hedgehog and, presumably, scanning sonars. Destroyer escorts armed with 3 in. guns were ultimately to have two fixed Hedgehogs. Some destroyer escorts were upgraded to these standards.

Mobilization also applied to aircraft. In 1950–51, naval aviation funds, including for patrol aircraft, were cut more than 30 percent, including a reduction in active patrol squadrons from 30 to 20. At that time there were still 29 reserve VPs. The FY 50 figures already reflected a cut from planned FY 49 strength of 17 land-based and 15 seaplane patrol squadrons. VP squadron strength was 9 aircraft. Patrol plane strength included 17 P4M-1 Mercators, which were used to gather electronic intelligence and thus had no direct ASW role. There were 2 blimp squadrons (ZPs). The FY 51 aviation plan (1 December 1950) showed 19 land patrol squadrons and 9 seaplane patrol squadrons (respectively, 171 and 72 aircraft), and 4 rather than 2 blimp squadrons (16

A group of destroyer escorts moored off San Diego on 4 June 1955 shows the limited ASW upgrade applied to many ships: a trainable Hedgehog, a modified bridge, and (not visible) a scanning sonar. *Spangler* (DE 696) has the extended sonar room of a high-bridge ship. Two ships (DE 509 *Formoe* and DE 368 *C. J. Doyle*) were not modified. DE 697 *George* was a heavily modified TE (*Buckley*)-class destroyer escort, her bridge completely rebuilt. DE 414 *Le Ray Wilson* has an open bridge built out around her original low bridge, which would have been similar to that of DE 368 alongside. All of these ships were steam (hence relatively fast) destroyer escorts. None of the ships with trainable Hedgehogs was part of the SCB 63/63A project. The policy seems to have been to plan installation on board all of the fast steam destroyer escorts with 5 in. guns (hence with superfiring twin 40 mm guns): TEVs, WGTs, and the TEs that had been rearmed with 5 in. guns (including the wartime radar pickets). Ships actually rearmed were: DE 217, 218, 219, 341, 351, 359, 361, 365, 366, 369, 406, 414, 416, 418, 422, 447, 534, 535, 679, 680, 684, 685, 696, 697, 700, and 708—a total of 26. RICK DAVIS

aircraft). As of 1 April 1953 there were 34 patrol squadrons (including seaplanes) and 15 carrier air ASW squadrons. A January 1952 version of the FY 52 aviation plan included provision of ASW helicopter squadrons (HS-3 and HS-4, later HS-1 and HS-3) to operate from escort carriers. The blimp squadron complement was increased from four to five aircraft. By 1955 the number of ASW helicopter squadrons had doubled. In 1959 the U.S. Navy had 20 squadrons of land-based patrol aircraft, 10 squadrons of ASW seaplanes, 10 ASW air groups (for 9 CVS), and 2 blimp squadrons. The main changes over the next years were elimination of the blimps in 1962 and elimination of the seaplanes in 1966. The U.S. maritime patrol force dwarfed that of RAF Coastal Command, which in 1949 had three squadrons, with a fourth being planned and a fifth expected in 1954.

ASW IN THE KOREAN WAR

ASW was a very small part of the naval war in Korea. Although the Soviets had a substantial Pacific submarine fleet, it apparently did not intervene.[1] Orders for U.S. naval forces were to attack and drive off any unidentified submarines they encountered; unidentified submarines in position to attack U.S. forces were to be considered hostile.[2] Apparently it was soon obvious that Soviet submarines would not engage the fleet operating off Korea (some Soviet bombers did probe U.S. task force defenses, however). The two ASW escort carriers in the Pacific, *Sicily* and *Badoeng Strait*, were ordered to Japan, but they were assigned to their alternate roles as Marine Corps support carriers. A carrier group formed of *Valley Forge* and the newly arrived *Philippine Sea* (CV 47) was assigned mainly to close air support, with hunter-killer operations as its lowest priority.

The fleet carriers had ASW detachments on board: Skyraiders AD-4W (or AD-4E) and AD-4N aircraft. All of these aircraft had alternative roles considered more urgent than ASW in Korean waters. The AD-4W, with its APS-20 radar, was used mainly for airborne early warning to counter the possible threat of North Korean air attack. To further complicate the situation, the sonobuoy dispensers carried by the AD aircraft could handle only the new SSQ-2 buoy, which was not yet available (an adequate quantity became available early in 1951).

Other ASW forces were two patrol squadrons and three destroyer squadrons. However, comments at the 1949 Undersea Warfare Conference indicate that the patrol squadrons (one of Mariners and the other of Privateers) probably had not yet been modified for ASW.

A western Pacific HUK group was formed on 23 January 1951 comprising one eight-destroyer division and the escort carrier *Bairoko*. The lack of destroyers and their marginal sonar performance led to a recommendation to use helicopters with dipping sonars, which were then under development. The main ASW weapon was the wartime Mk 54 depth bomb. Mk 24 torpedoes were available but were not normally carried on patrol. The ASW aircraft were Avengers: TBM-3W radar aircraft and TBM-3S attack aircraft (typically 12 TBM-3S and 6 TBM-3W). Mobilization accelerated modification of shore-based aircraft to ASW configurations, so that squadrons available in the Far East in 1951 had PBM-5S and PB4Y-2S aircraft; there were also HUK teams of P2V-3 and P2V-3W Neptunes.

Demonstrated limitations of the escort carriers employed in HUK operations off Korea led to repeated strong recommendations that fleet carriers replace them. The fleet carrier *Valley Forge* was successfully tested in the ASW role off the U.S. West Coast. On 8 August 1953 five fleet carriers, two of them active, were redesignated ASW carriers (CVS), two more active carriers being added on 1 January 1954.[3]

THE ESCORT PROBLEM AGAIN

Many existing NATO escorts were inadequate. Having gone down a step to produce Type 14, in June 1950 the assistant chief of Britain's Naval Staff asked his Director of Naval Construction (DNC) to go down another step, to what was soon called a third-rate frigate.[4] With the world's largest shipbuilding industry, the British assumed that in an emergency they would be building most NATO escorts, with U.S. effort being devoted to merchant ships, which were simpler. The Director of the Tactical and Staff Duties Division (DTSD) of the Naval Staff proposed adopting standards of performance for the different categories of NATO ships rather than detailed requirements. He suggested first-, second-, and third-rates, with speeds and endurances of, respectively, 22 kts and 4,000 nm; 19 kts and 3,000 nm; and 14 kts and 2,000 nm, all speeds being "deep and dirty." The first-rate ship would have a weapon equivalent to Squid; the second-rate, Squid or Hedgehog; and the third-rate, depth charges. First-rates should be able to contribute to the antiair defense of a convoy; a second-rate would be able to defend only herself while in company with other escorts. For that she needed only a multiple close-range mounting with visual gun direction.

DNC sketched two alternative designs, and in November 1950 the British offered standards for NATO frigates, all of them built to lower (hence more affordable) standards than their high-end Type 12 (*Whitby* class). The proposed first-rate was based on the third-rate sketched the previous June. The proposed second-rate was a 19 kt ship corresponding to projected modernization of existing British Loch- and Castle-class frigates. On this basis a 14 kt third-rate frigate armed with depth charges was defined, but the Director of Torpedoes and ASW (DTASW) considered it "folly" for any NATO country to build one. Presumably the definition was intended to swell NATO frigate numbers by including older, less capable ships.

Building for lower speed would make ships affordable; DTASW pointed out that not all enemy submarines would be fast, and that even fast submarines would not always be able to go fast. The need to search for fast submarines could be overcome by numbers—the more escorts available, the less maximum speed they would need to protect a convoy adequately. However, it would be difficult to provide the desired Atlantic endurance (4,500 nm at 12 kts) in a small hull. DTSD, who was responsible for assembling the requirements to which British warships were designed, commented that the endurance requirement might not apply to the French, Italian, and Dutch navies, which were concerned mainly with coastal convoy protection and with trans-Mediterranean convoys. Smaller frigates might be much more affordable for them. In the event, none of these countries built British-designed frigates.

Late in 1950, Controller asked DNC for a small A/S frigate that would cost about a third as much as a first-rate. He envisaged it as a natural follow-on to the second-rate Type 14, which would cost about a third as much as a first-rate (Type 12), compared with half as much for Type 14. There would be two versions: Type 17 for ASW and Type 42 as a coastal gunboat. The ASW version would need an endurance of 4,500 nm at 12 kts, the coastal gunboat 2,000 nm at 12 kts.

In theory, the fuel weight saved would go into power, so that the Type 42 would make 22 kts, the Type 17 only 19. In fact, 22 kts required 8,200 SHP on a single shaft (7,800 on two shafts) compared with 4,700 and 4,500 for 19 kts. As in World War II warships, providing the right combination of power and fuel economy in a package suited to mass production was a problem. Splitting the program into two different power plants would further complicate the problem.

Finding a relatively powerful yet compact and economical (for range) power plant was not simple. A/S frigates had to be able at the least to cross the Atlantic without refueling. That defined an irreducible minimum size and cost. Even with her 4,500 nm margin, a Type 14 compelled to use high speed to deal with heavy and prolonged attacks on a convoy would have to be refueled at sea. The proposed Type 17 would have to refuel twice. She would be a worse sea-keeper, though she could keep the sea at a higher speed than a slow convoy. Her worse seakeeping would probably reduce her sonar performance compared with Type 14's. Director of Navigation and Direction showed that the criterion of a 17-day Atlantic crossing at 12 kts equated to an endurance of 5,000 nm, which should increase to 5,500 to allow for periods at higher speed and a safe reserve.

The British were introducing the lightweight Deltic diesel, but it probably would not be available in quantity early in a war (Deltics did power many British minesweepers later). Production of the heavier Admiralty Standard Range diesels, which powered the first-generation postwar British frigates, could not be expanded quickly enough (this was not a problem in the United States, where diesels were considered the obvious choice of mass-production escorts). As in World War II, existing low-technology steam plants did not offer the desired endurance for either type. DNC ended up with using 5,000 SHP geared turbines: one turbine, one boiler, and one shaft for Type 17; two turbines, two boilers, and two shafts for Type 42. The boilers would be downgraded versions of the Y.101 type in Type 14 frigates. Type 17's endurance would have to be reduced to 3,000 nm at 12 kts. Type 42's endurance would be only 1,300 nm (rather than 2,000) at 12 kts. These reductions were not really a surprise: Type 14 devoted about 40 percent of its deep displacement to machinery plus fuel, but the Type 17 could spare only about 30 percent. No doubt, specially designed machinery could be designed for Type 42, but the whole point of the exercise was to avoid that sort of bottleneck. It was understood that giving Type 17 a speed better than 19 kts would "greatly improve" her ASW capability. A Type 14 plant (Y.101) derated from 15,000 to 12,500 SHP emerged as a viable alternative.

Draft Staff Requirements were issued on 9 February 1951 in hopes that ships could be laid down in 1953–54, but they were "to be built under an emergency programme only." The requirement for an endurance of 4,500 nm at 12 kts was revived. As in the earlier studies, the ASW battery would be a single Limbo and four homing torpedoes. As in a Type 14, gun armament would be limited to close-in antiaircraft, in this case one twin and one single Bofors. As sketched, the ship was not very much smaller than a Type 14: about 1,050 tons and 275 rather than 300 ft long. If the ship were any larger, she would be little less expensive than a Type 14. The British Ship Design Policy Committee concluded that the whole idea of an inexpensive third-rate ASW frigate had been illusory, largely because machinery was such a problem. The committee proposed dropping not only the third-rate ASW frigate but also the second-rate (Type 14) as soon as the current program had run. A simple new escort should be designed. That never happened, and in mid-November 1951 the Naval Staff abandoned Type 17 altogether and decided to build more Type 14s instead.

The idea of a standard NATO ASW frigate survived, as did the requirement for 4,500 nm endurance at 12 kts. In January 1953 SACLANT announced that he needed three times as many escorts as he would have at the outbreak of war. He planned for 8 through escorts per Atlantic convoy, backed by a total of 10 support groups (each comprising 1 light carrier and 3 escorts). Convoys in focal areas would have 2 more escorts. In the Channel, the estimated requirement was 3 escorts per convoy plus 15 support groups (3 escorts each, the supporting aircraft coming from land bases).

NATO now had a Standardization Committee intended to develop common designs for various weapons. In the spring of 1953 the committee announced a NATO frigate project, proposals for which would be submitted in midyear.[5] France, Italy, the Netherlands, and the UK were invited to submit proposals. Speed had to be at least 25 kts, and the ship had to be armed with an ahead-throwing weapon and homing torpedoes. Merchant ship speed, hence convoy speed, was rising: the new U.S. Mariners made 20 kts, and even war-built Victory ships made 14–16 kts. If the frigate spent an hour attacking a submarine, it needed enough speed to chase down the convoy. That might require twice the convoy's speed. New escorts had to be faster than submarines, and they might have to outrun 23–24 kt acoustic torpedoes.

A third-rate escort would have the same endurance but a speed of 23 kts and would be essentially a better-armed version of the wartime U.S. steam destroyer escort.

The British offered a "Utility" frigate, apparently Type 14. It is not clear whether the Dutch or the Italians offered designs, although the Italians were then building their four *Canopo*-class frigates (which lacked the desired endurance). The French offered the austere E54 escort then in the design stage. Like Type 14, it was an attempt to fit much of the battery of the first-rate frigate (E 52) into a much less expensive hull, but with similar seakeeping. Like the E52, it would have 57 mm guns (but fewer of them), ASW rockets, and ASW torpedoes. The ASW rocket launcher (the French equivalent to Limbo) was wanted for surprise attacks on submarines, the homing torpedoes for deliberate attacks. In July 1953 the project was recast: the ship should displace no more than 1,000 tons, make at least 25 kts, and be armed with ASW weapons but only self-defense antiaircraft weapons (40 mm guns, as in Type 14, rather than 57 mm).

The French were caught between the projected needs of Atlantic ASW and the current needs of their colonial empire. They developed a parallel project for an "*aviso*." The vessel did not have to be quite as fast, but it needed sufficient endurance to steam from France to, for example, Dakar in West Africa. In peacetime it would carry a few troops, and it would have a gun suitable for shore bombardment. In wartime it would be used for ASW. It seemed unlikely that the long endurance could be reconciled with high speed. The French did not have the sort of high-temperature high-pressure steam plant that made it possible to combine long endurance with relatively high speed in a Type 12 or a U.S. *Dealey*. At least initially the *aviso-escorteur* was expected to meet NATO third-class frigate requirements: 23 kts, 4,500 nm at 12 kts, 1 rocket launcher (48 rockets), and 4 ASW torpedoes. There would be 2 torpedo tubes. The French design showed 60 rockets rather than 48, 2 twin 57 mm antiaircraft guns, and two 20 mm guns. For the Union française (French empire) role, one of the twin 57 mm mountings could be replaced by one 100 mm gun. Sonars in the ASW role would be one search sonar (DUBV-1) and one attack sonar (DUBA-1).

The French parliament voted funds for three E54s, though the design was criticized because it did not take into account the needs of the Union française. Members of the parliament doubted that it would be worthwhile on such a small displacement. The navy minister offered to reexamine the E54 design to give it a better ASW armament. He hoped that the United States would pay for the ships through its Off-Shore Purchases (OSP) program. By this time, however, the United States was no longer willing to finance such construction. The French saw no point in spending their very limited money on an emergency ship—much as the British had decided two years earlier. Instead the French built the *Commandante Rivière*–class *avisos-escorteurs*, in effect rejecting the NATO requirements for a very large escort fleet.

The Royal Canadian Navy also became interested in a second-class escort, planning ten *Vancouver*-class frigates.[6] Not only were the Canadians impressed by SACLANT's call for more escorts, but they were also concerned that their new *St. Laurent* class was too expensive and took too long to build. New ships were justified partly as replacements for the existing *Prestonian*-class frigates converted from

The *Dealey* class might be considered the U.S. equivalent of the British *Whitby*, designed slightly later. The Weapon Alfa abaft the forward twin 3 in./50 was their main battery, corresponding to the British Limbo, but mounted forward. This is USS *Lester* before her upgrades with DASH and triple lightweight torpedo tubes. The design began as a more modern version of the World War II 173-ft PC. It emerged substantially faster and better armed than World War II destroyer escorts. Characteristics were approved on 8 February 1951. World War II experience of warship production problems caused by limited gear-cutting capacity explains the choice of a single screw, but they were still capable of 27 kts. The design was attractive enough for Norway and Portugal to build it in modified form. NHHC

The four *Claude Jones*-class destroyer escorts were a U.S. approach to designing second-rate frigates suitable for wartime mass production. USS *John R. Perry* is shown off the Charleston Naval Shipyard, 23 May 1959. She has a single stern depth charge track, but her main ASW weapons are her triple torpedo tubes and her two Hedgehogs forward. The design was complicated by the requirement for a speed of 21.5 kts coupled with an extraordinary range (initially 12,000 nm at 12 kts). It was solved by coupling four diesels to a single shaft, enabling the ship to run on fewer diesels if desired. In the end, the ship cost only 10 percent less than a *Dealey*, and it was much slower and less capable. Its main virtue was that it could be built in large numbers in an emergency (thanks to its diesel power plant), but by the time the four ships were completed, that need was no longer envisaged. USNI

war-built Rivers. Flag Officer Atlantic Coast strongly supported the project at the 6 May 1953 RCN Senior Officers' Conference. Plans called for 10 Canadian ships and 3 more for NATO allies, plus another 17 propulsion systems, the long lead items in any further program.

The new ships would have been 315 ft long, displacing 1,700 tons, compared with 2,800 (fully loaded) for a *St. Laurent*. A single-shaft steam plant would have driven the ships at 24 kts (compared with 28 for a St. Laurent). Endurance would have been the NATO standard of 4,500 nm at 12 kts. ASW armament would have been double Limbo and four U.S.-type Mk 32 homing torpedoes; gun armament would have been one twin 3 in./50 mounting (apparently originally a twin 4 in.) forward and four 40 mm. Complement would have been 167, versus 249 for a *St. Laurent*. Sonars would have been a U.S. SQS-10 scanning type (QHB), a British Type 170B attack sonar to control Limbo, and a Type 162 to search the bottom. The design included space for a future long-range sonar. Unlike the *St. Laurent*, this ship would have an open bridge.

The Canadian Chiefs of Staff approved this plan, except that the number of propulsion systems was reduced. Permission to begin the final design was given in February 1954. At this time estimated cost was $12.5 million compared with $21 million for a *St. Laurent*. About six months later the Naval Staff began to doubt the value of the ship, and it was dropped from the 1955–56 program. In May 1955 the Sea-Air Committee argued that *Vancouver* lacked the endurance, ASW capability, and air capability to face modern Soviet submarines and aircraft, and it lacked the margin for upgrades as well. Follow-on *St. Laurent*–class frigates were built instead of the

Vancouvers, although the formal decision to do so was not taken until 1957.

Only the United States seems to have built anything like the NATO mobilization frigate, in the form of the four *Claude Jones*–class ships. They stemmed from the first report of the new (1954) Standing Committee on the Long-Range Shipbuilding and Conversion Plan, chaired by RADM W. G. Schindler. Its May 1954 report argued that construction of general-purpose destroyers should be abandoned in favor of three types: fast task force escorts (the DL 6 class, which became the *Farraguts*), fast ASW ships, and mass-production escorts. The preferred fast escort was a converted 1,620–1,630-ton destroyer (DDC). None of this worked as planned. The prototype DDC conversion was dropped. The low-cost escort cost only 10 percent less than the more capable *Dealey* then being built. A diesel (rather than steam) power plant was adopted in hopes that the ships could be mass-produced in an emergency. That limited speed to only about 21 kts, below the standard NATO was then using for third-rate escorts. By the time the four ships were completed, moreover, the U.S. Navy no longer wanted masses of escorts for use in wartime.

U.S. LONG-RANGE PLANNING

After the Korean War the U.S. Navy shifted toward emphasizing the offensive side of sea power, at the time centered on carrier task forces. Given their high speed, the main threat they faced was air attack. Defending against that entailed a huge investment in fighters and then in antiaircraft missiles, which in turn required expensive fast surface ships. It was less and less possible to replace the mass of war-built destroyers, let alone destroyer escorts.

The new naval orientation was evident in the 1955 report of the new Long Range Objectives Group (LRO), which was intended to rethink the overall shape of the U.S. Navy in view of the shift toward missiles and nuclear weapons. The current fleet was inescapably shaped by earlier decisions—particularly the World War II building program—that expressed earlier priorities and strategy. By looking a decade ahead, the LRO could escape some of that constraint. Its first report emphasized strategic nuclear striking power in the form of task forces built around converted battleships (BBGs) armed with supersonic nuclear cruise missiles (Regulus II). The new supercarriers, which otherwise would provide the fleet with nuclear punch, increasingly seemed to be too expensive, largely because they had to be able to launch large nuclear bombers. It seemed that the cost of carriers could be cut drastically by replacing the bombers with Regulus missiles, and their fighters by vertical takeoff airplanes, which seemed to be on the verge of practicality.

Soviet submarines were a threat to this kind of force projection, not least because they could certainly intercept the slower replenishment groups on which the fast task forces would depend. By 1955 the Navy had been experimenting successfully for several years with SSK operations in the Arctic. The LRO therefore expected the future U.S. Navy to deploy a 50-submarine barrier near the Soviet coast, an idea that became a constant in U.S. ASW planning.

A second barrier would be deployed farther out to sea, out of reach of most Soviet surface forces. In the 1955 report, this second barrier in the Atlantic would be built around 12 helicopter/VSTOL carriers (CVHGs) equipped with long-range active and passive sonar. They would operate VTOL (vertical takeoff and landing) fighters and ASW helicopters, and would have Terrier missiles for self-defense. The fighters would deal with low-flying Soviet bombers. A Pacific barrier would employ surface ships equipped with long-range sonars. Hopefully the barriers would make it unnecessary to replace the aging escorts on anything like a one-to-one basis: the study called for only 75 new ocean escorts.

The 1955 plan was unaffordable, but CNO ADM Arleigh Burke valued the look into the future. In 1959 the LRO produced a new plan intended to be more realistic. By the time it was being written, hopes for VTOL airplanes had faded and Regulus was being cut back in favor of the Polaris ballistic missile. It was also clear that attack carriers would be wanted for nonstrategic strikes, so they had to stay. The barriers were still valued as a way of diluting the submarine threat in the open ocean, but by this time the threat of Soviet strategic submarines was being taken seriously. The backup barrier therefore moved toward the U.S. coast, where it could deal with them. In describing the planned force Admiral Burke emphasized that the ASW force would help protect the United States against missile-launching submarines: "While our main defense against such attack must be our own invulnerable deterrent forces, and while a *total* defense against *any* form of nuclear attack is impossible, we cannot, and do not intend, to leave any enemy a free shot at us." The antisubmarine force could give excellent warning of submarine attack and could destroy many of the attacking submarines. By this time SOSUS was promising the ability to seek out enemy submarines in the open ocean.

With the Polaris program well underway, the projected U.S. force included 50 missile submarines, 40 of them armed with Polaris ballistic missiles. About a dozen would be armed with small precision missiles that could also be used tactically.

About 400 of the planned 900 combatants would be devoted to ASW, including 75 submarines, 65 of them nuclear. Escorts were needed for the attack carriers and the nine ASW carriers as well as the underway replenishment groups and amphibious groups. In a confidential discussion of the new program distributed within the Navy, Burke also mentioned convoys, but by that time they were no longer significant. A summary of the plan showed 72 destroyers, 182 ocean escorts, and 60 ocean pickets (YAGRs and DERs). Nearly half the ASW ships would carry antiair missiles, and all would have ASW helicopters or missiles. Most would be built around advanced types of sonar, and some might be of radically new design. The ships and submarines would be backed by about 1,300 aircraft, out of a total of 7,000: 500 long-range, 400 short-range (for carriers), and 400 helicopters. By comparison, 1,100 aircraft would be fighters, attack planes, and Marine assault transports. A total of 400 reconnaissance aircraft included 70 long-range seaplanes for mining and reconnaissance, presumably the Seamaster program then envisaged. The Navy would operate about its current long-range patrol plane force, amounting to about 30 long-range patrol aircraft squadrons (including those in the Naval Reserve).

The plan required 254 escorts and destroyers, few of which were being built. The Navy could not afford simply to replace the war-built destroyers (it was accepted that the destroyer escorts could not be replaced). The pace of new construction hardly sufficed. An attempt to design a new mass-production destroyer failed altogether. The choice was to defer the problem by modernizing existing ships under the Fleet Rehabilitation and Modernization (FRAM) program conceived in 1958, which applied to all *Sumner*- and *Gearing*-class destroyers and to three *Fletcher*-class DDEs. There were two versions, FRAM I and the less elaborate FRAM II. FRAM I applied to *Gearing*-class destroyers, which had enough space between their stacks to take an ASROC launcher (they received the associated SQS-23 sonar). They were also given the drone antisubmarine helicopter (DASH) and triple lightweight torpedo tubes. *Sumners*, which lacked sufficient space for ASROC, were given DASH and their hull sonars upgraded with variable-depth sonar (VDS) aft. Other existing destroyers and escorts were given DASH and, in some cases, upgraded sonars. In theory FRAM offered ships another five to eight years of useful life, perhaps up to the late 1960s. There were also FRAM upgrades to ASW carriers and diesel submarines. FRAM was considered so urgent that it was given a priority like that of the Polaris program.

SONAR AND WEAPONS

Through the 1950s, the most important U.S. surface ship sonar was SQS-4 and its upgrades, operating at about half the frequency of its predecessor QHB. The next step down brought a "reliable range" of 10,000 yds for SQS-23. The range linked it with the ASROC ASW weapon and also with DASH. The next step down in frequency and up in range was the SQS-26 of the 1970s.

A ship detecting a submarine beyond the range of ahead-thrown weapons needed some means of attacking. Initially the Mk 35 homing torpedo seems

The standoff weapon: the Rocket-Assisted Torpedo (RAT) on a twin 5 in. gun house. The torpedo payload was a modified Mk 43. ASROC was a much-enlarged RAT (RAT C) carrying a Mk 44 (later Mk 46) torpedo or a nuclear depth bomb. USNI

to have been considered a solution, included in the armament of DDEs. The British similarly planned to install torpedo tubes on board their Type 12 and Type 14 escorts. None of the long-range homing torpedoes seemed satisfactory, not least because they would take so long to reach the submarine the ship was tracking.

In 1950 the Hartwell group urged development of rocket-propelled standoff weapons to give surface ships standoff ranges consistent with those of new-generation sonars. The Naval Ordnance Test Station (NOTS) at Pasadena developed standoff weapons under the generic designation RAT (rocket-assisted torpedo).

After successful tests in 1952–53 using Mk 24 torpedoes, in 1953 NOTS proposed a RAT B carrying a modified Mk 43 torpedo (EX-5, a modified Mk 43-0). The launcher was a pair of arms that could be fitted to the sides of the standard twin 5 in./38 gun mount, standard in *Sumner*- and *Gearing*-class destroyers. Alternatively, a twin launcher could replace a twin 3 in./50 antiaircraft gun. RAT B's range of 1,500–5,000 yds was consistent with that of the new SQS-4 sonar. It was first tested on board the destroyer *de Haven* in July 1954, using a deck launcher for ten firings. In 1956 the Atlantic Fleet Destroyer Force pressed for mass installation of RAT (Mk 110) with ten reloads, to replace all Hedgehogs. RAT was also specified for the FY 57 missile destroyer (*Charles F. Adams* class). The torpedo's homing would counter submarine evasion while it was in flight. However, it turned out that the error margin inherent in RAT exceeded the homing range of the torpedo.

The alternative was a rocket-assisted nuclear depth bomb, proposed in March 1953 by the Naval Ordnance Laboratory (White Oak). Initially NOL envisaged a depth bomb projected using a Talos missile booster and a jet-assisted takeoff (JATO) rocket, but it proved possible to shrink that.

By 1956 NOTS was offering the longer-range RAT C, which could lift the new Mk 44 homing torpedo out to 10,000 yds—consistent with the "reliable range" attributed to the projected SQS-23 sonar. RAT C could also lift the nuclear depth bomb being developed by NOL. Initially RAT C seemed too inaccurate to deliver a torpedo properly, so its fin

A standard ASROC box launcher on board the destroyer *H. J. Thomas* on completion of her FRAM I refit at Mare Island, 6 August 1965. Reloads were stowed in the box alongside the after funnel and the launcher control cabin. The new funnel cap was intended to protect the ship's boilers from nuclear blast. NHHC

An ASROC firing by USS *Charles F. Adams*, 18 November 1960. Although only a single rocket is being fired, the two-rocket section of the launcher elevates to fire it. NHHC

USS *Agerholm* fires the one and only nuclear ASROC test shot, 11 May 1962. NHHC

area was increased, reducing the Mk 112 launcher to eight rockets. In this form RAT C became ASROC, the A/S Rocket, a standard U.S. Navy standoff weapon.[7]

In 1956, the Atlantic Destroyer Force feared that RAT would be too massive for installation on board existing *Gearing*-class destroyers. These ships were scheduled to receive the long-range SQS-4 sonar, but not anything beyond the current dual Hedgehog and new lightweight torpedoes. The force proposed adapting a drone helicopter. Experiments beginning in February 1957 using a small, manned Bell 47 (HUL) showed that it could take off and land on the large destroyer *Mitscher*. In August 1957 CNO approved development of DASH. In April 1958 the U.S. Navy contracted with the Gyrodyne Corporation for a torpedo drone version of its lightweight Rotorcycle (YRON-1), DSN-1 (later redesignated QH-50). The specification required that the drone control system be capable of placing it within 200 yds of a plotted submarine position at a range of 10,000 yds. Specified payload was a single Mk 43 torpedo, which the initial production drone could barely lift. However, the turbine-powered QH-50C could carry two Mk 44 torpedoes or one Mk 46 or one nuclear depth bomb. The follow-on QH-50D could lift two Mk 46 torpedoes. QH-50C deliveries began in November 1962. By that time many ships had been modified with DASH facilities. For a time in 1960 DASH seemed preferable to ASROC, at least theoretically. Classification of sonar targets was difficult, so that the expendable ASROC might often be wasted. A recoverable weapon—the torpedo under

An early DASH carries its Mk 44 torpedo payload. DASH was the world's first armed unmanned aerial vehicle (UAV). It was powered by a turboshaft engine (its air intake and exhaust are visible), with counter-rotating rotors for stability. The object at the rear of the torpedo is its parachute. Later versions of DASH with a more powerful engine could lift two torpedoes. Effective range was limited by a ship's radar horizon, as the ship had to track the helicopter in order to control it. Concurrent development of the helicopter and its control system proved unfortunate. DASH was conceived in 1957 as a means of offering greater freedom of action under hazardous conditions; CNO issued Development Objective AS-04504-2 on 21 August 1957, and in June 1958 decided that the first operational helicopters would be needed in FY 62. Fleet deliveries of the operational QH-50C were to begin after July 1963. The program was affected by a summer 1960 decision to ban aviation gasoline (which was to power the initial DSN-2 version), so this version was canceled in favor of the turbine QH-50C. The first drone flight was made on 12 August 1960, and the first flight from a ship (*Hazelwood*) on 7 December 1960. Of a total of 750 drones made, through April 1966 losses amounted to 231 QH-50Cs and 131 QH-50Ds. Atlantic Fleet data showed that DASH system effectiveness consistently exceeded that of ASROC, but the rate of loss was considered unacceptable. USNI

a DASH—seemed preferable. Too, the helicopter promised more accurate delivery.

DASH was the first armed unmanned air vehicle, and as such it had unanticipated problems. The operator in CIC had only radar tracking as feedback; some DASH drones flew back to their ships upside down. DASH also proved unreliable in service. By 1968 one DASH was lost, on average, for every 80 flight hours. More than half of the 746 drones built were lost at sea. The Navy's request for continued production was deleted from the FY 68 budget. Yet the Japanese made DASH work, and a spotter version (Snoopy DASH) proved successful in Vietnam. Gyrodyne argued in 1967 that the 150-hour mean time between failures (154 flights) was lower than the expected failure rate that had justified production in the first place. This expected failure rate had made DASH competitive economically with ASROC—in combat but not in training, when very few ASROCs were fired.

With DASH in trouble, in 1966 a program for a manned light helicopter was created. It would equip the large fleet of new escorts (*Garcia*, *Brooke*, and *Knox* classes) then planned or under construction; as many as 100 ships were envisaged. This new Light Airborne ASW Vehicle (LAAV) was particularly needed to

DASH takes off from a FRAM destroyer carrying its Mk 44 torpedo, controlled by the operator just abaft its hangar. Unfortunately, there was no feedback from the drone as it returned, and it sometimes approached a ship upside down. The standard DASH installation consisted of a hangar and the small helipad shown. Initially the requirement was 25 hours between loss, but the contract requirement (which was not met) was greater. USNI

A LAMPS I helicopter on board the destroyer escort *Harold E. Holt* for trials, 1974. The hangar and pad designed to support DASH have been considerably enlarged. The box launcher abaft the helicopter pad is for the Basic Point Defense Missile System, whose radar director is atop the hangar. NHHC

A LAMPS I helicopter refuels from a ship. Note the MAD bird on the pylon. When operating, it would be towed well behind the helicopter to limit the distorting effect of the helicopter's metal mass. This helicopter carries auxiliary fuel tanks instead of lightweight torpedoes, and its sonobuoy ports are not visible. LAMPS I was intended to operate within 50 nm of a ship (for about an hour). Its acquisition sensors, the sonobuoys, read out their data directly to the ship, which commanded the helicopter into position to attack. The requirement for shipboard processing amounted to a limitation to line-of-sight operation—that is, first convergence zone. The helicopter carried a mix of 15 preprogrammed SSQ-41 (LOFAR) and SSQ-47 (CASS) buoys. Its AKT-22 link to the ship carried sonobuoy data (unprocessed) from up to four buoys, whose data the ship could process using two two-channel SQS-54 LAVA (Low Frequency Acoustic Vernier Analyzer) processors. Also visible is the shallow radome of the LN-66 surface search radar, which justified the helicopter's designation as a multipurpose, not only ASW, platform. The helicopter also carried one or two Mk 46 torpedoes and eight Mk 25 marine markers. For example, it could drop a maker on a MAD contact and use a series of such markers to indicate submarine course and speed. NARA

localize and attack submarines detected at convergence zone by the ships' SQS-26 sonars.

It was soon apparent that an ASW role alone would not justify the new program. After the Israeli destroyer *Eilat* was sunk by Egyptian Styx missiles fired from beyond the horizon, LAAV was rethought as the Light Airborne Multipurpose System (LAMPS), capable of detecting ships beyond the horizon using a lightweight radar. The radar also enabled the helicopter to spot for gunfire. For its primary ASW role, it carried sonobuoys and a lightweight MAD (ASQ-81). The outputs of the buoys were processed on board the parent ship. The project was urgent enough that an existing helicopter, Kaman's UH-2 Seasprite, was adapted. Even so, progress was slow. In mid-1970 the UH-2 was being used to prove LAMPS concepts, but no LAMPS I airframe had been chosen. Four HH-2s were modified, two for ASW and two for missile defense, for tests on board the missile destroyer *Truxtun* scheduled for that September. In March 1970 the *Belknap* class and *Truxtun* had been chosen for initial tests because they could accommodate the 12,000 lb HH-2D helicopter without major modification.

By this time the program was considered urgent. Early in 1971 the Kaman H-2D was formally chosen as the initial LAMPS helicopter; that March CNO ADM Elmo R. Zumwalt Jr. announced that 115 UH-2Cs and UH-2Ds would be converted into SH-2 LAMPS. The first 20 would become SH-2Ds, the rest SH-2Fs with improved rotors, avionics, and relocated tail wheels. The 95 suitable destroyer escorts would all require enlarged hangars. The first SH-2D flew on 16 March 1971. It had a Canadian Marconi LN-66 radar (the only suitable one available); the ASQ-81 MAD; 15 LOFAR or active sonobuoys, 4 of which could be processed by the parent ship via a data link; an ALR-54 threat receiver; marine markers; a tactical navigation system; and provision to carry lightweight torpedoes. For its final attack, it carried a paper range recorder sufficient for omnidirectional active buoys or, presumably, one directional buoy. There was no onboard computer and therefore no system enabling the helicopter to operate independently of the parent ship. The first unit was assigned to the missile frigate (cruiser) *Belknap* on 7 December 1971.

The LAMPS helicopters could detect and track ship targets, extending a ship's effective antiship missile range against a destroyer from 20 nm to 50–60 nm. The electronic support measures (ESM) horizon was extended from 30 nm to 60 nm, using the ESM array on board the helicopter. Mk 46 torpedoes could be delivered at convergence zone range (70,000 yds) rather than at ASROC range (10,000 yds). Typically the helicopter dropped one omnidirectional buoy to confirm the presence of a target, then flew a circle around that datum, just beyond the indicated range, to find the target.

The initial LAMPS was considered the first step in a larger program. A proposed LAMPS II (to operate out to convergence zone) was canceled in favor of a much more capable LAMPS III. Work was delayed when Congress cut the LAMPS program in 1971. In September 1973 OpNav changed the program, envisaging 100 nm missions (second or even third convergence zones), which in turn required a three-man crew and an onboard sonobuoy processor. Presumably this change was related to plans to deploy towed arrays, which were demonstrating second convergence zone capability. Meanwhile the Army was developing a new utility helicopter (UTTAS) with much the performance the Navy wanted.

In October 1973 CNO formally called for a LAMPS III capable of remaining on station for two hours at a range of 100 nm (i.e., at the second or third convergence zone) while carrying two Mk 46 torpedoes. The helicopter would need better long-range navigation, more sonobuoys (up to 25, including 4 DIFAR), and onboard acoustic processing. However, like LAMPS I, the longer-range helicopter was still intended primarily for direct operation from a ship's CIC, its sensors controlled by an acoustic-sensor operator (ASO) in the sonar room; a remote-radar operator (REMRO) in CIC (who received the helicopter's radar output via data link, and who could directly control radar range and operating mode); an electronic warfare operator (EWO) in CIC; and the air-tactical control officer (ATACO). The helicopter carried a pilot; an airborne tactical officer (ATO), who would take over from the ATACO as directed (e.g., when the helicopter passed

A LAMPS III (SH-60B) helicopter over the frigate *McInerney*, which has another SH-60B on her stern helipad, March 1981. The helicopter shows the radome of its surface search radar and smaller radomes covering ESM antennas. The stub below the national insignia could carry a lightweight torpedo. Sonobuoy chutes were on the other side. The helicopter carried 25 buoys, though typically it would use 12 on a mission. It was the redetection/prosecution sensor associated with the SQR-19 towed array. NHHC

out of radio range); and a sensor operator (SO). The line-of-sight data link would be duplexed for voice and digital data, with low sidelobes for electronic counter-countermeasures (ECCM). As with LAMPS I, the final attack sensor was the ASQ-81 towed MAD.

In September 1974 the Army's H-60 UTTAS was formally chosen for both LAMPS III (SH-60B) and as the replacement carrier ASW helicopter.

When the British considered which helicopter to buy to support towed-array frigates (they chose Merlin), they evaluated LAMPS III, which was then entering service. They considered the system adequate for the first convergence zone but thought that its longer-range endurance would be inadequate for a medium-range active/passive mission and for relocation and attack, presumably at the second convergence zone. They liked its high speed (140 kts) but considered it severely limited by its cabin height (4.5 ft) and its volume (less than 30 percent of that of a Sea King), which would make it impossible to fit a dipping sonar and a sonobuoy system. U.S. frigate-towed arrays gave the parent ship the flexibility to turn and to carry out the speed changes required to operate the SH-60 without losing contact for any lengthy period. As with LAMPS I, data was linked to the parent ship, the helicopter having only limited autonomous capability. The tailwheel configuration and the inability to swivel or to deal with significant out-of-wind conditions imposed deck limits. The British argued that the helicopter was limited by undesirable features related to its Army parentage. It is not clear how much of this was intended to push the choice to the British-produced EH-101 Merlin.

As foreign navies deployed lower-frequency sonars, they too wanted standoff weapons. The British adapted the Wasp light helicopter as a kind of manned DASH that they called MATCH (Medium Range Torpedo-Carrying Helicopter). MATCH originated with the conclusion that no frigate was likely to be fast enough to get within Limbo range of a fast submarine. Instead of keying escort speed to submarine speed, it should be set by the speed of the ships to be escorted. In 1956 First Sea Lord ADM Sir Louis Mountbatten strongly favored a lightweight helicopter carrying lightweight

A Wasp helicopter (of the MATCH system) on board a British Type 81 frigate, the first class with an integral helipad. The helicopter was a manned equivalent to the U.S. DASH drone. Requirements were developed in 1959. The helicopter was initially to carry two Mk 43 Mod 3 torpedoes, later Mk 44; weights allowed for 200 lbs for a detection device, which did not materialize (the file shows interest in both an acoustic device and MAD). The Admiralty memorandum justifying the program pointed out that asdic range already exceeded shipboard weapon range; the problem would worsen with the introduction of Type 177. Air-flight weapons (ASROC, the French Malafon, and Ikara) all suffered from the drawback that they were one-shot weapons with significant reload delay. After tests of a Fairey Ultra-Light helicopter on board the frigate *Grenville*, a specification for an ASW version was issued in August 1958. By this time Saunders-Roe was offering the considerably more capable P.531; the specification was reissued in October 1958 to call for the ability to carry two lightweight torpedoes rather than one. Three development helicopters were delivered in October–November 1959, and were soon being evaluated on the frigate *Undaunted*. Westland absorbed the Saunders-Roe helicopter division, the developed version of the P.531 being produced as the Westland Wasp. It could carry two Mk 44 or Mk 46 lightweight torpedoes or one British WE 177A nuclear depth bomb. The first production order was placed in September 1961. A helicopter flying at 300 ft could be directed at a range of up to 9 nm. By April 1960 the MATCH program was being conducted in parallel with British procurement of Mk 44 lightweight torpedoes. The *Leander*-class frigates were designed to accommodate MATCH. A special display console enabled a controller to see both the location of the helicopter and the position of the submarine as indicated by sonar. USNI

homing torpedoes, which the ship would vector onto the target based on sonar and radar data merged in her Action Information Center (AIC) using a special combination display. The Royal Navy accepted that the helicopter would be unable to fly in some weather, possibly because at this time its attention was shifting to the Third World, where weather conditions were milder. A Fairey Ultra-Light helicopter was successfully tested from HMS *Grenville*. The Royal Navy bought the somewhat larger Saunders-Roe Wasp, which could lift two torpedoes rather than one. Helicopter direction was first tested in September–December 1959. The manned helicopter offered flexibility. Its crew could spot for gunfire, and the helicopter could launch light missiles (the French SS 12) against surface targets.

The Royal Navy still wanted an all-weather standoff weapon. It had rejected ASROC in the late 1950s. In 1959, without any input from Britain, the Australians began developing the rocket Ikara missile, designed to carry a homing torpedo 20,000 yds, which was about the expected limit of sonar range. Ikara was command-guided, so one ship could launch it on the basis of another ship's sonar data. Minimum range was 3,000 yds. Typical reaction time against a target at 10,000 yds was 2 minutes. In 1962 the Royal Navy decided to adopt Ikara under the codename Blue Duck.[8] The project study was formally approved by the Navy Board in 1963. An operational requirement (NSR 7668) was formally approved in May 1963 and was endorsed by the Operational Requirements Committee on 5 November 1964. At this point the British planned to use a new digital fire control system embodied in their ADAWS command and control system, which had a ship-to-ship data link. Ikara would arm the new carrier, its Type 82 escorts (six of which were planned), and eight Ikara destroyers. In 1966 the Royal Navy decided not to arm Ikara with nuclear depth bombs. The Ikara missile was modified to deliver the Mk 46 torpedo the Royal Navy adopted in the 1970s, but a major redesign would have been needed to deliver the heavier Stingray of the 1980s. Ikara was retired in favor of helicopters, which were more flexible.

By the time Ikara was in development, the Royal Navy had undergone the 1966 and 1967 Defence Reviews, which emphasized the need to economize. The Royal Navy would operate a high-low mix of ships, consisting of relatively lightly armed ships operating

An Ikara missile on its launcher, displayed at the Royal Australian Navy Museum at Garden Island, Sydney. The dark object under the airframe is a torpedo. The Australian requirement was stated in the late 1950s, prototype development beginning in 1959. U.S.-British-Australian trials were conducted by HMAS *Stuart* in 1963–64. Unlike ASROC, Ikara was command-guided, with a range of 20,000 yds (i.e., twice ASROC). It carried a transponder, and it dropped its torpedo on command. The Royal Navy argued that, given its command guidance, Ikara could be used by a group of ships most of which had effective sonars but lacked the missile. They would transmit target data to the Ikara ship or ships via a data link, enabling it to vector the missile against a target that the firing ship might be unable to detect. In 1960 the Royal Navy evaluated Ikara alongside other standoff ASW weapons: MATCH, DASH, ASROC, and the French Malafon. The decision was eventually made to adopt Ikara. In the late 1980s, however, both the Royal Australian Navy and the Royal Navy gave up Ikara, leaving shipboard helicopters as their only standoff ASW weapons. They justified the decision based on maintenance problems of the launcher and loader. DR. RAYMOND CHEUNG

HMS *Dido* was fitted with the Ikara system, its launcher in the protective "zareba" forward (which replaced the twin 4.5 in. gun). Weight compensation included removal of the air search radar, justified on the ground that Ikara ships would operate in mixed groups. Note the triple torpedo tubes and the VDS aft. The ship retained a single Limbo, whose well is visible at the after end of the flight deck. An antiaircraft upgrade included two single Bofors guns (one of which is visible) and Seacat. USNI

singly or in groups, backed by more capable ships. In November 1966 the Board of Admiralty decided on the shape of the future fleet. It cut the Type 82 to a single ship and decided that priority should go to anti-air warfare (AAW) ships (Type 42 destroyers). Frigates and destroyers would be armed with MATCH helicopters directed by duct sonars, but they would be backed by cruisers carrying Sea King helicopters. Ikara would be deployed widely enough to deter enemy submarine operations. In January 1968 the expected 1972 Soviet submarine fleet would include 35–50 new attack submarines with speeds over 30 kts. All Soviet submarines would have torpedo ranges up to 20,000 yds. They would probably have to come up into the duct briefly in order to classify targets and obtain accurate fire control data. That in turn would provide Royal Navy ships with an opportunity to detect and attack them, providing attacks could be prompt enough. Hence Ikara. It could be fired as soon as a target was classified, and it could be readied to fire again in 40 seconds. Given the new emphasis on NATO cooperation, the Royal Navy argued that Ikara should be deployed in sufficient numbers to contribute effectively to deterring Soviet submarine operations in the Atlantic.

In 1967 limits on British ship design capacity made it unlikely that entirely new destroyers could enter service before 1977–78. The navy argued that Ikara had to be fielded much sooner because it was so significant in the newly central NATO context. It proposed that to match the originally planned time scale, Ikara should be installed on board existing *Leander*-class frigates on their first long refit. A simplified system could carry the UK version of the Mk 44 lightweight torpedo and also the Mk 31 then in development (and subsequently canceled), with potential to deliver NASR 7511 (Stingray) when it was introduced in the 1970s (in fact Stingray was both late and too heavy for Ikara to deliver it). The *Leanders* would need a new handling system; early in 1968 they were expected to carry 16 missiles, compared with 24 in *Bristol*. The handling system was simplified by eliminating the capacity for nuclear depth bombs and for the Mk 46 torpedo (subsequently reversed). Ikara was seen as an all-weather urgent attack weapon, capable of firing salvoes and thus able to keep attacking. The aim was a 30 percent or better kill probability in any one attack. A conversion program was attractive because it would defer work on a new Ikara destroyer into the 1980s.

At this point five more *Leanders* were planned (FSA 41–FSA 45). Installation of Ikara would set back the first ship by about nine months and the next two by about six months. Alternatively, the first five long-refit *Leanders* could be modified, extending their refits from 18 to 24 months. Design capacity limited the choice to one option or the other, and the long-refit option was attractive because it could be extended to more ships without the need for additional design work. Too, modifying the last five ships would preclude an earlier shift from *Leanders* to a planned RN/RAN frigate (which did not happen in that form). To provide sufficient space, either the 4.5 in. gun or MATCH would have to go. MATCH was seen as a complementary ASW weapon, and it also provided a reconnaissance/strike capacity. Removal of the gun was accepted reluctantly, but there was no other way to get Ikara to sea in numbers. At this point the Admiralty Board did not envisage more than 8 Ikara *Leanders* out of a total of 68 destroyers and frigates, so the loss of surface firepower was acceptable.

Because the ability to fire Ikara on the basis of information from other ships was so valuable, the Ikara *Leander* needed an automated command/control system with a data link: ADAWS. The ships would be given Type 184M all-around scanning sonars in place of their Type 177s. At this point, in 1968, the British hoped that their system would be exportable. It came close to stated Dutch requirements, and it was thought to be potentially attractive to Argentina and New Zealand. The Operational Requirements Committee was unhappy with the reduced number of missiles. The navy countered by pointing out the quick reaction of the system, which could be fired immediately on classification of a target, closing a datum at 360 kts.

The Operational Requirements Committee had compared Ikara to MATCH in 1964, when it endorsed Ikara initially. MATCH could not be considered a quick-reaction weapon. Even at the highest degree of readiness the helicopter took at least two minutes to take off and then close the target at 90 kts. That assumed that the helicopter had not been configured for any other role, although that might not be the case. The 1967 Supplementary Statement on Defence had emphasized the value of helicopters for surface strike and reconnaissance. The eight Ikara *Leanders* were later considered Batch I of that class.

A Lynx HAS.2 helicopter flies over the modified *Leander*-class frigate *Sirius*. Compared with the Wasp, the Lynx offered greater range and carrying capacity. However, it had no ASW sensors, only a Seaspray nose radar. There was provision for sonobuoys and a data link (for shipboard processing) and MAD, but none was fitted. The helicopter could carry two Mk 44 torpedoes or one Stingray or one WE 177A nuclear depth bomb. The French version of the Lynx carried a dipping sonar. An abortive "uprated Mk 4," which would have entered service in 1984, was initially to have had up to eight sonobuoys (Barra and CAMBS) with an onboard processor. When that was delayed, a dipping sonar was substituted, the U.S. AQS-18 being chosen. The program was dropped in favor of concentrating on the Sea King replacement, which became the Merlin (EH-101). The frigate shows a reel for a variable-depth sonar right aft. The triple torpedo tubes forward, at the after end of her superstructure, were considered a replacement for the Limbo urgent-attack weapon. On her after superstructure are rudimentary antimissile defense weapons: a quadruple Seacat launcher and the Corvus countermeasures launcher. Jammers line the vertical sides of her tower mast. Exocets have replaced her 4.5 in. gun. The helicopter was considered a means of replacing the gun, since it carried short-range air-to-surface missiles. USNI

The British replaced their Wasp helicopters with Lynxes, which could lift the projected NASR 7511 lightweight torpedo (which became Stingray). Like its predecessor Wasp, Lynx had no onboard ASW sensors. Instead the improved performance went into antiship capability in the form of a surface search radar. The later HAS 8 version had a central tactical system and a trainable forward-looking infrared sensor (FLIR). Lynxes were armed with the more capable Sea Skua antiship missile. Lynx was a joint

Anglo-French program. The French navy fitted its Lynxes with DUBA 4 dipping sonars, but reportedly the Royal Navy could not afford them.

DIRECT SUPPORT

In 1954 CDR J. H. Dolan of the U.S. Navy told the British TAS Conference that the Navy had taken the revolutionary step of trying to integrate submarines with its surface ASW forces.[9] Submarines could reliably detect snorkeling submarines at ranges of 35–45 nm, but they lacked weapons to follow up. They also seemed better equipped to classify what they detected (as submarine or nonsubmarine) than surface ships.

The U.S. Navy had already tried many combinations, including an SSK with one or two patrol planes, with three destroyers, with four destroyers and patrol planes, and with HUK groups including carriers. In one case, a Neptune was vectored by the SSK 40 nm at night at 200 ft altitude for a visual, radar-off attack. In another, a 60-mile vector of a Neptune resulted in MAD contact and a successful attack. Alternatively, the SSK could be vectored to attack. On one occasion an ASW aircraft that had expended its weapons made contact and cued the SSK to snorkel for 30 nm and make an undetected attack.

As in the past, communication and identification were difficult, forcing airplanes and submarines to rely on complex and unreliable techniques. Submarines had to transmit DF signals, use vertically directed lights at night, or paint their decks orange in the hope that they would be recognized as friendly. Sometimes submarines used sonobuoys to relay information to aircraft monitoring them. Alternatively, a submarine could rise to periscope depth to use her snorkel whip radio antenna. A submarine could also use her underwater telephone to relay data via a destroyer. In mid-1954 the submarine squadron in Norfolk was experimenting with a UHF buoy, which would permit secure and reliable communication and positive identification. A destroyer operating with an SSK had to stay off the direct line between the SSK and the target so that her own sound would not mask that of the target. For that matter, on hearing the destroyer, the target would probably stop snorkeling and go deep, breaking contact with the SSK. Dolan said that "adherence to doctrine, good planning, and adequate briefing of participants does much to improve a basically inadequate system."

CNO considered that the promise of coordinated operations involving SSKs justified including such operations in A/S exercises. The operating forces were equally enthusiastic. CinCPacFleet wrote that "coordination has produced ASW effectiveness which far exceeds the ASW capability of either the SSK or the aircraft operating independently." The Pacific submarine force commander claimed results of major tactical and strategic significance.

The new ASW tactical development unit, Task Group Alfa, included two Guppy submarines, *Cubera* and *Sea Leopard*, which were usually used as targets. However, in the fall of 1958 CinCLantFleet reported substantial improvement of tactical coordination between surface, air, subsurface, and SOSUS elements. However, surface-submarine cooperation did not last, at least in the U.S. Navy.

ASW AIRCRAFT

By the late 1940s a new generation of aircraft adapted specially for ASW was being developed. The U.S. Navy had two large land-based patrol bombers that might be considered candidates for adaptation: the Lockheed P2V Neptune and the Martin P4M Mercator. The Navy also considered adapting the much larger Air Force B-29 Superfortress (as the P2B-1). Full adaptation entailed installation of the APS-20 snorkel-detection radar, a shorter-range ASW attack radar (APS-31 or -33), a searchlight, a sonobuoy receiver, and provisions to launch a homing torpedo and to drop depth charges. Rockets were wanted to attack a surfaced submarine. The Neptune was chosen.[10] By 1948 the only problem was to accommodate the big APS-20 as well as the planned Mk 35 homing torpedo. Initially (in the P2V-3) there was not enough engine power to handle the weight and drag of the APS-20, so that the Navy had to make do with hunter-killer teams of radar and strike aircraft. The advent of a more powerful version of the R-3350 engine made it possible to produce a single-package airplane, the

The Lockheed P2V Neptune was the standard U.S. land-based patrol bomber of the 1950s and early 1960s; it also became a de facto NATO standard. The bow, dorsal, and tail turrets clearly show this P2V-4's bomber heritage. Not visible is the radome of the big APS-20 snorkel-search radar. The usual combination of attack radar and searchlight are, however, visible: an APS-33 radar in the port tip tank and a searchlight in the starboard tip tank. In later aircraft the tail turret was replaced by a MAD in a large fairing, and the other two turrets were also eliminated, the upper one replaced by a skylight and the nose turret by an observation station. USNI

P2V-4 (P-2D in later nomenclature) version, first flown on 14 November 1949 and delivered between January 1950 and March 1951. This airplane had tip tanks, with the nose of the starboard tank accommodating the searchlight needed for close-in delivery of ASW weapons against a surfaced submarine at night (the port tip tank of the next version contained an APS-31 attack radar in its nose).

Reflecting their bomber origin, most Neptunes were built with a nose turret, a dorsal turret, and a tail turret. When operating near enemy shores they were subject to attack; for example, a VP-6 Neptune flying near Vladivostok was shot down by Soviet fighters on 6 November 1951. After the Korean War, however, the Neptunes following up HF/DF and later SOSUS cues remained farther out to sea and were unlikely to face enemy fighters, making their defensive armament dead weight. The three turrets were eliminated, and a MAD tail eventually replaced the tail turret, a clear observer's nose replaced the nose turret, and a skylight over the ASW space replaced the dorsal turret. Meanwhile the aircraft's ASW sensing package became more elaborate. P2V-4 added six fuselage sonobuoy chutes and a sonobuoy operator. In U.S. aircraft of the follow-on P2V-5 version, the ASW team, up to then split by the wing spar, was brought together in a row facing to starboard, making

The U.S. Navy continued to use big seaplanes alongside its land-based patrol planes, valuing their inherent mobility thanks to seaplane tenders. Here a P5M Marlin is hoisted on board the seaplane tender *Pine Island*. The seaplane employed the new type of high length-to-beam-ratio hulls BuAer sponsored for better sea performance. Sonobuoys were launched from inside the hull, but weapons were carried in the engine nacelles and underwing. The boom atop the T-tail is for MAD (ASQ-8); the bulb at the after end of the hull is for a navigational system (APN-122). USNI

it possible for one officer to supervise all of them and coordinate what they saw. This lengthwise arrangement persisted in later U.S. land-based ASW aircraft.

The Neptune became a de facto NATO standard; the only country other than the United States producing a long-range land-based maritime patrol airplane in the 1950s was the United Kingdom. Production through 1962, including aircraft for the Military Assistance Program, totaled 1,051, not counting another 130 built by Kawasaki in Japan under license.

In addition to land-based maritime patrol aircraft, through the 1950s and into the 1960s the U.S. Navy maintained a large force of seaplane patrol bombers—initially wartime PBM Mariners and then their replacement P5M Marlins. Compared with land-based aircraft, seaplanes were inherently mobile (using their tenders). The Marlin was a direct descendant of the Mariner exploiting BuAer research on longer, narrower hulls for better seakeeping. Like the Neptune, the Marlin was a general-purpose patrol airplane adapted for ASW. It retained only the tail turret of its prototype. A big radome (for APS-44, later APS-80 radar) replaced the projecting nose turret, and the dorsal turret was eliminated. Sonobuoys and weapons could be carried in bays under the two engine nacelles, and later aircraft had MAD (ASQ-8). The seaplanes were all retired in 1966. France, the only foreign user, received ten of them. A 1955–56 project for a replacement (the Convair P6Y, which reached only the mock-up stage) died.

The only other country developing a long-range maritime patrol airplane in the 1940s was the United Kingdom, with its Shackleton.[11] Developing an almost completely fresh design allowed Coastal Command to rethink internal arrangement; it concluded that a long-range maritime patrol airplane should function almost like a ship, with a captain as system supervisor. The specification therefore called for a tactical station manned by the captain, the normal navigator, the tactical navigator, and the sonobuoy operator. They were apart from the two pilots and the bomb-aimer. Other crew stations were for the radar operator, the wireless operator, and the gunner manning the mid-upper turret. Initially the project was unpopular, to the point that in January 1949 an American exchange officer was told that the manufacturer, Avro, could not meet the desired specification and was demanding further concessions in performance "which would make the aircraft unsuitable for Coastal Command's needs." He suspected that the British were fishing for U.S. offers, having been encouraged (correctly) by the outcome of the U.S. 1948 elections to think that Congress would

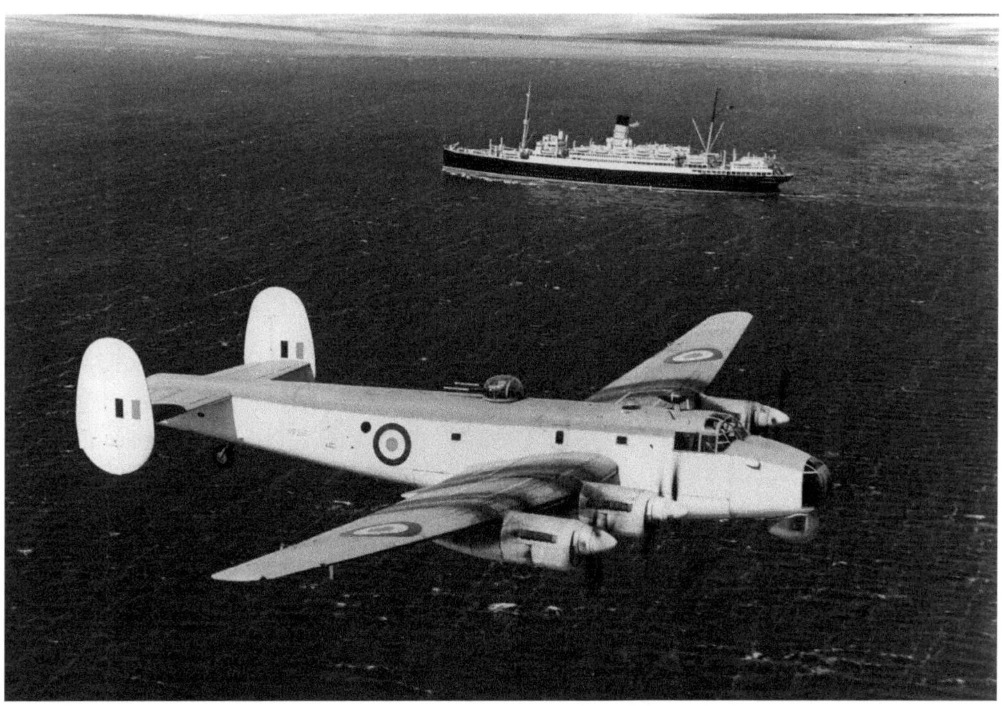

The Shackleton MR 1 was the first airplane designed specifically for maritime reconnaissance, including ASW. It used some parts of the existing Lincoln bomber and the Tudor airliner, but it was an entirely new design. Note the Coastal Command low-visibility livery, similar to that used by U.S. Navy Atlantic patrol bombers during World War II. This one is VP256. TONY BUTTLER

support Western Union rearmament. Without satisfactory new aircraft, the RAF would have to depend on a dwindling supply of adapted war-built Lancaster bombers. U.S. P2V-2s did visit Britain in May 1949 (Operation Paddle), but they were not well received. They had not been adapted for ASW, and their crews were apparently not familiar enough with them to explain either their electronics or their electrical system. The chief of BuAer had to write a long memo to the British explaining matters. Ultimately Coastal Command received 52 Neptunes, 2 of them for evaluation and 4 for AEW experiments (the APS-20 radar had been conceived for AEW).

The same equipment used by long-range shore-based ASW aircraft could be taken on board carriers, but initially it was considered too heavy for a carrier airplane. Once the APS-20 had demonstrated its ability to detect a snorkel at a usable range, AEW Avengers were given the alternative role of ASW hunters. They were to cue TBM-3S "killers," which would use searchlights, short-range radars, and sonobuoys. The next step was a more capable carrier ASW airplane, Grumman's AF Guardian. Its search radar version was AF-2W (the attacker was AF-2S).[12] The Guardians derived from the last wartime Grumman torpedo bomber, the TB3F, redesigned in 1946 to eliminate the originally planned jet engine supplementing the piston engine.

A hunter-killer pair of Guardian (AF) aircraft. The AF-2W hunter is in the foreground, with the big radome of its APS-20 snorkel-detection radar. The AF-2S killer is in the background, showing the nacelle of its APS-31 radar under its starboard wing. A corresponding nacelle under the port wing carried a searchlight, the radar-searchlight combination used to attack a surfaced submarine. The vertical data link antennas that enabled the two airplanes to coordinate are visible on both. Neither airplane shows the later ESM radome over and just abaft the cockpit. USNI

Parallel versions of the Douglas A-1 Skyraider were developed as well, initially as AD-3W and AD-3S. The production versions, operational in 1950, were the AD-4N and AD-4W. They were intended to operate as detachments on board fleet carriers for self-defense. The hunter carried an APS-20; the killer had a shorter-range APS-31 enabling it to home on the submarine and a searchlight to illuminate it for the attack. Both killer aircraft carried sonobuoys in pods developed by Douglas for the Skyraider. As its designation indicated, the AD-4N had an alternative night attack mission. The Skyraiders were the beginning of specialized ASW detachments for attack carrier self-defense. A Skyraider detachment was the first carrier air ASW unit on the scene at the outbreak of war in Korea. During the Korean War they were usually diverted for other, more urgent missions. Unfortunately, too, the sonobuoy pods could accommodate only the new SSQ-2 omnidirectional sonobuoy, which was not available in quantity until early in 1951. In Korea, AD-4Ns were supplemented by F4U-5NL Corsair fighters, which could carry one Mk 54 and four or six rockets.

After side-by-side evaluation of both types of aircraft in May 1950, BuAer decided that neither showed an overwhelming advantage and both had some disadvantages. The Guardian had an extra crewman; the Skyraider had to carry its torpedo externally (BuAer was then developing a cocoon to protect it against cold temperatures). The Guardian searchlight installation was satisfactory. The Skyraider seemed to be better for low-visibility flying. The Skyraider system, which used automatic control through its automatic radar bombing system, had not yet been evaluated. The BuAer Military Requirements Division decided to buy both. Guardians were assigned to the *Commencement Bay* (CVE 105) class carriers then assigned to ASW. Skyraiders went to large-deck fast carriers. Once the Korean War broke out, however, Skyraiders were used as attack aircraft.

The AF-2W had provision for an APR-9B radar search receiver and an APA-70 radar homing receiver, which were considerably more capable than the search receivers installed on board Avengers.[13] APR-9 occupied a small, upright cylinder above the fuselage; the homing receiver was in the wings. Later a much larger APA-69 homing receiver and the APR-9 were both mounted in a large, streamlined radome. Production was slow; the first Guardian squadron (VS 871) to get APR-9 and ASA-70 did not get them until 30 April 1953, at the end of the war. In postwar exercises, about half the positive submarine contacts these aircraft obtained began with ESM detections.

Postwar practice in Guardian ships was for the guppy with its APS-20 to fly a search pattern, cueing the accompanying "scrapper" when it detected a possible snorkel. The scrapper would descend to 500–700 ft and accelerate to 180 kts until his APS-31 radar picked up a contact. At 1.5–2 nm from the target, the scrapper bombardier would turn on his searchlight. Once he saw the target, he could aim his attack using a periscope protruding under the airplane. If there was no visual contact, the airplane dropped a sonobuoy and smoke marker on the datum estimated from radar contact, then dropped a pattern of sonobuoys and smoke markers around it 2,000 yds from the datum (as in World War II). The bombardier would monitor the sonobuoys, seeking cavitation sounds, which he could compare in intensity to estimate submarine course. He estimated the target's speed by counting propeller turns.

About 1953–54 standard CVE practice was to fly two search teams at night, often with an additional search plane as a radio relay to maintain contact with the search teams when they were outside a ship's radio range. Search altitude was set by the ability of the APS-20 to detect snorkels under existing conditions; it was usually 1,000–1,500 ft. Typically, search aircraft flew out to 40–60 nm from the ship at 135–140 kts. The scrapper attack aircraft was considerably heavier than the guppy search plane, so it normally used the carrier's H-IV (to port) catapult, the guppy flying from the shorter H-II.

By 1949 BuAer wanted a single-package airplane and pursued a single-package version of the Guardian, which emerged as the AF-3S. In the AF-3S and the larger single-package Grumman S2F Tracker, the key concession to buy compactness was to replace the APS-31A attack radar with a similar radar with a higher-gain (i.e., larger) antenna, which would not match APS-20 performance. AF-3S therefore had a larger wing pod carrying the higher-gain APS-31B. Its other new feature was a retractable MAD boom (ASQ-8). Forty were made, but the concessions were not adequate, and the airplanes were assigned to ASW attack squadrons as scrappers. The squadrons liked the MAD sensor. It enabled an AF-3S to create a barrier ahead of a moving force. To make up for the weight of the MAD and the heavier (more capable) radar, the bombsight periscope, the HF radio, and the flasher and the K-25 camera had to be given up, and the load of expendables (such as sonobuoys) had to be reduced. Endurance was reduced from six to five hours.

BuAer wanted to amalgamate the hunter and killer functions in a single airplane. While it backed development of the single-package Grumman Tracker, it also supported development of the single-package AF-3S shown here. Compared to the earlier AF-2S killer, its wing-mounted APS-31 radar had a larger (i.e., higher-gain) antenna, and it added a MAD boom. This combination was not quite enough, so the AF-3S was employed as a killer ("scrapper"). The prototype is shown. USNI

Meanwhile BuAer conducted a design contest (OS-117, issued on 20 January 1950) for a new single-package airplane.[14] Although the airplane had to be able to operate from *Commencement Bay*–class escort carriers, in fact it operated only from much larger ships, mainly *Essex*-class carriers converted for ASW.[15] Although single-engine designs were acceptable, many all-weather pilots considered twin engines mandatory. BuAer evaluators noted that a twin-engine craft "undoubtedly [had] a psychological value far beyond the true statistical advantage it possesses in safety." Therefore BuAer would choose a single-engine design only if no twin-engine design either could meet the requirements or offered inadequate margins for already planned electronic growth. That was not the case. None of the proffered designs was entirely satisfactory, but Grumman's was closest to what BuAer wanted.

As with the AF-3S, the key concession to carrier operation was the radar, an APS-33 (an all-round equivalent to the APS-31) in a retractable "dustbin." The other sensors were the same MAD as in the AF-3S and sonobuoys (2 directional SSQ-1 and 18 nondirectional SSQ-2). No searchlight seems to have been required, but airplanes had to accommodate 60 flares. The weapons were a Mk 41 torpedo and six 5 in. rockets. The specification required a 1,000 lb electronics weight margin for "electronic and airplane changes." Grumman won primarily because it offered much better engine-out (single engine) performance and the highest speed and climb rate. It had better

The single-package carrier ASW airplane: a Grumman S2F Tracker lands on the carrier *Valley Forge* (CVS 45) in 1961. She shows the usual searchlight under her starboard wing. Her radar "dustbin"—used for both snorkel search and attack—has been retracted. She had both a bomb bay and underwing stubs. Her retracted MAD boom is visible under her tail. NHHC

A Royal Australian Navy S-2E Tracker displays its sensors: an extended radar dustbin, and the MAD stinger. This version eliminated the big ESM radome of earlier ones, ALD-2 antennas being incorporated in extended wingtips. The APS-38 radar of earlier versions was replaced by APS-88. The ASQ-8 MAD was replaced by ASQ-10, on a longer boom. A sniffer on the port side is not visible here. The fuselage was lengthened to accommodate LOFAR processing (AQA-3). Sonobuoy capacity was doubled. The small ventral radome covers a Doppler navigation antenna. DR. JOSEF STRACZEK

The Royal Navy adapted its two-seat Firefly fighter as a carrier ASW airplane. This Firefly AS.6 has just landed on board HMS *Eagle* during the NATO Mainbrace exercise, 23 September 1952. The four underwing objects are sonobuoys. The other ASW sensor was an ASH (U.S. APS-4) radar, which occupied one of the two wing nacelles (that on the opposite wing was a fuel tank, to balance it). There were also marine markers. Antisubmarine weapons were two depth charges or eight rockets or two Dealer B homing torpedoes. The airplane was cramped enough that its sonobuoy receiver was an alternative to its normal radio set. The later Firefly AS.7 had a crew of three rather than two. Although it was much disliked, it was ordered into production to keep the Fairey company alive pending production of the desired carrier ASW airplane, the Gannet. USNI

spotting on a flight deck and a more flexible interior arrangement. The novel sonobuoy stowage in the nacelles was "one of the best submitted." Grumman's design became the S2F Tracker in a new VS (for ASW) series (the SF designation had already been used for a scout built by Grumman in the 1930s). The most apparent changes from the proposal were provision of a searchlight on the starboard wing and the addition of an ESM (called ECM) mast atop the cockpit carrying an APR-9 radome. Production aircraft had APS-38, an improved version of the APS-33 radar specified for the competition.[16]

The analogous British single-package carrier ASW airplane was the Fairey Gannet. Like the Shackleton, it had a long bomb bay, presumably sized by the requirement to carry the Pentane homing torpedo. Although that was not obvious, it was twin-engined, its contraprops driven by two separate engines. After the 1957 Defence Review the Royal Navy drastically slashed its carrier force. In view of the new strategy emphasizing out-of-area operations, it could no longer afford ASW aircraft that required full carrier decks. Helicopters made it possible to move the ASW function off the strike carriers onto specialized helicopter ships. When the planned helicopter carriers were abandoned, the helicopters moved on board the fleet carriers. The Gannet survived as an airborne early warning platform equipped with an APS-20 radar. It is not clear whether it or its predecessor, the Skyraider AEW Mk 1, was used to search for snorkels as well as for low-flying enemy aircraft.

Until the drastic shift in British defense policy in 1957, the Royal Navy planned to convert merchant ships to escort carriers in wartime, drawing on the very numerous British merchant fleet. These ships would have relatively small flight decks and might well lack catapults, so a special austere ASW aircraft was needed for wartime production: the Fairey Seamew.

It was understood that helicopters equipped with dipping sonars could counteract the new submarines with their long-range homing and pattern-running torpedoes, but for several years helicopter technology did not provide enough lifting power and hovering capacity. Work on U.S. ASW helicopters was suspended accordingly (U.S. Navy helicopter development was not affected, because there were other urgent helicopter roles). The ASW role was particularly difficult because it required helicopters to hover for a great proportion of each flight, as much as 60 percent. Hovering consumed considerably more fuel per minute than cruising flight, though it was less expensive in wind.

The third Fairey Gannet prototype (WE 488) with her dustbin radar antenna housing (for an ASV 19 radar) extended. The large bomb bay forward of the dustbin housed two homing torpedoes and sonobuoys. Rockets could be carried underwing. There was no MAD stinger, nor was there a searchlight. Although the Gannet appears to have a single engine, she was powered by two turboprops driving contraprops. Gannets were exported to Australia, West Germany, and Indonesia, in the latter two based ashore. DR. JOSEF STRACZEK

In August 1949 the BuAer rotary wing group announced that technology was now ripe for ASW.[17] The helicopters now entering production, such as the twin-rotor HUP-1, had been designed four years earlier, yet demonstrated what was needed. The fraction of gross weight available for payload had already increased from 17 to 35 percent. A helicopter could therefore carry a useful load of 5,000–6,000 lbs, which would be enough for 4 hours of cruising flight, a crew of 3 (pilot, copilot, and sonar operator), a sonar, and 1,000–1,200 lbs of armament. The BuAer Armament Division was interested in a short lightweight homing torpedo similar to the new Mk 41. A helicopter/airship version would not need long range and ruggedness (for high-speed drops), hence might weigh 500–600 lbs; a helicopter could carry two of them. This 14,000–16,000 lb helicopter could operate from carriers, merchant ships, and modified LSTs. The Navy had already gained considerable experience operating helicopters from ships' decks. Helicopters had hovered at night, in bad visibility, in winds up to 45 kts (gusts up to 60). The commercial Los Angeles Airways carried mail in its helicopters at night, suspending operations only when visibility in its landing and takeoff areas was less than one-eighth mile. A helicopter sonar screen could operate 20 nm from a fleet or convoy. A helicopter contacting a target could fly out to its immediate vicinity and drop its torpedo.

To test helicopter attack potential, the Advanced Weapons Unit at Key West developed a prototype lightweight homing weapon from the abortive wartime Mk 30 torpedo. At this time the Operational Development Force had been assigned to evaluate tactics employing a dipping sonar (AQS-1) using an HRP helicopter. By January 1950 it was extending that to include attack; it wanted to install a weapon rack for the new ultra-lightweight torpedo.

On 30 January 1950 BuAer requested bids for an ASW helicopter (to specification OS-120); they were received on 15 April, shortly before the Low Report urgently recommended work on just such an aircraft. Bell won with what became the twin-rotor HSL, carrying an AQS-4 dipping sonar.[18]

Seeking an interim capability, the Navy activated its first operational helicopter squadron (HS-1) in October 1951 at Key West using ex–Marine Corps HRS helicopters redesignated HO4S, nominally observation helicopters. They proved underpowered in the hot and humid Key West conditions. The preferred alternative, an ASW version of the HUP the Navy had already bought, appeared in 1953, as the Korean War wound down. It also proved underpowered, hence unreliable in the hover, and was rejected. The Navy modified Marine Corps HRS (S-55) helicopters as HO4S-3s to equip its ASW helicopter squadrons on an urgent basis. They were just in time for the final period of the Korean War. Their performance was marginal, so the new helicopter could not carry both a dipping sonar and torpedoes.

An HSS-1N helicopter shows its AQS-4 dipping sonar, May 1962.
NHHC

BuAer had concerns about buying the Bell helicopter. Bell had never built anything so large, and it had never built a twin-rotor helicopter. Normally BuAer would have bought a second, less innovative helicopter as insurance, but when that was proposed in July 1950 there was barely enough money to buy the HSL. In 1951 the situation was quite different, the Korean War fiscal spigot having been turned on. BuAer began negotiations to buy the runner-up in the 1950 competition, Bell's S-58, an enlarged S-55.[19] It flew in June 1954, a year after the seriously delayed HSL. By that time the Navy was already cutting back the Bell contract from the original plan to buy 300; eventually only 50 were made. The new Sikorsky helicopter, designated HSS (Seabat), offered twice the payload of the HO4S. Although it was designed to carry sonar, MAD, and torpedoes, in practice the Seabat was too small for single-package operation. Typically, it dipped a sonar and cued surface ships. The U.S. Navy first gained true single-package capability with Sikorsky's S-61 Sea King (HSS-2, later SH-3).

The HO4S, HSL, and HSS were all conceived to operate from carriers. The Royal Canadian Navy followed a very different path, concentrating on operating big helicopters from its destroyers.[20] In September 1953 HMCS *Iroquois* and *Athabaskan* participated in a HUK exercise involving helicopters, which maintained contact with a submarine initiated by a long-range airplane until a surface attack unit could arrive. The CO of *Iroquois* considered this the most important advance in ASW since World War II.[21] Some RCN officers saw helicopters as force multipliers that would make up for the large number of escorts SACLANT demanded at that time. A special Naval Staff committee appointed to study requirements for helicopters proposed in 1953 that future RCN destroyers (DDEs) be equipped to control screening helicopters.

For Canada, the important consequence of the strategic shift sponsored by the U.S. Navy in the mid-1950s was the need to prosecute SOSUS contacts hundreds of miles to seaward. Land-based aircraft could gain and maintain contact, but it took ships to prosecute contacts. In its sustained battle with the RCAF over ASW airpower, the Royal Canadian Navy argued that sea-based aircraft could maintain contact with submarines offshore more efficiently than land-based ones. Ships alone could not suffice, due to their limited speed. In 1954 the RCN produced a Staff Requirement for what it called a 1960 carrier-based helicopter, with a range of 150 nm and a maximum speed of 100 kts, to carry one torpedo for screening and three for search/attack. It had to have anti-icing and an autopilot. Initially these helicopters were to operate from the navy's carrier. Although it did not quite meet the requirement, the Sikorsky S-58 (HSS-1) was chosen, and the Naval Board approved purchase of an initial 12. This project died in a battle with the RCAF, which argued that it should control all Canadian military aircraft. The RCN was left with an experimental unit of less-capable S-55 (HO4S-3) helicopters.

With only one carrier, the Canadians became interested in operating big helicopters from smaller ships. By early 1956 the approved Staff Requirement for the Improved *Restigouche*-class escort (DDE) called for her to be able to operate a helicopter. In September the frigate HMCS *Buckingham* was fitted with a helipad, though not with servicing facilities. Another pad was installed on board the larger frigate *St. Laurent,* comparable with the planned Improved *Restigouche*. By this time the RCN was already working on a means of rapidly securing a helicopter landing on a pitching and rolling ship. The RCN quickly decided that a helicopter would be a permanent addition to the ship rather than a temporary one which the ship would refuel and rearm.

RCN plans for a new generation of missile-armed escorts (*Mackenzie* class) died because the ships were deemed unaffordable. That left the helicopter as the only candidate for major improvement of new

The Royal Canadian Navy was unique in developing a means for its frigates to operate big ASW helicopters that other navies assigned to carriers. This Sea King sits on the big helicopter platform of HMCS *Skeena*, one of the original *St. Laurent* class. Modernization included installation of a big SQS-505 variable-depth sonar and elimination of the ship's Limbos. She was fitted with triple lightweight torpedo tubes instead as urgent-attack weapons. Other navies adopted versions of the Canadian Beartrap system, which made it possible to handle a large helicopter on a lively ship. USNI

escorts. As a stopgap, the RCN decided to build an improved version of the existing *Restigouche*, the most advanced available ASW ship it had. By 1958 the RCN was choosing between the Sikorsky S-61 (which became the Sea King) and the Kaman HU2K (which became the UH-2 and then LAMPS I). Presumably in response to a Canadian suggestion, in July 1958 the Working Panel on Surface Systems of the U.S.-UK-Canada Tripartite Technical Cooperation Committee recommended that A/S helicopters be integrated into ASW weapons systems in 1960–66 to fulfill the long-range attack requirement. Late in 1958 the Annual Review of the NATO North Atlantic Council urged Canada to make shipboard helicopters its primary ASW system. In late April 1959 the Canadian treasury approved changes to the last two improved *Restigouches*, which became the *Annapolis* class. Given the limited space for landing and storage, the new helicopter had to be a single-package type. Specifications approved in September 1959 required it to carry a dipping sonar and MAD plus two Mk 44 torpedoes or a 1,200 lb nuclear depth bomb. Initially the Canadians favored the HU2K, but through 1961 the price of this helicopter (which could not deliver the nuclear depth bomb) escalated to the point that they withdrew their offer to Kaman. The RCN had initially rejected the S-61 because of its size and complexity, but the U.S. Navy had adopted it (as the HSS-2) even though it required a longer flight deck and a larger hangar, and their evaluation team was enthusiastic. By the time the first three helicopters were delivered, the RCN had a satisfactory haul-down system, the key to operating the big helicopter in rough weather. Simplified equivalents are used by other navies, including the U.S. Navy.

STRATEGIC ASW

Before the Korean War broke out, the U.S. defense establishment was preoccupied with the threat of an emerging Soviet strategic bomber force armed with nuclear weapons. A series of radar lines across Canada covered the most likely routes Soviet bombers would take. Because Soviet bombers might end-run these radars, the U.S. Navy became involved in continental defense, commissioning radar picket ships and operating long-range airborne early warning aircraft (modified Lockheed Constellations) and even radar-equipped warning blimps. These contributions to strategic air defense naturally morphed into naval concern with strategic undersea defense against the Soviet missile submarines that appeared in the late 1950s.

The outer line of pickets comprised converted destroyer escorts, nearly all of them diesel craft with inherently long endurance. The draft characteristics described the radar picket force as an Offshore Continental Barrier and noted that pickets would be required in considerable numbers in the event of hostilities. Conversion of diesel destroyer escorts thus seemed to be the best option; 81 FMRs were available. The first two pickets were FMR-type destroyer escorts converted under the FY 49 program (project SCB 46, characteristics approved 28 February 1950: DER 316 and DER 317).[22] The first-draft characteristics were dated 25 May 1948.[23] BuShips forwarded an initial study dated 29 July 1949. The Ship Characteristics Board decided the electronic characteristics in 1949–50. The lead conversion yard, Mare Island, received initial conversion plans in February 1950.

ASW did not figure as a primary or secondary task for the pickets, at least explicitly, so the initial tentative characteristics showed only a fixed single Hedgehog (but also a QHB scanning sonar). However, the 1949 sketch design showed a single trainable Mk 15 Hedgehog, supplemented by a single depth charge track (not previously wanted). The Mk 15 was requested by the BuOrd representative on the SCB at a 27 April 1949 meeting, which decided that it should be investigated although it had not been included in the formal characteristics. At this point the specified sonar was SQS-1 (equivalent to QHB) plus a sonaramic receiver (QXB, a broadband sonar intercept set) already desired. Revised characteristics (15 June 1949) added the long-range SQR-1 passive sonar. Presumably topweight required the initially planned battery of two single 5 in./38s to be replaced by two single rapid-fire 3 in./50s.

Haverfield (DER-393) was a standard radar picket conversion, except that in this photograph her 3 in./50 guns are not enclosed in fiberglass covers. Note the trainable Mk 15 Hedgehog in No. 2 position. USNI

Another four DERs followed under the FY 51 program (DERs 142, 244, 318, and 333). Further conversions followed, initially under the FY 54 program: SCB 46A for 2 steam destroyer escorts that had been retained incomplete since World War II (FY 54) and 28 FMRs converted under the FY 54–57 programs. Preliminary characteristics for the main conversion program, for the FY 54 and later programs, were issued 5 November 1953 (SCB Memo 201-53) and added launching and stowage facilities for three Mk 32 torpedoes or an equivalent weight of Mk 43 torpedoes. The draft showed SQS-4, but BuShips recommended substituting SQS-1, which required less space and seemed adequate. BuShips suggested reserving space and weight for later installation of SQS-4 (low frequency). The DER 316 class already had SQS-4; the proposal was for SQS-4 with DMCC (Dual Maintenance of Close Contact, the ability to depress the beam to maintain contact with a deep submarine). Characteristics were approved on 13 January 1954.

In 1961, with the system at its height, one DER was constantly on station between Greenland and Iceland, with a radar picket airplane (WV-2) also constantly on station, and another on station half the time between Iceland and the UK. In April 1960, due to budget restrictions, the DER force was cut back, but stations were retained between Iceland and the Faroe Islands and between the Faroes and Scotland. In the Pacific, two DERs were on search and rescue duty, but also participated fully in the early warning system.[24] The radar barriers were finally stood down in 1965, because it was clear that the main threat to North America was Soviet missiles, not aircraft. While the barrier was active, it was backed by 16 Liberty ships fitted with long-range radar but not with armament: *Guardian*-class YAGRs, the initial Y indicating ships assigned to naval districts (typically yard craft, though these were not).

In 1954 the U.S. Navy became intensely interested in countering Soviet submarines armed with long-range missiles. By this time it appeared that the United States would soon be able to defend itself effectively against a Soviet long-range bomber attack thanks to radars, interceptors, and surface-to-air missiles linked to a new computer command-and-control system.

Defense against submarine-launched missiles would be a different proposition. Their time of flight would be so short that they would probably could not be intercepted. A 1,000 nm sea-launched ballistic missile would fly for only 10 minutes before reaching its target, and warning time might be no more than 5–6 minutes. At this time SAC based its plans on a 15-minute warning of an ICBM attack, the missiles' flight time being 30–45 minutes. Too, 32 of the 38 U.S. cities with populations over 500,000 were in a

500 nm wide U.S. coastal strip vulnerable to attack from offshore. It seemed that only ASW could solve this problem. The Canadians agreed that an offshore barrier was necessary. In 1958, SACLANT created a North American ASW command specifically to maintain barrier forces about 500 nm offshore. Previously the British had exerted pressure to preclude such a development (which would weaken trade protection), but they were presumably resigned to it once they rejected the previous emphasis on a global war.

The perceived threat of Soviet submarine-launched missiles preceded the better-known threat of land-based ICBMs dramatized by Sputnik in 1957. The latter caused President Dwight D. Eisenhower to create the Gaither Committee to investigate the problem of surprise attack. Among its conclusions was that SOSUS could help warn of such an attack, the theory being that the Soviets would surge submarines to sea as preparation for their attack.

Although Khrushchev had been touting submarine-based missiles as part of his Revolution in Military Affairs, the threat did not become tangible for some time. On 26 January 1954, a Soviet directive was issued to arm long-range submarines with ballistic missiles (the Volna program). The existing R-11 Scud was chosen. On 16 September 1955 the Zulu-class submarine *B 67* fired the world's first submarine-launched ballistic missile. Five more Zulus were converted to become the basis for the first Soviet purpose-built ballistic missile submarines, Project 629 (Golf class), which fired the R-13 (NATO SS-N-4) ballistic missile. Whiskey-class submarines were later converted to fire P-5 (NATO SS-N-3) cruise missiles. Both P-5 and R-13 had to be surface-launched. Estimated ranges were 220 nm for SS-N-3 and 350 nm for SS-N-4. The latter was later replaced by the longer-range R-21 (SS-N-5), which could be fired from underwater.

A mid-1957 British report on Soviet R & D noted that "unidentified structures have been seen on Soviet submarines, one possible explanation of which could be that the Russians are experimenting with guided weapons launched from submarines."[25] Soon there were numerous reports of submarines with hangars on deck, with launching ramps, and with airplane-like objects on deck. U.S. naval intelligence estimated that in 1957 the Soviets had a 500 nm cruise missile (i.e., Regulus), and that by 1962 they might have a 1,000 nm missile (equivalent to the contemporary, but abortive, U.S. Regulus II). The first Zulu V, with its two ballistic missile tubes, was seen at Rosta in 1956; a good photograph that showed the missile tubes was obtained in May 1959.

As the threat became more widely known, the U.S. public increasingly demanded strategic ASW. Both Congress and the Air Force agreed, the latter due to the threat missiles posed to the bomber deterrent. Op-05 (ADM Robert Pirie) complained to CNO Admiral Burke that the Navy was being pushed more and more toward a defensive stance, with demands to solve the problem of "not enough ASW." Solutions included extending SOSUS with new elements, including open-ocean sensing buoys. There was, however, a fundamental problem. No matter how well Soviet missile submarines could be tracked, they could not legitimately be attacked until they began firing missiles—or at least were known to be preparing to do so.

The first Soviet strategic submarines had to surface and raise their ballistic missiles (SS-N-4/R-13) to the top of the sail before firing. The associated Echo-class cruise missile submarine had to surface to fire and had to elevate her missile tubes visibly. However, a submarine could fire the next-generation ballistic missile (SS-N-5/R-21) while submerged.[26] R-21 went to sea on board modified versions of the earlier diesel and nuclear ballistic missile submarines. The only warning would be sounds associated with preparation to fire, which might be heard by a trailing U.S. submarine or surface ship. U.S. ASW forces might try to limit damage to the United States by destroying the submarine before she could fire all her missiles. Thus, one proposal was to search for the ionization trail left by the first missile, automatically trace it back to its apparent source, then fire a large nuclear weapon at that datum.

Initially the Soviets sent few submarines into the open sea, so a SOSUS observation of a large sortie could constitute strategic warning. In 1960 the Atlantic and Pacific fleets each reported fewer than ten submarine contacts that could be considered positive.[27] Within a few years, however, the Soviets were routinely sending much of their submarine fleet to sea.[28]

SOSUS revealed that initially the Soviets deployed their submarines in what they thought were secure holding areas well off the U.S. coast. By 1967 Soviet diesel ballistic missile submarines were routinely patrolling from Petropavlovsk to holding areas 500–900 nm from Hawai'i. Another holding area was equidistant from Hawai'i and the West Coast (1,000 nm either way), and another was 800 nm from the Pacific Northwest. Each holding area (there were others in the Atlantic) was about two days' transit from a launching area. When Yankee-class submarines (Project 667A) began carrying the 1,300 nm SS-N-6 (R-27) missile, the existing holding areas became potential launch areas, and new holding areas were set up farther from the U.S. coast. Movement from these areas to within range of targets would constitute strategic warning, though not enough to justify preemptively attacking the Soviet submarines.

When the U.S. Navy approached the British in 1967 to set up a SOSUS array off the UK coast, it explained that the associated arrays were needed to detect Soviet missile submarines hiding on the far side of the Mid-Atlantic Ridge and invisible to SOSUS arrays on the U.S. side of the Atlantic. The British apparently accepted that explanation, and the SOSUS station at Brawdy in Wales opened in 1974.

Even if Soviet missile submarines could not be prevented from firing, their initial launches could be reported well before the first missiles struck. A procedure for flash reporting submarine-launched ballistic missile (SLBM) launches via the SLBM voice reporting net involving ASW forces and the SAC was set up in 1967 and then tested regularly. Although the chance of spotting a single launch might be small, an all-out Soviet attack would involve so many submarines and missiles that some launches would almost certainly be seen and reported, triggering retaliation.[29]

Submarine navigation was an important if obscure factor: the missile had to be aimed from a known point. The Soviets apparently did not have a U.S.-style ship's inertial navigation system (SINS) until the Project 667AU (nine late Yankee class, distinguishable by a bulge on the upper part of their sail) for buoys to contact Soviet navigational satellites to update their inertial systems; later all other 667As were similarly fitted.[30] Without SINS, the submarines had to rely on something else, most likely transponder beacons, but possibly recognizable parts of the sea bottom. The missiles themselves had to be set to hit particular targets. Lacking sophisticated computers, the Soviets hardwired early versions of their strategic attack missiles to hit specific targets. That is evident because they applied designations to missiles with more flexible targeting. R-27 (SS-N-6) seems to have been the first Soviet naval strategic missile whose launching ship had a firing (presumably multitargeting) computer.

PUTTING IT TOGETHER: THE CUBAN MISSILE CRISIS

Until the Cuban crisis, the U.S. Navy seems to have associated ASW with general war, which would entail hunting down Soviet submarines intending to attack ocean shipping and also hunting down Soviet missile submarines. The U.S. Navy expected the fast aircraft carriers that would fight around the Eurasian periphery (by far the most likely scenario) largely immune to submarine attack thanks to their speed and the inability of Third World enemies to find them well out to sea. Soviet submarine transfers to China, Egypt, and Indonesia apparently did not affect this conclusion, although they certainly were taken into account by the Royal Navy, which pursued a roughly similar strategy.

The Cuban crisis demonstrated that Soviet submarines could have an impact even in limited war: the four Foxtrots (Project 641) that reached Cuban waters could have attacked U.S. naval forces offshore. An important part of the deterrent effect of the quarantine around Cuba was the location, identification, and continuous tracking of the Soviet submarines in the western Atlantic. The crisis immediately increased the R & D budget for ASW, which was already 20–25 percent of the total Navy R & D budget, by 20 percent.[31]

The Cuban crisis was the first time since World War II that the U.S. Navy deployed its ASW forces against real targets. By this time the Julie and Jezebel sonobuoy systems were considered mature, but Jezebel contacts were sometimes confused, partly, it was said, due to interference in sonobuoy readout, presumably meaning in lofargrams. As a fully passive sensor, LOFAR could not necessarily distinguish between the relatively weak but regular sound of a submarine and broadband noise.

SOSUS relied on a library of Soviet submarine signatures. Prior to the crisis, nearly all of the detections that constituted the library were of surfaced submarines.[32] Worse, at least as late as 1961, Navy schools were teaching that some Soviet submarines used four-cycle diesels (with distinctive firing rates detectable by SOSUS), whereas before 1963 the standard engine was the two-cycle 37D. When SOSUS data for the years prior to October 1962 was evaluated in 1965, only a single valid detection was found, a previously unidentified Foxtrot detected by the Able and Sugar arrays off Cape Hatteras. It was making 14 kts on the surface, returning to the Northern Fleet. Contact was held at night for several hours before the submarine retired out of range.

SOSUS identification was further complicated by the Soviets' practice of snorkeling on only one engine, using the rudder to compensate for the off-axis propeller thrust. Distorted water flow to the single propeller created a high level of cavitation. The source of that cavitation was not apparent until the U.S. naval quarantine sharply reduced commercial shipping near Cuba. Contacts that were left were evaluated as Soviet submarines mainly because there were no other plausible candidates. A Neptune (P2V) sent to investigate one of the unusual sound sources spotted a snorkel. When it dropped a sonobuoy, it received the same puzzling acoustic signature that SOSUS had detected. The identification of the anomalous signatures remained controversial until 1963–64, when the array in Norway detected numerous Soviet diesel-electric submarines deploying and returning to the Northern Fleet base.

The Foxtrot deployment to Cuba was the first time the Soviets deployed any substantial submarine force to the western Atlantic. Apparently six Foxtrots deployed well in advance of the crisis, and four of them remained in the western Atlantic.[33] The Soviets tried to deploy the Foxtrots covertly, as they handled the other aspects of the military buildup in Cuba.

As the Cuban situation deteriorated, U.S. strike and amphibious forces moved to the Caribbean and air ASW support was provided. On 24 October a submarine-aircraft barrier employing Canadian and U.S. ASW aircraft and ten U.S. submarines was established at Argentia in Nova Scotia to detect submarines as far forward as possible. Overall, the ASW effort required four ASW carriers (CVS), whose onboard ASW aircraft (VS) flew about 50 percent more hours than the land-based patrol aircraft (VP).

Twenty-nine submarine contacts were investigated, of which six were positively identified as submarines (another four were probables, and five were possibles). ASW Task Group Alfa identified a submarine seen on the surface as a Foxtrot. On 31 October units forced a Soviet submarine to surface after maintaining sonar contact for 35 hours. Another Foxtrot (with a different number on her sail) was forced to the surface on 3 November. This last submarine was under surveillance from 24 October through 17 November. On the other hand, 450 flight-hours were spent on one contact erroneously classified due to interference in sonobuoy output (not further explained). For a time, it appeared that the nuclear carrier *Enterprise* was being trailed by a Soviet nuclear submarine (in fact, the Soviets never deployed their new nuclear submarines to the Caribbean). In recent years some Russians have claimed that one of the four Foxtrots was undetected and unprosecuted.

THE GREAT FEAR

By 1960, SOSUS was the great hope of U.S. ASW. The British had argued from the beginning that submarines could be silenced to the point that SOSUS would become completely ineffective. They were already working to silence their new class of diesel submarines. In 1959 the first such boat, HMS *Porpoise*, came to the United States and demonstrated what silencing could mean: detection range shrank to perhaps 30 nm. As yet SOSUS had not observed Soviet diesel submarines, and there was no certainty that the Soviets would not do the same thing, neutralizing SOSUS almost completely.[34]

The main experimental active systems in 1960 were Lorad, Artemis, Atlantis, and Fishbowl. Lorad was a low-frequency shipboard system that was expected to detect submarines out to a range of about 20 nm. It was tested on board the submarine *Baya*, which had a transmitter in her bow and receivers on long arms that extended to each side.

Artemis was a more ambitious fixed bistatic system ultimately intended to replace SOSUS, with a hundredfold increase (compared with Lorad) in both transducer dimensions and radiated power. Work on Artemis began in 1958. It was intended to employ large bottom-mounted transducers near a shore power supply and large receivers several hundred miles away, also near a land terminal. As a step toward developing Artemis, the tanker *Mission Capistrano* was equipped with a massive sonar transducer (a quarter-scale version of Artemis) in her after oil tank.[35] The goal was to be able to detect a submarine at a range of 500 nm, similar to what was expected of SOSUS in the late 1950s. By the time the test installation was ready, SOSUS was demonstrating really-long-range detection.

Atlantis was a cabled grid of receiver hydrophones spaced 15 nm apart in a surveillance area, each with 2 channels (50–150 and 350–450 Hz). All cable runs would be laid in water 2,000 fathoms deep or deeper, off the continental shelf. The low-frequency receivers would be used for lofargrams. The higher-frequency receivers would receive active acoustic signals provided by Artemis transmitters. Signals from the large transducers would be reflected from submarine and surface targets into the nearest two or three hydrophones. In 1959 it appeared that this technique could be carried out 1,000 nm from a shore station, and that it could localize a surfaced or submerged target to within a mile. Shadow areas would be covered by moored buoys with limited detection capacity.

Fishbowl was an upward-scanning active sonar.

The Soviets did not silence their submarines, so these projects were not pursued, but the fear remained that at some point the Soviets would silence their submarine fleet and render SOSUS impotent. That was one reason SOSUS was so carefully guarded.

8
THE NUCLEAR REVOLUTION

THE CRISIS

NUCLEAR SUBMARINES were nothing like their diesel forebears. The difference went far deeper than a few additional knots of maximum speed. The underwater endurance of a diesel submarine was limited by battery capacity, so her commander always wanted to keep some in reserve. He would attack a convoy at a speed (about 12 kts) at which he enjoyed substantial (typically four-hour) endurance, saving some capacity for a burst of high speed to get out of trouble. The finite battery capacity also explained the most effective ASW tactic: holding down a submarine to exhaustion. ASW aircraft had snorkel-detecting radars because a submarine would have to snorkel to replenish her battery after getting clear of a target group.

A nuclear submarine could sustain high speed indefinitely. She could make the fullest use of her speed both when attacking and when withdrawing after revealing her presence. She could never be held down. The only viable ASW tactic was to destroy her directly. The U.S. Navy's initial exercises against the first nuclear submarine, USS *Nautilus*, demonstrated how drastically ASW had changed.

In a 1955 exercise, *Nautilus* successfully overtook a hunter-killer group making 16–18 kts. The HUK

The nuclear revolution: a Soviet November-class nuclear attack submarine. She was threatening enough when credited with a speed of 20 or 22 kts, but in fact she was a lot faster, with a rated speed of 28 kts. Initially these submarines were not run at maximum speed due to problems with some of their eight steam generators. In 1968 a November showed her high maximum speed when she chased the nuclear carrier *Enterprise*, en route to the Far East for the Vietnam War. This one, *K 8*, was suffering a fire off Cape Finisterre; she sank on 8 April 1970. USNI was unable to detect her at a distance. Fast escorts could track her in low sea states, but they could not hit her because she could maneuver away while weapons were in the air. The existing Mk 43 Mod 1 homing torpedo consistently acquired *Nautilus* but was not fast enough to hit her. The exercise showed a five-to-one exchange rate, but later it was estimated that about eight surface ships (including the ASW carrier) would be killed for each *Nautilus* sunk. *Nautilus* could, moreover, dive quickly under the layer, drastically reducing surface ship sonar range, and still keep track of the surface ship. She could then pop up to fire.

In 1956 exercises, *Nautilus* could evade even when the surface force started with contact. Starting at 5 kts at a depth of 150 ft, she evaded destroyers and escorts at 1,000 yds range (sometimes 4,000 yds). In more than 26 attacks, the attackers made no hits at all using Hedgehogs, Weapon A, and Mk 35 and Mk 43 torpedoes. All but once *Nautilus* broke contact. One time a destroyer held her for an hour, but otherwise she was held for an average of four and a half minutes. In one exercise three destroyer escorts fired 31 sub-caliber Weapon A rounds at her in the five or six minutes before she broke contact. She could not hit them either, because her fire control computer (Mk 101) could not produce a solution quickly enough. Of seven torpedo attacks, she got only two hits, one of them by sneaking up after breaking contact. Generally, *Nautilus* found herself running away from surface ships, but lacking stern torpedo tubes could not fire at her attackers.

Nautilus was tried against a HUK group over a 24-hour period. She avoided cavitation by running at 5 kts, and MAD proved useless against her. The HUK group claimed seven kills using Betty nuclear depth bombs, but on analysis all proved false. Similarly, claims of MAD contacts and disappearing radar contacts proved untrue. The claimed kills were 40–50 nm from where the submarine actually was. The destroyers suffered severe false target problems, *Wilkinson* (with the new SQS-4) less so.

The fleet got similarly disappointing results against the fast and very maneuverable *Albacore*, in 1955 credited with a speed of 26 kts. She also enjoyed rapid acceleration and reduced cavitation. *Albacore* generally broke contact by diving under the layer. A submarine with unlimited underwater endurance would find that even easier to do.

Initial exercises with *Albacore* were described as more like dogfights than the deliberate, plodding, often stereotyped attack runs on the sluggish pre-nuclear submarines. The ASW destroyer *Sarsfield* used her scanning sonar to keep contact and attack when possible; she started her attack runs at 27 kts, maintaining speed to counter last-minute maneuvers by the submarine, trying to fire her Mk15 Hedgehog dead ahead, but training it as necessary. Upon firing, she tried to maintain contact, usually by turning in the direction of the submarine's last course. Then she tried to open the range sufficiently for another attack, though reacquisition was the main objective. To preclude a counterattack (*Albacore* had no torpedo tubes but could presumably simulate a shot) *Sarsfield* maneuvered violently at high speed, the maneuvers not affecting her scanning sonar. *Sarsfield* had done very well against a Guppy II submarine (55 percent per run, 44 percent per firing pass, compared with the usual 30 percent by a major ASW ship).

Experience with *Nautilus* showed that she could not only escape ASW surface ships with her superior speed, she could also sink them. This realization spurred interest in a fast nuclear attack submarine, which materialized as USS *Skipjack*. The U.S. Navy apparently soon concluded that future nuclear submarines might well make 30 kts, although *Nautilus* was limited to about 23, and initially it seemed that *Skipjack* would make only about 26 (she was later credited with substantially higher speeds, and for a time was considered the fastest submarine in the world). With ships that fast as targets, the fleet needed sonars that could function at the highest possible speeds.

It was widely accepted that ASW escorts and torpedoes needed a 50 percent speed advantage over their quarry. That was practicable when fighting a 15–20 kt diesel-electric submarine, but not for a nuclear submarine that might be able to sustain 25 or even 30 kts. The need for speed encouraged interest in unconventional ships, particularly large hydrofoils. In fact, only aircraft, including helicopters, would be fast enough. An alternative possibility was long-range sensing to guide long-range standoff weapons. Weapons were also a problem, as torpedo seekers would have to overcome the noise generated by water flowing over them at high speed (as well as the noise generated by the motors or engines driving the torpedo at high speed and the propellers).

LOOKING FOR SOLUTIONS: PROJECT NOBSKA

Recalling the value of the Hartwell study, in 1956 CNO ADM Arleigh Burke ordered a study of the impact of the new kind of submarine. The Project Nobska participants were told that the United States might soon face advanced-design true submarines armed with long- and short-range missiles capable of attacking ships at sea as well as the U.S. coasts.[1] Told that their recommendations could not be implemented in less than five years, the participants concentrated on the period between 1960 and 1970. In addition to general war, the group considered limited war in which the continental United States would not be attacked; in that case the Navy might have to deploy ASW forces anywhere in the world. The United States already faced the largest submarine fleet in history, and the Soviets were credited with the ability to build 100 submarines each year. The United States might soon face a significant number of true submarines capable of launching missiles against U.S. shipping and the U.S. coasts. As yet there was no definite information on a Soviet nuclear submarine program, but the Nobska group assumed that was coming. The new submarines would make it far easier for the Soviets to undertake a strong submarine offensive, despite their unfavorable geography.

Op-93, which was responsible for long-range studies, saw the coming SQS-26 sonar as a vital counter to the new fast submarines. An operational version should be designed at once, and all future escorts should have it. Aircraft would be needed to exploit the new sonar's expected convergence zone capability, since the range to a target in the convergence zone was uncertain; helicopters (three or four per convoy) would probably be best. On this basis Op-93 killed the currently planned SQS-26 ship, the Sonar Scout, in favor of a conventional ship. Op-93 also considered SQS-26 essential for strike force ASW. Its prototype ended up on board a large destroyer.

Probably the single most important Nobska recommendation was to build small nuclear submarines, each armed with two or three ballistic missiles, which

The water is so clear that part of the huge SQS-26 sonar dome is visible beneath the bow of the destroyer escort USS *Lockwood*. *Knox*-class ships were notoriously wet. This photo shows two of the cures, a bow bulwark and a spray strake. The ship's Basic Point Defense Missile System, itself an add-on, has been replaced by a Phalanx point-defense system right aft. The ship has the enlarged hangar and helicopter pad of the LAMPS I system (note the landing circle on the pad and the control box on the after side of the hangar). NHHC

would present an enemy with unique military and political challenges. The group noted that "not of small value is the drain that the added ASW requirement would put upon the resources of [the] enemy." A missile that weighed no more than 30,000 lbs would be compatible with a submarine of moderate size and complexity. Given a warhead that could be available by 1965, a missile of this size could reach 1,000–1,500 nm, the latter distance sufficient to hit most targets in the Soviet Union. One of the participants in the study, Dr. Edward Teller, seems to have made the crucial connection between a lighter weapon and the size of the missile, asking, "Why use a 1956 warhead in a 1965 missile?" A yield under 1 MT would suffice for most targets of naval interest. Nobska thought it would also be worthwhile to develop a shorter-range nuclear missile that could be fired from a torpedo tube.

The Navy was already designing a big ballistic missile submarine to launch three massive Army-developed Jupiter liquid-fueled ballistic missiles. The Navy treated strategic attack as a secondary mission, to be executed mainly by aircraft. It could not hope to match the number of strategic bombers wielded by the Air Force. The Nobska group's conclusions turned the Navy toward far more compact missiles. In 1957, responding to an Air Force request to enlarge the national nuclear weapons stockpile, President Eisenhower had the Army and the Navy evaluate the nation's current nuclear war plan. The result was shocking. It turned out that much of the Air Force's nuclear arsenal was unneeded overkill; an affordable force of strategic submarines would suffice. Using Nobska's smaller missile, the Navy could multiply the missiles on board strategic (Polaris) submarines without requiring an unaffordable number of them.[2]

As the first Polaris submarines were deployed, the Soviets felt constrained to emphasize ASW, to the point that in one case they designated a ship (Kresta I) clearly designed to attack carriers (the previous emphasis, because they carried nuclear weapons) as large ASW ships. It took the U.S. Navy some years to accept that the Soviets were now emphasizing strategic ASW. Beginning in 1970 the Soviets deployed Krivak (Project 1135) class frigates. They were designated

The key Nobska recommendation: a U.S. strategic missile submarine. USS *Patrick Henry* is shown, newly completed, on 22 July 1960. The Nobska panel emphasized that the existence of such submarines would draw the Soviets into spending heavily (and futilely) on strategic ASW. It was soon evident that a force of about 40 such submarines was a sufficient U.S. deterrent. Initially the U.S. naval leadership had assumed that no naval nuclear force could match the massive tonnage wielded by SAC. Analysis ordered by President Eisenhower revealed that all that tonnage amounted to massive overkill. NHHC

as large ASW ships, but U.S. analysts assumed that their big missile tubes contained an unknown anti-ship missile. Only after about five years did the Navy accept that instead of being a carrier-killer the missile was really an ASW standoff missile, a kind of Soviet Ikara (though far more massive).

The Nobska group's torpedo tube tactical nuclear missile was soon rethought as the ASW submarine rocket (SUBROC), which could take advantage of the range of the new passive sonar. In January 1957 two parallel studies were underway, due for completion by March 1958: Stinger at the Naval Ordnance Laboratory (White Oak) and Marlin at the Naval Ordnance Test Station (China Lake). Stinger became the SUBROC underwater-to-underwater missile. The Naval Ordnance Laboratory at White Oak pointed out as early as July 1956 that Stinger would be so fast that it could be fired based on a single-ping range. The alternative ASTOR nuclear warhead torpedo seemed to require much more elaborate target tracking.[3]

The Nobska committee considered current (1956) U.S. ASW defense negligible and pressed for immediate production of an effective long-range submarine detection system. SOSUS should be completed, and recent work on long-range active sonar should be exploited. A very numerous Soviet submarine force in the North Atlantic could saturate SOSUS and might even use explosives to jam the system. Nobska was aware of Britain's claims that snorkelers could be silenced well enough to defeat SOSUS, and it estimated that in 1965 a quiet SSN could do even better. Hence SOSUS had to be backed by a long-range (hence low-frequency) active detection system. It would be more difficult to jam, and it might achieve half-degree accuracy at 30 nm. The committee's report included a diagram of a possible barrier showing passive bottom arrays connected to moored buoys carrying active sonars and linked by radio to explosive echo-ranging aircraft up to 100 nm away. Aircraft cued by the combination of SOSUS and long-range active systems should use explosive echo-ranging to attack submarines with nuclear depth bombs (Lulus). Nobska suggested buying 500–1,000 Lulus by 1959.

Explosive echo-ranging seemed so promising that attempts were made to apply it to submarines (as SEER) and destroyers (as DEER). In effect, SEER was the distant forerunner of bistatic submarine sonar systems. The DEER concept included a proposal to use small VSTOL drones (ducted fans) to distribute the hydrophones.

Convoy tactics would likely fail in the event of an attack, so Nobska proposed a strategic shift: setting up barriers to keep enemy submarines out of the open ocean. There would be barriers off the U.S. coasts and also abroad; the committee suggested an Iceland barrier, a Skaggerak barrier, a Gibraltar barrier, a Japan barrier, a mid-Pacific barrier, and peripheral barriers. Barriers would also deal with the threat of Soviet strategic missile submarines. The barrier concept was important into the 1960s, but its expected wartime value declined as the Soviets deployed more and more of their submarines in peacetime. Barriers would still destroy submarines returning from patrol, but in that case it would be some months before they destroyed most Soviet submarines.

Nobska proposed a new generation of specialized ASW ships, their development tied to construction of experimental ships. The Navy did build the experimental escort ship USS *Glover*. Attempts to adopt an exotic hull form failed, and *Glover* emerged as a modified version of the production destroyer escort, the *Garcia* class. A manned sonar buoy was proposed but not built as such.

An ASTOR (Mk 45) torpedo on board a torpedo retriever. Note the studs along the length of the torpedo to support its 19 in.-diameter in a standard 21 in. tube, enabling it to swim silently out of the tube. The propellers are covered by a protective shield; this torpedo had conventional contraprops, not a pump-jet. USNI

Mk 45 employed a silent form of ranging to create a quiet weapon system: PUFFS (Passive Underwater Fire Control Feasibility Study), which employed a row of three transducers to measure the curvature of the wave front created by a target submarine. On modernized diesel submarines, PUFFS employed three big sonar domes, as here on board USS *Pickerel*. She was subject to FRAM modernization, which included sound-mounting one of her three diesels. Nuclear submarines had a form of PUFFS employing much less prominent transducers. USNI

The Navy was already developing the Naval Tactical Data System (NTDS) to link ships and aircraft, mainly for fleet air defense. Nobka pressed for an ASW equivalent, which emerged a decade later as the ASW Ship Command and Control System (ASWSCCS).[4]

Nobska was struck by how little information the Navy had concerning Soviet submarines. An acoustic intelligence program (AIDA) began in 1958 with specially configured aircraft. Its objects were to obtain guidance for new surveillance and detection systems, to confirm suitability of current equipment, to assess the characteristics of Soviet submarines, and to study the character and direction of Soviet underwater acoustic investigations. Aircraft were favored because they did not suffer from the self-noise of other platforms, could get close to their subjects, and could identify them. Their work made it possible to modernize the catalog of Jezebel (LOFAR) signatures. This work was extended to submarine surveillance in 1963.

A new generation of faster torpedoes was essential, despite the problem of operating sonars effectively at higher speeds. Given those problems, the Nobska committee recommended an interim submarine weapon: a nuclear torpedo that might be available within two years (July 1958) and might marry the existing Lulu nuclear depth bomb to 100–200 existing torpedoes. BuOrd was already developing a nuclear torpedo, which emerged as ASTOR (Mk 45), so it rejected the Nobska idea.

Nobska proposed reversing the previous decision not to develop a very fast (40–45 kt) homing torpedo to counter *Albacore*. The group pointed out that there was an urgent need for a fast lightweight homing torpedo, because without it, airplanes would be unable to attack very fast submarines.

Nobska called for two new torpedoes: a much faster lightweight (searching at 15 kts, acquiring a target at 1,500 yds, and homing at 45 kts, with a running range of 8,000 yds) and a new heavyweight (acquisition range 2,000 yds, range 35,000 yds at 45 kts). In each case the slow search was accepted to get around the torpedo self-noise and flow problems. Because the Mk 44 designed to attack fast submarines was not yet in service, Nobska had to compare what it wanted with the slower Mk 43. The two new Nobska torpedoes emerged as the lightweight Mk 46 and the heavy Mk 48. Both proved difficult to develop, not entering service until 1968 and 1971, respectively. Until then, U.S aircraft and surface ships were armed with Mk 44 lightweight torpedoes adapted to 20-kt targets; U.S. submarines had Mk 37s with an effective speed of about 24 kts.

Mk 46 torpedoes on deck, identifiable by their slightly raked tail fins. AUTHOR'S COLLECTION

A Mk 48 ADCAP 6AT torpedo dwarfs a Mk 54 lightweight torpedo behind it at the 2006 Euronaval arms show. The horizontal plate and the plate wrapped around the upper part of the torpedo, toward its nose, are the antennas of the standoff active magnetic fuze (with separate plates creating the triggering magnetic field and detecting its distortion by the target ship). The torpedo's pump-jet is visible at its rear end. Mk 54 is the length of standard U.S. lightweight torpedoes. AUTHOR'S COLLECTION

Nobska led to a decision that henceforth all U.S. nuclear attack submarines would have a primary ASW function, hence would have long-range sonars and be silenced. The first of the new breed was USS *Thresher*. Although considerably larger than her *Skipjack* predecessor, she was nearly as fast, thanks to very careful hydrodynamic design, including a drastically smaller sail placed farther forward than on previous nuclear submarines. *Thresher* was also much deeper diving than earlier submarines, though not nearly deep enough diving to reach the deep sound channel. That sort of performance was an unmet goal of U.S. and British submarine designers in the 1960s. NHHC

An indirect consequence of the Nobska study, not at all explicit in it, was that henceforth all U.S. nuclear attack submarines would have a major ASW role. The existing *Skipjack* class was redesigned as the quiet *Thresher* class. The very-long-range BQQ-2 submarine sonar was already in the design stage. The existing Mk 37 torpedo was not compatible with long-range sonar; BQQ-2 was paired with the nuclear SUBROC ASW missile.

Nobska proposed a submarine that could exploit the deep sound channel (4,000 ft down). A quick study showed that such a submarine would displace 7,500 tons and would have very little capability, but the idea recurred later.

SSN VERSUS SSN

By the late 1950s nuclear submarines seemed to be the best antidotes to enemy nuclear submarines. In 1957, in the wake of the Nobska group's study of the implications of the new technology (see below), CNO Admiral Burke ordered that all future nuclear attack submarines should be adapted for the ASW mission. The first affected was the *Thresher* (later *Permit*) class, which was specially silenced (rafted, among other things) and had the whole bow devoted to a huge sonar (the torpedo tubes were moved aft).

Once the U.S. Navy had a second SSN (*Seawolf*) it could try SSN versus SSN experiments to see whether submarines could do any better than surface and air ASW craft. It was found that they could detect each other outside torpedo range when at speed, but they could not hit each other due to limited torpedo range. When they detected each other within torpedo range, their fire control computers were not fast enough to enable either to shoot before the other was off and away. A 1958 exercise (*Nautilus*, *Seawolf*, and *Swordfish*) showed that an SSN could close another SSN using her active and passive sonars (active when the target submarine went quiet) but could not kill her because the Mk 27 torpedo lacked sufficient range. Sometimes the submarine hit herself. A 1961 exercise (*Nautilus* vs. *Skipjack*) began, for the first time, with a detection, which was to be pursued to a realistic kill using the new Mk 37 torpedo. There were no hits. *Skipjack* was considerably quieter than earlier U.S. nuclear submarines, and she was difficult to hold, get accurate bearings on, track, and hit. In one encounter, the target submarine was ordered to cavitate, so she was immediately detected—but the submarine following was always counter-detected well before she could shoot. Classification proved very difficult. Later in 1961 the first really quiet U.S. nuclear submarines were used in the tests, first *Tullibee* and then *Thresher. Tullibee* detected *Nautilus* at 20,000 and 30,000 yds, but in this test *Nautilus* was very noisy at 6 kts due to a bent propeller blade. *Tullibee* had to run shallow and cavitate for *Nautilus* to detect her. In a September 1961 exercise with *Skate* and *Tullibee*, *Thresher* completely outclassed the earlier-generation *Skate*. At low speed *Tullibee* detected her at 30,000 yds. Using an active ping, she ranged on *Skate* at 36,000 yds. *Thresher* could outrun the Mk 37 torpedo and go deep. When deep and at high speed she seemed practically invulnerable.

SSNs were almost never hit by practice torpedoes. Ranges were often miserably short. The main lesson of these exercises was that to be effective against other submarines—an increasingly vital role—SSNs needed a fast long-range torpedo, which did not yet exist. It emerged more than a decade later as Mk 48.

Occasionally an SSN was able to hit another SSN: in an Atlantic test, 12 torpedoes were fired for 1 hit, in the Pacific 28 shots yielded 5 hits. To some extent the problem was insufficient time spent in submarine-versus-submarine exercises, as the SSNs were badly needed to exercise against surface and air forces—which could not hit them at all.

The early tests raised some questions that would keep recurring. Were barriers, already adopted to counter conventional submarines, the best way to use SSNs against Soviet SSNs? Might the enemy try rushing them, to saturate them? Once the Soviets had large numbers of SSNs, would barriers remain viable?

Within a few years the situation was transformed. The U.S. Navy had sophisticated, quiet nuclear submarines optimized for ASW. By 1966 ASW strategy envisaged barriers of such submarines across the likely routes from Soviet bases to their wartime operating areas. One-on-one exercises suggested that the quiet new submarines could achieve very high exchange rates. OSD's Director of Defense Research and Engineering, Dr. John S. Foster, wondered whether the Soviets could find a way around the new strategy. He assigned the problem to the Defense Science Board (DSB). Foster was particularly concerned that the Soviets might use antibarrier tactics such as massing submarines, and he feared that they would succeed by thinking very differently from their U.S. counterparts.

By late 1968 the U.S. Navy's Submarine Development Group Two had been running submarine-versus-submarine exercises for about two years and had amassed considerable data for the DSB task force to analyze. The exercises showed that the outcome

Soviet Foxtrot (Project 641) class submarines were probably the last of the big anti-SLOC program of the 1950s (the later Tangos [Project 641B] were probably intended mainly for ASW). In effect Foxtrots were much-improved versions of the long-range Zulu (Project 611) of the initial postwar program. This one shows the big HF radio antenna common to many Soviet submarines of this era. A total of 160 such submarines was planned, but only 58 were built for the Soviet navy, some being transferred abroad. The first two were completed in November 1959. Production for the Soviet navy was completed in 1971. Others were built for export. Initial units had the same 37D engine as Zulus, hence much the same acoustic signature, but Project 641K submarines had faster-running Type 2D-42 diesels. This one was caught on the surface by a Nimrod in May 1987. The bulb-shaped object on the mast fed a radar warning receiver. The "windows" on the sail were ports for a sonar intercept receiver. USNI

This Romeo (Project 633) was the companion to Foxtrots, just as Whiskeys complemented Zulus. In contrast to the early postwar program, Romeos were not built in large numbers, although initially it seems that many were planned. The Soviets built 20 before selling parts and tooling to the Chinese, who built these submarines as their Project 033. There was also unauthorized production in North Korea. The long faired periscope shears suggest that these boats had longer periscopes than Whiskeys, giving them a deeper periscope depth. Their Staff Requirements (TTZs) were drawn up in 1952–53, but production was not ordered until 1956. The engines were the same 37Ds as those in Whiskeys. USNI

depended on relative noise levels, sonar performance, and weapon effectiveness, of which the first two factors favored U.S. submarines so much that even coordinated enemy tactics involving up to five submarines made little difference. However, these advantages might fail if the Soviets could saturate target-handling or weapon capability.

The working group assumed that except when on battery, Soviet non-nuclear submarines were much noisier than a *Permit*; that their sonars were equivalent to the BQR-2B that equipped older U.S. submarines; and that their torpedoes were no better than the U.S. Mk 37. Their conclusions were sensitive to such assumptions. For example, 14–20 dB quieting of a transiting submarine (equivalent to the difference between the very loud *Seawolf* and the reasonably quiet *Skipjack*) would reduce the exchange rate between single adversaries from 5:1 to 2.8:1. Atypical sonar conditions could even the odds by drastically reducing U.S. sonar range. Given the right tactics, submarines with *Skipjack* characteristics but with improved sonar, an effective weapon, and appropriate cooperative tactics could be formidable opponents for a *Permit*. U.S. exercises had not gone beyond the first-shot stage, but multishot engagements had been modeled.

Of 656 exercises SubDevGru 2 had conducted over the past two or three years, 521 had been one-on-one. The targets were the loud SSN *Seawolf*, a diesel submarine, and the relatively quiet *Skipjack*. *Skipjack* was considered the more dangerous opponent. In 79 engagements, the defending SSN killed her 14 times, but she killed the defender 5 times. *Seawolf* failed to kill the defender even once in 111 encounters. The diesel submarine was more effective than *Seawolf*, the detection rate falling from 86 to 65 percent. The diesel submarine was detected 115 times out of 178 opportunities. Out of 68 attacks, 42 were considered accurate. The exchange rate was 10:1. Kill probability with a single Mk 37 Mod 0 torpedo was 65 percent, compared with 45 percent against *Seawolf* but only 8 percent against *Skipjack*. The defending *Permit* was also faced with pairs of diesel transiters. The measured exchange rate fell from 10:1 to 6:1, but that may have been misleading since there were 178 engagements with 1 transiter but only 24 against 2. Compared with a *Skipjack*, the diesel submarine was easier to detect (0.65 vs. 0.49) but more difficult to attack (0.64 vs. 0.79). Experienced submarine officers said the diesel submarine was more difficult to track and close (and it was not defenseless).

Exercise results from the past few years revealed that quiet U.S. submarines could often get into position to attack Soviet submarines trying to get through the barriers, but their weapons were grossly inadequate. A U.S. SSN could detect enemy submarines far beyond counter-detection range, but she had to close to within a few thousand yards to fire. Getting there passively often took hours. In 16 of 24 attempted Mk 37 Mod 1 torpedo firings in one exercise, the torpedo failed to function. Later, 10 out of 11 functioned, but the target (a *Skipjack*) detected and evaded 5 of them. In another exercise, only 7 of 40 attempted launches succeeded. The torpedo was being modified, for example to avoid surface capture, but it was grossly unsatisfactory. The DSB task force considered the torpedo problem one of the most critical facing the submarine force.

The coming Mk 48 might solve the problem, but the task force preferred entirely new weapons. Mk 48 was a general-purpose weapon that fit neither the long-range nor the short-range role. For close engagements the task force recommended an adapted Mk 46 lightweight torpedo. For longer ranges it wanted a fire-and-forget weapon. It liked a non-nuclear derivative of SUBROC carrying a Mk 46 torpedo. Such a weapon was already under consideration as an unsolicited proposal. As described by the DSB Task Force, it could be fired at any depth down to 200 ft and at a maximum speed of 5 kts, with a range of 30,000–40,000 yds (a note in the body of the text referred to convergence zone range, 60,000–65,000 yds). The weapon would be fired on the basis of passive sonar data. Current methods were not good enough (about 20 percent accurate), so some new passive technique was needed (target motion analysis, not PUFFS, a passive fire control system that could triangulate the target based on noise it emitted). As with SUBROC, once a passive solution had been reached, the submarine could refine it using a single ping or a few pings.

The DSB predicted (correctly) that the Soviets would operate their diesel submarines on the surface instead of snorkeling whenever they were not likely to be seen by aircraft. They would thus radiate less of their machinery noise into the water. On the surface they would be difficult to detect or classify. To deal with them, the DSB wanted an effective but inexpensive torpedo to attack surface targets at ranges out to 10,000 yds. The board thought the simplest approach would be to adapt the existing Mk 45 nuclear torpedo (ASTOR) with midcourse guidance by wire and passive homing. This idea had already been studied at length. In 1968 the task force estimated that it could be ready in two years. Instead the Navy provided Mk 48 with a surface option. In the longer term, the task force proposed a long-range SUBROC-like missile with an air-homing warhead. The U.S. Navy eventually got much of this in the form of the Sub-Harpoon, though it was developed to meet a British requirement.

The great problem was saturation: the possibility that the Soviets would throw a large number of submarines at one submarine in a barrier, hoping that most would get through. In one calculation, there was a very high probability that if the Soviets flooded part of a barrier with 200 noisy submarines, no more than 10 of them would be sunk (assuming the surviving transiters took advantage of the hole created).

Exercises against the diesel submarine were somewhat inconclusive. In one, the results were about even, but under poor sonar conditions the defending *Permit*-class SSN did much better, as her BQQ-2 sonar greatly outperformed the BQR-2B. Exercises also showed that, up to a point, massing noisy slow submarines against quieter more mobile ones (*Permits*) resulted in greater *Permit* effectiveness because there were more targets. An experiment in which a submarine was escorted by a surface ship showed that the noise of the surface ship would tend to reveal the submarine.

DIRECT SUPPORT

Particularly if she operated under the layer, a submarine might enjoy very long sonar range. As early as November 1959 SubLant proposed coordinating U.S. SSNs with HUKs, task groups, and convoys to kill enemy submarines. This idea recurred in the 1960 U.S. ASW summary. The submarine would search ahead under the layer. The new BQQ-2 sonar might see a 10 kt snorkeler at 130 nm and a fast SSN at about 190 nm. The new submarine *Thresher* was built with two features associated with combined submarine–surface ship operation. Her fire control system enabled her to accept contacts from another unit. She also had an experimental secure communications link, to be used with either surface ships or other submarines. Work on underwater communication was badly set back when *Thresher* was lost, because the underwater communications development team was lost with her.

At this time the notional ASW task group consisted of an ASW carrier (CVS), six destroyers, and two SSNs. It was intended to sanitize an area in which a carrier task group or other high-value formation might operate. The combination was expected to have a 30 percent probability of detecting a quiet submarine operating below the layer, and an 80 percent chance of detecting a noisy submarine such as a snorkeler. Destroyer screens of high-value units such as carriers were intended to have a 50 percent probability of detecting a conventional submarine, which was hardly adequate in the face of the new nuclear submarines. It was therefore proposed that each deployed fleet include one ASW task group to supplement its destroyer screen. Screens had to be augmented when ships operated for an extended period or in the face of a high threat.

This seems to have been the last gasp of the U.S. Navy's interest in close cooperation between submarines and surface ships for some time. However, the Royal Navy became intensely interested in such coordination because in HMS *Dreadnought* it had an extremely effective long-range sonar coupled to very ineffective torpedoes. The British therefore sought to use the longer-range weapons of the surface group to exploit what the submarine could see, using a near-vertical link through a surface ship variable-depth sonar. The link ship had to track the SSN below her, which she had to protect from accidental attacks. The combination was unwieldy, but it interested the U.S. Navy as U.S. interest in submarine support of surface forces revived in the mid-1960s. The U.S. Navy noted that in a 1965 British convoy exercise (Aigrette), *Dreadnought* detected 24 of 25 valid targets. In March 1965 CinCLantFleet, who had opposed using nuclear submarines to support task groups, reported British progress using *Dreadnought*, which had been briefed at the February 1965 NATO ASW conference at Londonderry.

The emergence of Soviet SSNs was behind America's new interest in coordinating surface ships and submarines. Although initially the Soviet SSNs could not reach high speeds, it appears that the U.S. nuclear submarine community assumed correctly that they would (the British did not). About 1964 ADM Hyman Rickover began to agitate for construction of U.S. nuclear submarines fast enough to escort carriers. In 1966 there were several proposals for a submarine version of the surface ship computer tactical system, which would be linked to the surface force in some way (which was the main problem). At this point the new submarine was often called the AGSSN. In theory it was an experimental auxiliary submarine, but it was clearly the favored replacement for the current *Sturgeon* class. The difference between the U.S. and British concepts was that the U.S. Navy planned to place its submarine escort well in the van of the surface force, where it was unlikely to run afoul of surface escorts.

About this time a Pacific Fleet submarine officer gave a lecture describing direct support by three SSNs, two of them 15–30 nm ahead of a convoy to avoid interference from friendly surface ships, with a safe lane between them so that they would not attack each other. Another SSN would trail the formation to deal with an enemy SSN overtaking the convoy, a distinct possibility. The SSNs would range on each other to keep station. It appeared that adding more SSNs would give only a marginal advantage, but fewer were not acceptable. The ongoing major fleet escort study showed that the optimum for a convoy was two or three SSNs and five to eight surface escorts. Destroyers were needed to deal with leakers from ahead (a quiet conventional submarine could penetrate). They would protect the wings of the formation, and they could prosecute SSN contacts in sectors the escorting SSN could not reach. Poor underwater communication stymied initial U.S. experiments with direct support in the late 1960s.

During the October 1973 confrontation in the Middle East, the Sixth Fleet's need for better protection against Soviet nuclear submarines led to a new effort to develop direct support tactics. A distinction was made between integrated support and associated support. In the former, the SSN was controlled by the OTC of the supported force; in the latter, the submarine was controlled by the Submarine Operating Authority. The latter was responsible for deconflicting submarines operating in any area. The OTC had greater flexibility when operating a submarine in integrated support, but integrated support made much greater demands on the most difficult part of direct support: rapid communication to keep the submarine at the right distance from the protected surface force and to verify its identification when challenged by friendly ASW forces, particularly aircraft. Without rapid and effective communication, integrated support was impossible. On the other hand, associated support depended on scheduled communications, which precluded quick tactical responses. Very simple messages (such as "come to depth to receive a message") could also be sent by sequences of practice depth charges and keyed sonar signals. Late in the Cold War the U.S. Navy added a bell-ringer feature to SQS-26/53 sonars, by means of which submarines within range could be notified that they had to prepare to receive radio messages. Messages from submarines involved similar awkwardness, the favored technique in the 1970s being to release a submarine-launched one-way transmission (SLOT) buoy. It was generally vital to minimize communication with the submarine.

The closer the supporting submarine was to the supported force, the more it was affected by the noise of the formation and by the sound of its sonars. It was soon found that a submarine would find passive sonar search difficult within 5 nm of a passive surface ship or 15 nm of an active SQS-26 ship. The same noise sources would drown out target sounds when the submarine was within 15 nm of a passive ship or within 30 nm of an SQS-26 ship, or in the convergence zone annulus of the supported force. On the other hand, although in theory a Charlie could fire its missile from 30 nm away, it would have to close within 20 nm to be sure of identifying its target passively. Supporting submarines would generally sprint and drift; the faster the force, the less time the submarines would have to listen (drift).

Communications problems caused the U.S. Navy to abandon close support in the 1980s, and eventually to detach SSNs altogether from battle groups. SSNs could still help battle groups by forming a defensive perimeter, normally less than 50 nm from the task group center. Alternatively, SSNs could search for and engage enemy submarines well ahead of the protected force (60–100 nm) along a projected route. SSNs could search for and engage an enemy inside or adjacent to a fixed geographic area before the task force operated there, or could form a barrier between the task force and the enemy. There were also surveillance-aided operations, such as interception of enemy submarines based on SOSUS fixes, trailing, and delousing (detecting shadowing enemy submarines). Trailing and delousing were considered Cold War rather than Hot War operations.

NEWLY SIGNIFICANT SOSUS

Nuclear submarines made SOSUS a far more successful and significant intelligence asset. In contrast to diesel submarines, which snorkeled only intermittently, nuclear submarines had to run their machinery continuously. Pumps and turbogenerators, for example, could not be turned off, at least not in early nuclear submarines.[5] Soviet First Generation nuclear submarines were particularly noisy, but all early nuclear submarines made significant noise. This generic weakness was apparently first revealed when the submarine *Nautilus* was tracked for several days.[6] At 20 kts she could be detected by one station at a range of 1,200 nm, but not she was not picked up at any useful range when running at 9 kts. In 1961, SOSUS tracked the prototype SSBN, USS *George Washington*, on her first deterrent patrol during her entire trip across the Atlantic to the United Kingdom.[7] *George Washington* had not been specially silenced because the Polaris program was considered so urgent that she was built on the hull of a *Skipjack*-class attack submarine.[8] The U.S. Navy adopted the British method of silencing machinery by rafting it (rigidly mounting it on a structure resiliently mounted on the submarine hull). This technique was kept secret, but it was likely to occur to any Soviet engineer intent on silencing Soviet submarines.

The great question was when the Soviets would silence their submarines to the point that long-range detection (SOSUS) would no longer be possible. Post–Cold War accounts of Soviet acoustic research indicated that the Soviets were well aware of sources of noise. The very considerable lags between such knowledge and its use seem to be attributable to the great inertia of the Soviet industrial system, which emphasized quantity over quality. In 1967, for example, a Western paper on the prospects of SOSUS argued that the Soviets had adopted noisy twin screws because of problems operating in ice. The author did not know that the single-screw design that became the second-generation Victor class had been adopted in 1959. He was unaware of that because it would be nearly a decade until the first Victor was completed.

Early upgrades to SOSUS included digital spectrum analysis, introduced in 1962, which was said to increase effective range by 50 percent.[9] Typically, the detection range increased from 600 to 1,100 nm. Stations were later fitted for automatic signal detection, which presumably made it possible to exploit weaker signals received from greater distances. The range improvement suggested that at some point SOSUS arrays on both sides of the Atlantic could cover the whole ocean, with dramatic consequences for hunting submarines. Detection range in the Atlantic was ultimately limited by the Mid-Atlantic Ridge and by seamounts that blocked sound propagation across the ocean. Given the greater effective range, it was worthwhile to extend SOSUS coverage to the far side of the Atlantic by completing arrays monitored from Brawdy in Wales. Full SOSUS coverage would make it possible to change ASW strategy in the Atlantic from reliance mainly on barriers across Soviet submarine routes into the Atlantic to interceptions based on SOSUS detections.

The Navy was also working on a next-generation Trident system intended to replace SOSUS if Soviet silencing defeated the existing system. Trident seems to have been a double line of upward-looking sensors that could pick up the flow noise of moving submarines, a signature impossible to suppress. It presaged the much later Fixed Distributed System (FDS). Work was also continuing on active systems such as Artemis.

When the U.S. Navy deployed fast single-screw nuclear submarines in 1959, a new source of submarine noise was discovered: blade rate. Previously, propeller noise had been associated with cavitation (noise created when bubbles caused by the propeller collapse). As a submarine operates deeper, water pressure increases, and cavitation disappears. Blade rate, a low-frequency sound caused by the propeller's disturbance of the water flowing over the after end of the submarine, is entirely different. That flow pattern reflects the shape of the after end, with its planes and rudders. As a propeller blade enters the wake of, say, the vertical rudder, it vibrates. This vibration travels up the propeller shaft and through

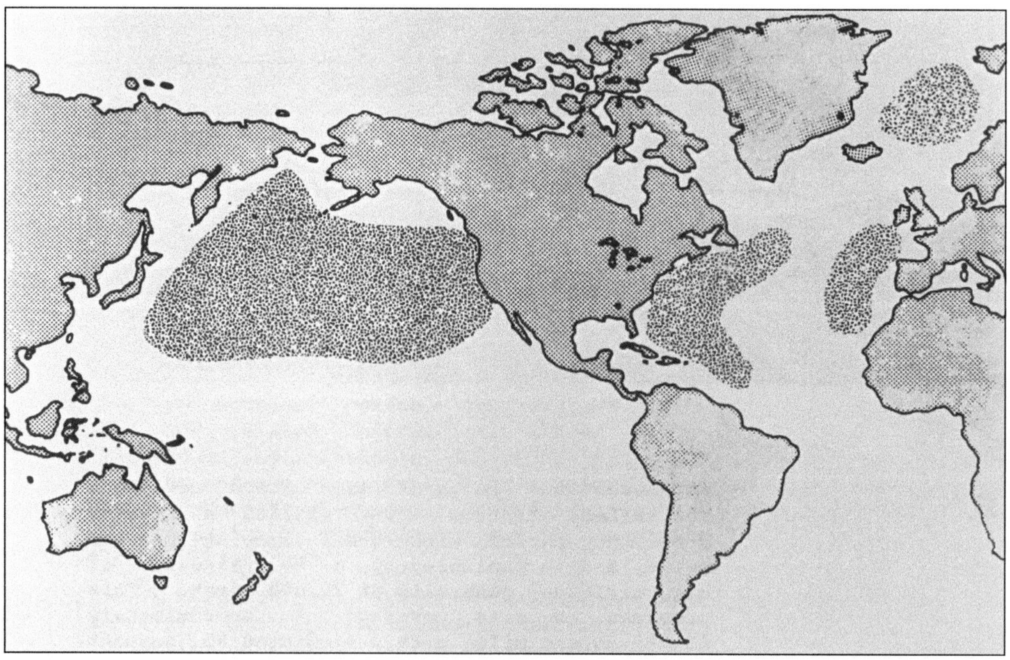

Worldwide SOSUS coverage about 1975, including the Backscratch array on the eastern side of the Atlantic, the coverage north of Iceland, and the Pacific coverage extended by the array in the Aleutians (given the inclusion of the British array, the date on the map must be too early). Note that there is no coverage in the Mediterranean or in the Indian Ocean. The source is the declassified Cross Report; originally this map was Secret.
AUTHOR'S COLLECTION, FROM NHHC

USNS *Mission Capistrano* was modified for developmental tests of the Artemis long-range active sonar system. The stack visible in her after well deck served a diesel generator. USNI

the hull to be radiated mainly out the bow, in the direction the submarine is traveling (also out the stern, to a lesser extent). For example, at 200 RPM, each blade hits the wake of a stern plane or rudder 4 times each time the propeller turns, 800 times each minute. That is multiplied by the number of blades, so that a 4-bladed propeller will vibrate 3,200 times each minute, a frequency of 53.3 Hz. Blade rate was typically detected because it modulated a submarine's flow noise. Demodulating that noise (a form of analysis called DEMON) revealed blade rate and its low harmonics.

Blade rate was a terrible surprise. The U.S. Navy was building a large fleet of silenced single-screw submarines, both SSBNs and attack submarines. Now it seemed they would not be quiet at all. The Navy began a crash program to find some cure, bringing in British acoustics experts to help. Two solutions were found. The U.S. Navy adopted the simplest of the two and reshaped propellers so that they entered the wakes more gradually. The new propeller blades were scythe shaped; they can now often be seen in photos. This solution could easily be applied to the large number of single-screw submarines already being built, the caveat being that the propellers had to be precisely machined. The alternative was to replace the propeller with a pump-jet, which would smooth out water flow before ejecting water from the after end. The British adopted pump-jets for their second-generation nuclear submarines, and they feature in current U.S. and British submarines.

Blade rate was an important intelligence secret. We now know that the Soviets discovered it about 1964–65, but that did not help them, because they lacked the ability to produce a large number of precision-machined propellers. The situation changed in the 1980s when Norwegian and Japanese firms sold the Soviets the computer-controlled milling machines they needed to make quiet propellers. To explain why that sale was so devastating a violation of NATO export controls, the Reagan administration revealed the nature of blade rate.

The first three full SOSUS systems were installed in 1954; the associated NAVFACs were at Ramey AFB Puerto Rico in September (operational in February 1955), at Grand Turk in October, and at San Salvador (in the Bahamas) in December.[10] Sites installed in 1955 were at Canadian Forces Station Shelburne (Nova Scotia); Nantucket, Cape May, New Jersey; and Bermuda. In 1956 NAVFACs were added at Cape Hatteras and Antigua. At this time the evaluation center in Norfolk was supplemented by one in New York. In 1957 the array at Eleuthera was raised to fully functioning status, and another Atlantic station was added at Barbados. That year the first Pacific station, at San Nicholas Island, California, entered service. Experiments with a shallow-water system began (they ended in 1964). Four more Pacific NAVFACs completed the initial Pacific system (a planned station on Vancouver Island in Canada was canceled due to a change in Canadian government): Pacific Beach, Washington; Coos Head, Oregon; and Point Sur and Centreville Beach, California. The planned Vancouver NAVFAC was replaced by additional processing at Pacific Beach. In 1959 an additional system was installed at Argentia, Newfoundland, to surveil the approaches to Hudson Bay. This curved shallow-water array was deactivated in 1968. According to a Naval War College lecture, in 1965 there were 12 Atlantic NAVFACs and 18 Atlantic arrays, including 2 each at Argentia, Cape Hatteras, and Eleuthera; and 3 terminating at Bermuda. A shallow-water complex of arrays between Newfoundland and the 1,000-fathom curve about 125 nm to the south protected the inshore side of the system. A station on the Norwegian coast became operational in 1963. A NAVFAC in Iceland (Keflavík) became operational in September 1966.

In 1967 a British mission discussing a U.S. offer of a SOSUS system was told that each NAVFAC monitored 1 or 2 arrays of 40 to 48 hydrophones, mostly in deep water off the continental shelf. The British had to be told about the importance of discrete frequencies and signatures created by propulsion systems and power plants. Lewes operated at 10–310 Hz, but its processor was limited to 10–150 Hz, using filters to divide the spectrum into 0.2 Hz slices. When a more detailed examination of one beam (searchlight mode) was wanted, the output of that beam could be run through another frequency analyzer covering 150–300 or 300–410 Hz, though Lewes could not use the latter range effectively due to the limited response of the hydrophones. Vernier recorders were a new feature. They made it possible to examine beam outputs over selected bands 50 Hz wide, with a filter band width of 0.1 Hz, to give a much-expanded frequency scale. Bearing could be ambiguous because beams formed on either side of the array gave a mirror image effect. In most cases arrays were sited so that acoustic propagation (hence range) on one side was reduced, for example by shallow water. Further ambiguity could usually be resolved by consideration of the plot. To filter out non-submarines, the station maintained a shipping plot of the whole western Atlantic area based on sitreps from the evaluation center. Skilled and experienced watchkeepers interpreted lofargrams, selecting harmonically related families of spectral lines, such as those arising from blade rate and diesel firing rates. Lewes maintained a comprehensive library of lofargrams. Contacts considered of interest were reported to the evaluation center. Upgrades to SOSUS included the ability to transmit grams of interest directly to the evaluation center.

A planned phased expansion of SOSUS, approved in 1964, emphasized improving cross-fix capabilities off the U.S. East and West coasts. This program was canceled in March 1965 in favor of a crash program focused on forward areas. New arrays would cover Soviet submarine transit areas rather than patrol areas, on the theory that once on patrol, submarines would be quieter, and hence more difficult to detect.[11] The possibility of forward-area surveillance was apparently demonstrated by the spectacular performance of an array off Adak installed in 1962, which could look into the Petropavlovsk area. Submarines in transit could be detected without resorting completely to the detailed signatures obtainable from submarines on patrol. Some of that signature data seems not yet to have been collected. Secretary of Defense McNamara approved the new program in the fall of 1965. It began with the Keflavík array. The program envisaged two new arrays each year through 1970, plus new R & D funds. A second array, looking out toward Petropavlovsk, was emplaced at Adak. The other initial arrays seem to have been in the Philippine Sea (NAVFAC Guam, in operation 1968) and at Midway (operational 1968). The primacy of the two Pacific installations suggests a connection with the Vietnam War and the need to deal with any possible submarine interference with the Seventh Fleet.

By 1960 the U.S. Navy was hoping to be able to detect and shadow a reasonable number of submarines entering a zone 1,000–1,500 nm from the coast, and to be able to destroy shadowed submarines if necessary.[12] For the moment, SOSUS covered potential submarine missile launch areas off the U.S. coasts, but not always the holding areas in which Soviet strategic submarines normally operated. U.S. intelligence initially credited the SS-N-3 strategic cruise missile with a range of 220 nm, and its stablemate the SS-N-4 with a range of 350 nm; its successor, SS-N-5 (introduced in 1963), was credited with 700 nm.

In the mid-1960s the United States seems to have offered SOSUS to Great Britain and Japan.[13] When Congress debated extending Caesar (SOSUS) overseas in 1967, it was assured that the system would not escape U.S. control. The British accepted the U.S. offer, resulting in the NAVFAC at Brawdy. A British team visited the United States in June 1967 and was allowed to inspect the NAVFAC at Lewes, Delaware.[14] A 1978 report for the House Committee on International Relations referred to a SOSUS array "between Japan and Korea." This was presumably a Japanese array.[15] Many accounts of Mediterranean ASW mention that U.S. forces there knew when Soviet submarines entered and left the sea, but that they were not tracked once they were in the Mediterranean. That suggests the existence of some type of acoustic fence in or just outside the Strait of Gibraltar.

As an indication of what SOSUS could do in the Atlantic, in 1968 the system reported 467,677 contacts. Sifted, that equated to 4,755 submarine contacts: 1,716 U.S. nuclear submarines, 2,451 U.S. diesel submarines, 303 Soviet nuclear submarines, and 285 Soviet diesel submarines.[16] Of 244 patrol plane (VP) sorties flown in 1967–68, 169 contacted Soviet nuclear submarines. When based on SOSUS data, contact was made 81 percent of the time. If SOSUS was cold for 24 hours prior to arrival at the datum, the patrol plane was unable to make contact at all. During this period the first observed transit of the Denmark Strait by a Soviet nuclear submarine was made, at a lower speed than a transit in the open ocean or in the Norwegian Sea. It would seem that the Soviets were aware of some sort of Western listening system and assumed its array(s) conducted surveillance in the GIUK Gap chokepoint. In 1969, the number of unidentified submarine contacts increased to 787, of which 425 were classified as certain or probable (and only 93 as non-submarines); of those, 258 were in the eastern Atlantic, including the Norwegian Sea; 84 were in the Mediterranean; and 23 were in the Ocean subarea.

Soviet submarine numbers in the Mediterranean increased sharply after the Six-Day War. During the 18 months between July 1967 and December 1968 it was estimated that there were 19 Soviet nuclear submarine deployments (11 Echo IIs, 8 Novembers) to the Mediterranean and 56 diesel submarine deployments (20 Whiskeys, 25 Foxtrots, 7 Romeos, 4 Zulus).[17] The Echo IIs patrolled in the Tyrrhenian Sea and the western Mediterranean, the Novembers in the eastern Mediterranean and east of Sicily.

Lacking SOSUS or any other long-range detection system in the Mediterranean, the Sixth Fleet relied on sweeps by HUK groups. Policy prior to the Six-Day War was unaggressive, but after the war, when there was a sense that the Soviet Mediterranean squadron might fight, all submarines not American or known to be friendly would be prosecuted to locate and identify potential threats. An intensive effort by the *Essex* HUK group, supported by Task Force 60 destroyers, Task Force 69 submarines, and patrol aircraft between 21 July and 10 September 1967, positively identified 13 Soviet submarines. HUK groups were particularly effective because they normally searched out to a radius of about 100 nm 24 hours a day. This was far more than the one and a half VP squadrons could do, with their average of 1.4 aircraft airborne all the time. OEG later estimated that one HUK group in the Mediterranean was equivalent to three VP squadrons in surveillance of evasive submarines in the Mediterranean, and to more than 12 squadrons if continuous, rather than daily, surveillance was wanted.

TACTICAL LOFAR

In 1954, with SOSUS showing great promise, BuAer asked Bell Labs to develop LOFAR sonobuoys and a

LOFAR localization technique. Bell Labs was already working on the problem. Its 1 October 1953 report stated that LOFAR information alone had successfully directed an airplane to a submarine snorkeling at an unknown bearing 105 nm from the Bermuda test SOSUS array. The airplane dropped modified buoys, which it monitored using a LOFAR analyzer.[18] The January 1954 Jezebel progress report included a test run against HMS Auriga, a P5M dropping several buoys and applying LOFAR analysis to the best buoy in the water. The outputs of the other buoys were taped for later processing. At high speed on batteries the submarine was detected at 8 nm. When snorkeling, she was detected at 28 nm by the same buoy. A broadband buoy detected harmonics of her blade rate. By this time Bell Labs had proposed a technique employing a circular sonobuoy pattern about 50 nm in diameter, the airplane applying LOFAR analysis to each buoy in turn for two to three minutes. The buoy detected maximum frequency when it was nearly directly ahead of the submarine, and minimum when it was nearly astern. Buoys abeam the submarine registered the average of the two. Given an approximate position, the airplane could create a smaller circle that should place the submarine within 1 or 2 nm.

Tests at the Naval Air Development Center began in 1955.[19] The first attempt to detect a submarine, off Bermuda on 19–24 January 1957, was considered successful. The system prototype was installed on board a P2V-7 on 9 July 1957, and it began test flights off New Jersey on 25 July. Evaluation was completed in January 1958, after which trials were conducted by the Operational Evaluation Force. The system became operational late in 1958. After that Lockheed fitted P2V-5s with Jezebel; P2V-7s received it either as a retrofit or during production, the latter as an Aircraft Service Change (beginning in April 1960). A snorkeling or fast-running nuclear submarine could be detected at 20–40 nm; a quiet nuclear submarine could be detected at 10 nm based on the noise of her cooling system.[20] A range of 125 nm was cited in one case. Such ranges made airborne LOFAR an effective way to search the probable submarine areas defined by SOSUS. To the extent that it could reveal details of a submarine signature, it could also solve the classification problem. A special data link (Heifer) enabled an airplane using LOFAR to send its signature data back to a shore center for comparison with a signature found by SOSUS. It also proved useful for one airplane relieving another, as the reliever could be sure that it was detecting the same signature.

Unfortunately, buoy LOFAR was nondirectional. Given its extraordinarily long range, it was difficult for an airplane to close in on a submarine; in the mid-1960s buoy LOFAR was credited with no better than a 5 percent chance of leading to an attack. In the mid-1950s Bell Labs had conceived two solutions. One, a directional buoy, was beyond the capacity of current technology. The other, CODAR (correlation detection), compared times of arrival of a broadband event picked up by two buoys close to each other. Correlation offered reasonable directional accuracy around the line projecting from between the two buoys, albeit ambiguous as to which side of the line between them. This ambiguity could be cleared up by planting a second pair of buoys and monitoring them. In practice CODAR was ineffective, because the airplane had to lay the pairs of buoys very precisely (which meant from low altitude). Even then it had to reacquire the target, which could be a lengthy process. Because it was a broadband technique, CODAR could not match LOFAR range.

Given their limited size, active sonobuoys could not pack much power. They were eclipsed by an alternative technique, explosive echo-ranging ("Julie"), which was developed by the Naval Air Development Center at Warminster, Pennsylvania. A practice depth charge exploded near the passive buoy would create a broadband pulse of energy, which would echo off a submarine.[21] Julie entered fleet service in 1957. It demanded higher-precision air navigation, as the distance between explosive source and buoy had to be known. Neptunes (P2V) were given special amplifiers that could drive the dual-track pen recorders installed for MAD. The high energy of the explosive pulse created reverberation problems, and the system could be used only in deep water (otherwise the

A U.S. Navy P-3C from VP-24. It is identifiable as a P-3C by the mass of preloaded sonobuoy tubes abaft the wing, in the area outlined in black. The sword antennas abaft them were used to locate buoys by interferometry. The airplane had an unusual surface search radar with coordinated nose and tail antennas (the tail antenna is under the boom for the ASQ-81 MAD). The airplane was entirely unarmed, which raised uncomfortable questions as the Soviets prepared to deploy carriers armed with fighters to reinforce the "combat stability" of their Northern and Pacific Fleet strategic submarines. USNI

reflection from the submarine would be drowned out by reflection off the sea bottom). Too, Julie generally involved only a single buoy, so it did not offer the desired location information. Even so, this capability was retained, and at the end of the Cold War Julie was revived as enhanced echo-ranging. The first buoy used with Julie was SSQ-23, a low-frequency (100–3,000 Hz) device. During May 1956, Atlantic Fleet carrier ASW aircraft (S2F-1 and S2F-2 Trackers) began a retrofit program in which they received interim Julie systems; about 46 had been equipped by October 1958. By 1960, 75 of 131 Atlantic Fleet Trackers had interim Julie equipment, the last scheduled to reach the fleet by December 1961.

After a new ASW office (Op-95) was created in OpNav in 1964 to concentrate the Navy's ASW projects, one of its first directives was to solve the LOFAR localization problem on a crash basis.[22] Ultimately, the solution was directional LOFAR buoys, pairs of which could locate a target. Such buoys required much more processing. The buoy multiplexed data from arrays at right angles with an omnidirectional array for comparison. That in turn required a digital analyzer, because the usual paper readouts from multiple LOFAR sensors on each buoy could not be compared. The new directional (DIFAR) buoys had magnetic compasses. Given quick direction-finding, DIFAR buoys had a much better chance of catching a conventional submarine snorkeling intermittently on a fast cycle. Unlike the precision-laid CODAR, DIFAR buoys could be laid from greater heights. To speed development, parallel contracts were let to Magnavox and Sanders Associates for quantities of prototype SSQ-53 buoys and associated digital AQA-7 processors. As a result of the crash program, SSQ-53 entered service in 1969. Further localization was done by an active buoy or by explosive echo-ranging, the active buoy being preferred.

The U.S. Navy became interested in having DASH drones on board destroyers lay buoy fields, a technique called Destroyer Jezebel (DesJez). As DESLOC, the idea was the subject of a 28 August 1962 Tentative Specific Operational Requirement. Potential range was 30 nm on a loud nuclear submarine. The destroyer would use an airplane-type AQA-3 analyzer, receiving buoy data via a 30 nm link. Buoys would be recoverable, laid by either the ship or her DASH helicopter. The idea had recently been tested by the Operational Development Force, using both LOFAR and CODAR processing (there were no directional LOFAR buoys yet). In tests, a QH-50D equipped with three long-range tanks flew to a sonar contact at 4,700 ft, dropped to sea level to launch eight buoys, loitered while the ship processed their output, and then dropped a Mk 44 torpedo. Maximum monitoring range was 12 nm. With long-range tanks, a QH-50D could reach 50 nm from a ship; it had a four-hour endurance.[23]

AIRPLANES

The combination of SOSUS and successful LOFAR buoys seems to have inspired a new view of land-based patrol aircraft. Given the reach of SOSUS, it was worthwhile to be able to span the Atlantic. A formal requirement (TS-146) issued in May 1957 called for a long-range airplane adapted from an existing

P-3s could cover the whole Atlantic. The main role of the ASW carriers changed to direct protection of strike carriers. USS *Randolph* (CV 15) is shown on 26 August 1966. Among the Trackers spotted forward is one E-1B (formerly WF-2) Tracer airborne early warning or snorkel search airplane. Another two are abaft the island, opposite three Sea King helicopters. NHHC

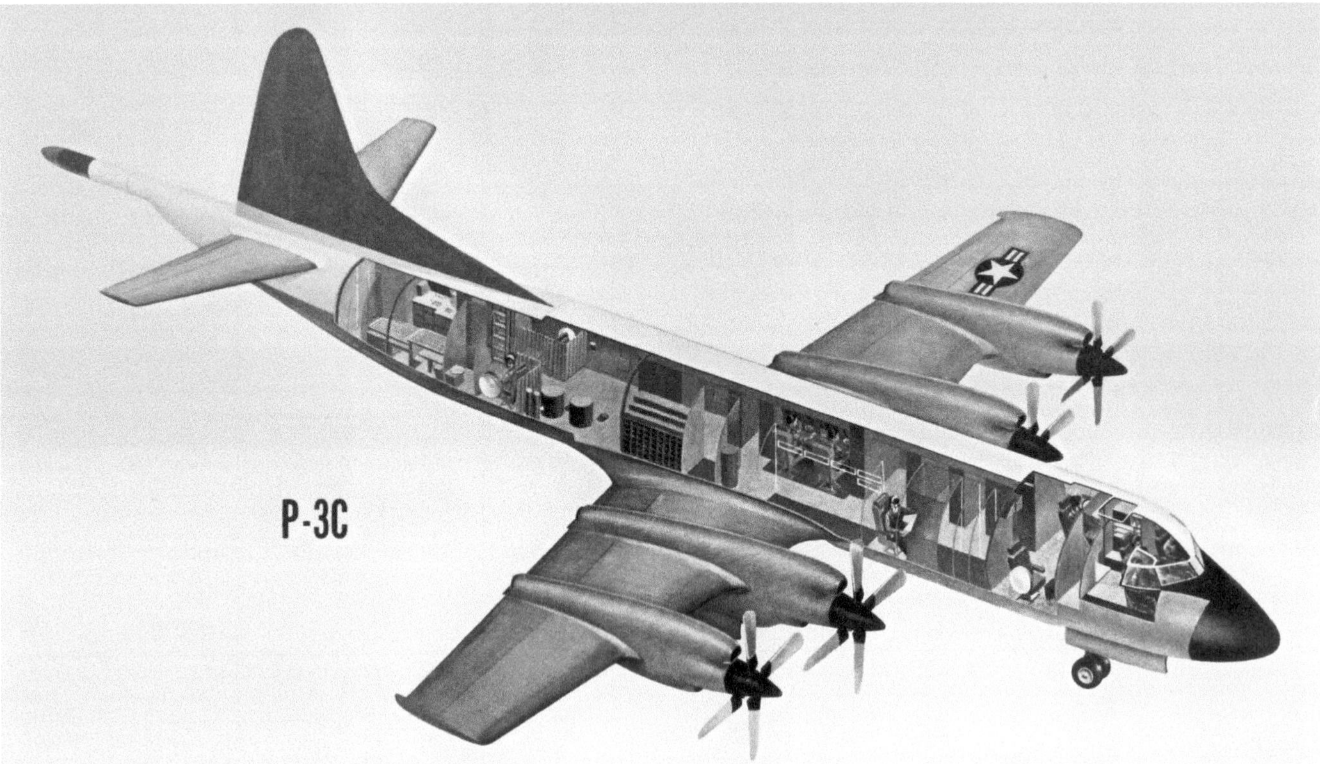

The P-3C was one of the first two maritime patrol aircraft with a computerized data system (the other was the Nimrod). In earlier P-3s, the tactical coordinator (TACCO) typically stood or sat facing a row of LOFAR analyzers, coordinating the work of the operators with the pilots and building a tactical picture in his head. In the P-3C shown here, there are only two operators, who enter their data into a central computer system. The TACCO sits separately in a seat facing forward. Given a central computer, the TACCO can devote most of his time to tactical decisions. The sensor operators detect targets (using radar and sonobuoy processors), providing elements of the tactical picture the TACCO built up in the airplane's central computer. Automated navigation eliminated the lengthy mechanical calculations formerly needed to determine relative sonobuoy positions. However, the system did not track sonobuoys as they drifted in the water, so the airplane had to overfly them periodically to verify their positions. Given the tactical picture, the central computer could compare the likely results of alternative tactics. The A-NEW system (literally "a new" system) was conceived in July 1960 at the Naval Air Development Center (NADC) in Johnsville, Pennsylvania. The first version was flight-tested on board a P-3A early in 1964; the P-3C entered production in 1969. In 1967 it was estimated that a P-3C would be two to three times as effective as a P-3B, but also that it would have as many electronic parts as all 144 P-3Bs. Buoys were selected by the computer, so the airplane had 48 external preloaded chutes, instead of the 9 chutes of earlier P-3s, whose buoys were all stowed internally. A P-3C also had reloadable chutes for buoys stowed internally. AUTHOR'S COLLECTION

airliner. Lockheed and Douglas submitted proposals. Lockheed won in April 1958 with an airplane based on its Electra turboprop airliner. The airplane was initially designated P3V-1, but it did not fly until the designation system had been revised; it became the P-3 Orion. Unlike its predecessors, the P-3 did not have a massive under-body radome. Instead, it had an unusual combination of coordinated nose and tail antennas (for an APS-80 radar) feeding a common display. It had the same in-line arrangement of acoustic and other sensor operators as the last version of the Neptune, making it possible for a tactical coordinator (TACCO) to see all the inputs and coordinate action based on them.

The first production contract was awarded in October 1960. An aerodynamic prototype flew in August 1958, and the first full prototype flew on 25 November 1959. The first preproduction airplane flew on 15 April 1961. Before that, in July 1960 the Naval Air Development Center (NADC) at Johnsville, Pennsylvania, proposed an automated tactical system, developed as A-NEW (literally, a new ASW system). The first Mod 1 version was flight-tested on board a P-3A early in 1964. The interim version was reportedly so successful that the fleet tried hard to bring it into production. It was incorporated in the P-3C, which was approved for production in August 1967. The first production prototype flew on 18 September 1968, roughly contemporary with the British Nimrod, which also had a computerized tactical system. In both airplanes, the computer system kept track of the tactical situation and enabled a TACCO to see the outputs of all the airplane sensors. Digitization made it possible to upgrade the P-3C fleet significantly.

Roughly contemporary with the Orion project was the European Atlantic, intended to replace the Neptune, which was serving with the Canadians, the French, and the Dutch. On 14 December 1956 the NATO Council called for a Neptune replacement, which was envisaged as a pan-European (NATO) project. In 1957 the council issued a specification (NMBR.2). National versions would have a common airframe and power plant (two turboprops) but their own equipment. On 28 October 1958 the French Breguet 1150 was selected by the NATO group of experts, a choice endorsed later by the NATO Armaments Committee. The resulting Atlantic was operated by France, West Germany, Belgium, the Netherlands, and Italy. The United States contributed

The Breguet Atlantic was conceived in the late 1950s to meet a NATO specification for a Neptune replacement. One is shown at a NATO ASW meet alongside two of the alternatives, a U.S. P-3C at left and a British Nimrod in the background. The prominent scoop under the nose fed a "sniffer." The radome above the vertical tail was for ESM. NARA

funds under the Mutual Weapons Development Program but never considered operating the airplane; it wanted much greater endurance. Like the P-3, the Atlantic had AQA-3 LOFAR sonobuoy processors. The later Atlantique combined the original airframe with a new computerized combat system, hence might be likened to the P-3C with its A-NEW.

The Canadians chose their own new maritime patrol airplane, the very-long-endurance Argus, which was based loosely on the British Brittania airliner.

Britain's replacement of the Shackleton was a drawn-out process that eventually resulted in the Nimrod, which used a much-modified Comet airliner

A Royal Canadian Air Force CP-107 Argus maritime patrol aircraft. Like the earlier Shackleton, it was intended to replace war-built Lancaster bombers adapted for maritime reconnaissance. The basic design was proposed unsuccessfully by Bristol as a Shackleton replacement. The final design employed parts of the Bristol Brittania airliner, but the airplane was powered by turbo-compound (piston) engines rather than the turboprops of the airliner, which gave it the longest endurance (not range) of its time. Patrols initially lasted 24 hours, but were reduced due to budget limitations to 18 and then 16 hours. The record flight time was over 32 hours, and the airplane landed with more than 2 hours left in its tanks. Typical cruising speed was 180 kts. The chin blister of this Mk 2 contained a British ASV.21 radar (as in the original version of the Nimrod). The blister above the cockpit was intended for an ESM D/F antenna, which was never fitted. The blister under the fuselage abaft the bomb bay was for an ALR-8 ESM antenna. This set included an airborne DF system (APA-69A) and a pulse analyzer. The MAD in the boom aft was ASQ-10, with a rated range of 1,800 ft. The airplane also had an ASR-3 sniffer (removed by 1978). The Argus was the first foreign airplane to get the U.S. AQA-3 LOFAR/CODAR processor. Note the searchlight. The last operational Argus flight was on 10 November 1980. USNI

A Nimrod MR.1 maritime patrol aircraft investigates a Soviet *Kiev*-class VTOL carrier. The carrier was advertised as part of an effort to improve the "combat stability" of submarines, meaning to beat off Western air and surface ASW. The Nimrod shows the searchlight in its starboard wing external tank. Not evident here is its ASR 21 radar. The big tail boom carries MAD (ASQ-10A), and the top of the vertical fin carries ESM antennas. Nimrod development was protracted. The original Air Staff target was forwarded to the Ministry of Aviation in April 1963; the new airplane was wanted by 1970, but acquisition was delayed because the RAF was buying so many other airplanes at the same time. At the earliest, it could not be produced before 1973. The Air Staff considered adapting an existing airplane; it looked at the developmental Breguet Atlantic as well as the U.S. Orion and adapted Trident and VC-10 airliners. In February 1965 ministers decided to reject all four aircraft in favor of a heavily modified version of the Comet 4 airliner. The short time available precluded adding any major new equipment. The new airplane improved on the existing Shackleton by adding Jezebel processors and a degree of automatic processing. A major upgrade was planned for the 1970s (Mk 2). Replacement of CODAR was first priority, DIFAR being the obvious (but rejected) candidate (Barra was adopted). The project included a short-range active/passive buoy to relocate a target and support an attack. Nimrod first flew on 23 May 1967 and entered service on 2 October 1969, becoming the first maritime patrol aircraft with a central computer system (the P-3C was slightly later). USNI

airframe. Flying slightly before the P-3C, it became the first maritime patrol airplane with a central tactical system computer. Its least satisfactory feature was its sonobuoy system, which comprised the existing short-range Sonics Mk 1C of the Shackleton and a LOFAR system that had no directional capability. The British later rejected the U.S. directional system (DIFAR) introduced in the late 1960s in favor of the Barra buoy developed in collaboration with the Australians.

A Sea King (HSS-2, later SH-3H) helicopter dips its AQS-13 sonar body. U.S. Sea Kings were assigned to aircraft carriers; they were not carried by smaller surface ships. The object under the nose is the housing for the antenna of the APN-130 Doppler radar altimeter, which was coupled to the autopilot to enable the helicopter to hover stably. The object under the belly is the radome of an LN-66HP surface search radar, both for snorkel detection and to detect enemy missile craft. The box abaft the radar contains decoys. Not visible is the ASQ-81 MAD bird on the other side of the helicopter. The helicopter is recognizable as the upgraded version because it shows the circular radome of its LN66HP radar. All SH-3s had an eight-channel Jezebel relay; none had onboard LOFAR processing capability. SH-3H had two channels to process active sonobuoys. In 1969 the Atlantic Fleet ASW command considered the single-sensor SH-3 obsolescent in the face of nuclear submarines; it also criticized the helicopter's slow dash speed and short radius. Command recommended fitting the helicopters with the LN-66 radar then being adopted for LAMPS I, and repeated a recommendation to add MAD as a localization sensor to complement the sonar. Without it, fixed-wing airplanes had to be diverted to classify sonar contacts using their MADs. At this time ASQ-81 was being incorporated in the new LAMPS helicopter. Although SH-3 production ended in FY 67, the helicopter remained in service, in modified versions, into the 1990s. USNI

A Royal Australian Navy Wessex HAS.31 dips its AQS-10 sonar from a bell-shaped underbelly housing. It was similar to the Royal Navy's Wessex AS.1 except for its uprated gas turbine and its U.S.-type dipping sonar. The Royal Navy issued a requirement for a single-rotor heavy ASW helicopter in December 1955; Westland offered a version of the Sikorsky S-58 (used by the U.S. Navy as its HSS-1) in February 1956. The prototype, a modified U.S.-built aircraft, flew in August 1956. In British service the Wessex replaced fixed-wing Gannet ASW aircraft. Its British Type 194 variable-depth sonar gave it a search speed of only up to 17 kts, but it was expected that new U.S. dipping sonars like the one shown here would increase that to 30 kts. Compared to a fixed-wing airplane, the helicopter would find it appreciably easier to find and fix a submarine. The fixed-wing airplane would cover a wider search area and had a radar, but it was argued that ASW helicopters would be supported by airborne early warning aircraft. Fixed-wing aircraft would run out of sonobuoys, whereas the helicopter had a dipping sonar which it would retain. From 1968 on, Australian Wessexes were upgraded to HAS 3 standard as HAS.31Bs. DR. JOSEF STRACZEK

A Royal Navy Wessex HAS.3 had the tactical system (including the dorsal radar) of the Sea King but the limited endurance of the earlier Wessex HAS.1. It did have an upgraded gas turbine (1,600 vs. 1,450 SHP). It was the first British helicopter capable of carrying a nuclear depth charge, the British-developed WE 177A (aka the 600 lb MC bomb). Reported yields were 0.5 kT for shallow water (down to 130 ft) and 10 kT for deeper water. When asked to dispense with the WE 177A, the Royal Navy argued that it alone offered a sufficient chance of destroying a fast deep-diving nuclear submarine like an Alfa. The WE 177A looked like a finned bomb (the RAF used other versions as tactical nuclear weapons); diameter was 16.5 in. and length 112.25 in. This version of the Wessex became operational in October 1967, but beginning in 1970 it was retired from all ships other than County-class missile destroyers. AUTHOR'S COLLECTION

WEAPONS

The developmental SQS-26 promised a reliable direct-path sonar range of 20,000 yds, twice that of the earlier SQS-23. Prospective ways of exploiting it were Extended-Range ASROC (ERA) and an extended-range version of DASH. To achieve sufficient accuracy at more than twice the range, ERA was given an autopilot. It was to have entered service in the early 1970s. Although it was tested in 1967 it was canceled, probably a victim of budget cuts during the Vietnam War. DASH range could be stretched using LORELI, a transponder buoy that a destroyer could track relative to the target, but this capability was never implemented due to inherent DASH problems.

The Nobska group's proposal for a shorter-range submarine nuclear bombardment missile evolved into the most exotic ASW weapon of the 1960s, SUBROC (submarine rocket). SUBROC actually predated Nobska, having been proposed by BuOrd about 1953 as part of the ALIEX study conducted after adoption of the Betty (ALIAS) nuclear depth bomb. In the spring of 1955 CDR J. D. Miller, who commanded the Pacific Fleet Regulus submarines, proposed a solid-fuel torpedo tube bombardment missile; OpNav issued an operational requirement (AS-08701) for a submerged-launch missile system in April 1956, before Nobska. The project died when BuOrd pointed out that underwater fire control at the range envisaged would be impossible. Development money became available in 1958 when the Triton bombardment missile was canceled. That year the feasibility study for the ASW version was completed. The first full-scale tests followed in 1961, and eight final tests (seven successes) in 1963. The missile's bombardment origin showed in its inertial guidance and its nuclear warhead (and its retention of an airburst capability). Development of the long-range BQQ-2 submarine sonar, with convergence zone range, made it a viable ASW weapon.

A BQQ-2 submarine would detect its target passively, using her fire control computer and manual plotting to deduce target position, course, and speed. She would ping once to determine the range, then fire. The missile was so fast that ping-stealing would not save the target. The powerful nuclear warhead (130 kT) would make up for any target maneuver executed while SUBROC was in the air. In 1960 SUBROC was considered essential as a means of SSBN self-defense. On that basis it survived a proposal to cancel it in favor of less expensive shorter-range conventional weapons (including a SUBROC carrying a lightweight homing torpedo).

An alternative version would have had a Mk 46 lightweight torpedo as warhead. The search radius of the torpedo would be 1,500 yds (but the estimated lethal radius of the nuclear warhead was 3,500 yds). This search radius would limit the effective range of a non-nuclear SUBROC to 20 nm. Even then it would have a lower kill probability than the nuclear version. CNO Admiral Burke doubted that the non-nuclear version could be developed alongside the nuclear one; he wanted the U.S. Navy to concentrate on the nuclear

commander of Pacific ASW forces argued that the logic of convoys no longer applied.[31] Once the Soviets had effective surveillance systems, they could find large convoys; adopting convoying would no longer frustrate submarines hunting for ships. Too, outside Japan, nowhere in the western Pacific could accommodate a 150-ship convoy at this time.

The antiship missile threat was dramatized when Egyptian missile boats sank the Israeli destroyer *Eilat* in 1967. In 1968 NATO held a major convoy exercise, Silver Tower, to test tactics against Soviet missile submarines and ships. It seems to have been the first such NATO exercise, and the results were depressing. Two carriers and some underway replenishment ships were sunk at the cost of two Soviet submarines. To keep attackers out of missile range, it seemed that area and distant support should have priority over close support (convoy). That led to proposals for "safe routing" and "sanitized lanes." Among other things, Silver Tower indicated that decreasing escort numbers made the direct defense of shipping (by convoying) less and less feasible.

The Soviets demonstrated the predicted SOSS-guided interception capability at the end of 1967.[32] On patrol between 4 December 1967 and 19 January 1968, a November (designated XRay-5 by the U.S. submarine tracking system) left Petropavlovsk, transited the Aleutians, and appeared off the U.S. West Coast. There she coordinated with the AGI (intelligence-gathering ship) *Gavril Sarychev*. This was the first such coordination observed in the Pacific. The submarine attempted to intercept the carrier *Enterprise*, which was en route to Vietnam. The carrier's escort, the nuclear frigate *Truxtun*, diverted to intercept the submarine, gained sonar contact, and coached in a P-3 that surprised her on the surface at dusk. COMASWFORPAC commented that the coordination with the AGI, as well as the near interception of the carrier, demonstrated "very advanced mission planning and an effective submarine command and control and support broadcast system." Maneuvers by both the submarine and the AGI seemed to imply at least a general knowledge of the West Coast SOSUS network. The Soviets were apparently aware of the impending departure of the carrier for the Far East. There were other attempts to intercept *Enterprise* at this time (N 285 on 5 January 1968, and N4/5 on about 12 January).

When U.S. carrier forces mined Haiphong Harbor in 1972, the Soviets sent one Echo I (*K 45*) and two Echo II (*K 57* and *K 184*) submarines from Petropavlovsk to a holding area not far from the Tonkin Gulf, where the U.S. strike carrier force was operating. The Soviet operation was revealed when *K 184* was trailed

USS *Guardfish*, shown here, successfully trailed the Soviet Echo II submarine *K 184* from near her home base at Petropavlovsk to nearly within range of U.S. carriers in the Tonkin Gulf. Revealing this deployment enabled President Richard M. Nixon to demand that the Soviets withdraw their missile submarine force. Revelation of the Soviet deployment caused a serious missile scare in the Pacific Fleet and led to urgent development of better point defense missiles. USNI

down to the holding area by USS *Guardfish* (SSN 612), which had picked her up off Petropavlovsk. *Guardfish* maintained her trail for 27 days, and detected the other two Soviet submarines as well. In Moscow for talks on strategic arms limitation, President Nixon raised the deployment with General Secretary Leonid Brezhnev, who agreed to withdraw the three submarines *Guardfish* had detected (another Echo II, *K 189*, may also have been involved).[33]

The Soviet operation emphasized the importance of the relationship between ASW and fleet air defense, because the three Soviet submarines posed a missile threat to the U.S. carriers in the Tonkin Gulf. Knowledge of their presence seems to have caused a near panic; carriers stationed Marines carrying Redeye missiles in the catwalks alongside their flight decks. The existing Sea Sparrow point-defense missile system could not have reacted quickly enough to destroy an incoming supersonic Soviet missile. On the other hand, to use their missiles the Soviet submarines would normally have relied on targeting information provided by a Bear radar reconnaissance aircraft, and the presence of a Bear—if it was detected—would probably have provided some warning. The visible outcome of the incident was U.S. acquisition of the much more effective (and much more automated) NATO Sea Sparrow system, and also development of the fire-and-forget RAM system.

TACTICAL COUNTERMEASURES

The Pacific Fleet sponsored a program of exercises to develop countermeasures to Soviet missile submarines called UPTIDE (United Pacific Fleet Project for Tactical Improvement and Data Extraction).[34] UPTIDE focused on protecting the carrier rather than locating and destroying the submarines threatening her. It therefore concentrated on the submarine's targeting problem, in a way that recalled the ASW discussion of deceptive tactics such as Haystack. The Soviets' long-range antiship missile threat could be dated from the appearance of the SS-N-3a (P-6) Shaddock missile on board Echo II–class submarines and Kynda-class cruisers in 1962. These missiles were typically fired on the basis of radar data from a Bear-D, but for the purposes of the exercise it was assumed that the Soviet submarine would have to find targets by herself. That already applied to the shorter-range SS-N-7 (P-40) missile on board Charlie (Project 670) submarines.

UPTIDE tested deceptive tactics intended to deny targeting data to Soviet submarines and thus prevent them from firing at a carrier. Deception emphasized passive sensors, because the big SQS-26 would be an unmistakable beacon. Initially, the goal of the exercises was to demonstrate the continuing value of ASW carriers, which were losing their open-ocean search role now that P-3 Orions could cover the entire ocean. The exercise series was very much the idea of VADM E. P. Aurand. The series was initiated by the Pacific Fleet in June 1968 and assigned to ASW FORPAC (which had been established in 1957), the Pacific Fleet center for ASW tactical guidance. Aurand had commanded a Pacific Fleet HUK group in 1965–67 before commanding ASW FORPAC. As a HUK commander, Aurand believed strongly in the CVS concept. He considered training exercises unrealistic because they were designed to bring forces into contact with attacking submarines, whereas in wartime the object would be to avoid such contact. He was also aware that past exercises had shown that the likelihood of detecting, localizing, and killing an SSN was only 10 percent. The focus had to change to evasion based on the submarine's reliance on her passive sensors to do her own detection, classification, tracking, and localization of targets.

Aurand called this "the thinking man's approach to ASW." The submarine had to follow a chain of requirements, so small reductions in the probability of each could have considerable consequences. For example, strict electromagnetic and acoustic emission control could complicate the submarine's task of picking the carrier out of a formation. Acoustic deception devices and practices (such as trailing two of the four shafts) could help. The mere knowledge, on the submarine's part, that such devices and practices existed could help; so could acoustic jamming. Using passive sonars could force down the submarine's speed and make her less likely to classify her target. On this basis, Aurand wanted tactical towed arrays. Aurand made time and resources available for UPTIDE by arguing that the previously scheduled exercises against simulated ballistic missile submarines would merely duplicate work already being done in the Atlantic.

Phase I of UPTIDE was three HUK ASW exercises (HUKASWEXes), on 12–16 January, 3–10 February, and 25 February–10 March 1969. Phase II was UPTIDE 2B through 2E (2A was canceled), beginning with 2B on 22 September–1 October 1969, then 2C on 18 January–7 February 1970, 2D on 12–16 October 1970, and 2E on 15–25 January 1971. Phase III was UPTIDE 3A (1–7 October 1971) and 3B (30 October–8 November 1972). A planned UPTIDE 4 series was canceled.

Phase I tested conventional ASW tactics: ASW Warfare Group 1 (built around the carrier *Kearsage*) and Desron 23 were opposed by two submarines notionally armed with cruise missiles: two diesel submarines in UPTIDE 1-69, and then two *Skipjack*-class SSNs in the others. The key metric was the survival time of the carrier. In 144 exercise-hours, the submarines launched 3 torpedo attacks and 19 missile firings (78 missiles); 87 percent of the missiles were judged to have met the parameters to acquire the carrier. Carrier survival time was nine hours. The submarines' most useful source of information was beaconing by the SQS-26 sonar, which they could detect at ranges of 150–225 nm. Beaconing enabled the submarines to fire from outside the ships' detection range. Broadband radiated noise, such as propeller noise, was much more difficult to classify.

Phase II examined dispersion, emission control, simulation of the carrier by the escorts, and active acoustic deception. Its target force followed a fixed path. In UPTIDE 2-B, *Hornet* and Desron 31 faced one nuclear submarine and one diesel submarine. The dispersed carrier formation steamed at below cavitation speed, the escorts attempting to lure the submarines away. The carrier avoided identification for the entire seven days of UPTIDE 2-B. Submarines spent all their time in unproductive searches. *Sculpin* used strategic information to fire eight missiles, two of which were considered to meet acquisition parameters. The HUK group commander considered the result an unprecedented success. UPTIDE 2C introduced an acoustic deception device (tergiversator), which emitted carrier and escort propulsion noises and simulated sonar transmissions. It was employed by a minesweeper. The submarine *Plunger* fired a missile at the minesweeper after listening for three hours and concluding she was a high-value unit. *Hornet* was undetected for 87 hours. UPTIDE 2D added more acoustic devices, including a buoy (SADSAC) and Nymph's Voices. They masked the carrier *Ticonderoga* for 93 hours, inducing the 2 SSNs in the exercise to launch 9 of their 16 missiles in 2 of their 4 attacks. Sixth Fleet was impressed enough to request acoustic deception devices of its own. They became standard. In UPTIDE 2E, *Ticonderoga* evaded detection altogether for 140 hours. When the carrier reverted to standard tactics for the last day of the exercise, she was targeted for missile attacks within 4.5 hours. In the whole of Phase II, fewer than a third of the missiles simulated were considered to have been able to acquire their targets.

However, the carrier force was unable to capitalize on the submarines' need to expose their sensors to find the carrier. They never exploited their sensor capability, and when they used their SQS-26 sonars, submarines were attracted to them; on eight occasions escorts were sunk.

Phase III used a new measure of effectiveness: miles safely traveled. The enemy was given integrated surface, submarine, and air threats. New capabilities were land-based patrol aircraft, towed arrays, and destroyer helicopters. Acoustic deception devices were used extensively. In UPTIDE 3A, the group managed 86 percent of its 900 nm path without a successful submarine attack, but it was detected and attacked after three of the five acoustic deception devices shielding the carrier broke down. Surface raiders armed with missiles were a considerable problem. Conclusions included the need to provide the high-value unit with a means of reducing its acoustic emissions. Acoustic deception devices were much more difficult to move and position when the carrier steamed more freely. With better sensors there were many more false targets, so better means of correlating and classifying contacts were needed.

UPTIDE 3B added further new capabilities, which included two squadrons of land-based patrol aircraft, a ship towing an array (TASS), SOSUS and HF/DF tactical support, and a DLG with a LAMPS helicopter. A *Hamilton*-class Coast Guard cutter also participated.[35] The high-value unit was the carrier *Ticonderoga*. The target group attempted six transits of the exercise area, three of them successful; performance was better than in 3A, the first UPTIDE transit scenario. The successful transits were made against the submarine *Sculpin*; those against the much quieter *Drum* failed. The exercises demonstrated the need for current intelligence concerning the opposition submarine. Information about *Sculpin* was adequate and timely. It was used in conjunction with data on the detectability of the high-value unit. Against *Drum*, there was less information, and it was either indeterminate or too late. During the last transit in the exercise *Drum* activated an acoustic projector (to simulate a louder Soviet submarine), which made it possible to maintain a usable submarine plot. Dispersion and acoustic deception made it difficult for the submarine to find the high-value target, but ships had to be close enough for mutual support (for example, against missiles) and for effective command and control.

Among other things, the exercise demonstrated the value of data links. The first transit tested a widely dispersed force, with two of the escorts operating one convergence zone apart. Shore- and ship-based ASW aircraft simultaneously gained LOFAR contact on *Sculpin*, coordinated by the ASW Classification and Analysis Center (ASCAC) on board the carrier. A P-3C began to direct the carrier-borne S-2s on the basis of its DIFAR contacts. An S-2 attacked a MAD contact (*Sculpin*) gained at a DIFAR fix to which it had been vectored by the P-3C. *Sculpin* was ruled out of action after a series of simulated attacks. The exercise resumed with *Drum*, the object being to test her ability to distinguish between acoustic deception devices and the carrier. SOSUS detected *Drum* and placed her on a line of bearing crossing astern of the carrier. Unfortunately, it took nearly eight hours for this information to reach the officer in tactical command. About three hours earlier *Drum* gained a positive sonar contact which she classified correctly as a warship (later the range was estimated as 58 nm). When *Ticonderoga* accelerated and changed course for flight operations, *Drum* positively classified the contact as a large warship and began to approach. She visually identified *Ticonderoga* and correctly classified her sonar contact, even though the carrier was operating on two screws in hopes that the submarine would identify her as a destroyer. *Drum* attacked (constructively) with six missiles. Although helicopters immediately retaliated, *Ticonderoga* was ruled out of action. Later it was learned that two of three scheduled deception device operations to the north of the carrier had aborted due to equipment malfunctions, and the intended degree of deception did not exist.

The second transit tested more concentrated forces and deception devices. Three of the escorts were placed one convergence zone apart. One of the others, *Mahan*, towed an acoustic deception device while playing 12- and 15-kt carrier tapes, spending one hour out of four playing a 20-kt tape and operating her radar deception device. The others of this latter group also played carrier tapes. The TASS ship was placed in the center of the transit area. *Ticonderoga* transited at 13 kts. *Sculpin* was picked up by a P-3C and ultimately successfully attacked. SOSUS detected what was thought to be *Drum*. Plans called for *Ticonderoga* to divert once a consistent contact could be maintained, then proceed around an acoustic deception device and transit between two lanes of such devices. *Drum* was not attacked, but she never attacked the carrier either.

In the third transit, the carrier was intended to pass through overlapping fields of influence generated by rows of acoustic deception devices, ship-towed ones on one side and a combination of ship-towed and helicopter-deployed devices on the other side. Two ships formed a "wolf trap" unit on one side. Both P-3s and S-3s gained contact on what turned out to be *Sculpin* during the first 90 minutes of the transit (there was also an invalid contact). *Sculpin* was sufficiently deceived by the decoys that she had to try to identify ships visually, eventually putting two out of three out of action. *Drum* was not involved.

In transit 4, *Ticonderoga* transited at high speed (21 kts) with one plane guard, covered by deception devices all around them. There were also expendable buoys simulating SQS-26 sonars. Escorts conducted SQS-26 sonar sweeps. *Drum* tried to use HF/DF information to find the high-value unit, but this information was not applicable to the exercise, having been collected earlier. A P-3 that gained second convergence zone contact on *Sculpin* managed to convert that to direct path. Since *Sculpin* was no longer considered a threat at this point, the airplane was ordered to break off and search for *Drum*. *Drum* gained contact with what proved to be *Ticonderoga*. The latter received SOSUS messages indicating that *Drum* was astern, so she ordered deployment of decoys including SQS-26 simulation buoys, as well as a lane of LOFAR buoys. *Drum* visually identified *Ticonderoga* from ten miles astern and launched simulated SS-N-7 missiles. This attack was ruled a success. Because *Drum* attacked from astern, she was not detected. As before, SOSUS information was not timely. Problems included the known poor performance of the AQA-4 buoy processor on board the S-2E carrier ASW aircraft.

Transit 5 exposed the carrier to both submarines, with the carrier passing down a lane of acoustic decoys including those towed by some of the ships. Two ships swept one side of the lane using SQS-26 sonars. When *Ticonderoga* and her close escort *Roark* entered the exercise area, *Sculpin* was detected about 100 nm away by a P-3. *Drum* apparently managed another unopposed approach from astern of the carrier, achieving hits with simulated SS-N-7s at a range of 12 nm.

A sixth transit was made possible by the early termination of transit 5. This time *Drum* operated her sound projector. The carrier again sought to run down a lane of deception devices, her aircraft laying down LOFAR fields ahead and astern. As in previous exercises, the towed array surveillance ship was assigned to operate independently for maximum detection range. SOSUS and TASS continuously provided submarine locations. There was at least one false contact.

By the spring of 1973, money was tight. ASW FORPAC was disbanded, and Aurand retired. The focus shifted to training. An UPTIDE IV that would have been focused on sea control was canceled. UPTIDE doubtless contributed a great deal to the shift away from high-powered active sonar toward towed arrays, and also to increasing interest in acoustic deception. Various forms of deception were very much in evidence in the big forward fleet exercises of the 1980s. These trends took time to become evident.

THE SYSTEM IN ACTION: 1973

In 1973 VADM Fred C. Bennett, USN, described an encounter with a Soviet November-class submarine that was en route from the Kola Peninsula to Cienfuegos, Cuba, between 11 January and 14 February.[36] The episode illustrates the ways U.S. systems could track a Soviet submarine. The November was first detected by the station at Keflavík on 12 January. She was tracked all the way through the Norwegian Sea, but on 17 January contact was lost in the Faroes-Shetland Gap. This was quite typical. The area north of the gap was a favorite ASW hunting ground because there was no heavy shipping concentration and because the water was isothermal. It was ideal not only for SOSUS, but for other acoustic sensors. Conditions were so good that it was estimated, based on all-source intelligence, that SOSUS detected all outbound nuclear Soviet submarines as they passed through this area. Diesels were more difficult; only 85 percent of deploying diesel submarines were detected.

As the November proceeded through the Norwegian Sea, the Navy did not fly a P-3 out to track her directly. Tracking was considered extremely expensive, and sonobuoys were then in short supply. In the late 1970s someone suggested that the Soviets could exhaust the U.S. supply by compelling the U.S. Navy to track their submarines for several months. At this time, too, SOSUS eastern Atlantic coverage was limited pending construction of the Backscratch arrays terminating at Brawdy, Wales. North Atlantic SOSUS coverage, terminating at Argentia and Brawdy, was at that time programmed for completion in 1977 to provide coverage south of Iceland and west of northern Europe.

In 1973 that was well into the future. During the latter part of January, a Soviet surface force appeared. It was eventually tied to the November. The surface unit, based on a Kresta-class missile cruiser, was spotted by a British Shackleton flying from Bally Kelly, and the next day a submarine tender was spotted joining the group. HF/DF and special intelligence indicated that the Kresta had deployed from the Northern Fleet about 17 January; the tender and a tanker had deployed somewhat earlier. The Soviet surface force headed into the Atlantic and arrived north of the Azores, where it was spotted on 28 January by a British aircraft out of Plymouth.

At this point SOSUS could identify the submarine only by its acoustic signature, meaning its type of machinery. It was evaluated as either a November or a Hotel-class SSBN, as both had much the same machinery. It was assumed that if it was an SSN it was headed to the Mediterranean. If an SSBN, it was assumed headed for a western Atlantic patrol station. On 25 January the SOSUS station at Argentia reacquired the submarine, which it held intermittently through the 26th. It had gained contact because a sound "window" opened that enabled the array to see eastward of the Mid-Atlantic Ridge. Given tracking by Argentia, a P-3 based at Lajes in Portugal gained and held LOFAR sonobuoy contact for 57 minutes on 26 January. The next day a Canadian Argus flying from Lajos to Greenwood, Canada, gained LOFAR contact, which it converted into radar contact; running in, it sighted a submarine that had broached and remained partly surfaced for 16 minutes. The light was poor, but the resulting photo was identified as a November. The submarine then submerged, and apparently remained submerged for the rest of its trip. It was now clear that the November was following a track similar to that taken by a November deploying to the Caribbean in the summer of 1969. That gave it special significance, so the P-3 squadron at Bermuda was ordered to search the area defined by SOSUS. The November was found on 30 January, using sonobuoys, 240 miles northeast of Bermuda, but contact was lost on 31 January. Sea conditions limited sonobuoy range to 2–5 nm and also reduced the effective SOSUS range. Too, the November was running at only 6 kts, at which she was relatively quiet because she was in turboelectric mode with relatively quiet turbogenerators.

SOSUS regained contact on 4 February. Meanwhile the Soviet surface group proceeded westward into the Atlantic, the ships separating to rendezvous south of Bermuda on 5 February, joined by the submarine the next day. SOSUS held the November through 7 February, the submarine running at 8–18 kts. A P-3 localized the submarine on 7 February, spotting her ECM mast and periscope. Three frigates were vectored into the area. Early in the morning of 10 February, USS *Calcaterra* made contact at 6,400 yds and maintained it for 8 minutes (and for 15 minutes later that day). She continued to ping on the November and saw her surface outside Cienfuegos. At this time *Calcaterra* held the record in the Atlantic for sonar contact on Soviet nuclear submarines, a total of six and a half hours.

9

FIGHTING A NEW SOVIET SUBMARINE GENERATION

THROUGH THE 1960S, those privy to the SOSUS secret nervously awaited a new generation of quieter Soviet submarines. In April 1964, for example, the director of ASW research and development pointed out that it was surprising the Soviets had failed to silence their SSNs. If they quadrupled their spending on individual submarines, they could match U.S. silencing, with dramatic impact on U.S. ASW. The prospect of Soviet silencing justified continued U.S. Navy interest in longer-range active sonars. However, any submarine with speed beyond 30–35 kts and a diving depth of 1,500–2,500 ft would be very expensive for the Soviets.

The obvious counter to quiet enemy submarines was to revert to a combination of direct protection of shipping (convoys) and barriers across likely routes between Soviet bases and patrol areas in the open sea. About 1969 the Long-Range Objectives Group envisaged 20 convoys at sea at any one time. In 1971–72 it was assumed that after 200–400 ships arrived in Europe in the first three months, NATO would run five 100-ship convoys to Europe each month as well as 300 ships per month to Japan. The Japanese requirement was not expressed in a number of convoys, because Japanese ports at that time were not considered sufficient to handle a 100-ship convoy. In the 1973 SEAMIX I study NATO ran seven convoys per month. SEAPLAN 2000 of 1978, written during the Carter administration, assumed ten convoys per month worldwide. SEAWAR 85 of 1980 assumed five pulses, each comprising three 50-ship convoys, running simultaneously. Studies generally showed that a convoy needed 8–17 escorts to fend off Soviet submarines.

The Victor (Project 671) class was one of the new quieter generation of Soviet submarines. USNI

The U.S. fleet suffered a double blow in the 1960s. First, the urgent Polaris program displaced new surface ship and carrier construction, presumably including a projected class of new ASW carriers (CVS); then the Vietnam War wore out many ASW ships, including ASW carriers and FRAM destroyers. After Vietnam, money remained very tight. Despite a steep decline in overall fleet strength, however, it proved possible to limit the loss of ASW escorts. At the end of FY 58, the active fleet included 245 destroyers, plus 32 destroyer escorts and 36 radar picket conversions. At the end of FY 65, that had declined to 231 destroyers, 21 destroyer escorts (1 of them old), and 21 radar pickets used mainly for patrols off the Vietnamese coast (the warning/barrier force was inactivated in the summer of 1965). Another 13 destroyers used for Naval Reserve training could be activated in an emergency. As of 1964 the projected FY 70 fleet included 240 destroyers. In fact, the FY 73 fleet totaled 242 surface combatants (including cruisers). In FY 73 four *Essex*-class carriers remained, all operating as

The FRAM (Fleet Rehabilitation and Modernization) program was an attempt to stave off the escort numbers crisis that would strike when World War II–built destroyers had to be discarded. USS *Wilson* is shown off the Philadelphia Naval Shipyard, 21 November 1963, after her FRAM refit. Her No. 2 twin 5 in. mount has been replaced by a pair of triple lightweight torpedo tubes, and an ASROC box launcher is set between her funnels, which have new caps intended to protect her boilers from nuclear winds. Aft she shows the hangar and pad for a DASH drone. Not visible is her new SQS-23 keel sonar. USNI

An attempt to deal with block obsolescence: the *Spruance*-class destroyer *Hewitt* maneuvers on trials in late 1976. Visible ASW features were the ASROC box forward (close to the reloads in the superstructure) and the helicopter pad and hangar aft. The usual triple torpedo tubes are hidden in the hull near the helipad. Not visible at all is the computer command system incorporating the ASW elements developed in the 1960s. The threat of fast submarines, which could threaten carriers, figured in the OSD analysis (the Major Fleet Escort Study) that led to the *Spruance* program and to the end of *Knox*-class production. Beyond pairing three or four air-defense ships with a carrier, it was less expensive to add pure ASW ships, as long as they could sustain 30 kts or more (a carrier's vulnerability increased sharply if she was required to run at less than 28 kts). By this time the Maritime Administration was predicting that in the 1970s merchant ship speed would be approaching 30 kts. A DX/DXG Development Plan was drawn up in October 1967, the DXG element meaning that ships would be designed to be completed either as air defense destroyers (DXGs) or as pure missile ships. The U.S. Navy never ordered the air defense version, but it took over four ships ordered by the Shah of Iran as its *Kidd* class, and the *Ticonderoga*-class cruisers were built on *Spruance* hulls. NHHC

The carrier *Kitty Hawk* departs San Diego for a western Pacific cruise, 25 October 1977. Her status as a multipurpose CV rather than an attack carrier (CVA) is indicated by the row of S-3 ASW aircraft alongside and in front of her island. The ASCAC installed to coordinate her ASW functions is not visible. NHHC

CVSs. There was no prospect of replacing them on a one-for-one basis.

When he appointed ADM Elmo Zumwalt Jr. as CNO in July 1970, President Nixon promised that money saved by retiring worn-out ships would go into recapitalizing the fleet. Zumwalt later said that he hoped to bring the fleet back to its pre-Vietnam size of about 900 active ships, although he thought 600 was a more realistic number. As a former chief of the Long-Range Objectives Group, Zumwalt had been made aware of the prospect that Soviet silencing would defeat SOSUS. In that case the long-range patrol aircraft would become much less effective. The Navy would have to run convoys, which would need organic air cover. Strike carriers would also need organic air cover, even though the *Essex*-class CVSs that were escorting them were wearing out. Convoy operations would translate into a desperate need for more ASW surface ships.

Zumwalt's approach to the strike carrier ASW problem was to move CVS aircraft onto them and redesignate them general-purpose carriers (CV) rather than attack carriers (CVA). At the same time, Zumwalt argued that the carriers should be able to change their air wings as circumstances dictated. The change was more than cosmetic: the carriers were given Tactical Support Centers (TSCs) to support operations by the new S-3 Viking ASW airplane, which replaced the earlier S-2 Tracker. The new centers derived from the successful TSCs used to support P-3 operations ashore. Given the rising threat of Soviet torpedo submarines in the Mediterranean (and the difficult sonar conditions there), the Atlantic Fleet had already embarked dipping sonar helicopters on two carriers assigned to the Mediterranean.[1] However, fixed-wing ASW aircraft added much more: an area capability to deal with Soviet submarines armed

The S-3 Viking replaced the S-2 Tracker when the big carriers were converted to multipurpose CVs. This S-3B was an upgraded S-3A adapted to broader missions, including antiship attack using the Harpoon missiles shown. To support them it had an ALR-76 ESM system, which employed the wingtip antennas shown. The APS-116 snorkel (and periscope) detection radar was replaced by an APS-137. The acoustic system of the S-3A was a version of the P-3C system; the S-3B had a UYS-1 acoustic processor in place of the earlier AQA-7. The airplane shows the preloaded sonobuoy chutes and interferometer antennas of the P-3C, but the upgraded system could track buoys in the water, eliminating any need to keep overflying them. USNI

with antiship missiles. *Saratoga*, the first CV, tested the concept in the Mediterranean in 1971. Instead of a TSC, she (and *Kitty Hawk*) initially had the ASW Classification and Analysis Centers (ASCACs) developed to support the S-2 Trackers of the earlier ASW carriers. To give the carriers greater ASW capacity, an ASW Tacair pod was developed for use by attack aircraft (A-6 and A-7). An airplane could carry two pods with four buoys each plus a fixed-channel Multi-Channel Jezebel Repeater (MCJR) link back to the carrier. It is not clear whether the pods were ever used.

Convoy air support was a different matter, requiring an inexpensive helicopter carrier. As many as 20 convoys might be at sea at any one time in a general war, but not all of them would require organic air cover; Zumwalt envisaged a force of 12 helicopter carriers. If they could be made inexpensive enough, eight might be bought instead of one large-deck carrier. The SQS-26 sonar demanded aircraft to prosecute targets at convergence zone range. To distinguish the convoy support ship from a carrier, it was initially called a DHK or DH. In 1967 RADM Thomas D. Davies proposed a small ship carrying six helicopters. In August 1969, LRO-81, the last in the series, included a 25 kt 12,000–14,000-ton ship with fleet speed accommodating 12 VTOL aircraft or helicopters, 6 of them in a hangar. Another study emphasizing the increased capability of airborne sensors and weapons over those on ships included a 5,850-ton "air-capable ship" with 24 kt speed that could accommodate 6 heavy helicopters (SH-3D). Davies proposed it when CNO Zumwalt issued an Action Sheet on the air-capable ship on 6 July 1970, soon after taking office. Zumwalt also pressed to make as many surface ships as possible air-capable and to accelerate development of the LAMPS helicopter. The helicopter sea control ship morphed into a larger project to move aircraft from expensive large-deck carriers to numerous smaller ones. As a surface sailor, Zumwalt was impressed by the growing Soviet antiship missile arsenal, which might wipe out the Navy's few capital ships. Initial money for the necessary VSTOLs came from reallocation of the NAVAIR budget; a proposed replacement land-based patrol airplane (VPX) was canceled in 1971.

The official model of the proposed Sea Control Ship. The aircraft are Sea King helicopters and AV-8B Harriers. In trials, the Harriers laid sonobuoy barriers and the helicopters prosecuted contacts the sonobuoys revealed. No such ship was ever built, but the Spanish navy built a version as its *Principe de Asturias*. NHHC

Zumwalt hoped that the air-capable ship, which was renamed the Sea Control Ship, would be approved by the end of FY 72, with eight ships being authorized in FY 73 for completion between 1978 and 1982. The helicopter carrier *Guam* was employed for a demonstration between late 1972 and April 1974.[2] Her primary means of detecting submarines at a distance was lanes of sonobuoys laid ahead of her 4–10 nm apart in lines about 30–35 nm from the ship by AV-8A Harriers borrowed from the Marines, the only fast VSTOL aircraft available. Contacts were investigated by Sea King helicopters, which could monitor eight buoys at a time. As the force advanced, more buoys were laid ahead of those in the water. By proceeding at more than 15 kts, the force built around the Sea Control Ship would force any Soviet submarine to approach at high (hearable) speed.

The ship had an ASCAC containing two analog LOFAR analyzers (AQA-5) and one digital analyzer (AQA-7) and a plot, with lines to the ship's CIC. Although the AQA-7 could handle directional LOFAR buoys, only nondirectional buoys were used in the test. During initial trials against a fast SSN, the submarine was detected on five of nine runs (more than 55 percent). Localization was achieved on a quarter of these runs (about 15 percent of all runs).

The considerable area in which LOFAR might detect a contact was too large for localization by helicopter dipping sonar.

The program proceeded, although the Operational Evaluation Force found the existing aircraft only marginally effective. For example, *Guam* failed to keep the required two helicopters continuously airborne. The ship's Harriers could not protect her against the agreed 1980 threat of Soviet antiship missiles carried on board bombers and surface ships.

Even so, two Sea Control Ship detail design contracts were awarded in January 1973, and long-lead funding was included in the FY 74 budget. By early 1973 the estimated price of the ship had risen considerably. On 30 January 1974 the Navy was authorized to proceed with the design of the lead ship, but only on the condition that it demonstrate its adequacy. Further *Guam* trials were encouraging but not conclusive. Opponents saw the project as a way of cutting back on expensive—but effective and flexible—full-deck carriers. Delays killed the project; after Admiral Zumwalt left in 1974, the Sea Control Ship was superseded by a project for a more capable VTOL Support Ship (VSS), which never materialized. It morphed into a small carrier, which died when its projected high-performance VSTOL aircraft failed to materialize. The only Sea Control Ship to be built was Spain's *Principe de Asturias*.

In the end, Zumwalt's vision of an air-capable fleet survived only in the helicopter facilities installed on board numerous escort ships.

The *Knox*-class frigate *Bowen* shows her telescoping helicopter hangar and her enlarged helipad, with a Phalanx point-defense system on her fantail. The door in her transom covers a VDS fish serving as depressor for a towed array. The ship processed passive (LOFAR) data the helicopter collected from her sonobuoys, the ship directing the helicopter to attack on that basis. The helicopter processed the active sonobuoys she used to localize the submarine. Evaluation by the Operational Test and Evaluation Force was that LAMPS I was only marginally effective against a modern nuclear submarine but *was* effective against a diesel submarine. Some of the early problems were cured after the initial evaluation. The towed array was a different story. During the operational evaluation of the Interim Sea Control Ship in the western Atlantic in April 1974, a *Knox*-class frigate using an interim towed array consistently gained submarine contact at 29–38 nm under predicted convergence zone conditions. It was predicted that under favorable conditions the array would have a 50 percent probability of detecting a second-generation Soviet nuclear submarine at the second convergence zone, and third-generation submarines at the first convergence zone. The situation was actually better than that because it took the Soviets some time to produce submarines with predicted third-generation silencing. USNI

Admiral Zumwalt's attempt to close the escort numbers gap: the *Perry*-class frigate *Thach* in the late 1980s, with a LAMPS III helicopter on her helicopter pad. Not visible is her main ASW sensor, a towed array streamed through her transom. She had a standard missile system capable, in theory, of reacting automatically to a Soviet sea-skimming missile such as the SS-N-7 or -9 fired by a Charlie-class submarine. The hope was that the ships could be built in single-purpose versions, for ASW, AAW, or surface warfare, as the low end of a high-low mix. It turned out that the Harpoon antiship missile and the Standard antiaircraft missile could be fired from the same launcher, reducing options to AAW and ASW. The AAW option was bought first, on the ground that the Navy already had a large number of specialized ASW ships built under earlier programs. Admiral Zumwalt wanted 50 FFG 7s, built at the rate of 8 each year. As costs rose, Congress became skeptical, and annual programs were cut, but in the end the class totaled 50. Sisters were built in Australia, Spain, and Taiwan. NHHC

When Zumwalt became CNO the Navy was engaged in buying *Spruance*-class destroyers, which in effect succeeded the SQS-26 frigates (destroyer escorts) of the 1960s. Zumwalt considered the *Spruances* too expensive. Instead, he wanted frigates cheap enough to build in larger numbers: the *Perry* class. Unfortunately, low cost meant limited capability, including a short-range sonar. The program was saved by the advent of towed arrays and by the frigates' capacity for big helicopters (two SH-60B).

ASW STRATEGY

The Soviets had built a huge submarine fleet. What were their plans for it in war? At the beginning of the 1970s, a British official handbook laid out the specific wartime objectives of the Soviet navy as then understood in the West: to contribute to the strategic deterrent (by providing and protecting SSBNs); to contribute to defense against Polaris; to counter Western strike fleets (i.e., carrier groups); to defend the offshore zone and support land operations (including assault operations); and to provide surveillance of Western maritime forces. "In war, their capability for interdiction of sea lines of communication [SLOCs] would be the greater, though always subject to fulfilling roles more fundamental to Soviet defense." Attacks on SLOCs "would have little effect on the outcome of a general war, but the mere threat could have a considerable influence on a limited war or confrontation situation." In 1975 the CIA argued that the Soviets were so fixated on nuclear-capable forces that they would concentrate initially on Western carriers and SSBNs. Conversely, they would emphasize protection of their own SSBNs. But while Western submarines were already successfully trailing Soviet SSBNs, Soviet submarines had enjoyed no success against Western SSBNs. The Soviets might mount limited anti-SLOC attacks in order to force NATO to disperse naval forces.[3]

This analysis is much like the one that would later underlie the successful U.S. Maritime Strategy.

The Yankee- (Project 667A) class ballistic missile submarine was the rough Soviet equivalent of the first U.S. SSBNs, though most units lacked the computers needed to keep missiles on target as the submarine moved. From a Western ASW point of view, the important thing about the Yankee was that the Soviets became obsessed with protecting their SSBNs during the initial non-nuclear phase of a war. The NATO nickname presumably reflected the broad similarity between this type of submarine and U.S. SSBNs, although details (such as twin screws in the Yankee) were radically different. The Yankee design was the basis of the Delta class with its much-longer-range missiles. The advent of the Yankee seems to have generated intense U.S. interest in placing SOSUS arrays on the far side of the Atlantic, since normal Soviet practice would have been to place Yankee holding areas there. The result was Project Backscratch, under which arrays were laid in the eastern Atlantic and monitored from RAF Brawdy in Wales. This Yankee seems to have been photographed about 1976. NHHC

Yankee submarine *K 219* in distress off the U.S. East Coast, about 590 nm northeast of Bermuda. On 3 October 1986 a missile tube exploded (note the cover blown open) when a seal in the hatch cover failed while she was submerged. High-pressure seawater ruptured the missile fuel tanks, causing a fire and explosion. She had suffered another missile tube explosion the previous year due to seawater leaking into a tube and reacting with missile fuel residue there. That tube was disabled and welded shut. The second explosion breached the submarine's pressure hull. She took on water and sank about 250 m before eventually reaching the surface on battery power, her reactor having been shut down. A nearby Soviet freighter attempted to tow her home, but that proved impossible as the flooding continued. She sank on 6 October 1986. Note that her sail planes are vertical, presumably because her hydraulics failed. U.S. NAVY

During the 1970s, however, Western views of likely Soviet wartime behavior changed radically back and forth, with enormous implications for ASW thinking. There were also many competing views of the requirements for such a war. For example, the U.S. Army and OSD (Office of the Secretary of Defense) planned on a protracted (at least 90-day) war, to which end OSD sponsored the POMCUS program to amass sufficient stocks for eight divisions; the personnel would be flown in when demanded. It seems, however, that consumables still had to be shipped in; in 1978 SACLANT planned to move 11 million tons across the Atlantic at the beginning of a crisis or war. A million men would be flown in at the same time.

In 1969 NATO formally adopted a planning concept in which a war would be the culmination of a three-month period of tension, followed by a month of much greater tension, and then by a week of non-nuclear war, followed by nuclear escalation. Presumably the Soviets would need a month to assemble the force for the European assault. In 1979, although in theory the British still adhered to the 1969 guidelines, the Royal Navy planned to maintain a North Atlantic sea lifeline associated with a more protracted war (Defence Minister Hugh Nott later used this disparity to justify drastic cuts in British surface forces). To what extent would the West need the North Atlantic SLOC to keep a NATO army fighting in Europe? The two main NATO navies could certainly envisage circumstances in which a war would be protracted, and hence in which shipping had to be protected.

U.S. hopes that a non-nuclear war in Europe could be protracted seem to have declined sharply after about 1973 and then revived in 1977–78.[4] The most obvious possible trigger was a major study of a NATO European war requested by President Jimmy Carter soon after he entered office. The study's authors concluded that, properly supplied, NATO could hold out against a full-scale Soviet attack for about two months (pessimists could point out that the same study showed that NATO would give up about a third of West Germany while holding out). The hope was that the Soviets would not want to initiate nuclear warfare; they would wait until NATO did so in accord with its own doctrine. Carter's secretary of defense, Dr. Harold Brown, saw the potential for effective protracted NATO resistance in new microcomputer technology, which could make possible non-nuclear systems sufficient to deal with a massed Soviet army. By 1984 Marshal Nikolai Ogarkov, chief of the Soviet General Staff, was telling his leadership that he could not guarantee victory unless the Soviets had comparable technology. The Soviets too began to write about a possible protracted European war.

In 1977 the British Joint Intelligence Committee placed SLOC interdiction third in Soviet naval priorities, after anti-SSBN and anti–Strike Fleet. At this time the main role of the Strike Fleet Atlantic, built around the strike carriers, was to prevent Soviet naval forces from breaking out through the GIUK Gap. Dissenters suggested that the manifest Soviet failure to solve the anti-SSBN problem might well have elevated the anti-SLOC role.

That year British and U.S. officers worked on a joint concept of operations (CONOPS) as a way of evaluating the impact of changes in command-and-control technology, particularly after the British withdrew from strike carrier operations when they retired HMS *Ark Royal*.[5] The concepts paper contrasted the earlier emphasis on SLOC protection with an emerging U.S. view that the Soviets would emphasize attacks against SSBNs and carrier strike groups, leaving residual forces available for secondary naval roles such as SLOC attack. Thus, the introductory chapter (written by the British) argued that the Soviets had learned the vital World War II lesson that the U-boats had almost won the war, and also that they understood that a secure SLOC might well deprive them of victory in a future European war. A revised CONOPS was needed mainly because of the dramatically increased threat to Allied shipping in the North Atlantic, which was worsened by the decline of the two main Western navies, particularly the Royal Navy. Challenged for evidence of Soviet interest in anti-SLOC warfare, the writer could only quote a statement made by Admiral Gorshkov in his book *Seapower of the State* (1976) that "the Soviet Fleet must have a war-winning potential on the high seas in order to counter the oceanic strategy of imperialism." The British took this to mean the "vital historic trade routes that sustain the West."[6] The final version of the paper argued that the offensive posture of the Warsaw Pact and SACEUR's dependence on prompt reinforcement and resupply from North America "have combined to restore 'sea control' as the fundamental and prerequisite function of US and UK maritime forces."

Director of Royal Navy Plans cautioned that the paper read too much as though World War III would be a rerun of World War II. Current assessments were that a new war, if it came, would be too short for America's "industrial might" or even resupply to count (except perhaps for military stocks). Major reinforcement from the United States could not get underway for at least three weeks, which might be too long.

The authors assumed that the Strike Fleet built around two U.S. carriers would be a dominant part of the future Battle of the Atlantic. Given sufficient warning, it would block any attempt by the Soviet surface fleet to break out of the GIUK Gap or to secure the Gap by seizing Iceland and northern Norway. The British were already committed to help defend northern Norway, working with a Dutch amphibious force. The writers distinguished between a short warning scenario (5 days or less) and a longer-term one (the Soviets decided to attack in 31 days and the alliance began to mobilize in 24, the "31/24" scenario). The short-warning scenario was daunting because U.S. forces, particularly a mobilized Strike Fleet Atlantic, would not reach the GIUK Gap before war broke out. On the other hand, to maintain surprise, the Soviets could not deploy much of their fleet into the GIUK Gap. NATO would seek to establish barriers to keep the bulk of the Soviet fleet, particularly its submarines, out of the North Atlantic. Before that could be done, some Soviet forces could surge through the Gap. Sudden Soviet mobilization would preclude assembly of forces to capture NATO facilities flanking the GIUK Gap; the Soviets would opt simply to destroy them. However, the Soviets would not need a very strong force to seize lightly defended Iceland.

By the time Strike Fleet Atlantic formed, substantial Soviet forces would have passed through the Gap; the Strike Fleet would have to take a southerly route and then fight its way north. "The transit of the Strike Fleet can be regarded as the center piece of [this] sea control campaign," the authors explained. The Soviets would be compelled to try to destroy the Strike Fleet, which in turn would be the principal means of destroying the Soviet submarines, surface ships, and bombers that otherwise would be free to attack NATO shipping.

In the long-warning scenario, NATO forces would be deployed well before the Soviets were ready to strike. NATO might stave off the war by deterring the Soviets as NATO forces trailed and marked (for destruction) Soviet surface warships. The Royal Navy made this point repeatedly in the late 1970s and early 1980s as it argued against proposed cuts to its surface force.

Planners assumed that during the run-up to war the Soviets would focus on the security of their strategic nuclear force (SSBNs); deploying blocking forces to the Norwegian Sea; maintaining surveillance of NATO carriers and submarines (including SSBNs);

Through the 1970s and early 1980s the Soviets built modified versions of their Yankee-class SSBNs armed with longer-range missiles that could be fired at North America from bastions—waters they expected to control. This Delta II (Project 667BD) is recognizable by the absence of a towed-array deployment tube on her vertical tail fin and a missile section lower than that of later Deltas. She was armed with 16 R-29U (SS-N-8) missiles. The modified Yankees had the same power plant as the Yankee, hence the same vulnerability to detection. USNI

Tango (Project 641B) was an improved Foxtrot adapted for ASW. Her totally reshaped bow accommodated a large Rubikon transducer, and she was adapted to fire wire-guided TEST-71M torpedoes. Compared with the Foxtrot, she had a hull better adapted to submerged operation, and greater storage battery capacity. Tango was more automated and better silenced as well, with some equipment being sound-mounted and the hull covered in anechoic material. Some piping was also covered in sound-deadening material. Tango was credited in the West with the ability to fire the SS-N-15 SUBROC-like missile, but that seems not to have been the case, at least as the submarines were designed. Eighteen were built. All units served with the Northern Fleet, although the first was seen at the 1973 Sevastopol naval review. USNI

and positioning forces to interdict SLOCs. The Soviet navy's most important task would be to establish safe havens (bastions) for their SSBNs in the Barents and White seas, and to deploy SSBNs there. The havens would probably be protected by mine barriers and by Soviet general-purpose naval forces. However, as NATO reinforced Europe, the Soviets would deploy submarines across the Atlantic SLOCs. An estimated division of Soviet naval forces showed an ASW blocking force in the Norwegian Sea assigned to protect the SSBN bastion (including 20 SSNs and 5 SSGNs) and an Atlantic surveillance/shadowing force including 30 SSGNs/SSGs, the latter intended to attack carrier battle groups. The estimated SLOC interdiction force would total 23 submarines (diesel and nuclear). This scheme did not show any Soviet submarines assigned the anti-SSBN role.

This analysis hardly justified a claim that the Soviets had learned the anti-SLOC lesson of World War II. It was more consistent with intelligence assessments that emphasized the anti-SSBN and anticarrier missions and assigned what was left over to possible SLOC attacks. The section on Allied objectives solved that problem by arguing that during the run-up to a possible war NATO had both to deter the Soviets and to avoid undue provocation. According to the joint RN-USN concept paper, to avoid provocation (i.e., triggering the perhaps imminent war) carrier battle groups would be kept out of striking range of the Soviet Union (south of Iceland) and SSNs away from Soviet SSBN havens.

The joint paper retained a Royal Navy paragraph arguing that SSBNs probably would not be attacked, but Director of Naval Warfare (RN) doubted that they could be distinguished from attack submarines.

In 1978 the British Maritime Tactical School conducted a study of how reinforcement shipping could be protected in wartime.[7] Presumably the assumed number (42) of opposing Soviet SSNs would be limited by their other assignments, although the threat to shipping could be considerably increased if the Soviets considered it more important. Evaluation of the Soviet submarine threat was complicated by the fact that, given the 2,500 nm range of Polaris and Poseidon missiles, Western SSBNs (hence Soviet anti-SSBN submarines) would be operating in much the same area as reinforcement shipping in the eastern Atlantic and around the Iberian Peninsula. In theory, then, the threat to shipping could change quickly and significantly if those submarines were reassigned.

NATO policy was to rush as much materiel as possible to Europe before the war actually began, not least because once convoys were instituted ships would spend much more time loading and unloading. The Tactical School assumed that convoys would run at 13.5 kts, but by this time more than half the large merchant ships were capable of more than 20 kts. The report apparently did not mention new merchant ship developments such as RO-ROs (roll-on-roll-off) and containers, which might markedly reduce turnaround time in European ports.

The previous year the Maritime Tactical School had studied the impact of Soviet radar surveillance satellites, which would easily identify convoys and report their course and (by deduction) speed. The only remaining virtue of convoying would be to concentrate ASW ships where the submarines would have to come. Since none of the escorts had substantial antiair capability, convoys might be wiped out by Soviet missile bombers. *If* it was accepted that attacks on shipping were the third Soviet priority, this threat might be ignored, and the problem limited to torpedo submarines.

Like the Germans, the Soviets required global submarine command and control for their widespread fleet. Western navies were surprised that they relied on HF radio, which could be received only by antennas above the surface. The Soviets converted some of their Golf-class ballistic missile submarines to HF relays (Project 629R) like this one. They apparently never activated the German Goliath VLF transmitter, whose signals could have been received by a submerged submarine. USNI

Through the 1970s and 1980s the U.S. Navy developed means of confusing the Soviet ocean surveillance system. The deployment of Tomahawk land-attack missiles in the 1980s was a case in point. It was known that the Soviets felt compelled to track every U.S. or British warship capable of launching a nuclear attack against the Soviet Union, which generally meant carriers and strategic submarines. Placing Tomahawks on board major surface combatants considerably expanded that number. USS *Virginia* is shown off Cape Henry, Virginia, 7 November 1985, with two Mk 44 armored box launchers for Tomahawks on her fantail. Each accommodated four missiles. NHHC

It seemed that there were only enough British escorts to form two convoy escort groups, barely enough for reinforcement shipping. However, wartime Europe would also need economic shipping, which could not be convoyed. The Maritime Tactical School became interested in defended lanes through which ships could run independently. It seemed that the new towed arrays would make that possible. A diagram showed a barrier up-threat containing towed array ships and submarines in 100 × 100 nm boxes. On the other side would be an acoustic separation zone, with a 20 nm traffic lane beyond it. Then would come a 20 nm wide separation zone, and then a 20 nm lane for traffic going the other way. Maritime patrol aircraft would prosecute contacts found by the towed array ships. The RAF protested that its limited number of aircraft could better be employed running down SOSUS contacts. That is probably why the first British escorts intended primarily to operate towed arrays, the Type 23 (Duke class), were associated with a big organic helicopter (the EH-101 Merlin). Critics of the concept pointed out that the Soviets would be sure to attack towed-array ships—which is why the nominally low-end Type 23 was given the best existing British point defense missile, Sea Wolf. Towed arrays could not be relied upon to detect diesel submarines, but these submarines would not be a threat to merchantmen capable of 20 kts or more.

In June 1978 a captain in British naval intelligence wrote that the Northern Fleet had 38 possible anti-SLOC submarines, of which 6–8 would be in the Mediterranean at current peacetime rates. The Soviets would probably set up a submarine barrier north of the GIUK Gap. Taking homeland and SSBN defense into account, few would be available for tasks such as interdiction, replacement of casualties, and offensive mining.

As the CONOPS paper had argued, operating between the Soviets north of the GIUK Gap and the

The British Duke class (Type 23) was conceived to exploit the potential of towed arrays, using an unusually large EH-101 Merlin helicopter; note the big hangar and helicopter pad. The design emphasized silencing, with the ships having a diesel-electric propulsion mode. Because they would operate far from other ships (to exploit their silencing), they were given the vertically launched Seawolf, a sophisticated self-defense missile system, the launchers for which are visible just abaft the 4.5 in. gun. The Harpoon antiship missiles were another self-defense feature, usable given the range of the helicopter. The 4.5 in. gun was added after the Falklands Conflict showed that British frigates had to be able to fight in scenarios other than the Cold War. The lead ship HMS *Norfolk* is shown. USNI

Great Circle shipping route to the south, the Strike Group would attract (and destroy) Soviet submarine and air forces. If it were held back in the western Atlantic or moved into the Mediterranean, however, the Soviets might feel free to release many more submarines against the SLOCs.

Few of those involved in evaluating the study seem to have had access to hard intelligence. One of them asked for evidence beyond statements by Admiral Gorshkov and others that the Soviets would concentrate their nuclear submarines against Western SSBNs and carrier groups rather than shipping. In response, the British Joint Intelligence Committee wrote that this issue had been reviewed regularly. Its conclusion was based not only on extensive Soviet writings, but also on the structure and implementation of their command/control and the size and shape of their fleet and its weapons. The number and character of Soviet submarine torpedoes seemed poorly adapted to a sustained campaign against merchant ships.[8] Recent intelligence indicated that 75–80 percent of the torpedoes on board Victor-class submarines were antisubmarine, as were 70 percent of those on Charlies. The Soviets had no dual-purpose torpedoes. About 25 percent of Soviet torpedoes seemed to be nuclear, hence would not be used in the assumed non-nuclear war. This high nuclear loadout would be consistent with the assumed primary mission of attacking high-value units. Given limited torpedo loads, the priority targets might well be fast combatants and their logistics support ships.

By this time Soviet naval exercises were being closely observed. The Soviets rarely if ever practiced attacking convoyed merchant ships (except for amphibious groups). Their normal practice of firing large salvoes of long-range torpedoes suggested that they expected to attack high-value units rather than merchant ships, against which every torpedo should count. The British contrasted the Soviets' practice with Western submariners' concentration on penetrating screens to fire from short range. Long-range shots would be substantially less likely to hit. Even at 2,000 yds, the estimated hit probability of a single unguided torpedo was only 0.5, and less against a fast, maneuvering target. The estimated wartime hitting rate would be no better than 25 percent. It was assumed that submarines released to attack reinforcement shipping would not be allowed to use more than half their torpedoes for that purpose.

Numbers available for an anti-SLOC campaign would be further reduced, it was thought, by low operational availability due to poor quality control and inefficiencies in design, production, administration, and bloc obsolescence. Probably only 60 percent of the Soviet force could conduct operations in an emergency, and with varying degrees of effectiveness. Geography would force Soviet submarines back and forth through chokepoints, where they would suffer heavy attrition. They would have limited time on station due to the distance from their bases to convoy routes—about 3,000 nm if convoys were routed south to extend transit time (and to avoid heavy air attack).

The sheer distance between Soviet bases and the North Atlantic shipping routes was also likely to make an interdiction campaign difficult. A diesel submarine averaging 5 kts (i.e., snorkeling) would take 54 days in transit. If, as in World War II U-boats, the Soviets averaged 25 days' turnaround, their submarines would spend only 12.5 percent of the time on station. However, it had to be admitted that if somehow the threat of SSBNs or the Strike Fleet were to be reduced sharply, the Soviets might shift priorities and release forces currently assigned to counter the Strike Fleet. This JIC analysis closely matched U.S. and Canadian views expressed at the June 1978 Canada/UK/U.S. CANUKUS naval intelligence conference.

If relatively few submarines were available for the anti-SLOC mission, they would be unlikely to achieve much against the large number of merchant ships at sea, counting only military resupply ships. During the first 30 days, 300–400 ships would be en route at all times. During the next 30, more than 500 more ships would depart for Europe. The Soviets would have to distinguish ships carrying military cargo from the mass of commercial ships.

The Admiralty trade office pointed out that as merchant ships became faster, convoying became less attractive. Pressure to complete planned reinforcement as quickly as possible would militate against any measure that reduced the speed of the merchant ships or the rate at which they could unload. Moreover, by this time it was clear that a nuclear submarine encountering a convoy could hide underneath it, more or less undetectable, while firing torpedoes up to sink most of the ships. Post–World War II exercises showed consistently that it would be difficult or impossible to deal with submarines submerged under groups of ships such as convoys.

Given its currently assumed priorities, the Northern Fleet could assign only 11 percent of its long-range diesel submarines: 5 Foxtrots. They could sink 36 ships. If all available long-range submarines other than Victors were used (12 Foxtrots, 2 Tangos, 5 Novembers, and 2 Zulus), 107 ships could be sunk. It would take all submarines other than SSBNs in the Northern Fleet (6 Charlies, 8 Echo IIs, 7 Julietts, 5 Novembers, 12 Victors, 22 Foxtrots, 2 Tangos, and 2 Zulus) to sink 276 ships.

The most prominent skeptic about Soviet deployments seems to have been the chief British submariner (FOSM), who thought that the intelligence service was downplaying the Soviet submarine threat to reinforcement shipping, neglecting scenarios in which such reinforcement was the major threat to Soviet ground forces. At the outset, U.S. and British submarines should be devoted to tracking and marking (making it possible to engage) Soviet submarines, which he considered the greatest threat. That would be far better than doing so when the threat actually developed later on. Another writer pointed out that "once an SSN has closed a target, the more ships that are present, the happier it is likely to be."

The U.S. intelligence establishment agreed that the two primary Soviet naval missions would be the destruction of U.S. SSBNs and carriers, because both presented nuclear threats (the U.S. Navy argued that the Soviets might also execute an anti-SLOC campaign). For example, in 1979 the CIA issued a summary estimate of Warsaw Pact forces facing NATO, the first such comprehensive estimate since 1971.[9] It dealt primarily with conventional forces. Given the need for unfettered naval operations from Northern Fleet bases, the Soviets would seek to establish control of the Norwegian and Barents seas and their approaches as soon as possible. This operation would probably absorb most Northern Fleet submarines and virtually all surface forces and aircraft. Some submarines might operate farther into the Atlantic to prevent the transit of NATO carriers and amphibious forces and to divert NATO naval strength. The Soviets would try to neutralize Western SSBNs near their bases and in the Norwegian Sea before they could launch. Operations in the north would have to be concurrent with the beginning of a ground attack in central Europe. Attacks on U.S. naval forces in the Mediterranean would be mounted at the same time. The primary targets would be SSBNs and carriers.

There was disagreement in the U.S. intelligence community as to the likely scale of attacks on resupply shipping, which it was apparently assumed would be carried out by reserve submarines. The CIA estimated that no Soviet reserve submarines would be brought to combat readiness in less than 90 days, but the DIA and naval intelligence estimated that 6–10 reserve submarines could be brought to combat status in 30 days, and a total of 25–30 in 90 days. CIA, NSA, and State Department intelligence all argued that the Soviets would not be likely to attempt a serious SLOC campaign unless they had previously defeated NATO carrier and amphibious forces without losing submarines. NSA further believed that the extent and degree of an anti-SLOC campaign would be largely scenario dependent. In a prolonged crisis where the outcome was in serious doubt, SLOC interdiction in advance of a conflict would increase.

DIA and the Navy argued that the Soviets considered the SLOCs so significant, and their submarine inventory large enough, that they would devote a considerable number of submarines to SLOC interdiction. CIA, NSA, and State argued that the Soviets were unlikely to mount serious attacks on SLOCs until attrition (or Western strategy) had substantially reduced the threat from NATO nuclear strike and projection forces. The Soviets would commit large forces to counter NATO carriers in or near areas like the Norwegian Sea and the Mediterranean. They would lack the numbers for simultaneous strong antifleet and antishipping operations. DIA and Naval Intelligence argued instead that the timing and extent of Soviet interdiction would depend more on the disposition and tactics of NATO naval forces and upon Soviet intentions and expectations in the course of the war than on the prior achievement of other Soviet naval tasks. For example, an early stalemate or a NATO decision not to deploy carriers and amphibious forces immediately into the Norwegian Sea would lead the Soviets to mount a substantial interdiction effort early in a conflict when NATO would be convoying critical war materiel, including elements of U.S. divisions and their equipment, to Europe.

CIA and State estimated that the Soviets' ability to attack merchant ships in the open ocean would be significantly constrained by submarine torpedo loads, lack of replenishment opportunities for the submarines, turnaround time, long transits, combat attrition, and limited target information. DIA and the Navy countered that these limitations were sensitive to the timing, manner, and level at which hostilities began, and were not sufficient to prevent the Soviets from mounting a significant SLOC threat. DIA argued that torpedo capacity was understated, that two 40 cm × 4.5 m torpedoes probably could be substituted for each of up to six of the 53 cm torpedoes in most classes. The Navy agreed that Soviet nuclear and missile submarines had limited numbers of torpedoes but argued that long-range diesel submarines had sufficient numbers on board.[10] We now know that several times the Soviets designed replenishment submarines that could have enabled their attack submarines to remain at sea, in one case going so far as to order one (which was canceled in favor of a ballistic missile submarine).[11]

This argument was aside from disagreements as to how many submarines the Soviets could devote to disrupting the North Atlantic SLOC. CIA, NSA, and State Department intelligence all estimated that about ten submarines would be available, some of them also having minelaying or port reconnaissance missions. DIA and Navy argued that typically twice that many would be positioned astride NATO SLOCs to attack both merchant ships and warships. Remarkably, there were no references to the lack of dual-purpose Soviet torpedoes or to standard Soviet submarine tactics involving large salvoes fired at long range. In 1978 new intelligence did show that Soviet submarines had substantially smaller torpedo loadouts than had been assumed.

DIA and naval intelligence insisted that the evident technical limitations of Soviet sensors and weapons could be offset in part by the coordination of ships and aircraft the Soviets practiced. CIA and NSA argued that Soviet intelligence would find it difficult to identify high-value targets among the many low-value ships in ocean traffic; finding one would be a matter of luck. DIA and the Navy argued that the Soviets would have clandestine and technical access to information enabling them to direct attacks against the more valuable convoys and—with greater difficulty—individual ships. This seems to have been a good bet, based on the Soviets' access to U.S. naval communications—which the other intelligence agencies apparently doubted.

The issue was the classic intelligence dilemma: should the key be known enemy intentions (at the beginning of a war) or enemy capabilities? At the outset the Soviets might well concentrate their naval efforts against carrier battle groups and Western SSBNs, but if their initial thrust ashore bogged down, what would keep them from turning naval resources against the reinforcement shipping keeping the NATO army alive, hence holding them back on land? Few prospective convoy escorts had even a slight chance of beating off an attack by Soviet antiship missiles. NATO was only beginning to equip escorts with short-range antimissile systems.

For some years Soviet naval writers had hinted that the most important Soviet wartime naval role would be to protect the SSBN force until it had to be used. Unlike other naval roles, this one was imposed by the Soviet leadership, which considered the submarines survivable (and defensible) in a way that land-based missiles were not. In a series of articles published in 1973, ADM Sergei Gorshkov, the head of the Soviet navy, claimed that SSBNs were more effective deterrents than land-based missiles because of their great survivability. SSBNs finally gave the Soviet navy the independent role it had long sought. Its attempts to carve out any other independent wartime role, symbolized by creation of a wartime oceanic theater of operations (TVD), failed altogether. After reading Soviet statements about SSBNs and their significance, in 1973 Brad Dismukes of CNA (Center for Naval Analyses) wrote that "at the least we should include pro- and anti-SSBN scenarios in our general purpose force planning, or run the risk of structuring a force which might be ill-suited to the most important war-fighting tasks it may be called on to carry out."

In its draft of the Navy's input into a new NIE on the Soviet Union (NIE 11-15-75), CNA argued that the Soviets seemed to be adopting a strategic "fleet in being" concept for at least a portion of their SSBN force. It was no great step to envisage the Soviets creating havens (bastions) to protect their SSBNs until they were ordered to fire their missiles. The appearance of the Delta class, whose missiles could be fired from within Soviet-dominated areas, presumably demonstrated the validity of this idea. The perception that the Soviets planned to protect their SSBNs in bastions was probably not controversial. For example, a 1976 CNA scenario for a future maritime war (for the Sea War 1985 study) included the creation by the Soviets of an SSBN haven in the Barents Sea surrounded by minefields and also defended by submarines and other naval forces. In the scenario, the need to create this haven at the beginning of the war motivated the Soviets to seize Iceland and northern Norway. This scenario was included in the papers written for the new U.S.-British naval CONOPS, so it cannot have been particularly exotic.

CNA inferred that the Soviets would allocate general-purpose forces to protect their SSBNs during the initial stages of a war, and would continue to do so if the war became prolonged. Such protection would entail a fundamental change in Soviet naval emphasis. If enough strength was earmarked to protect the bastions, little or none would be left for the role most threatening to the West—the attack on the transatlantic SLOC. At least initially, the Navy resisted this analysis. There may have been skepticism that the Soviets had a motive for protecting SSBNs in bastions; Western SSBNs, after all, were expected to look after themselves. Few were aware that the Soviets had a reason to fear that their SSBNs were vulnerable in the open ocean, because few in Western navies were aware of how successful U.S. and British submariners had become in trailing Soviet SSBNs. Successful trailing could easily become successful engagement, meaning that the Soviets could quite realistically doubt that their SSBNs could survive in the open ocean. They had to be protected by other Soviet naval forces. U.S. and British policymakers seemed unaware of the successful trailing program; they preferred to imagine that all SSBNs were inherently survivable, hence that the existence of SSBNs guaranteed strategic stability. These same policymakers were generally unaware that the Soviets were doing their best to trail Western SSBNs so that they could be destroyed at the outbreak of war.

U.S. naval officers were impressed by the growth of the Soviet fleet and by the prowess it demonstrated in, for example, the spectacular Okean exercises in 1970 and 1975, which showed off the Soviets' anticarrier tactics. Enormous resources were clearly being poured into a global Soviet navy, which seemed to be designed to overcome Western sea control and to deny the West the ability to project sea power. U.S. naval officers contrasted the rapidly growing Soviet fleet with their own aging, shrinking fleet. It seemed impossible that the Soviets would turn from an offensively oriented fleet to one tasked primarily with protecting the SSBNs.

Although it was not appreciated in the West, the Soviets greatly overestimated the power and competence of the Western naval forces, particularly the U.S. Navy. For example, late in the 1960s the Soviets believed that Western ASW had largely solved the problem of attacks against convoys by nuclear submarines.[12] This would have been news to anyone in NATO. The Soviets justified this view on the ground that submerged submarines could now be detected at ranges up to 30 km. The Soviet paper continued, "if you add to this the ASW helicopters and ASW aircraft that had appeared, then one can hardly claim that it was easier for nuclear subs to attack a modern convoy than it was during conditions that existed in World War II. The creation by the end of the 1960s of an echeloned comprehensive ASW system with a stationary system for illuminating the situation [presumably SOSUS] made the deployment of nuclear submarines more difficult (in the event of a war)."

By the late 1970s the SSBN bastion idea was apparently widely accepted in the West. It appeared that the Soviets envisaged a two-phase war. During the initial, non-nuclear phase they would try to tip the nuclear balance. On land, they would try to destroy NATO nuclear weapons while safeguarding their own. Western strategists had long assumed that at sea they would try to destroy NATO's nuclear SSBNs and strike carriers. Safeguarding their own SSBNs was the natural complement to such offensive action. The Soviets would assume, moreover, that the first Western priority would be to hunt down Soviet SSBNs in their bastions.

The controversial idea raised by CNA was that the Soviets would devote much or all of their general-purpose fleet to defending the bastions rather than to (for example) destroying carrier battle groups. Analysts reading Soviet writings could point to Admiral Gorshkov's justification for investing in increasingly expensive surface combatants: he had to ensure the "combat stability" of the submarines. This idea that the pro-SSBN role would absorb nearly all Soviet naval resources seemed insane to U.S. naval officers, who could not understand why the Soviets would use all of their expensive fleet to shield their SSBNs. Credible (and sensitive) intelligence that became available from 1979 onward gave Americans insight into how the Soviets thought, which turned out to be radically different from their own way of thinking.[13]

"How the Soviets thought" seems to have been a distinctly Soviet method of analysis that was amazingly conservative by Western standards. Applying it would leave little or nothing for anything but the primary role of safeguarding the Soviet strategic submarines. Probably the new sensitive information included details of Soviet naval war plans. Western officers would find Soviets' evaluation of what they had to do, and with what resources, difficult to believe, but examples of this sort of nominally objective ("scientific") analysis crop up in many post-Soviet accounts.[14] A cynic might link this overestimation of what was needed to accomplish various approved missions as a way of justifying the scale of the Soviet military-industrial complex, which had overriding political power. Remember that Khrushchev was deposed for demanding a military revolution that affected many of the politically connected managers in the system by forcing them to change what their factories were making, and thus cutting their bonuses (and to some extent their jobs). Late in the Cold War it became apparent that maintaining the rate of production was a primary driver of Soviet military programming. Soviet political forces favored military spending, whereas in the West the pressure went the other way.

Before gaining this understanding, the U.S. Navy was already deemphasizing SLOC protection on the basis of the understood Soviet emphasis on destroying nuclear threats: the Soviets were expected to mount an anticarrier campaign while trying to destroy Western SSBNs. Little would be left over to attack the SLOC. The most startling proof that U.S. naval thinking was changing came in May 1979.[15] During USN-RN plans and policy talks on 8–9 May 1979, VADM William P. Crowe, then chief of plans and policy, told the British that in wartime, contrary to the declared U.S. view on employment of the Strike Fleet, the U.S. Navy was considering moving the two ready Atlantic carriers into the Mediterranean to reinforce the Sixth Fleet. Crowe pointed out that of the ten attack carriers nominally assigned to NATO, three would have to come from the Pacific; but that was no longer possible. Crowe referred to the need to convince Japan to stand with the United States. Given the impossibility of taking carriers from the Pacific, the U.S. Navy had concluded that operations must be sequential rather than simultaneous: "Since the weight of attack is assessed as more likely to be greater on the Southern Flank than the Northern Flank, and since success over the Soviet Mediterranean Squadron would immediately secure the Mediterranean Sea, the USN delegation implied that the clear choice would be to assemble an adequate force in the Mediterranean to achieve early domination there; this force might well include all the four US carriers in Availability Category A1." In the Atlantic, NATO would rely on surface action groups, submarines, and ASW aircraft for a holding action across the GIUK Gap. There would be no carrier operations in the Atlantic until either carriers in a lower state of readiness were brought forward or carriers were released from the Mediterranean.

Crowe's statement suggests that the U.S. Navy had now accepted that the Soviets would hold back the bulk of their attack submarines while concentrating on the anti-SSBN campaign, protecting the bastions protecting their own SSBNs, and operations against carriers wherever they were. Placing the carriers initially in the Mediterranean would, among other things, enable them to destroy anticarrier forces there without exposing them to Northern Fleet anticarrier forces—which would be held back until the carriers came within range. Presumably it was believed that the bombers, too, would be withheld until the carriers had been dealt with. Crowe's statement was a first step toward the U.S. Maritime Strategy of the 1980s, which is said to have exploited new sensitive intelligence showing how the Soviets thought.

As for SACLANT, by 1981 his stated plan was to run ships independently at first. He would institute convoys only if losses mounted unacceptably. That could be read as acceptance that initially the Soviets would not be assigning many (perhaps any) submarines to interdict SLOCs. Until they mounted a concerted attack on the SLOCs, independent shipping would be far more efficient, and safer, than convoys. That was aside from the inherent problems convoys would face given space surveillance and submarines capable of hiding beneath them.

FIGHTING AN ASW WAR

In the early 1970s U.S. analyses of ASW needs were based not on the evolving strategic picture, but rather on the needs of a renewed Battle of the Atlantic. These analyses, based on large-scale simulations, shaped strategic developments through the 1970s.

To produce the FY 73 Program Objective Memorandum (POM-73), which would guide near-term force development, in 1971 the U.S. Navy conducted a massive simulation of an Atlantic ASW battle under the designation NARAC-G.[16] The context was a protracted non-nuclear war, the basic assumption being that the Soviets would avoid using nuclear weapons to avoid escalation. This was a standard assumption at the time. By about 1976 the West appreciated the considerable scale of Soviet naval nuclear weapons, but analysts assumed that they would be loath to use them.

In the simulation, the war began with a Soviet invasion of NATO Europe. Given how important shipping to Europe would be at the outset, the most logical Soviet naval strategy would be to surge as many submarines into the Atlantic as possible—65 percent of the Soviet submarine force—before the war in Europe began. When war began, they might well already be outside areas covered by SOSUS.

Early shipping losses would be horrific: in the first month of the war, 46 percent of all merchant ships sent to sea would be sunk. After that, losses would decline as Soviet submarines were sunk, down to 19 percent of merchant ships in the second month and 15 percent in the third, for an overall loss of 26 percent of the merchant ships sent to sea. In the first month ASW forces would kill 27 percent of Soviet submarines in the Atlantic inventory, then 18 and then 12 percent, for a net 57 percent in three months. Earlier studies had concluded as much, but this time merchant ship losses would be far worse and more Soviet submarines would survive.

The question was, which investments should be made in the near term to alleviate the situation? A study of the prospective 1981 situation was included in a 2 December 1971 list of studies to be carried out in 1971–72. The postulated 1981 U.S. force was already practically set by the Five-Year Defense Plan (FYDP). The scenario was much the same as for NARAC-G.

In the scenario, a Soviet attack on NATO Europe was the first phase of what would be a global war; for example, the Soviets were expected to attack South Korea on D + 30. The global scenario highlighted the inadequacy of U.S. naval forces unable to defend areas outside Western Europe. This scenario was, moreover, somewhat watered down by assuming that the Soviets would not attack the NATO flanks as they struck the central front in Germany, even though they could do so without seriously degrading their central front offensive. It was assumed that they would limit their military objectives to reduce the danger of nuclear escalation. Iceland was assumed secure throughout, and carriers were not needed to defend it and southern Norway. Those assumptions made it possible to assign five carriers to defend convoys against air and submarine attack, with "a major positive effect on the results." NATO allies and the French (who were no longer part of the NATO military alliance) were expected to contribute their own forces plus more than 100 merchant ships to supplement the declining U.S. flag fleet to take U.S. reinforcements and supplies to Europe.

The two main differences between the 1973 and 1981 war threats were that the Soviets would have third-generation quiet submarines (equivalent to the U.S. *Permit/Sturgeon*) and that they would have Backfire (Tu-22M3) bombers with sufficient reach to attack convoys previously safe in the mid-Atlantic.

NARAC-G assumed that in the Atlantic, the Soviets would deploy 28 first-generation submarines (E and N classes), 24 second-generation (C, V, and P classes), and 89 diesel submarines, a total of 141 submarines. SEAMIX I assumed that the force would consist of 23 first-generation, 38 second-generation, and 20 third-generation nuclear submarines as well as 26 diesel submarines, a total of 107. Although third-generation submarines would account for only 19 percent of the 1981 force, they were expected to account for 41 percent of ship sinkings.

The SEAMIX I results were better than those of NARAC-G, but hardly reassuring. Despite some highly favorable assumptions, 15–38 percent of Allied merchant ships would be lost. The United States and NATO would continue to control the western Mediterranean basin, but the ability to maintain occasional shipments to Greece and Turkey would be marginal. The Soviets would almost certainly seize control of the Turkish Straits. To some extent the Soviets' ability to fight in the Mediterranean was tied to their base in Egypt, from which they would soon be ejected. Without that base, the Soviet fleet in the Mediterranean could not be sustained, nor could Soviet submarines there be resupplied.

War plans called for U.S. naval forces in the Pacific to be drawn down heavily to fight in the Atlantic. The U.S. naval forces remaining there would have to be withdrawn to the eastern Pacific to protect shipping between Alaska, Hawai'i, and ports in the continental United States. The study does not seem to have taken Sino-Soviet tensions into account (the Soviets and the Chinese had fought on their border in 1969). The Soviets were expected to deploy forces to the Indian Ocean to cut the flow of oil from the Persian Gulf, Libya, and the eastern Mediterranean. As a result, not taking shipping losses into account, the United States and Western Europe would have only 45 percent of their peacetime oil supply.

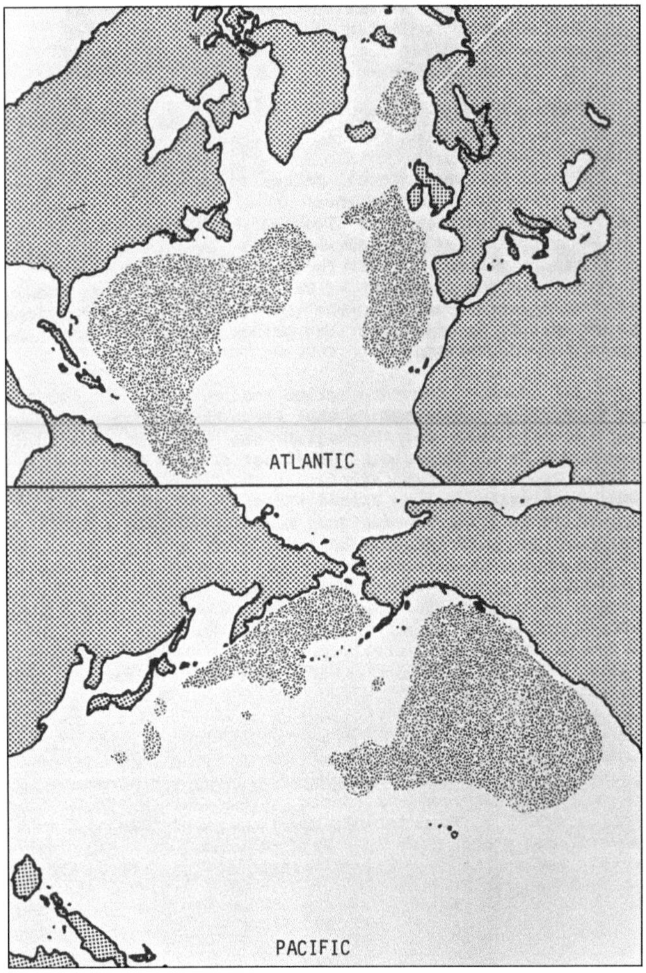

SOSUS coverage in alerted searchlight mode about 1977 against a quieter Soviet submarine, from the declassified Cross Report. FROM A DECLASSIFIED ORIGINAL IN NHHC

USNS *Bold* (T-AGOS 12) was a SURTASS ship. Note the large satellite antenna radome; the data collected were transmitted ashore for most processing. She is shown about 1993. NHHC

Improvements in ASW would reduce the Soviet submarine fleet more quickly, but their best and quietest vessels would survive. Second- and third-generation nuclear submarines would be more difficult to detect passively but would return the same target echoes as first-generation submarines, so surface escorts would kill more of them.

Current programs envisaged substantial improvement to SOSUS, extending coverage to the patrol areas Soviet submarines would have to occupy in wartime. The current Phase II upgrade would extend the existing network and repair and replace existing arrays. New arrays would cover the entrance to the Mediterranean at Gibraltar and the egress from the SSBN base at Guam, the latter to prevent the Soviets from stationing submarines there to trail U.S. SSBNs. Soviet SSBN holding areas in the North Atlantic would be covered. The SOSUS "probability" areas would be made smaller, hence easier to search, and target capacity would be increased sixfold (translated as an 8 dB improvement in recognition differential, which indicated the required signal-to-noise ratio). The upgrade was included in a list of near-term possibilities used in a 1977 Naval War College series of near-term budget alternatives.[17]

There were also ship-towed equivalents to SOSUS arrays.[18] Development of the full-scale Surveillance Towed Array Sensor System (SURTASS) began in 1976, technical evaluation being completed in March 1978; the first full system test, including the satellite acoustic data link, was conducted in July 1979. Contracts for the first three SURTASS ships were let on 26 September 1980, with the first ship, USNS *Stalwart*, entering service late in 1984. *Stalwart* began the first operational SURTASS patrol on 12 January 1985. In August 1985 she became the first SURTASS ship to operate in the Mediterranean, an area of particular interest because no SOSUS arrays had been placed there. Because the surveillance ships could change SOSUS coverage simply by moving, they drastically reduced the Soviets' understanding of the fixed parts of the SOSUS system. At his trial for espionage, John Walker was accused of having revealed SOSUS array locations to the

Soviets, who published maps supposedly showing array locations.

To support operations outside the Atlantic, the Navy planned a small-scale deployable SOSUS-like system at the time called MSS (Moored Surveillance System). As then envisaged, MSS would be deployed by aircraft, surface ships, and submarines and would report via satellite or signals processed by ASW aircraft. Like SURTASS, it was expected to be ready in the 1980s. That date turned out to be overoptimistic. MSS became the air-deployed RDSS (Rapidly Deployable Surveillance System), which was canceled in 1986. The concept has since been revived, for example as a covertly deployable surveillance system.

The shipboard ASW situation was not so favorable. With the death of Extended-Range ASROC, LAMPS I helicopters were the only surface ship standoff weapon with a range beyond 10,000 yds. They could prosecute targets in the first convergence zone, but too few of these helicopters were in service. In the early 1970s it seemed unlikely that the number could be made adequate before 1985. The main surface ship and air ASW weapon, the Mk 46 torpedo, had limited passive detection and localization capacity and inadequate capability against anechoic coatings (the situation would be alleviated by the near-term improvement program [NEARTIP] for the Mk 46 torpedo). ASW command and control was inadequate, as was the ability to integrate ASW sensors. It was difficult to localize targets detected at long range. Torpedo defense was inadequate, as it depended completely on the Nixie towed decoy (the Soviet wake-following threat was not yet known).

The bright spot in surface ASW was the tactical towed array, due to enter service in 1977. The array would enable surface ships to detect reliably submarines out to at least the first convergence zone. Instead of announcing its presence with loudly pinging sonars like SQS-26, a towed-array surface force could quietly detect submarines at great distances (in a British convoy study, up to 100 nm away).

Mining was clearly important, but the mine stockpile was inadequate. Here too, however, there was a bright spot: the Captor rising mine.

Aircraft were marginal ASW assets mainly because their sensors were inadequate against patrol-quiet submarines and because of the Mk 46 problems. When operating in cluttered areas, they needed more sonobuoy channels so that they could deploy more buoys: the new ASCL (Advanced Sonobuoy Communications Link) would triple available buoy channels. LAMPS I and the SH-3 had inadequate tactical navigation and ASW data management capability. Beginning in 1975, the SH-3 would be given a new tactical navigation system (APW-123). However, LAMPS I could not accommodate much more equipment; its limits would not be overcome until the follow-on LAMPS III entered service in 1983. The Mk 46 problem would finally disappear as the Advanced Light Weight Torpedo (ALWT) entered service.

Submarines needed better communications if they were to be used in direct support.

The Soviets introduced a new ship detection system in the form of radar satellites (RORSATs). They were conceived in 1959 by Vladimir Chelomei (who was responsible for the long-range antiship missiles) as the sensing element of an anticarrier system whose weapon would be Chelomei's IRBM (the satellite would be launched by a Chelomei-designed rocket). The authorizing decree was issued on 1 August 1961. The project covered both the active-radar US-A and a passive US-P. Chelomei lost the program for this *upravleniye sputnik* (reconnaissance satellite) after Khrushchev fell from power. Vehicle flight tests (without the radar) were conducted beginning in 1965. Flight tests of the complete satellite, powered by a nuclear reactor, began in 1973, and the satellite system was formally accepted for military service in 1975. It had the worst reliability of any Soviet space system. According to post–Cold War information, it was too rarely available to provide useful data. Satellites were typically launched in pairs; there were also passive satellites. They were powered by small nuclear reactors because large solar arrays would have caused drag even in the extremely thin atmosphere in which they operated. U.S. analysts estimated that the satellite radars could detect medium and large ships except in rain or high seas. Not knowing that the system was seriously defective, Western observers saw it as the final blow to the ability of convoys to escape in the vastness of the ocean. Given its distinctive formation, a convoy would surely be readily recognized on the image produced by a radar satellite. A British officer wrote that every six hours Moscow would have the current position of any convoy and could deduce convoy heading and speed. In the 1980s the satellite reconnaissance problem was considered so serious that the U.S. Navy considered adopting an antisatellite weapon. The idea was scrapped because it was thought (probably incorrectly) that the Soviets had many reserve satellites available for quick launch.

Soviet espionage success was a less tangible factor. John Walker and his accomplice Jerry Whitworth were both communications petty officers. Walker served at SACLANT, while Whitworth had various roles, including one on board the ship that at one time during the Vietnam War distributed keyword lists in the Far East. The two compromised the wiring manuals for all current U.S. Navy code machines, including modifications to machines compromised when the North Koreans seized the intelligence-gathering ship *Pueblo*. Whitworth once apparently supplied the Soviets with 90 days of code keys, presumably wanted by the Soviets to reconstruct the way such keys were created. At his trial, Walker was charged with having revealed the holding areas of U.S. Polaris submarines and a map of SOSUS arrays. Walker began working for the Soviets in 1968 and was not caught until 1985. By that time, he had retired from the Navy after bringing Whitworth into his operation to keep it going. Later he recruited his brother-in-law (who supplied almost nothing useful) and his son.

The Soviets seem to have been quite willing to risk exposing Walker by using his product operationally. By the early 1970s the U.S. Navy was aware that Soviet ships were turning up where they were not expected to be, presumably in Polaris holding areas. It was assumed that the Soviets had some exotic wide-area submarine sensor, and through the 1970s considerable effort went into finding out what it might be. Repeated analyses of Soviet ASW, now largely declassified, showed that nothing like that was known, but speculation was rife nevertheless. As SACLANT, ADM Isaac Kidd concluded that there was a spy in his command, but polygraphing (which reportedly included Walker) failed to unearth him. At least one senior intelligence officer thought (correctly) that the Soviets were reading U.S. naval communications, but NSA, which developed the cipher machines, rejected the possibility. The effects of Walker's treachery may have waned in the early 1980s, as demonstrated by the surprise with which the Soviets reacted to U.S. exercises in the north at that time. They manifested no great surprise when the Pacific Fleet tried a similar exercise off Petropavlovsk, however, presumably because Walker's son was on board the flagship *Nimitz* (he was caught shortly afterward).

Soviet submarine silencing in the late 1970s and beyond is often attributed to the Walkers. But that ignores the fact that the Alfa, designed much earlier, embodied considerable silencing. It also ignores the pronounced sluggishness of the Soviet naval industrial sector. The decision to adopt rafting, which became visible about 1975, might well have been made in the mid-1960s.

In SEAMIX I, 33 percent of merchant ships were lost in the first month. After that losses fell to 9 and 5 percent, respectively, over the two next months, for an overall loss of 14 percent. Under NARAC-G assumptions of 70 percent rather than 65 percent predeployment of Soviet submarines, the initial loss rose to 35 percent and the net loss over three months to 16 percent. Forces projected for 1981 would kill 47 percent of Soviet submarines in their Atlantic inventory in the first month, then 18 and 5 percent, the last figure reflecting the dearth of submarines operating in the Atlantic at that time. Ship losses were still unacceptably high.

Further analysis showed how important maritime patrol aircraft were. Halving their effectiveness would increase sinkings by nearly a third. Aircraft cued by SOSUS would account for 77 percent of kills of first-generation submarines (but 53 percent of second generation and 34 percent of third generation; they would also account for 63 percent of sunk diesel submarines). Nuclear submarines would account for only 11 percent of sunken first-generation submarines, 26 percent of second generation, and 24 percent of third generation, plus 17 percent of sunken diesels.

The two new features of the 1981 scenario were towed arrays and the Captor rising mine. Helicopters vectored on the basis of towed array contacts would account for 5 percent of sinkings of first-generation submarines, 11 percent of second generation, 23 percent of third generation, and 5 percent of diesels. Captors were credited with 7 percent of first-generation submarines sunk, 10 percent of second generation, 19 percent of third generation, and 15 percent of diesels. The more survivable third-generation submarines would make more transits of the Captor barrier laid across the GIUK Gap, so more of them would be sunk by these mines, which had both passive and active sensors and would be less sensitive to quieting.

In SEAMIX I, 65 of the 107 submarines with which the Soviets began the war were be sunk. In fact, in 1981 the Soviets had a total of 19 first-generation nuclear attack submarines (5 Echo I converted to torpedo submarines and 12 surviving Novembers) and 39 second-generation nuclear submarines (22 Victor I and II plus 17 Charlie I and II) divided between their Northern and Pacific fleets. The first-generation total considerably exceeded the assumed figure in NARAC-G and even the more moderate one used in SEAMIX I. The second-generation figure roughly matched the Northern Fleet total in SEAMIX I. The first third-generation submarine was the prototype Victor III in December 1977. It did not meet U.S. expectations. By the end of 1981 there were four such submarines. There were also diesel submarines, the survivors of large programs of the 1950 and 1960s.

The SEAMIX I analysts tried three alternative means of protecting shipping: independent routing, close convoys for selected shipping, and expanded convoys. As the World War II experience had shown, independent routing was worst. Close convoying would reduce losses by 19 percent, but it reduced average cumulative delivery by 21 percent due to assembly and dispersion delays. Too, as the British had pointed out in the 1940s, a submarine could use a close convoy as cover and fire up at the ships while evading detection in the surrounding noise. Denying the submarine that haven by expanding a convoy might reduce shipping losses by up to 29 percent. With ships as much as 10 nm apart, a submarine would be exposed after sinking a merchant ship. This new tactic might reduce losses by as much as 17 percent. The expanded convoy had to be tested in practice.

Clearing the Atlantic of all but essential shipping on the outbreak of war would alert the Soviets that any ships they detected were worthwhile targets. The SEAMIX I analysts asked whether it would help to encourage all shipping to continue, diluting the Soviet effort. The sea would become noisier, and ports would be congested, but the idea was deemed worth pursuing as a way of reducing early losses of vital cargoes.

Another possibility was a defended lane.[19] Each of two lanes in a 1972 Naval War College game consisted of a series of boxes that could be sanitized in turn on a sprint-and-drift basis by fast surface ships such as surface effect ships. These craft could be backed by maritime patrol aircraft, submarines, and other ASW forces. There would be no attempt to concentrate targets in convoys.

The Soviets' Backfire bombers were at least as threatening as their submarines, because they could fire their missiles from well beyond the range of air defense missiles. Even if convoy escorts shot down their missiles, the Backfires could fly back to base, reload, and return until the escorts exhausted their weapons. Current air defenses between Labrador and the United Kingdom were considered inadequate. According to the SEAMIX I summary, "Losses of shipping to the Backfire threat . . . could exceed losses to submarines. . . . [A] Soviet strategy which commits a large percentage of their best submarines deployed in the Atlantic against the carrier force might be feasible. Such a strategy might eventually pay off in terms of trans-Atlantic ships sunk if Backfires could achieve unopposed access to the sea lanes after the carrier strength was reduced below effective levels." The carriers were already needed to provide ASW cover to shipping and to deter Soviet surface operations in the open ocean. If the Soviets decided to attack them first, to reduce that protection, NATO ASW ships and aircraft would have to be redeployed to reinforce carrier ASW defenses, and land-based fighters would have to be deployed to strengthen air defenses in the GIUK Gap. That in turn would require access to bases in Greenland and Iceland.

The Navy authors of SEAMIX I pointed out that planned cuts in the carrier force level were incompatible with the likely increases in air and surface threats to the areas the carriers were expected to cover. Committing most carriers to the Atlantic would make it impossible to fight effectively in the Mediterranean and the Pacific.

The carriers needed a way to engage the Backfires before the bombers could launch their missiles. The SEAMIX I report advocated what was later called the Outer Air Battle: fighters should engage the bombers rather than the missiles they launched. That became practicable with the advent of the Aegis and other anti-saturation defensive systems.

It seemed likely that the situation would improve to marginal in 1981 and to adequate by 1985. Ship passive capability would improve markedly with completion of the SQQ-23 passive program (PAIR) and the advent of the tactical towed array, which was expected to reach long ranges at speeds of up to 20 kts. Ships were also being quieted. Rubber sonar dome windows would greatly improve SQS-26 capability.

In 1975 it seemed that the Navy would prevail in the western Mediterranean but not in the eastern Mediterranean and the Arabian Sea. Although the West would likely prevail in mid-Atlantic operations, the outcome would be uncertain; on the Atlantic sea lanes, the Soviets would prevail in the Norwegian Sea in both the near term (FY 78) and the far term (FY 87). Area ASW capability would become marginal as early as FY 76.

QUIETER SOVIET SUBMARINES

We now know that the Soviets tried (very unsuccessfully) to make their first-generation nuclear submarines reasonably quiet. Unsurprisingly, they continued this attempt with their second generation—the Charlie (Project 670A tactical missile submarine), Victor (Project 671 torpedo submarine), and Yankee (Project 667A strategic submarine). All had versions of the same second-generation power

A Charlie- (Project 670A) class missile submarine shows her blunt nose and, faintly, the outline of some of her missile doors. Charlie was the first Soviet nuclear submarine with a single reactor, probably adopted because it was conceived as a mass-production attack submarine to meet Khrushchev's demand for large numbers to attack Western sea-lanes. Its role changed as Khrushchev developed his nuclear Revolution in Military Affairs and the focus turned to attacking Western naval nuclear forces—initially strike carriers. The P-7 (SS-N-7) missile in the submarine's enlarged bow was credited with a range of 4–30 nm. Unlike the longer-range P-6 (SS-N-3), it was targeted from the submarine (often in conjunction with other submarines, which could provide other target bearings for triangulation). USNI

plant, which came to be called the CVY plant after the NATO class designations. The first of the group, the Victor-class attack submarine K 38, was delivered on 5 November 1967. The first Yankee, K 137, was completed on 6 November 1967. The first Charlie, K 43, was completed on 5 November 1967.

By June 1968 Western intelligence knew that six submarines of a new type existed, but little more was known.[20] On 14 December 1968, NAVFAC Keflavík first detected Soviet Victor- and Charlie-class submarines.[21] In 1969 all of the second-generation submarines were credited with speeds of up to 30 kts, compared with 25–26 kts for the November class (which was actually capable of 30 kts).

The Victor class was credited with improved self-noise, better sonar, and higher speed.[22] In 1972 the vessel was described as "a sophisticated attack submarine combining high speed and deep-diving ability with good sonar performance. It is probably the first credible ASW submarine produced by the Soviets."[23] U.S. Navy analysts interpreted the Victor class as a direct response to the threat of U.S. strategic submarines (Polaris boats). In 1971 Victor was (accurately) credited with a speed of 32 kts, making her the fastest operational submarine in the world (the nature of the Alfa class was not yet known). It now appears that the Victor design was started in 1959. At that time the most prominent strategic threat to the Soviet Union was U.S. aircraft carriers, and the new submarine was presumably conceived to attack them. The Victor's massive bow sonar offered ASW capability.

Because of her unusual (presumably hydrodynamic) shape, Charlie was credited provisionally with "improved self-noise characteristics, better sonar, and a higher speed than any previous Soviet nuclear submarine." In fact, the bulbous bow carried eight antiship missiles; by January 1970 they were assumed (correctly) to be short-range (35 nm) antiship missiles, perhaps (incorrectly) related to Styx. That made Charlie the most startling of the new submarines. The underwater-launched missile offered almost no warning as it approached at about 10 nm/minute at fairly low altitude: less than 3 minutes' warning from 30 nm away. In theory, new automatically triggered defensive missiles would solve that problem, but captains were leery of activating them.

In 1975, Charlie was considered the greatest single threat to U.S. carriers because, unlike the first-generation Echo, the vessel did not require support by any other platform. By this time almost every Soviet fleet exercise over the past few years had included some form of anticarrier strike. Realistic training began in 1971, and such exercises were held at least once a quarter from 1972 on. By 1975 the Soviets were training against Sixth Fleet carriers. The attackers were a Charlie and/or a diesel Juliett with longer-range missiles, and at least once two Charlies. It seemed likely that the attack would be mounted by two submarines, either a Charlie and a Victor or two Charlies. The Soviets were known to have a variety of underwater communications systems (it turned out that communication was integral to all integrated Soviet sonars). These systems incorporated distance-measuring equipment (DME). Two submarines could triangulate the target based on the measured distance between them and the bearing each had on the target. Typically, they loosely trailed the carrier from her rear flank and her rear. They could attack almost silently. It seems likely that the practice tactic described in 1975 had been observed many times in the Mediterranean. Observed range of the Charlie missile was 30 nm. If a Charlie and a Victor worked together, the missile attack would be followed by torpedoes from both submarines. If only one missile submarine was available, she could work with a surface tattletale, which would withdraw just before the strike. By 1975 the Soviets were using paired Charlies and Victors not only in the Mediterranean but also in the open ocean, with no surface ship support, relying on the Soviet submarine broadcast for initial location information. Charlie was actually considerably slower than Victor because she was powered by one rather than two reactors.

The new Soviet nuclear submarines were considerably quieter than the earlier rather loud ones. Initially Charlie- and Victor-class submarines were considered comparable to the *Skipjack*, the last U.S. submarine not designed specifically to be quiet enough for ASW. However, further experience showed that they were somewhere between *Nautilus* and the next class (*Skate*) in noisiness. Silencing reduced propeller blade rate frequency (propellers turned more slowly), cavitation noise (which had been a valuable aid to tracking first-generation submarines), and much of their broadband noise. However, the new submarines did have an important acoustic weakness. Instead of the DC generators of the earlier units, whose speed varied with that of the ship's turbines, these submarines had 50 Hz AC turbogenerators, their speeds carefully regulated.[24] The submarines thus generated a stable 50 Hz line. It or its harmonics (whole-number multiples), the strongest of which came at 300 Hz, were easy to detect (the 300 kHz line drove the designs of early towed arrays). Variations (due to Doppler) in the received frequency indicated the submarine's speed relative to the listener. Such variations could also indicate a submarine's maneuvers. No such certainty was available with first-generation Soviet nuclear submarines, as the frequency of their turbogenerators varied with turbine RPM. The 300 Hz line was particularly useful in the Mediterranean, where shipping noise blanketed the 50 Hz line.

The rub was that 300 Hz was also the fifth harmonic of the standard U.S. 60 Hz turbogenerator signature. In the late 1970s a U.S. frigate was lauded for keeping track of a Soviet submarine all the way down the U.S. continental shelf using her towed array. In fact, she had been tracking herself, her machinery noise being reflected off the sea bottom into her towed array.

The story of the new classes began with a 1958 meeting at the Malakhit design bureau, which had been formed to design the first Soviet nuclear submarine. At this time the Rubin design bureau dominated Soviet submarine design. A Malakhit engineer claimed that given a free hand, he could design a submarine as fast as a torpedo. His key ideas were to use a titanium pressure hull for light weight, to enable him to fit a lightweight reactor into the smallest possible hull. He would also save weight and volume by automating the submarine for minimum crewing and by adopting a single hull, as opposed to the double hulls the Soviet navy favored.

At this time Malakhit was designing a fast submarine intended to attack carrier strike groups using underwater-launched missiles (Amethyst system, SS-N-7 [P-70] missile): Project 661 (Papa).[25] To reach high speeds, Papa was given two 40,000 SHP reactors driving twin screws. Like the proposed superfast submarine, she had to trade pressure hull weight for the considerable weight of the reactors and turbines. The project was important enough to sponsor the development of titanium welding techniques for the new submarine's titanium pressure hull. This hull offered the submarine a diving depth of 400 m (1,312 ft), which considerably exceeded the 320 m of a Victor; but depth was not as important as high speed. The required speed was 38 kts, but as designed the submarine was expected to make 42. On trials in 1971 she actually made 44.7 kts, the highest known submarine submerged speed. However, at high speed she was very noisy, and the vibration damaged her hatches, her conning tower door, and a buoy fairing. The concept design was approved in May 1960. *K 162* was completed on 31 December 1969, but seems to have escaped Western intelligence. She was not reproduced.

In 1958 Malakhit submitted an unsolicited proposal for a single-hull attack submarine with a body-of-revolution hull and a single reactor driving a single screw, presumably inspired by the engineer's high-speed proposal.[26] Shortly afterward a design competition for four submarine classes was announced (apparently for the only time in Soviet practice): an SSBN (Project 667), a large SSN (Project 669), a small SSN for mass production (Project 670), and an ASW SSN (Project 671).

The ASW SSN was to be suitable for production at an inland yard (presumably Gorkiy), so displacement was limited to 2,000 metric tons, making it transportable along the canal from Gorkiy to the White Sea. It would be armed with 4 torpedo tubes (8 torpedoes), would have a speed of about 30 kts, and would dive to 300 m. The competitors were Rubin, Malakhit, and the Gorkiy design bureau (TsKB-112, later named Lazurit). Rubin was assigned Project 667 (and subsequent SSBNs). Its big Project 669 was dropped. Project 670 went to Gorkiy, where it was reconfigured as a cruise missile submarine (Charlie). As Malakhit would suggest for Alfa, the antiship and antisubmarine roles were fused in a single Project 671, which was assigned to Malakhit.

The single-screw configuration was unconventional. Malakhit overcame opposition by proving that it minimized noise and displacement (hence made maximum speed) compared with the Soviet navy's preferred twin-screw configuration. Malakhit argued that it was 30 percent more efficient than a conventional twin-screw design. Admiral Gorshkov approved the design "as an exception." Although the displacement of the new design was about 30 percent greater than that of a November, the two ships had about the same wetted surface—that is, about the same resistance. The greater diameter of the body-of-revolution hull accommodated a much larger sonar array. According to the Malakhit design history, a two-reactor plant was chosen to provide a margin for future modernization. The reactors were mounted side by side. The decision to separate the ship's service turbogenerator from the main machinery (i.e., it was not geared to the propeller shaft) provided Western submarine hunters with an invaluable bonus in the form of the stable 50 Hz line. Victors employed the ship system automation adopted for the more exotic fast submarine (Alfa), but not the combat system automation; in Soviet service they were called "semi-automats," whereas Alfas were "automats."

The very fast titanium submarine was conceived in 1958 and briefed within Malakhit in June 1959.[27] The design was enthusiastically supported by both Soviet naval chief Admiral Gorshkov and B. Ye. Butoma, at that time chairman of the state committee for the shipbuilding industry. The design study was completed in June 1960. The project was formally authorized on 23 June 1960, and the concept design completed on 31 December. Malakhit told the Soviet navy that the new submarine would be a technological leap; if it were built, someone would find a role for it. The Soviet state resolution authorizing this Project 705 described it as an ASW submarine, presumably as part of the emerging anti-Polaris program applied to

An Alfa- (Project 705) class submarine shows her blended hull form, without any flat upper deck. Alfa was the first Soviet nuclear submarine with a rafted power plant, and was so quiet that until the end of the Cold War Western analysts thought she had a turboelectric plant. USNI

virtually all Soviet naval projects at the time. Given the emphasis on ASW, Malakhit had to find a way to quiet the submarine to increase her sonar range. Maximum expected detection range was 20 cables (2 nm).

Malakhit stressed that a single hull would minimize wetted area, hence resistance, and also flow noise. But the Soviet navy rejected the single-hull concept in favor of greater reserve buoyancy. The double hull added 250 tons (more than 15 percent) to surface displacement; length increased by 8 m and beam by 1.3 m. Total submerged volume increased by more than 1,000 cubic meters (i.e., by more than 1,000 metric tons). That cost an estimated 2.5 kts in speed, reducing guaranteed speed to 40 kts (initially it was 45).[28] The Alfa's proposed crew was only 15, compared with the November's 104 (including 24 officers). That was more than doubled to 32—of whom 31 were officers (the sole enlisted man was the ship's cook). The operating depth had to be reduced from 500 to 400m (1,312 ft), the same as that of other Soviet nuclear submarines. However, the radical idea of automating the ship was pursued, all control being concentrated in a single console. For the first time in

Soviet practice, torpedoes were ejected hydraulically, so they could be fired at depth.

Because it was interested mainly in performance, Malakhit developed alternative versions for antiship operation (armed with the new 65 cm torpedo) and as an SSBN.[29] Malakhit later proposed that the antiship and ASW versions be combined, although it was not clear why very high (loud) speed was necessary for ASW (Malakhit claimed that the submarine could evade ASW torpedoes).

To achieve the performance it sought, Malakhit needed a compact steam plant generating 40,000 SHP at a specific weight of 12–14 kg/SHP. The designers ultimately adopted a liquid-metal reactor, which had been tested on board the last November-class submarine, *K 27* (Project 645).[30] *K 27* introduced the 50 Hz AC ship's service turbogenerator driven separately (and stably) rather than from the propeller shafts. A conventional pressurized-water reactor was heavier, and the saturated steam it produced entailed a turbine about 300 tons heavier. It was argued that because the primary loop of a liquid-metal reactor was not under high pressure (as in a pressurized-water reactor), a break in the primary loop would not be as dangerous. U.S. experience with the liquid-metal reactor in *Seawolf* showed that leaks would be common. The Soviets' experience with Project 645 was so bad that her reactor section was removed and dumped in the White Sea. Once the reactor went critical, its liquid metal had to be kept molten; if it froze, the steam generator would be ruined. The first of the Alfa class had to be cut in half after a freeze-up; cynics joked that she was simultaneously in service in the Baltic and Northern fleets. She was discarded in 1974, never having been operational.

Unlike other Soviet second-generation nuclear submarines, Project 705 employed 400 Hz generators. Presumably the much higher frequency was adopted for compactness, but it would also have made passive detection more difficult. It appears that Project 705 had the first rafted power plant in Soviet practice. The West knew nothing about it until a photograph of the rafted unit appeared in a Russian publication in 2002. Until then it was thought that the impressive silencing of the Alfa class was due to adoption of turboelectric propulsion—which would have been considerably heavier than the rafting actually employed. According to a Russian account, the required level of silencing was never attained.

The first unit, *K 64*, was built at Leningrad and became known to the West in 1971. It was accepted in December 1971, but the second unit did not follow until December 1977. By that time Victor III surpassed Alfa in everything but speed. By 1973 the program was so expensive that Butoma, by that time a senior defense program official, tried (initially unsuccessfully) to kill it. Seven Alfas were built; an eighth was scrapped on the slip. The Soviets did return to titanium in the 1980s, but they never built any more liquid-metal reactor submarines. Nor did they try to reach the speed of the Alfa.

To the West, Alfa was initially a puzzle, apparently contradicting the Russian adage that better is the enemy of good enough, supposedly a favorite aphorism of Admiral Gorshkov, the architect of the modern Soviet navy.[31] In 1969 a small new submarine, which turned out to be the prototype Alfa, appeared alongside the fitting-out quay at the Sudomekh submarine yard in Leningrad. A senior U.S. submarine analyst soon concluded that it was a new nuclear attack submarine with an estimated submerged displacement of 3,700 tons (actually 4,320 metric tons). It was the world's smallest nuclear submarine but had the usual high reserve buoyancy of Soviet submarines. Nearby was a previously unseen "highly reflective" pressure hull section. Reports of similarly highly reflective sections also associated them with the Severodvinsk submarine yard. The analyst concluded that the sections were made of titanium. Skeptics pointed out that titanium is very difficult to handle or to weld in air. It later turned out that the Soviets had built a special titanium-handling facility. The analyst concluded, and the Navy eventually agreed (mistakenly), that the Soviets had adopted titanium to enable their submarine to dive much deeper than any previous Soviet submarine. Titanium would also defeat MAD.

The compactness of the submarine seemed to indicate other advances: probably it was highly automated (it was), and it was powered by a new compact reactor. The U.S. Navy did not want to believe that it was powered by a liquid-metal reactor, knowing the dangers of such a power plant. The Navy also disliked the automation claim because computer control was considered inherently unsafe—and that was without knowing how limited Soviet computer technology was. Another puzzle was a visible joint down the side of the submarine's sail, which was correctly interpreted as an enclosure for a rescue sphere, again contrary to the usual Soviet disdain for personnel safety. Human intelligence collected as the new submarine was being built confirmed that she would have an unusually small crew, although figures varied. In 1981 a rescue sphere was seen being lowered into the area where an Alfa was fitting out. The diameter of the sphere gave an estimate of its capacity, hence the number in the Alfa crew. In March 1979 an Alfa running trials in the Barents Sea was monitored at over 40 kts. This performance far exceeded what would have been possible with a conventional pressurized-water reactor, so for a time there was speculation that the Soviets were using exotic means of reducing drag, most likely by ejecting polymers to smooth water flow.

Alfa had considerable impact on Western navies because it was believed (wrongly) that the titanium hull would enable the vessel to dive much deeper, and hence evade U.S. ASW weapons. In fact, the Soviets used the light weight of the titanium pressure hull to balance the weight of the powerful reactor. Diving depth was the same 320 m as in a Victor—but rated speed was 41 rather than 30 kts. If a torpedo needed a 30 percent speed edge over its submarine target, that seemed to demand a 53.3 kt torpedo. Mk 46 was clearly outclassed. It seemed that only the big Mk 48 could catch an Alfa. The U.S. Navy released requirements for an Advanced Light Weight Torpedo (ALWT) to deal with the new submarine. By 1972 the U.S. Navy was dissatisfied anyway with Mk 46, so its decision at that time to develop a new lightweight torpedo does not necessarily indicate agreement with the speculation about Alfa-class capabilities. In 1975 design requirements for the new torpedo included a range of 15,000 yds at 55 kts and a homing range of 3,000 yds. No depth requirement was reported at that time. By the time a competition was held, the requirement included the ability to dive very deep. Alternatives were a closed-cycle engine using lithium fuel (Mk 50) and an advanced battery (Mk 51); Mk 50 won. Ironically, by the time the Mk 50 was in mass production the Cold War was ending, and it was superseded by a much slower Mk 54 that incorporated its homing head and Mk 46 propulsion.

Alfa was first seen at about the same time as Papa and Bravo (Project 690). That they and Charlie, Victor, and Yankee could emerge in quick succession was terrifying evidence of the breadth of Soviet submarine design capability. In fact, Bravo was a target submarine, and there was only one Papa. On the other hand, there were three distinct Soviet submarine design organizations: Rubin, Malakhit, and the small office at Gorkiy that designed Charlie. Both Rubin and Malakhit were large organizations. The United States and Britain, in contrast, had only a single submarine design organization.

SOLVING THE LOW-FREQUENCY PROBLEM: TOWED ARRAYS

The prospect of really quiet deep-diving Soviet submarines seemed to demand new kinds of sensors that could see through the surface layer and yet achieve long ranges. In 1969–71 three new techniques were being explored: the reliable acoustic path (RAP), refracted surface reflection (effectively surface bounce), and the vertical linear array (VLA). RAP used a deep sensor looking up, spotting submarines in an overhead cone. If the sensor was deep enough, the cone could cover a considerable area at practicable submarine operating depths. VLA would focus beams in the vertical, to avoid both the surface and the bottom. Most shipping noise would be trapped in the layer. Initially it seemed that all of these sensors would be so elaborate (hence so expensive) that they would have to be recovered after use. Hopefully they would detect submarines at a considerable distance. Their ship or airplane or helicopter would dash to that place to attack before the submarine got much farther away. The idea was associated with a variety of platforms, including fast VTOL aircraft and big hydrofoils and surface effect ships. The surface craft would deploy deep RAP sonars using extremely fast winches (to get down and up from great depth) before sprinting. It turned out that some of the advantages claimed for the exotic sensors could actually be realized in new expendable sonobuoys.

A much simpler solution was found: the towed array, employing a shipboard form of LOFAR.[32] The idea of applying LOFAR techniques to ships was not new. In the spring of 1955 Bell Labs had become interested in what it called Mobile Listening, a shipboard application of LOFAR. Bell's scientists hoped that its range might be half of that achieved by the shore stations, which might have meant 200 nm (at this time SOSUS range was being given as 400 nm).

In the late 1950s Marvin Lasky of the Naval Research Laboratory began to experiment with arrays towed from submarines and surface ships. He resurrected work done during World War I with towed multi-element listening arrays. The offshore oil industry was using towed streamer arrays to detect echoes of underwater explosions. Lamont Geophysical Laboratory tried a ship-towed hydrophone in January 1959 as part of a project to monitor and reduce self-noise. The array employed 11 hydrophones on a cable as much as 3 miles long. At 100 Hz–10 kHz, at up to 10 kts it showed the lowest self-noise ever achieved, lower than the quietest submarines. Lasky contracted with the Chesapeake Instrument Company to build and test arrays. A Chesapeake Instruments research vessel tested a large-diameter array in the Bahamas in the summer of 1961. The first U.S. submarine tactical towed array appears to have been Towflex, built by Chesapeake at Lasky's request specifically for trials on board the experimental submarine *Albacore* to measure her near-field noise; it was tested in February 1962, probably as an ONR project (the array was also called TOAD). At this time *Albacore* was also testing a prototype DIMUS passive spherical sonar.

A 1964 National Security Industrial Association (NSIA) ASW study recommended that a submarine tow a complete 1,000 ft SOSUS array on a 10,000 ft cable. The following year NSIA proposed that the next submarine have a bow array and a towed array, their output processed by the BQQ-3 LOFAR spectrum analyzer.

An array towed by a surface ship had self-noise comparable to the best submarine sonars. Because array length was unconstrained by the ship's hull, a ship could focus low-frequency sounds. A Tentative Specific Operational Requirement (TSOR S22-32T) was issued on 6 December 1962 and, in revised form, on 15 January 1965. The array would be towed by destroyers and ships of opportunity. It could cover the flanks and rear of a moving force. The same idea was applied to submarines at about the same time, though in submarines such arrays became operational earlier.

As with SOSUS, signals received by the line of hydrophones on the array could be combined to form beams, which could be subjected to the usual Jezebel frequency analysis to detect targets. The beams of a towed array formed cones wrapped around the array. The bearings of nearby targets were ambiguous, but at long ranges the segments of the conical beams cut by the surface and the bottom (or the deepest possible submarine targets) were clear. Maneuvers or other techniques were needed to resolve right/left ambiguity.

It appears that the first U.S. naval use of towed arrays was by submarines gathering acoustic intelligence of Soviet submarines under the TUBA program. Initially, diesel-electric submarines used hull arrays and precision Bell Labs sound recorders (BQH-2).[33] The recorder was so sensitive that a submarine had to be stopped to use it. Since flow noise precluded detecting lower-frequency Soviet submarine noise, the U.S. Navy adopted towed arrays, which could operate outside most of the submarines' flow noise. When the British decided to adopt narrowband analysis in 1967, BQH-2C was a candidate analyzer. *Sturgeon*-class submarines were fitted with the first U.S. tactical towed arrays (as TUBA II), their bow sonars being used to pick up higher-frequency sounds. The TUBA II array (BQH-4) was optimized for precision recording and had no tactical role, but it was the basis for the later submarine tactical towed arrays. The design of the BQH-4 development model was frozen in March 1966, and production specifications were framed about September 1968, at which time plans called for ordering three sets about March 1969. A development model was evaluated in March 1968. It was made by both Hughes Aircraft and Chesapeake Instrument Company (which was later bought by Gould). The array offered a high data rate and had a moderate aperture (180 ft) and a large diameter (3 in.). Total array length, including the vibration-isolation modules, was 258 ft, and it was towed on a 2,600-ft cable; it listened between 10 Hz and 20 kHz. It was credited with excellent beam-forming but had poor self-noise at speed. As might be imagined in an intelligence device, it did not offer the sort of high sustained reliability a tactical array would need. TUBA II was installed on board about 15 *Sturgeon*-class submarines. By the fall of 1963 ONR was reporting that a new 450 ft array offered very promising LOFAR performance. The main problem was how to stow a longer array (perhaps 1,000 ft) on a 10,000 ft cable.

The towed sensor had an immediate tactical value: it could look aft from outside the flow noise of a submarine. That was clearly important for a strategic submarine. In July 1964 the U.S. Navy established a Specific Operational Requirement (SOR 23-27) for a submarine towed array. By the time the requirement was written, the array was also wanted for attack submarines. In addition to looking aft, it could offer greatly improved LOFAR classification capacity compared with the usual bow array; it could monitor the submarine's self-noise and thus help her to maintain silence; and it might also support better acoustic communication. The SOR envisaged quick development based on the existing TUBA II array. It should be operationally evaluated in FY 67 and installed not later than FY 68 in *Sturgeon*-class attack submarines and then beginning in FY 70 in SSBNs. About 150 arrays would be needed. The initial version was expected to detect a 15 kt ship in the first convergence zone in the Norwegian Sea. It seemed that it might even be able to detect a quiet submarine (like *Thresher*) at 30,000 yds (and a noisy fast one at twice that range). Because it would be streamed outside the submarine hull, the array had to be reliable, with at least 1,000 hours between failures (the goal was 2,000). The only problem was that the BQQ-3 spectrum analyzer used by the BQR-7 bow array did not properly cover the low-frequency range (10–150 Hz) the towed array was expected to explore (in fact, its output at higher frequencies, about 300 Hz, was more important).

Bell Labs and Chesapeake Instrument competed for the hindsight contract. They offered, respectively, an array stowed in a tube between the hull and the superstructure, with a range beyond 40,000 yds, tested on board USS *John Marshall*; and an array towed from the sail (with a range of about 10,000 yds), tested on board USS *Thomas Jefferson*. Chesapeake Instrument built the Bell array. The Chesapeake array went to sea in early December and the Bell array in late December 1962. Both were successful, but each had drawbacks. The Bell system was too expensive, while the Chesapeake array was towed above the missile tubes and might interfere with them. Bell's stowage arrangement was adopted. Bell Labs received the hindsight contract for what became the BQR-15 submarine array. The operational evaluation report on BQR-15 was issued in June 1970. It used a large-diameter (3.5 in.) array and transmitted its data via tested twisted-pair technology. It had a narrow aperture (104 ft) with fair self-noise at speed (it used four-element hydrophones). Uniform element spacing made for varying beamwidth at different azimuths. It was considered quite expensive ($400,000, the objective for an operational array being $150,000). Given its limited length, the hindsight array was most effective at 300 Hz rather than at the fundamental 50 Hz frequency of Soviet submarine machinery, as its beams were much narrower at the higher frequency. Conversely, Soviet second-generation submarines were loud enough at 300 Hz that a short, towed array was effective against them.

The Soviets were expected to silence their submarines. The first response to that seems to have been SITTAR (Submarine Integrated Tactical Towed

Array), intended to make up for the loss of broadband noise that had been exploited by BQR-15. SOR 23-40 was issued in tentative form on 7 October 1969 and then in complete form on 13 November 1970. The new array would detect tonals in the 10–1,500 Hz range at up to 15 kts better than the existing array could detect broadband above 1 kHz. The aperture was to give a beam width of 10 degrees at 50 Hz, the target tonal. It seemed that a 500 ft array streamed 1,000–2,000 ft astern would suffice. Unlike BQR-15, SITTAR would have a small diameter, ⅝ to 1.5 in. (the upper limit was set by a requirement that the array be deployable through a line wiper). This array would be reelable, that is, it could be retracted; it was to be streamed and recovered at up to 8 kts and a depth of 200 ft. Given its small diameter, it might be stowed by being wrapped around the submarine's main shaft.

In 1970 it was decided that the developmental digital sonar of the *Los Angeles* class (BQQ-5) would be designed to be compatible with a BQH-4-type towed array, its low-frequency receiver integrated into the new sonar. Initial tests completed in December 1972 showed that as modified the sonar could meet the narrowband tracking threshold against the projected 1975 Soviet threat as well as against the existing second-generation Soviet submarines. At this stage the envisaged array was about 240 ft long including a 186 ft aperture, with 3 in. diameter. Compared with BQH-4, it was expected to have lower self-noise and greater reliability. It could either clip on or be housed when not in use. In May 1973 each of the BQH-4 manufacturers, Hughes Aircraft and Chesapeake Instruments, was contacted for three prototype arrays for tests in FY 75; they were to be refurbished as the first BQQ-5 towed arrays.

Existing submarines had to be fitted with towed arrays. Because this was an urgent program, the U.S. Navy initially opted for clip-ons. The system was called STASS (Submarine Towed Array System, later BQR-25). Chesapeake Instruments received the first production contract in 1975, defeating Hughes Aircraft, whose array was less rugged and reliable, though quieter at speed. STASS was evaluated in the Pacific in March 1972. Later the array was housed in a long tube on the outside of the pressure hull, the towing cable being coiled around an internal winch. STASS employed the TB-16 array, developed under a 1973 contract for a dual-purpose array, for both

***Hartford* (SSN-768) shows the housing for her TB-16 towed array; her long, thin TB-23 was wound around inside her after ballast tank.** USNI

TUBA and tactical roles. The 240 ft array (186 ft acoustic aperture) was towed on a 2,000 ft cable.

Sturgeon-class submarines had a reelable installation. The array was stowed in a tube running outside the pressure hull; the reel, cable, and winch were located in the forward main ballast tank. The array was streamed out of a tube leading from the port horizontal stabilizer. Early *Los Angeles*–class submarines had permanently streamed (clip-on) arrays. As the submarine went to sea, a small boat would clip the array to the submarine's horizontal stabilizer. In service the array could be damaged during high-speed runs. *Los Angeles*–class submarines were therefore provided with reelable arrays, despite their drag.

The Soviets introduced rafts in Alfa and Victor II (Project 671RT), but at least in the latter the raft seems to have been ineffective. Western sources never touted it as a great silencing improvement over Victor I. The Victor II raft apparently did little or nothing to blank out the 50 Hz line. It appears that the first evidence of Soviet interest in silencing their turbogenerators came in 1975, when a newly refitted Yankee showed a 6 dB reduction in SSTG noise, presumably meaning the 50 Hz line. A second quieted Yankee was seen in 1978 (11 dB reduction) and a third in 1979 (8 dB).[34] The first Victor III (Project 671RTM), *K 524*, which also displayed turbogenerator silencing, was completed in December 1977. With this submarine the Soviets finally effectively reduced machinery tonal harmonics, so that the emphasis had to shift from 300 Hz to 50 Hz. The latter was more difficult to suppress but required a longer array to maintain gain and bearing precision (TB-23, the 4X array). Victor III seems to have benefited from much more careful acoustic measurements, possibly reflecting an understanding that the Western powers were relying heavily on narrowband detection. It used better machinery sound isolation mounts. In effect it was a half step between the second-generation Victor and the successor third-generation Akula (Project 971: Soviet name Shchuka-B, Victor III being Shchuka). The much longer U.S. 12X towed array was an attempt to maintain passive tracking of third-generation submarines by again increasing gain at low frequency, presumably to keep using the 50 Hz line.

A Victor III type submarine (Project 671RTM or 671RTMK) with the characteristic bulbous towed array housing; the array is wrapped up inside the housing and unreeled through the hole at the rear of the pod. The Soviets had rafted the power plant of the Victor II (Project 671RT), but without much success in silencing. Presumably that meant that they reduced broadband noise but not the narrowband noise critical to tracking Soviet submarines. The RT in the designations indicated that these submarines had two 65 cm torpedo tubes in place of two of their 53 cm tubes. All Victors could fire the Soviet SUBROC-equivalent (81R), but Victor III could also fire the RK-55 Granat (SS-N-21) cruise missile as well as the improved 83R and 84R versions and the Shkval torpedo. USNI

Victor III featured the first Soviet towed array, which was deployed from a pod atop its vertical fin. The nature of this array reflected the Soviets' changing understanding of silencing. It was reelable, because its designers initially assumed that it would detect a target's flow noise. The submarine using the array would have to slow dramatically to listen; it would necessarily employ sprint-and-drift tactics. After the array and its reeling machine had been designed and ordered, the Soviets discovered that Western towed arrays were picking up tonals, hence could function while the submarine was generating its own flow noise.[35]

Towed array techniques were also applied to surface ships. As the U.S. submarine array program began in earnest, there was also a less-advanced surface ship towed array project (ADO 22-23X). The first U.S. Tentative Specific Operational Requirement for a surface ship towed array seems to have been issued on 6 December 1962. As revised in January 1965 it stated that experiments had shown that a towed array could provide self-noise comparable to that of the best submarine sonars. Arrays were wanted both for destroyers and for temporary installation on board other ships for specific tasks, such as creating local barriers and detection on the flanks and rear of a task group or convoy. Arrays could also be used to gather acoustic intelligence. A feasibility study was wanted to explore ways to overcome restrictions in maximum speed and maneuver. The array would employ LOFAR processing. Initial results were so promising that it was considered very desirable to investigate a high-speed system, which would complement hull sonar.

When Bell Labs received the BQR-15 contract, Chesapeake Instruments was funded to do surface ship experiments. Chesapeake's initial approach was to stream two arrays 350 ft apart, their outputs being cross-correlated. The towing ship carried a LOFAR sonobuoy processor. In May 1965 this array was tested against submarines off Bermuda. A snorkeler was detected 67 nm away. In 1967 Dr. G. Sebastien of DDR&E suggested that towed arrays be used for submarine surveillance in the Mediterranean, where there was no prospect of deploying SOSUS. At this time Hughes and Litton were also making towed arrays. After all three towed array producers gave presentations, specifications were written for a Mediterranean surveillance system called the Interim Towed Array Surveillance System (ITASS, SQR-14). The first SQR-14 was produced under the FY 68 program; NAVSEA requested bids on 1 May 1968. In 1969 tests were run with a prototype submarine TUBA array. The program was to comprise three air-transportable systems that could be accommodated in vans.

The contract went to Chesapeake, which delivered the three systems to the fleet in September 1970, about 14 months late. SQR-14s were installed on board three destroyer escorts: *Hammerberg*, *Courtney*, and *Lester*. In effect this program brought SOSUS to the Mediterranean. The three ships accounted for

USS *Hammerberg* as a towed-array surveillance destroyer escort in the Mediterranean, 1972. The processing equipment for the array was housed in a van on her helicopter pad and in her helicopter hangar; presumably the gear on her fantail was the array itself. It was 730 ft long (including its drogue), towed on the end of a 7,500-ft cable. Weapon A was removed as weight compensation, but she retained her lightweight triple torpedo tubes. Of the three *Dealey*-class escorts assigned to this role (as Escort Squadron 8), she and *Lester* had SQS-23 sonars. *Courtney* had the smaller SQS-41, and consequently had more fuel and greater endurance. The measure of detection range on a snorkeling target was the effective detection range (EDR), defined as the range within which the ship should always detect an exposed target. In the first controlled exercises using the array, the greatest EDR was 84 nm in the southern Balearic Basin, falling to 63 nm in the Levantine Basin, 58 nm in the northern Balearic, 57 nm in the Ionian Basin, and 50 nm in the Tyrrhenian Sea. NHHC

The former MSO 521 (*Assurance*) as the experimental towed surveillance array ship AG-521. Note the trough right aft for the array. Towed array service wore out the three destroyer escorts, which were not designed to be maintained for short visits to forward depots, both ashore and on a tender. NHHC

about half of all Mediterranean submarine sightings, including visuals. Their success led the Sixth Fleet to set up the equivalent of a U.S. ASW tactical coordination center, with P-3s flying out of Sigonella prosecuting submarines found by the towed-array ships. Initially the ships focused on the 50 Hz line so successfully used by SOSUS. In the Mediterranean, however, it was blanked by shipping noise. The system succeeded when processing was shifted to the sixth harmonic of the line, 300 Hz. The destroyer escorts' success was attributed partly to the very low sidelobes of the SQR-14 array. Shipping noise in the Mediterranean is very directional, and sound propagation is excellent, so that a submarine operating between ship traffic was generally detected.

The old destroyer escorts were not ideal surveillance platforms, having spent only about 20 percent of their time at sea, and only half of that in surveillance. They were replaced by a pair of converted minesweepers (MSO 520 and MSO 521, *Alacrity* and *Assurance*). They were redesignated AG 520 and AG 521, their sweep gear removed and replaced by SQR-15 for tests.[36] Initial test results from these ships were impressive. In the May 1975 Surveillance-Aided Intercept exercise southwest of Bermuda, *Alacrity* detected the SSBN *Andrew Jackson* at 99 nm and the attack submarine *Pargo* at 124 nm. *Assurance* averaged 95 nm against *Jackson* and 117 nm against *Pargo*. Both submarines were simulating Soviet submarines with noise sources augmenting the sound of their very quiet rafted machinery. In this exercise, *Alacrity* averaged 55 percent contact-holding performance. Fractional holding time (the percentage of the time held when the target was within 200 nm) was 60 percent. For both TASS units, 28 percent of range estimates were within 5 nm, and 55 percent were within 20 nm. Bearing accuracy varied between targets that were roughly end-on and targets that were roughly broadside-on, the latter being more accurate. This performance was a radical improvement on past experience. Submarines were also tested using the new STASS arrays. Their average initial detection range was 105 nm, but one of them managed to reach out two convergence zones farther, in one case to 152 nm. The implication was that open-ocean areas previously considered safe for nuclear attack submarines could now be made quite hostile by a few submarines. The two towed-array AGs were stricken in 1977.

Six improved SQR-14As were ordered under the FY 72 program, three for the Atlantic on board frigates and three for the Pacific on board *Spruances*, as a

USS *McCloy* with her second-generation SQR-15 towed array during NATO exercise Northern Merger, 9 September 1974. She operated in the Atlantic and the Mediterranean; her sister *Bronstein*, similarly equipped, operated in the Pacific. The reel of her array is visible abaft the break of her forecastle. Conversion included removal of her after twin 3 in. mount and the captain's gig. Processing equipment (aircraft-type AQA-5s) was in a van in her hangar. Late in October 1983 she successfully tracked a Victor III, only to have the submarine snag her towed array cable. NHHC

The frigate *Albert David* is shown refueling in the Pacific en route to the RIMPAC exercise, 7 April 1975. She could be distinguished by her low hangar, not adapted to LAMPS but sufficient for towed-array processors. Not visible is the drum for her SQR-15 array, which deployed over her stern. Tracking Soviet Yankee-class SSBNs, she and her sister ship *Sample* (DE 1048) pioneered a technique for resolving array bearing ambiguity without maneuvering. They steamed along parallel courses 200 nm apart. Each communicated to the other via a P-3 using UHF radio, which the target submarine (and land bases) could not intercept. Once the submarine was localized, the P-3 could attack. NHHC

mobile surveillance capability. Arrays were upgraded to SQR-15 configuration in 1974. SQR-15 was adopted in preference to the 1,200 ft SQR-14B, which was intended for strategic and broad-area surveillance. The first were on board the two *Bronstein*-class frigates in 1975, and six were on board *Garcia*-class frigates (one was lost in 1983 when a Soviet Victor III class submarine cut through the tow cable with her propeller). Arrays were moved onto *Spruance*-class destroyers (three on each coast) in 1983–85. The electronics equipment was in a van, for cross-decking.

The success in the Mediterranean led to two different lines of development. DDR&E Jerry Cann feared that further work on SQR-14/15 would compromise either the surveillance mission or the tactical mission, so he split the program in two. One line of development was to extend surveillance range using a much longer array towed by a quiet ship: SURTASS (UQQ-2 sonar system). The ship would cruise at low, quiet speed, filling gaps in SOSUS coverage. It would use a seismic-type array, the kind used in oil exploration. The other was a tactical array for surface ships. In March 1973 DDR&E promulgated a plan for the next-generation TASS and TACTLASS (tactical towed array), envisaged as versions of SQR-14.

The first tactical towed array experiments, using one in an exercise, were conducted in 1973 using USS *Patterson* (DE 1061). She detected targets 13 times in the first convergence zone, and twice in the second zone.

The tactical program resulted first in the Interim Escort Towed Array (IETASS), SQR-18, and then the ETASS itself (SQR-19). An SOR was issued in FY 73 after an experimental array was bought in FY 72. The array was tested on board *Knox*-class frigates (initially USS *Moinester*) and a minesweeper, contracts for the first two IETASS being let in August 1974. The array was conceived as a simpler alternative to SQR-14/15. Established in FY 75, the ETASS program could be traced back to the initial FY 68 towed array project. Contracts for 2 IETASS were awarded in August 1974; in FY 80, 12 were retrofitted to *Knox*-class frigates, 4 more following in FY 81. Rapid development reportedly made for some

The *Knox*-class frigate USS *Pharris* shows the ship end of her towed array, emerging from the cut in her transom initially intended for a variable-depth sonar. She is shown at Portsmouth in the United Kingdom, 25 May 1978. Although she has the enlarged hangar and helipad for a LAMPS I helicopter, she lacks the point defense missile system of other upgraded frigates. NHHC

unreliability. The initial production version was towed from the VDS fish on board *Knox*-class frigates; it was first tested on board USS *Hewes* in May 1978. The improved SQR-18A(V)2 had a depressor, so it did not need the fish; it was towed by *Knox*-class frigates never fitted with VDS and by some *Perry*-class frigates. The 730 ft array included 32 vibration-isolated hydrophones (8 modular hydrophone sections). An upgraded SQR-18A (1981) used an 800 ft array. Its processor embodied ALICE (adaptive line-canceler and enhancer), which could cancel out the noise of a nearby ship.

During the Interim Sea Control Ship operational evaluation in the Atlantic in April 1974, a *Knox*-class frigate using the IETASS system consistently gained submarine contact at ranges of 29–38 nm during predicted convergence zone conditions. Naval authorities predicted that under favorable conditions IETASS would have a 50 percent probability of detecting second-generation nuclear submarines at the second convergence zone, and third generation at the first convergence zone.

The objectives of the SQR-19 program included detection out to the second and third convergence zones. It was to detect targets at an own-ship speed of up to 20 kts and to survive at about 30. Initially it was to have used a 490 ft array, but in the summer of 1979 the program was restructured to detect submarines at the second convergence zone when the ship was running at high speed. The array was lengthened to 800 ft, towed on a 5,600 ft cable, after 1983 experiments showed the need to go to much lower frequencies to deal with the silencing of new Soviet submarines. The acoustic aperture was doubled to 16 modules, and a fourth aperture (8 module) VLF section was added. The other 3 apertures were nested, consisting of 8 acoustic modules (4 LF, 2 MF, 2 HF) each about 40 ft long, consisting of 48 equally spaced hydrophones. The first array was installed on board USS *Moosebrugger* in November 1981 and was operationally evaluated in November and December 1982. A lengthened array was tested in FY 87. SQR-19 became the U.S. standard, part of the SQQ-89 suite. Full-scale production was approved on 26 December 1984. SQR-19(2X), a lengthened array to detect lower-frequency tonals, was tested in FY 87.

The fleet had to learn to use towed arrays. For example, a target directly under the array would be picked up in the same cone as one to either side. Surface ships learned that they had to maneuver to overcome this sort of error. Fortunately, in really deep water the conical beams intersected with the bottom (or with maximum submarine depth) far enough away that a ship could not mistake contacts for ones coming from directly below her.

STRATEGIC ASW

When the Soviets fielded the Yankee class with a longer-range submerged-fire SS-N-6 missile, the Atlantic Fleet was ordered to develop a countermeasure. This project supported a new round of boost-phase interception tests by the missile destroyer Barney (DDG 6) in May 1971.[37] As before, the destroyer was to maintain contact and firing position on a Soviet SSBN. Past tests had shown that the necessary level of ship readiness could be maintained. CNA studied counters to possible short-range strikes by Yankees intended to decapitate U.S. command and control (Blue Mail). The Operational Evaluation Group produced its own study, Operation Squeeze Play.[38]

That SOSUS had an 84 percent chance of detecting a Yankee passing through the Norwegian Sea was encouraging. However, an SSN operating in a barrier there had only a 32 percent chance of picking it up (net probability 27 percent). The SSN would almost certainly gain sensor contact once properly vectored (96 percent), and the probability of successfully trailing the SSBN once picked up was 60 percent; but multiplying out these figures gave a net probability of only 15.4 percent. Unfortunately, there was only a 33 percent chance that the single Mk 37 torpedo fired by the submarine would kill the SSBN: in the end, all the effort would offer a net kill probability of only 5 percent. If the trailing SSN had to wait until it heard the SSBN begin to fire, OEG estimated that on average the trailing SSN would destroy only 0.19 missiles out of the 16 on board.

This performance could be considerably improved. The poor probability that the SSN could pick up the SOSUS contact was due partly to poor positional accuracy and partly to the long time delay between SOSUS detection and notification to the SSN. These were solvable problems; the submariners were already working on precision navigation, and messages could already be passed reliably to U.S. SSBNs. The fast Mk 48 soon to enter service would largely solve the torpedo problem. Instead of being slower than the SSBN (only 22.5 kts), it would be a good deal faster. Because the trailer had to stay close (so that it could regain a lost contact), it could not use SUBROC, the best U.S. ASW weapon. It might even be too close to use ASTOR, and the OEG suspected that the process of securing the president's authority to use a nuclear weapon would be too slow. Patrol planes would do even worse, because localization was too inaccurate and (in the reactive case) too lengthy. OEG hoped that once directional LOFAR (DIFAR) was operational, aircraft could localize targets more quickly, though even then the aircraft would be too slow to interrupt SSBN missile firing.

This left trailing by missile-armed surface ships, which might shoot down missiles in the air. Exercises showed that the pings of an active sonar would warn the SSBN, enabling her to escape. Fortunately, a passive alternative, the towed array, had suddenly become available. In FY 70 tests, surface ships successfully covertly trailed a submarine simulating an SSBN, which failed to detect them. The exercise was conducted at night, because in daytime the submarine might see the tracking ship or ships. It is not clear to what extent the strategic ASW program accelerated work on the U.S. towed array. Nor is it clear to what extent the boost-phase tactic was practiced after 1971.

The main anti-SSBN effort seems to have been to trail deploying Soviet strategic submarines. Most trailing operations remain classified, so it is impossible to say when they began in earnest. That OEG was estimating the probability of trailing in 1971 suggests that the practice was already well established. British nuclear attack submarines also participated in the trailing program.[39]

Trailing turned out to be extremely important because it demonstrated to the Soviets that their strategic submarines could not operate safely in the open ocean. Although the operations were secret, apparently from time to time the Soviets were deliberately allowed to know that they could be trailed at will. They probably derived similar information from the coded naval messages they broke thanks to the Walker spy ring. It seems clear that the knowledge of successful trailing led the Soviets to concentrate their strategic submarines in bastions, which in turn deeply affected their naval strategy in the latter part of the Cold War.

Soviet SSBN operations were cautious. With all 34 Yankees completed, the Soviets maintained just 2 of them on each of 2 patrol stations, in the western Atlantic and the eastern Pacific, corresponding well with a U.S. estimate that they could keep about 15 percent of their force continuously on station. Observation of Yankee operations indicated that their patrols usually lasted 60 days. Typically, a Yankee would take 5 days to get to the open ocean and 9 days to reach a holding area. It would need at least 2 days on launch station to find its station and settle down navigation systems and to communicate with its home base.

The four deployed Yankees were initially supplemented by a single Delta in the Barents/Greenland Sea, with another expected in the Pacific late in 1975. Estimated maximum capacity for the Yankee force was 30 percent within firing range of the United States, given factors such as overhaul requirements and personnel deficiencies. A much larger percentage of the Delta force would be within range of the United States due to their missile's much greater range.

The extreme secrecy accorded trailing may have been due to fears that publicity would have killed it. Advocates of arms control typically saw strategic submarines as inherently invulnerable, hence a guarantee of strategic stability. Successful trailing brought that into question: strategic submarines were safe only to the extent that they could be properly designed and operated. Painfully aware that their submarines were vulnerable, from time to time the Soviets called for the establishment of strategic submarine sanctuary areas. In fact, once tension mounted sufficiently, such agreements would become worthless. The U.S. Navy successfully argued that the West would be surrendering an important advantage without gaining anything in return. Even so, President Jimmy Carter, who strongly advocated strategic arms control, seems to have been interested in the idea. The successor Reagan administration rejected several Soviet approaches to such arms control.

Attempts to trail, and if necessary destroy, strategic submarines were hardly one-sided. It now appears that the main mission of Victor-class submarines was strategic ASW. The Soviets seem never to have succeeded in the prerequisite—the ability to intercept and trail the submarines for any lengthy period. The strategic mission was cited to explain why Victors were so much faster than Novembers, but that difference was illusory; it just took a long time for Novembers to overcome teething problems so that they could reach their designed speeds.

10

END-GAME

IN THE 1980S the U.S. Navy proclaimed its new Maritime Strategy. At about the same time, the Soviets began to introduce a new generation of submarines that were finally silenced to something like U.S. levels, ultimately threatening to end-run long-range detection systems such as SOSUS. Although the two developments were not linked, the Maritime Strategy proved to be a viable counter to the quieter Soviet attack submarines. It was certainly not envisaged as such. Through the 1980s the U.S. Navy continued to invest heavily in measures it hoped would improve its ASW performance.

Despite its name, the Maritime Strategy was a way in which a fleet designed for a wide variety of operations, particularly on the Eurasian periphery, could fight and win in a particular scenario, a NATO war in Europe—the possibility that most concerned the Washington defense community. The European contingency was the least likely of all possibilities, because it was widely believed that it could escalate into a general nuclear war that would destroy both the Soviets and the West. The standard Maritime Strategy briefing package included, early on, a graph of destructive potential versus likelihood, the big war offering the worst destruction in the least likely case. President Eisenhower's reasoning in the 1950s still made excellent sense in the 1980s: no one would opt for such a war unless very badly cornered. The Eurasian periphery was still likely to be where the East-West conflict would play out, and that was where the U.S. Navy's mobile forces should be committed.

The Cold War end-game forced the Soviet fleet out of the open seas and into western and eastern bastions, where the Soviets thought they could defend strategic submarines such as this Typhoon (Project 941). Typhoon was unique among Soviet submarines in being built of several separate pressure hulls, necessary because her solid-fuel missiles were so massive—because Soviet solid fuel was so inefficient. USNI

On the other hand, the European scenario seems to have driven Soviet thinking. As it evolved, the Maritime Strategy exploited a Soviet obsession with nuclear weapons that went far beyond anything in the West. To the Soviets, the course of a big war would be determined by the evolving nuclear balance between East and West. That made the primary objects of Soviet forces to tip the balance in the Soviets' favor.

In the initial phase of the war, then, they would seek to destroy U.S. nuclear-armed naval forces and NATO nuclear forces such as bombers. The Soviets assumed that NATO would do the same, hence their decision to concentrate their strategic submarines in defended bastions.

Western ASW superiority was crucial in shaping the Soviets' perceptions and use of naval forces. Western nuclear submarines trailed the Soviets' ballistic missile submarines whenever they went to sea. This success continued even as the Soviets improved and quieted their strategic submarines, and as they developed the ability to fire missiles from under the Arctic ice. By the late 1970s, developing Western, particularly American, ASW made it less and less likely that the Soviets could destroy Western nuclear-armed warships such as strike carriers. This was an indirect form of shipping protection. To the extent that the Soviets doubted that they could do the essential naval nuclear tasks, they could not spare naval forces for roles such as shipping interdiction. Conversely, if it could beat off Soviet attempts to destroy it, the U.S.-led Strike Fleet in the Atlantic could support NATO ground forces by striking at the flanks of an advancing Soviet-led army.

The core idea of the Maritime Strategy was a return to classic naval thinking: the sea is vast, and it does not lend itself to creating defensive positions. Navies win by seeking out and destroying their enemies. This explicitly offensive mindset can be traced back to the Navy's reaction to the Carter administration (1977–81). President Carter was determined to cut the nation's budget to cure a very serious inflation problem. Given the sheer size of the defense budget, he had to cut defense funds. To do that, Carter dismissed naval power projection as an expensive irrelevancy. He would limit contingency planning to a NATO war in central Europe. U.S. naval planners, who had not previously spent much effort on what they considered a very unlikely European conflict, argued that the power-projection Navy would have a very important role in any European war. They found the Army's approach to such a war (which relegated the Navy to ensuring transatlantic transportation) unimaginative, devoid of any offensive mindset, and therefore ultimately defeatist.

The Carter administration assumed that forces sufficient for a NATO war could handle anything else. That was true of the Army and the Air Force in Europe, except for the problem of transporting and supporting an army elsewhere. It was not at all true of a Navy defined entirely by the need to resupply Europe. Remarkably, actual demonstrations of the Navy's argument—the revolution in Iran and the Soviet invasion of Afghanistan—had little impact on the administration's decision-making. Both events called forth naval responses, but they did not convince the Carter administration of the urgent need for more naval strike forces (built around more carriers).

U.S. Navy planners' examination of European scenarios had heretofore entailed a defensive posture and SLOC protection.[1] Scenarios generally involved formation of the Strike Fleet Atlantic, built around U.S. carriers of the Second Fleet. As the Soviets built up their force of antiship bombers based in the Kola Peninsula, as well as the submarines there, the early NATO concept of moving the Strike Fleet into the GIUK Gap was deemed less and less practical.[2] Yet the SLOC protection mission would be compromised if NATO lost control of the Gap through Soviet seizure of Iceland and northern Norway.

A classic convoy-based defense of shipping was less and less practicable, given declining ship and aircraft numbers. The OSD system analysts saw SOSUS-directed maritime patrol aircraft (mainly P-3Cs) as a solution, but whether that would or could work depended on whether the aircraft would be permitted to attack the submarines they found. Despite considerable investment in the system during the 1970s, SOSUS coverage of likely submarine patrol areas was limited. The patrol planes could not attack until war broke out, and they would be useless if the submarines vanished into the blank spaces in SOSUS coverage. Existing convoy escorts would be unable to deal with threats such as Soviet missile bombers and missile-armed submarines.

In the late 1970s, emerging naval technology was changing the situation. The new Aegis antiair system promised to defeat submarine- and bomber-launched missiles. The new towed arrays were enabling surface ships to detect nuclear submarines two or even three convergence zones, perhaps a hundred miles away. Shipboard helicopters could exploit this sort of detection. With the Mk 48 torpedo, U.S. SSNs could successfully engage fast Soviet submarines without having to resort to the nuclear SUBROC, use of which might not be approved early in a war. Given the new technology, the U.S. Navy could compel the Soviets to deal with its strong battle force before it could reach the SLOCs, breaking Soviet sea power in the process. This was classical naval strategy adapted to current conditions.

Once it had established maritime supremacy, the fleet could use its maritime mobility to counter the much less mobile Soviet army advancing through Europe. Attacks on the flanks of the Soviet army would force the Soviets to devote substantial forces to defend them. Soviet territory might also be threatened, again forcing the Soviets to dilute the force they would direct against NATO land forces on the central front. As the first SACEUR, General (later President) Eisenhower had envisaged much the same strategy 30 years earlier, describing Western Europe as a peninsula extending out from Eurasia. Proponents of the new strategy sometimes pointed to a book by Alfred Thayer Mahan, *The Problem of China*, which described a naval approach to a land problem, and thus demonstrated that the approach followed classic lines.

The Navy could also offer a true non-nuclear deterrent. It could force the Soviets to face the prospect of a continuing struggle even if their army managed to reach the English Channel. In that case the British Isles would almost certainly remain unconquered, and an intact North Atlantic SLOC would keep them alive. This had been the logic of the earliest NATO war plans.

The Navy's rejoinder to the Carter administration's cuts was *Seapower 2000*. Referring to the U.S. Navy of 2000 avoided a frontal attack on the administration, but in fact the study looked to the nearer future. It may have been the first to argue that the new naval technology would make it possible for the fleet to achieve "maritime air supremacy" in the north by destroying the limited force of Soviet naval long-range bombers.

At about the same time that *Seapower 2000* was being written, Pacific Fleet commander ADM Thomas B. Hayward became interested in the influence his force could exert on the Soviet strategic flank in the Far East. The Navy had long argued that any European war would inevitably extend to the Pacific because the Soviets had large naval forces there. That argument had stressed defense: forces should be maintained in the Pacific to limit losses there and to encourage Japan to retain its ties to the United States. Hayward looked at the situation very differently. The Soviets maintained large forces facing China; they had fought the Chinese on the Yalu River in 1969. Tying them down would impose "virtual attrition" on the Soviet forces in Europe. The Soviets would probably interpret U.S. air strikes in the Far East as possible precursors to a Chinese attack on the Soviet Far East. To demonstrate what he had in mind, Hayward had his pilots practice 500 nm strikes rather than the 100 nm strikes employed during the Vietnam War. Hayward's new Pacific concept was one reason why Admiral Crowe could not promise that Pacific carriers would move to European waters in an emergency.

Hayward's Sea Strike concept was widely briefed in the Navy and elsewhere in 1977–78. It engaged the Washington analytical community, which largely determined how defense resources were allocated. The briefing emphasized that the Soviets could move Far Eastern naval (including air) units to Europe more quickly than U.S. naval units, particularly carriers, could get there.

Hayward came to Washington in 1978 as a CNO well primed to encourage an offensive mindset among fleet commanders. As president in 1981, Ronald Reagan favored a more assertive policy toward the Soviet Union. His Secretary of the Navy, John Lehman, argued for the strategy embodied in *Seapower 2000*, which he had helped write. He strongly supported its proposal for a 600-ship fleet that included a more powerful carrier force. Lehman, who had written his PhD dissertation about the role of Congress in national defense, argued that the Navy had to back its request for the necessary funding with a coherent explanation of how it would use its forces in wartime. To do that, he had his staff in effect stitch together the basic war plans of the naval theater commanders. These plans fit together because they reflected the offensive approach championed by Admiral Hayward. The combined plans showed how the Navy expected to fight in a global war. Through the 1980s the combined plan was elaborated and integrated to create a unified U.S. Maritime Strategy—the Navy's view of how it could contribute to success in the sort of global war the Soviets might choose to fight.

The new approach incorporated a changed perception of the naval side of a European war. The option of concentrating the Atlantic carriers in the Mediterranean was abandoned. The focus returned to the Norwegian Sea. The target there was the vast Soviet base complex on the Kola Peninsula. This was a return to the earlier idea of attack at source. This time, however, the antiship force to be destroyed included, perhaps was even dominated by, Soviet naval missile bombers. Advocates of the new approach pointed out that the bombers and their expert crews could not quickly be replaced. They, not the surface ships and the submarines of the Northern Fleet, were the main Soviet battle fleet.

Initially, advocates of the Maritime Strategy did not see the Soviets' SSBN bastions as central to NATO strategy. The idea of forward operations was to draw out and destroy Soviet naval forces, such as missile-carrying bombers, to gain freedom of action. However, attacks on Soviet naval forces covering the SSBNs in their sanctuaries certainly would help reduce any Soviet threat to the SLOCs. Intelligence confirming the Soviets' emphasis on protecting their strategic submarines arrived slowly, and the conclusion was controversial. For example, when he was briefed on the newly understood Soviet bastion strategy in August 1981, Admiral Hayward doubted that the Soviets would operate their fleet in so bizarre a way. They were certainly deploying their general-purpose fleet aggressively in peacetime. By the spring of 1982, however, Hayward and other senior officers were convinced that in wartime the Soviets would concentrate on protecting SSBNs in the bastions. The new perceptions were embodied in the Navy Net Assessment approved in December 1981. The new consensus was that in wartime, Soviet naval general-purpose forces would be concentrated close to home waters not only to protect SSBNs but also to protect the homeland.

In addition to deploying a fleet to block the door to the bastion in the White Sea, the Soviets demonstrated that they could operate at least some of their SSBNs in the Arctic, surfacing in polynyas to fire. The U.S. Navy and the Royal Navy publicized their own ability to operate attack submarines in the ice, specifically to demonstrate that they could continue to pursue Soviet strategic submarines. The U.S. Navy conducted strike exercises in the far north to show that it could attack the Kola and the nearby bastion; it also ordered *Seawolf*-class submarines specially adapted to the bastion attack mission.

Forces in the NATO north were so crowded together that subtle distinctions between Soviet attacks to relieve pressure on the bastion there and

Carrier-killer: an Oscar- (Project 949) class cruise missile submarine shows the towed array tube atop her tail fin and, faintly, the outlines of the missile doors on her port side. With her SS-N-19 (P-700) missile system, she replaced the Echo I as a long-range antiship missile submarine. Unlike Echo II, she did not have to surface to receive targeting data or to launch missiles or to provide midcourse guidance for her missiles. She also carried more missiles, 24 rather than 8. The requirement that produced Oscar and its Granit missile system apparently was stated in 1969. USNI

Soviet attacks to destroy carrier battle groups because they presented nuclear threats were meaningless. The main import of the new information was that the Soviets were unlikely to release attack submarines to attack the transatlantic SLOC even in a protracted war.

U.S. Navy strategists were aware that the Soviets were obsessed with Western naval nuclear striking power. The Soviets tried to keep track of all U.S. (and British) ships armed with nuclear weapons that could be used against the USSR. The more such ships there were, the more difficulty the Soviets would have in tracking them. The U.S. Navy deployed Tomahawk cruise missiles, which could carry nuclear warheads, on board many of its surface combatants. It was impossible to tell whether a ship was armed with nuclear or non-nuclear Tomahawks. The Navy also demonstrated that carrier bombers could mount very-long-range nuclear strike missions, forcing the Soviets to engage the carriers at greater ranges, where attacks would be more difficult—and bombers could not be covered by Soviet fighters.

The Soviets' difficulties were further complicated by the advent, in the late 1970s, of the Trident C-4 missile, with a range of 3,800 nm rather than the earlier 2,500 nm. Soviet attack submarines hunting Tridents would have to be deployed far from NATO shipping routes. The Soviets were still not displaying any capability to hunt down Western SSBNs.

The U.S. *Seawolf*-class submarines designed in the 1980s were particularly adapted to the bastion-attack mission. Sinking a Soviet strategic submarine would create a "flaming datum" that might attract the Soviets. To limit that vulnerability, *Seawolf*'s designers gave her a much higher quiet speed than her predecessors had so that she could clear a flaming datum before Soviet submarines or other ASW forces turned up. She would be endangered mainly when entering and leaving the bastion, because only then could Soviet ASW forces be certain that she was hostile. She was given twice the torpedo capacity of earlier U.S. nuclear attack submarines so that she would not have to leave the bastions as often as earlier U.S. submarines. She also had much-improved sonars and larger-diameter torpedo tubes. The larger-diameter tubes could accommodate much larger torpedoes—but they could also enable the current torpedoes to swim out silently, again reducing the submarine's vulnerability. The general failure of underwater identification friend or foe (IFF) made it likely that Soviet ASW forces hunting U.S. submarines in the bastions would often find themselves killing Soviet submarines in error.

The Soviets used their submarines to respond to the deployment of NATO intermediate-range nuclear weapons in Europe, themselves intended to counter Soviet "Euro-strategic" weapons. The Soviets were concerned that these weapons, particularly Pershing precision missiles, could rapidly destroy their command-and-control center, Moscow. Beginning in May 1984 they deployed missile submarines (Deltas and Echo IIs) off the U.S. East Coast in positions from which they could quickly attack U.S. command and control using depressed-trajectory missiles. The Soviets called the deployments "analogous responses" to NATO's deployment of intermediate-range nuclear weapons in Europe. The new deployments did not affect the continuing pattern of keeping three Yankees off the U.S. East Coast and two off the West Coast.[3] This was in addition to the Soviet attack submarines operating off the U.S. SSBN bases. The "analogous" deployments ended in 1986, part of a more general concentration of Soviet submarines in home waters under the pressure exerted by the U.S. Maritime Strategy.

Naval missions can change dramatically in wartime. U.S. perceptions of the USSR's wartime intentions were shaped by critical sensitive intelligence information indicating what the Soviets planned to do. Soviet submarines were mostly well adapted to anticarrier and antisubmarine roles; by the 1980s the big fleet of diesel antishipping submarines had largely worn out. However, plans can change. The U.S. Navy itself was a case in point. Prior to December 1941 its submariners were trained to concentrate on the enemy's capital ships; they received no training for a campaign against enemy shipping. On 7 December 1941, however, they were ordered to "execute unrestricted submarine warfare against Japan." Despite faulty torpedoes and a lack of relevant training, they sank a considerable tonnage of Japanese merchant ships in 1942. What would happen during a protracted war in which a NATO army relying on the transatlantic SLOC managed to stall or even repel the Soviet army? Would the politically dominant Soviet army demand that submarines be released to attack the SLOC? It could be argued that only sustained pressure against the bastions, particularly the one in the Barents Sea, could ensure against such a shift.

NEW SOVIET SUBMARINES

As early as 1975, there was evidence that the Soviets were aware of the 50 Hz turbogenerator line that had been so useful to sonar detectors. Thanks to computer-controlled machine tools they obtained in Norway and Japan they were able to reduce blade rate by changing propellers on a large scale. They adopted anechoic coatings. Although not effective against low-frequency sonars, the coatings deadened echoes from the high-frequency sonars of homing torpedoes, greatly reducing their homing ranges. In the late 1970s the U.S. Navy began an urgent NEARTIP (near-term improvement program) to restore the efficacy of the Mk 46 lightweight torpedo, its standard aircraft, ship, and ASROC weapon.

After the collapse of the Soviet Union, a Soviet vice admiral wrote a book about Soviet warships in which he explained that the Soviets had begun ten-year programs in 1970.[4] The first five years were devoted to research on a chosen theme; the fruits of that research were put into production in the next five years. He claimed that the theme selected for 1970–75 was silencing of every ship signature. That would accord with the design of the first quiet Soviet submarine, Victor III.

A Sierra (Project 945) titanium-hulled submarine. Unlike attack submarines of the previous generation, she had a single-reactor power plant and a rescue chamber in her sail. The detailed design was done by the Lazurit (rather than Malakhit) design bureau to meet a requirement issued in March 1972. The stated role was to hunt and destroy enemy SSBNs and carrier strike groups. Project 945 had a double hull and a water-cooled reactor. It had two additional 65 cm torpedo tubes. She was quieter and could dive deeper. A modified (lengthened) Project 945A could carry the RK-55 Granat (SS-N-21) strategic cruise missile and was quieter than Project 945. All torpedo tubes were 53 cm in diameter. Given the high cost of the submarine, only four were built. These submarines drew too much water to be completed at their building yard at Gorkiy, so they were finished at Severodvinsk. Planned construction at the Far Eastern yard at Komsomolsk-na-Amur was impossible because there was no local source of titanium. USNI

The Soviets kept on producing new submarines: a new strategic type (Typhoon) in 1981, a new anti-ship cruise missile type (Oscar) in 1980, and a new attack type in 1984 (Sierra/Akula). The new designs came five to ten years before they were expected in the West. The new submarines did not achieve the expected high speed (45 kts, up from an earlier estimate of 38 kts, sustainable for eight hours); nor did they reach the expected 4,000 ft diving depth (Sierra and Akula had a normal operating depth of 480 m [1,574 ft] and a not-to-exceed depth of 600 m [1,968 ft]), only slightly deeper than those for Victors. The more impressive figures expected in the West shaped the U.S. ASW R & D program of the 1980s. The most important advance represented by the third generation was the missile system in the Oscar (Project 949 Granit), intended to replace the

An Akula (Project 971 Bars), the quieter successor to Victor III, intended primarily to launch cruise missiles. Like Victor, Akula was designed by the Malakhit design bureau as a steel equivalent to the titanium-hulled Sierra, to be built in more yards. It was the first Soviet multipurpose attack submarine, able to attack shore targets using the RK-55 Granat (SS-N-21) cruise missile. Like Sierra, it was quieter and could dive deeper. It had four 53 cm and four 65 cm torpedo tubes. The Russians considered it the quietest nuclear submarine they had built to date. Submarines were built both at Komsomolsk-na-Amur in the Far East and at Severodvinsk. USNI

old Echo II. Unlike its predecessor, it did not rely on a visible airplane to downlink its targeting data; it also did not have to remain surfaced to direct its missiles. Instead, it linked to a radar satellite and fired while submerged on a fire-and-forget basis. The impressive size of the Typhoon (Project 941 Akula) turned out to be a consequence of the decision by Defense Minister Dmitry Ustinov to adopt solid fuel for its missiles, Soviet solid fuel performance being far behind the West's. All these submarines were quieter than their predecessors.

It appears that Western navies were unaware of how long these projects had been underway before they reached the stage at which components became visible. The first knowledge of the Typhoon project in the West seems to have come in 1974, when Leonid Brezhnev warned President Gerald Ford that if the United States continued to pursue its Trident project, the Soviet Union would pursue Typhoon, which was undefined at the time. The first visible hint of Typhoon seems to have been the appearance of massive reduction gears outside the assembly building at Severodvinsk. Their function was not obvious; for a time there was speculation that a carrier was being built. The new attack submarines were conceived in 1972, shortly after the first Alfa was completed. Project 945 Barrakuda (Sierra) used the same titanium hull material but had a single conventional pressurized water reactor of a new type, tested on board a new Soviet nuclear icebreaker. Given the high cost of a titanium hull, the Soviets built an equivalent steel submarine, Project 971 (Schuka B, NATO Akula). The two differed in torpedo tube numbers and in torpedo numbers. The new reactor, which could produce up to 50,000 SHP, also powered the other third-generation submarines (two powered an Oscar and a Typhoon).

Only in another new attack submarine, *Komsomolets* (Project 685 Plavnik, NATO Mike), did the Soviets attempt to exploit titanium for deep diving; test depth was 1,000 m (3,280 ft). Her technical requirement was issued in 1966, but the technical project (i.e., the sketch design) was not approved until December 1974. Completed in October 1983 and running trials between August 1984 and July 1988, *Komsomolets* was clearly an experimental vessel. She was lost on 7 April 1989 due to an onboard fire. Her designer blamed an untrained crew for failing to operate her firefighting equipment properly (he claimed that she had been sent to sea with a scratch crew to satisfy a headquarters scheduling order).

Only a few of the new submarines had been completed by the time the Berlin Wall came down in 1989: six Typhoons (another was broken up on the slip), five Oscars, two Sierras, and seven Akulas.

There was also a new class of diesel submarines, Project 877 (Varshavyanka, NATO Kilo), the first of

The Lada- (Project 677) class submarine *St. Petersburg* at that city for the 2011 naval exhibition. She differs externally from the Kilo class in having sail planes and an upper vertical fin aft carrying the tube of a towed array. Internal differences include stronger steel and a digital combat system enabling the submarine to track five targets while engaging two. She is also said to be much quieter than a Kilo. Laid down in December 1997, she was not in service until May 2010. ALEXANDRE SHELDON-DUPLEIX

which was completed in 1982. The Soviet nickname suggests that Kilo was intended for Warsaw Pact navies, although only Poland and Romania received it (as did the Soviet client states Algeria, China, India, Iran, and Libya). Presumably because it was not considered an ocean submarine, Kilo had 45-day endurance (Tango's was 80 days, Foxtrot's 90, and Zulu's 75; Whiskey's was 30). Like the new nuclear submarines, Kilo would seem to reflect increased Soviet awareness of the threat of Western passive sonars. A submarine recharging her batteries is loudest when snorkeling, since all of the sound goes into the water. By the time Kilo was designed, Soviet submarines often recharged on the surface, despite having perfectly adequate snorkels. Of course, they had to be sure that no ships were nearby. The Soviets used a device they called Kodak because it took a radar snapshot of the surrounding area. Kodak had an omnidirectional radar transmitter in a dome, surrounded by cavity-backed spiral receiving antennas like those in many ESM arrays; in fact, until the end of the Cold War it was thought to be an ESM antenna. Kilo was also equipped to defend herself on the surface; she carried six man-portable antiaircraft missiles in her sail.

That Soviet submarines were armed with torpedo-tube missiles was apparently initially unknown. SS-N-15 had entered service in 1969; it armed the Victor II–class submarine (Project 671RT, the RT apparently meaning "rocket torpedo"). After 1980 Victor III (1977: Project 671RTM) could be armed with the improved PLUR 83R or PLUR 84R; later Soviet submarines (Projects 945 and 971) could be armed with the 65 cm PLUR 86R and PLUR 88R Veter missiles. Although in theory these were ASW weapons, their nuclear warheads offered them an impressive antiship capability. It is not clear when Western navies became aware of them.

There were also shocking torpedo developments. Soviet torpedoes were tested in the Caspian Sea, far from intelligence-gathering Western submarines. Until well into the 1980s, Western navies seem to have been unaware that Soviet antiship torpedoes used wake-following guidance, which was virtually impossible to decoy. The first indication that the Soviets used standoff (magnetic) exploders seems to have come from an unclassified Soviet book about torpedoes that mentioned such exploders as commonplace information.

The end of the Cold War prevented construction of a further Soviet submarine generation: Projects 881, 885, 955, and 957. Project 881 (Merkuriy) was intended to replace Oscar (Project 949/949A), using the new Bolide antiship missile, a longer-range development of the SS-N-19 missile of the Oscar class. Given the increased size of the new missile she carried, the submarine was huge, about the size of an SSBN; she was similar in outline to the Project 935 SSBN. Merkuriy was canceled in 1989, but the missile survived in a project to rearm Oscars.

HMS *Conqueror* in 1975.
RAN NAVAL HISTORICAL STAFF VIA DR. JOSEF STRACZEK

Project 885 Yasen or Granay, another cruise missile submarine, was based on the Alfa and Akula designs. Work on the design was scheduled to begin in 1977 and to be completed in 1985. The first of the class was begun in 1993, but due to the collapse of the Russian economy she was not completed until 2013. Unlike the much larger Project 881, this submarine was armed with the shorter-range Oniks missile, quad-packed in eight multipurpose tubes. The OK-650KPM reactor was derived from the reactors of the Akula, Sierra, and Oscar classes. This was the first Russian submarine to have the Irtysh-Amfora (MGK-600) sonar, with its spherical transducer, flank arrays, and towed array (without any nacelle).

Projects 935 and 955 were strategic submarines to replace Delta and Typhoon. Project 955 (Borey 1) was designed to carry the D-31 strategic missile system, and 935 (Borey 2) the D-35 system. D-31 was canceled, followed by its replacement SS-N-28 (D-19UTTKh) Bark system, which was initially intended for modernized Typhoons). Only Borey 1 was actually built, armed with the SS-N-32 Bulava missile. Although it was ready for production in the early 1990s, the first submarine was not laid down until 1996 and was not completed until 2013. The missile actually carried replaced the canceled ones, and the first unit incorporated parts of canceled Akula II and Oscar II submarines.

Project 935 was a parallel project to carry 12 single-warhead SLBMs. This missile was canceled, and in 1989–90 the submarine project was suspended pending further studies.

Project 957 Cedar was a companion attack submarine designed in 1980–89, with 20 being planned. The project was canceled in 1989 due to shipyard overload. It seems to have been a single-hull submarine with an unusually great length-to-beam ratio, specially silenced. It was intended to replace the Victor, Sierra (Project 945), and Akula (Project 971) classes, to be built in both steel and titanium (Project 957T) versions.

Project 833 was a small antisubmarine submarine based to some extent on Alfa, with a displacement of no more than 2,500 tons. Work stopped in the early 1990s.

THE FALKLANDS CONFLICT AND WESTERN ASW

The 1982 Falklands Conflict demonstrated what the U.S. Navy and the Royal Navy had been telling their governments for years: navies were needed to deal with contingencies far different from a hot NATO war. The war did have many features of the expected NATO naval war, but on a much smaller scale. In a big NATO war, the NATO Strike Fleet would have faced a combination of Soviet naval bombers wielding antiship missiles, submarines, and possibly Soviet surface ships. The British expected to mount an amphibious landing in northern Norway. In the fight for the Falklands, the British faced the Argentines' bomber- and ship-launched antiship missiles (Exocets), submarines, and a surface fleet. The British had to protect their task force and to support an opposed landing in the Falklands. Nothing like this had happened since 1945. The analogy made the war intensely interesting not only to the combatants but also to the United States and, presumably, to the Soviet Union.

Perhaps the most important lesson for the Western powers was that their focus on the Soviet threat had blinded them to the possibility of a war in the Third World fought against Western weapons and systems. The British submarine commander (flag officer submarines, or FOSM) later wrote that his commanders' knowledge of the Argentine force facing them was so poor that they were studying *Jane's* as they steamed south. Operators of submarines' radar intercept devices, which proved to be important, lacked data on the Western radars the Argentines operated.

The Royal Navy was considered the best NATO antisubmarine force. However, it was increasingly optimized against Soviet nuclear submarines, which its SSNs often trailed, and it was used to operating in nearby waters such as the Norwegian Sea. Abandoning operations "East of Suez" had drastically limited the Royal Navy's familiarity with shallow warm waters. There was no information about the signatures of the Argentine submarines. The British had adapted to the threat of Soviet nuclear submarines, with their consistent LOFAR signatures. On the other hand, the Cold War had provided the British with a sophisticated cryptologic capacity, which seems to have enabled them to read Argentine naval messages.[5] Once the United States declared support for the British it provided extensive assistance, including the fruits of satellite reconnaissance, which could tell them whether particular Argentine warships were at sea.

The British correctly assumed that the main threat to their task force was Argentine aircraft, some of them armed with Exocet missiles. They did have to reckon with Argentina's submarines: two German-built Type 209s and two much older U.S. Guppies. Only one of each type was operational: the Type 209 *San Luis* and the Guppy *Santa Fe*. Thus they had to consider that any of their most important surface units—the two carriers and the liner *Canberra*—might be torpedoed. They had no way to repair underwater damage so far from home. Finally, the Argentines had a carrier (*25 de Mayo*) and a surface fleet that included the large light cruiser *General Belgrano* and destroyers, some of them armed with Exocet missiles.

Unlike the Norwegian Sea, the area around the Falklands was not wired for sound, so the offensive ASW techniques that had guided U.S. and British ASW were not relevant. Too, the British did not have the sustained maritime patrol aircraft coverage they would have enjoyed in the Norwegian Sea. Their few Nimrods, which could spend very little time on station, were used mainly for surface search and to follow up intelligence information.

The underwater environment was very different from that of the Norwegian Sea. Temperature profiles were good, so that sonar performance was limited by background noise, either biological or from the shore or from the sea state. On one occasion sonobuoys laid southeast of the islands detected the British task force more than 60 nm away. It appeared that this performance could be attributed both to convergence zone conditions and to a downslope environment. The British moved the task force north specifically to operate in conditions less favorable to long-range detection.

Sea life was so dense that one officer wrote that "the sea is like soup." It was estimated that sea life could have made as much as a 10 dB contribution to volume reverberation, which would mask sonar echoes and sound from a submarine. Whales behaved much like submarines, even swimming at typical submarine speeds. Shoals of fish and krill could appear as submarines on active sonars. The same problems apparently affected the Argentines as well, even though in theory they should have been more familiar with their own waters. The British suffered a high false alarm rate; it fell as operators gained experience but rose again when fresh ships joined the force.

The continental shelf extended uniquely far to seaward off Argentina, making it possible for the Argentine submarine *San Luis* to bottom for extended periods. Weather on the shelf was predominantly mild and calm. However, at and to the east of the Falklands the sea turned rough and the weather cold. There were many free icebergs, unlike the Arctic pack ice in which British submarines had exercised. Much of the shallow water was filled with wrecks of old whaling ships, among which a bottomed submarine could hide from sonar.

The Argentines apparently considered the area inside the 50-fathom line (300 ft depth) a safe refuge from Royal Navy SSNs, but in fact British submarines spent extended periods there and at times dived in less than 27 fathoms (162 ft). In calm and unusually clear water they risked detection from the air when functioning as pickets. In some places bioluminescence in the water made night operations easy to detect. In very shallow water an SSN running at speed created easily seen surface disturbances.

Nuclear submarines offered the British the valuable ability to deploy without any publicity, hence without affecting diplomatic attempts to head off the crisis. On 29 March FOSM was ordered to prepare an SSN to go to the South Atlantic, to be followed by a second and then a third. The first two were sent south while the crisis was escalating and before the Argentines invaded the Falklands (on 2 April). HMS *Spartan* was recalled from an exercise and loaded in Gibraltar with weapons and stores from the diesel submarine *Oracle*. She sailed on 1 April and arrived in the Maritime Exclusion Zone (MEZ) on 12 April. HMS *Splendid* was withdrawn from operations in the Northern Fleet's exercise areas (she was trailing a Victor-class SSN) and sailed immediately (on 1 April); HMS *Conqueror* (which had a towed array) sailed on 4 April. Later two towed-array SSNs, HMS *Valiant* and HMS *Courageous*, arrived. *Valiant* and *Courageous* had improved spectrum analyzers. All had clip-on arrays, which could be damaged in water less than 100 fathoms (600 ft) deep. The more recent (and quieter) *Spartan* and *Splendid* and the surface ships all lacked towed arrays. The SSNs managed to operate far from home without much logistic support. The silencing rafts of the two *Swiftsures* suffered failures that could not be made good; *Spartan* was too noisy.

Satellite communication made it possible to control the submarines from the submarine force headquarters at Northwood in the United Kingdom. The staff at Northwood worked with the task force staff at sea in the flagship HMS *Hermes*, which had an embarked submarine staff officer. An important lesson was the inadequacy of the Joint Maritime Intelligence Center at Northwood, which could barely cope with the flood of messages it received. However, it was credited with a major contribution to the success of the attack by *Conqueror* against the Argentine cruiser *General Belgrano*. Overall, the British submarines enjoyed support broadly equivalent to what the SOSS provided to Soviet submarines.

To the surprise of the British, the Argentines had their own long-range sensor, demonstrated when they found and attacked the British task force in the open ocean en route to the Falklands. The Argentines had exploited British satellite communication links, using a Doppler technique developed at an Argentine university (similar techniques later became common). The British then had to assume that the Argentines knew approximately where their fleet was. The best they could do was to avoid Argentine submarines by exploiting water conditions that would be difficult for the Type 209. They also tried to exploit bad weather to avoid air attacks.

To avoid blue-on-blue problems, a task force ASW weapons-tight boundary was set up between the task force operating area and submarine patrol areas (SPAs), with a 30–50 nm buffer between the weapons-tight boundary and the nearest SPA. Northwood enforced the kind of water management developed for the GIUK Gap, assigning patrol areas to single submarines. Submarines were barred from crossing boundaries even in hot pursuit. That was essential because sometimes submarines detected each other: for a time *Valiant* held a towed-array (narrowband) contact that turned out to be *Splendid*. The British did not use their existing underwater IFF system.

The British declared an MEZ around the Falklands, which would be enforced by SSNs. Its 200 mile radius was a compromise between maintaining an area small enough to patrol but large enough not to

The two British VSTOL carriers were the only source of heavy ASW helicopters with their dipping sonars. They were also the only source of effective fighters, hence drew the most effective frigates (which had the best air defense missiles) around them. Photographed by VA-85 on 8 September 1991, HMS *Invincible* is in her Falklands Conflict configuration. The helicopter aft. is a Sea King. BAKER COLLECTION

restrict the SSNs (the MEZ was later expanded to a 300 mile Total Exclusion Zone). The rub was the need for positive identification. Small diesel-electric submarines such as Type 209s had weak signatures (and small sonar cross-sections). The frequencies they did emit might be below those best exploited by British submarine sensors.

The Argentines had two major surface warships—the light carrier *25 de Mayo* and the cruiser *General Belgrano*—each of which became the core of a task group. On 1 May the Argentine naval commander ordered an all-out attack on the British surface force. The carrier group was to launch at first light and search for and attack the British task force. A second task group operating south of the exclusion zone would attack British warships fleeing the air strike. A third task group (*Belgrano* and two destroyers) would attack British ships in the waters south of the Falklands. The destroyers in this group had Exocet missiles, hence were a more potent antiship threat than the cruiser. The Argentine operations order was decoded by GCHQ and passed to *Spartan* and *Splendid*.

The British considered *25 de Mayo*, with her strike aircraft, the most serious threat. The British fleet commander, ADM John Woodward, envisaged a worst-case scenario: the carrier would attack at first light with 30 aircraft, some of them armed with Exocets. The cruiser group, rushing north through the night, would fire Exocets from a different direction. He later said that this possibility made it imperative to derail the attack by stopping the *Belgrano* group. The War Cabinet was so impressed by the carrier threat that it ordered her sunk even if she was outside the MEZ. Using two SSNs, in different areas due to water management, was expected to allow for alternative carrier courses. By the time they received the message, the SSNs were more than 100 nm from the supposed position of the carrier. Neither submarine managed to reach the carrier, which never launched any aircraft (apparently she lacked sufficient wind over her deck).

Conqueror was ordered to intercept the *Belgrano* group. Although outside the exclusion zone, the cruiser was considered fair game because she was threatening the British task force. Prime Minister Margaret Thatcher later wrote that the War Cabinet considered the *Belgrano* group equivalent to the carrier group. All Argentine combatants in international waters became fair game.

Early on 1 May *Conqueror* began to trail the cruiser from beyond the horizon, using her clip-on towed array. Initially the cruiser group headed east toward the Falklands with two escorts. When the cruiser turned back into shallower water, the CO of *Conqueror* had to reckon with the possibility that the array might snag on the bottom and be pulled off, ending the submarine's long-range trailing capability. Reportedly the decision to allow *Conqueror* to attack *Belgrano* became urgent because the latter was heading into the shallower water of the Burdwood Bank.

Conqueror fired three straight-running (unguided) Mk 8 torpedoes at a range of 1,200 yds, making two hits and sinking *Belgrano*. A third torpedo that hit one of the two escorting destroyers apparently failed to explode. The destroyers failed to find, let alone attack, *Conqueror*, though some British media accounts said that she had been extensively depth-charged—as an attacking submarine might have been during World War II. *Conqueror* used Mk 8s because the guided Mk 24 had too small a warhead and was not considered reliable. The sinking apparently deterred the Argentine surface fleet from any further moves beyond the 12 mile limit. From that point on Admiral Woodward could concentrate on the air threat.

Splendid found the Argentine carrier in waters that Argentina claimed as territorial. In a War Cabinet debate, the Foreign Secretary argued that a sinking would derail sensitive negotiations with the United States and might outrage other Latin American governments and bring them in alongside the Argentines. These were not particularly persuasive arguments, and on 8 May the rules were amended: the carrier would not be attacked if she was more than 12 hours' steaming range from the task force. That was supposed to preclude carrier-launched attacks, but the Ministry of Defence pointed out that the attack range depended on the load of the attack aircraft; the carrier could fight from a much greater distance than some imagined. The rules were changed again to allow attacks on any Argentine warship or military airplane outside the 12 mile limit off the Argentine coast. By that time *Splendid* had to withdraw to repair one of her turbogenerators, shutting down her reactor and either operating on batteries or snorting. She found it impossible to snorkel in the heavy seas, could not surface, and had only barely enough battery power.

The SSNs also served as pickets, detecting air raids as they crossed the Argentine coast. Stationed off Rio Grande, *Valiant* made more than 300 reports, 100 of them over a period of 3.5 days. Aircraft were sometimes detected by sonar and sometimes visually by periscope. Communications intercepts were of limited value due to limitations of the equipment, lack of knowledge of the frequencies the Argentines were using, and the absence of Spanish linguists. The new Northoe II combined periscope and communications/ESM mast on board *Splendid* was considered very successful.

The Argentines had two functioning submarines: the German-built *San Luis* and the modernized U.S.-built Guppy *Santa Fe*.[6] The British apparently did not know that the other Type 209, *Salta*, was not operational, but U.S. satellite data showed clearly that she was at her home base (she was undergoing hull repairs to cure a serious noise problem). The Argentines sent both operational submarines to sea.

The old Guppy *Santa Fe* was sent to the outlying island of South Georgia with urgently needed supplies for the Argentine force that had seized it. En route she sent an HF message (which GCHQ decrypted) announcing her imminent arrival.[7] The British SSN *Conqueror*, which had delivered commandos to retake the island, was informed of her presence but failed to find her. Having landed her supplies, *Santa Fe* proceeded on the surface to recharge her battery and to make necessary repairs (her superstructure had been damaged in a storm), intending to dive once she reached deep water. She never made it.

By this time a small British task group intended to retake South Georgia was nearby. It included the missile destroyer *Antrim*, which had on board a Wessex helicopter. The Wessex dropped Mk 11 depth charges on either side of the submarine. The explosions temporarily knocked out all power and permanently damaged the submarine's air intake valve, making it impossible for her to snorkel (according to a British evaluation, she was rendered unable to dive). The Argentine submarine commander later told the British that he assumed that he would be attacked with ASW homing torpedoes. Knowing that they could not attack surface targets, he stayed on the surface. The Lynx helicopter from the Type 22 frigate *Brilliant* dropped a Mk 46 torpedo, presumably to hit the submarine if she dived on seeing the depth charges. Wasp helicopters attacked with AS 12 wire-guided missiles, damaging the submarine's sail. *Santa Fe* withdrew to the port in South Georgia, which was contested for

With few frigates, and the best of those assigned to AAW defense, the Royal Navy in the Falklands found itself relying heavily on helicopters, particularly Sea Kings like this one, operating from the two carriers. They had much more sophisticated tactical systems than their U.S. counterparts, built around a dipping sonar and the dorsal radar. The Royal Navy adopted a license-built version of the U.S. SH-3D after evaluating a variety of other helicopters offered in response to a 1962–63 specification. The weapons load was four Mk 44 (as here) or two Mk 46 or two Mk 44 plus a WE 177A depth bomb or two torpedoes and two Mk 11 depth bombs or a Mk 46 and a WE 177A. Although in theory the desired helicopter was a single-package type, the Royal Navy considered that a helicopter could not dip, retrieve its sonar, and then reach the indicated datum quickly enough to hit a fast deep submarine. Unlike the U.S. Navy's SH-3, the British Sea King HAS.2 shown had no MAD. Like its U.S. counterpart, it could drop and process sonobuoys (initially Jezebel only). Experience of false targets in the Falklands showed the need for MAD to help classify targets, although the presence of numerous wrecks complicated that. The improved HAS.5 (identifiable by its squatter radome) had a Sea Searcher radar and Barra buoy processor; some also had ASQ-81 MAD. HAS.6 introduced a new dipping sonar (Type 2069) and a new digital buoy processor. This Sea King HAS.2 is carrying four Mk 44 torpedoes, recognizable by their ring tails. The frigate is HMS *Lowestoft*. BLUE ENVOY COLLECTION

a time between British and Argentine marines. Once abandoned, *Santa Fe* flooded and bottomed.

The British had operated against (and been impressed by) German and Norwegian submarines comparable to the Argentine Type 209s. Training exercises suggested correctly that existing British sonars would perform poorly against such small targets.

The British had few escorts, and many of those they did have were devoted to air defense. A typical screen consisted of an inner screen of two or three frigates 7 nm from task force center, with three Sea Kings, each covering a 9 × 40–degree sector, 11 nm from task force center. They covered the sector forward of the mean line of advance, which was usually at 12 kts. Sea Kings were equipped with Type 195 dipping sonars and with a torpedo (Mk 46, but Mk 44 for shallow water) and a Mk 11 depth charge. For air defense, Type 22 frigates, which had both the best sonar (2016) and the best close-in air defense (Seawolf), were assigned to screen high-value ships against missile attack. ASW numbers were further reduced by detaching frigates to escort noncarrier task groups and as plane guards. That left the British relying heavily on dipping sonar Sea Kings flying from their carriers to form an outer ring of ASW defense.

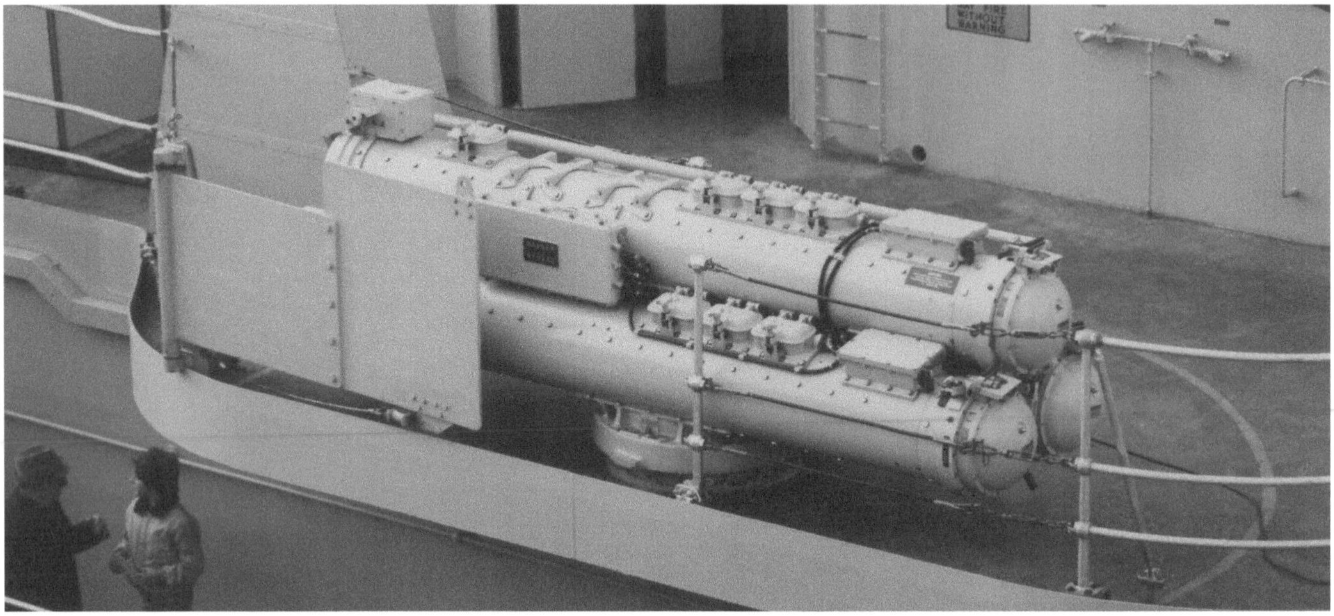

A British STWS-2 launcher for lightweight torpedoes, the standard Royal Navy urgent-attack system at the time of the Falklands Conflict. Postwar analysis of torpedo firing showed that far more weapons were fired than might have been expected, some of them before targets had even been classified. That was consistent with the experience of previous wars, in which crews often attacked on learning that a submarine might be present. The new feature in the Falklands was the high cost (and limited supply) of the torpedoes. Note the extemporized splinter plating to protect the torpedo warheads from bomb splinters and cannon shells. The Falklands Conflict illustrated the multidimensional character of a modern maritime war, in which ASW forces might well be subject to attack by aircraft and by surface ships armed with antiship missiles. The placard warning about firing without warning reflects the fact that the mounting was controlled from the operations room (CIC), not manually from a nearby position. USNI

The British read a 27 April Argentine message ordering *San Luis* to patrol a designated area north of the Falklands. *Spartan* was ordered to patrol a line to the west of the islands through which *San Luis* had to pass. She failed to detect the Argentine submarine even when *San Luis* closely approached her while snorkeling. It was later estimated that excessive noise reduced the range at which *Spartan* could detect *San Luis* to less than 1,000 yds.

Spartan and *Conqueror* both searched *for San Luis* during her transits to and from the islands, hoping that she would reveal herself either when snorting or perhaps when cavitating. *Spartan* searched passively, but *Conqueror* used her active sonar in intermittent discreet transmission mode. Neither made contact. It seemed that *Spartan* failed to detect the submarine in transit because she lacked a towed array (the absence of towed arrays was also blamed for the failure of both *Splendid* and *Spartan* to contact the Argentine carrier). She was also too noisy due to machinery problems.

When *San Luis* was ordered back from the Falklands to Argentina, British communications intelligence revealed the submarine's approximate route. *Valiant* was ordered to intercept her. *San Luis* was quieter than the SSN, and estimated detection ranges were short. *Valiant* was operating in the flat calm weather of the continental shelf, where biological noise abounded. A fishing trawler in the area often confused *Valiant*'s sound operator with bursts of cavitation, flutter, diesel signature (due to the firing rate of the diesel), and tonals. It must not have helped that specifics of the Type 209 signature had not been collected before the war. The computer combat direction system planned for later British SSNs would probably have helped disentangle the numerous contacts. *Valiant* could not definitely classify any of them. She was also frustrated by an intelligence failure. Intelligence indicated that *San Luis* would turn north, but the most promising contact turned south (as *San Luis* actually did, toward a different Argentine naval base).

This contact may not have been *San Luis* at all. *Valiant* tried to clarify the situation by going active, but that only further confused the situation due to the large number of contacts.[8]

There were two offensive ASW operations against *San Luis*. The first, mounted on 1 May, was based on intelligence indicating that *San Luis* would be patrolling the area the Argentines designated "Maria." *Brilliant*, *Yarmouth*, and three Sea Kings from *Hermes* were deployed from the task force. *San Luis*' sonar detected and tracked what seemed to be a British ship, but a quick look through the periscope showed nothing. She detected multiple sonar emissions, some identifiable as coming from British Type 184 sonars (which the Argentines also had). Some presumably came from British dipping sonars. She fired a torpedo on the basis of passive sonar tracking. The torpedo broke its guidance wire and failed to home on any ship. *San Luis* may have grossly underestimated the target range, possibly due to exceptional water conditions (which also affected the British). Her passive ranging sonar (a French DUUX-2) was unreliable at this time.

Brilliant reported a solid sonar contact 2 nm away and fired her two Mk 46 torpedoes. Both exploded soon after entering the sea (one or both may have hit a whale). A MAD search by a Lynx found nothing. As they were being hunted by ships and helicopters, the Argentines on board *San Luis* thought they heard a torpedo in the water; later it seemed to be approaching from astern. It appears to have come from a Lynx from HMS *Brilliant*. *San Luis* launched decoys, turned, and went deep. The torpedo was heard to pass. There were numerous explosions, presumably Limbos.

As the bottom was only 50 m (164 ft) deep, *San Luis* bottomed and remained there overnight. She heard repeated explosions identified as Limbo salvoes. The Argentines claimed that *San Luis* was counterattacked by the 2 frigates and 3 helicopters for about 20 hours before she broke off. *Yarmouth* and *Brilliant* and the helicopters expended considerable ordnance, but they did not claim that they heard any torpedoes, and no ship showed hull damage that might have been associated with a dud torpedo hit. Following orders to return to the task force, *Yarmouth* obtained an apparently solid sonar contact within the Maria area and attacked with Limbo; she classified the contact as a non-submarine.

In another incident at about the same time, *Invincible* was screened by the frigates *Alacrity*, *Arrow*, and *Plymouth* 70 nm from area Maria. All of the frigates reported torpedo sound (HE), but the Admiralty Underwater Weapons Establishment (AUWE) analysis found it unlikely that a modern torpedo such as SST-4 could have been detected simultaneously by all three. It seemed more likely that the noise came from a torpedo decoy. The incident was finally classified as non-sub after a helicopter spotted a school of dolphins.

It was impossible to confirm or disprove the Argentines' claim of having mounted an unsuccessful attack on a British warship. Given the considerable expenditure of weapons in the second incident, it was possible that *San Luis* thought she was under continuous attack.

The second offensive operation was mounted on 5–6 May near the wreck of the missile destroyer *Sheffield*. Jezebel buoys were laid in the northern approaches to the wreck. Presumably the British thought the Argentines would approach in hopes that British ships would congregate near the wreck, but there were no contacts.

On 5 May there was a possible Argentine torpedo attack on the carrier *Invincible*. A lookout reported a periscope, so one of the carrier's Sea Kings dropped a torpedo. The Type 21 frigates *Alacrity* and *Arrow* reported hearing a heavy torpedo and maneuvered violently to evade it. The frigate *Plymouth*, more than ten miles away, reported the same torpedo sound; she fired her Limbo. Later in the afternoon the missile destroyer *Glamorgan* reported a periscope and a sonar contact; it is not clear whether she fired a torpedo at it. A helicopter using a dipping sonar to investigate the contact determined that it was a semi-submerged rock. AUWE initially rejected the claim that a torpedo wake had been seen from *Invincible* but then relented: a torpedo on maximum-speed setting, running shallow, might well cavitate to create a visible wake.

It later emerged that on manual setting SST-4s did indeed run shallow. However, the apparent torpedo did not turn as the carrier maneuvered. It was later apparent that *San Luis* had been nowhere near the carrier at the time.

Also on 5 May, the Argentines reported that their carrier-based Trackers and Sea Kings attacked one (possibly two) British submarine 45 nm from Camerones Bay. No British submarine was anywhere near there at the time. The Argentines claimed that the incident began with a radar contact, followed by a sound contact by a Sea King.

The next day *Yarmouth* fired four Limbo salvoes at a submarine supposedly confirmed by a Sea King helicopter using its dipping sonar and MAD. A hunt by the frigates *Yarmouth* and *Brilliant* (the latter with a 2016 sonar) was interrupted by an air raid alert, but it involved attacks by homing torpedoes, Limbo, and helicopter-delivered depth charges. The target turned out to be a wreck. All of this action turned out to be well north of *San Luis*.

On 11 May a Wessex helicopter from the missile destroyer *Antrim*, dipping its sonar well ahead of a small convoy, detected what seemed to be a valid submarine contact, with the right apparent size and shape. A Mk 46 torpedo was dropped. The apparent submarine seemed to escape by, among other things, suddenly changing speed (up to 28 kts) and depth. It was officially classified as a whale, but the helicopter crew believed that the contact was a foreign submarine maintaining a secret presence. If the apparent evasion speed was accurate, she was an SSN. The British had considered the possibility that the Soviets might use their own submarines to watch the unfolding war. Their warning to other nations to avoid the war zone was unlikely to have deterred the Soviets (who monitored the war to some extent using their ocean surveillance satellites).

Also on 11 May, *San Luis* heard two fast warships. The frigates *Alacrity* and *Arrow* were returning to the carrier group, having transited Falkland Sound to see whether it was mined. They had just sunk an Argentine logistics ship. They may have detected *San Luis* but classified the sonar contact as a non-sub (a submerged rock). They soon accelerated to higher (sonar deaf) speed because they had to return to the carriers, with their air protection, before daybreak. *San Luis* launched two SST-4 antiship torpedoes in sequence at what seemed to be optimum range. One may have hit and damaged the torpedo decoy towed by *Arrow*.

The Argentines had their own scares. On 8 May, on hearing what was thought to be an approaching torpedo from a hostile submarine about 2,500 yds away, *San Luis* maneuvered and released decoys, then fired an antisubmarine torpedo (a Mk 37). The torpedo exploded, probably on hitting the bottom. There was no secondary explosion. A few hours later a similar sonar contact was detected. *San Luis* was too close to fire a torpedo at it, but she took a course to ram. At the last minute the contact was identified as krill.

After the unsuccessful attack on the two frigates, *San Luis* was ordered home. She had never approached the British carriers, the only targets whose destruction would have been decisive.

Due to numerous false alarms, the British task force expended a surprising amount of ordnance. A postwar official British account listed 235 ASW operations, probably nearly all against false targets.[9] On only three occasions was it highly probable that an Argentine submarine was present. Not counting numerous targets quickly classified in the sonar control room, the average time to classify a target to non-submarine was 25 minutes. In 13 cases it took more than an hour to classify a contact. Of eventual classifications as non-subs, 36 were biologicals, including 9 whales, which returned sonar echoes very similar to those of a small submarine. There was no indication that MAD either positively aided classification or reduced classification time. In several cases, weapons, particularly depth charges, were used to aid classification. To overcome the limitations imposed by the shallow waters, the British used air-delivered Mk 11 depth charges and Limbos. Set to 30 ft, the depth charge was considered an aid to classification. It was considered nearly useless against bottomed or near-bottomed submarines. According to a postwar evaluation, providing an alternative 120 ft

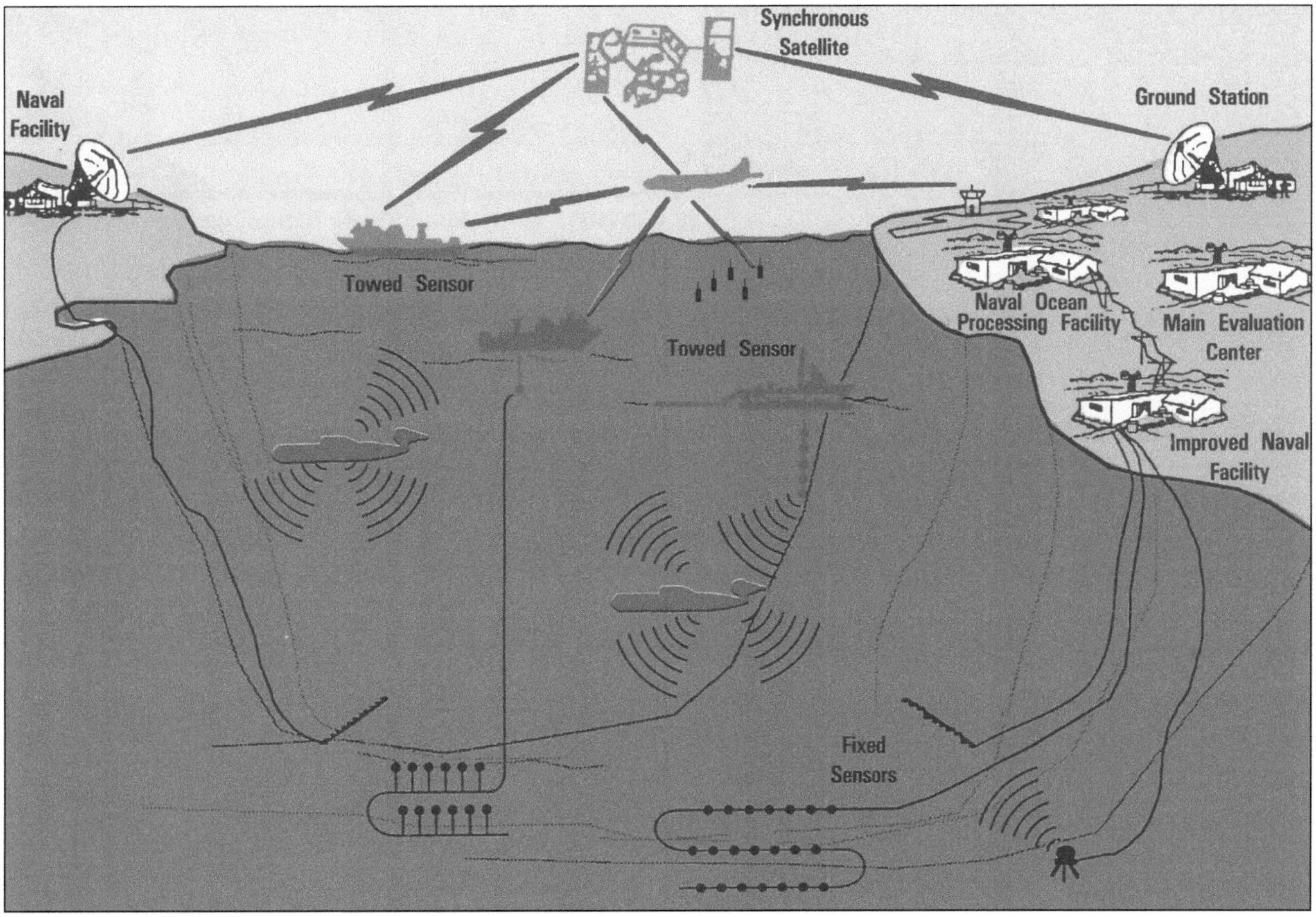

An advertising drawing of the elements of the Integrated Undersea Surveillance System (IUSS) shows the sorts of sensors envisaged late in the Cold War. The three strings of one of the fixed sensors represent the Fixed Distributed System (FDS), which would detect a submarine as it passed over different strings to measure its course and speed. The single string above and to the right represents a SOSUS array, looking at horizontally over the ocean, whereas FDS looks vertically. A double string of upward-looking sensors, to the left, seems to connect with a sensor on board a T-AGOS surveillance ship. The five floating sensors near the surface are presumably sonobuoys. CHRIS CARLSON

setting would make the Mk 11 far more effective in shallow water, particularly where there were many non-submarines.

The scale of British ASW ordnance expended was a reminder of the psychological impact of even a single submarine—hidden, difficult to find, and hard to kill. Wartime exigency would justify using up weapons on a scale far beyond peacetime estimates. The British estimated that they had expended 41 torpedoes (including 26 Mk 46s and 7 Mk 44s), 68 Mk 11 depth charges, and 125 Limbo projectiles. Postwar analysis showed that 10 torpedoes were fired before any target classification at all, and at least 6 of 10 following a possible submarine (POSSUB) classification. Three Mk 46s of 18 analyzed exploded prematurely, and 1 Mk 44 (of about 5) ran on the surface. Due to ambiguities, it was impossible to be certain of torpedo failure rates. Mk 46 failures were between 3 of 10 and 3 of 16, Mk 44 failures were between 1 of 7 and 1 of 3. The numbers involved were too small to make these figures meaningful.

As for classification by depth charge, after the war AUWE argued that "a Warsaw Pact submarine, lying silent and stationary for its own very good reasons and having a pretty good idea of how close we can drop a depth charge, will not be sufficiently disturbed by a few unpleasant explosions to change its behavior pattern. The risk of damage is not high and a submarine that does not react will be classified as a non-sub." Several ships wanted depth charges as stop-gap shallow-water weapons, but during the war virtually all ASW incidents occurred outside the 100 fathom line.

British sonars, other than Type 2016, seemed ill-adapted to detect the Argentine submarines. Postwar analysis showed that whales were not mistaken for submarines as often as had been reported, partly because their target strength was significantly smaller. Ships using the most common British hull sonar, Type 184, detected whales four times, compared to once each for Type 177 and Type 2016. Three ships had 2016, 2 had 177M, and 17 had 184M. Passive capability was very limited.

Of contacts assessed as certain submarines (CERTSUBs), one was biological (not a whale), one was a bottom feature, and one was not assessed. By 1982 the British had decided that Type 184 was missing many (larger) targets because it was difficult for its operator to stay alert enough to react to fleeting indications on the CRT. The follow-on Type 2016 was intended specifically to overcome this problem by using computer-assisted detection. Unfortunately, that alteration left the operator without any ability to listen to echoes, which would have helped distinguish a real submarine from surrounding clutter. Whales and shoals of fish regularly gave a high weight of evidence and a submarine-like image. On the other hand, the computer (and an experienced operator) would have mistakenly classified the SSN *Courageous* as marine life. Type 2016 also rejected bottom contacts out of hand as non-submarines; but the enemy could (and did) bottom. Even so, Type 2016 was considered a very useful classifier, particularly when combined with MAD. According to a postwar Royal Navy critique, it "eased the difficult problem of sorting the whales from the Type 209s, or, rather it should be said left us convinced the contacts were all whales."

Operators reported the sound of rigid inflatables (Geminis), towed decoys (Unifoxers), and even own-ship noise as torpedoes. One report would trigger others; according to the postwar lessons-learned report, "only the absence of a flaming datum allowed the [ASW commander] to take a detached view and cool rapidly escalating whale hunts. There is a need to teach sonar operators what torpedo [noise] (all types) sounds like when heard through the sonar."

All three towed-array SSNs deployed without the broadband analyzers that would have given them much-improved ability to detect and classify surface ships. A Type 209 on motors was so quiet that the standard British SSN sonar (Type 2001) would not detect her passively outside 1 nm. An SSN with the existing towed array would likely detect a snorting Type 209 from about 10 nm; but such opportunities would be brief and fleeting. Initial detection of a quiet diesel submarine would as likely be broadband as narrowband. Only two of the three towed array SSNs had wideband analyzers. A trial conducted soon after the war using the SSN *Sceptre* showed that their initial detection ranges would have been about doubled had the towed-array SSNs been fitted with sonar signal acoustic processors (SSAPs). *Conqueror* was effectively limited to only a few selected threat frequencies, greatly reducing her ability to recognize relevant signatures. The British concluded that broadening low-frequency towed-array response to include the 3–60 Hz spectrum would have enabled detection of *San Luis* at much greater ranges. The next-generation Type 2026 and Type 2046 towed arrays (to enter service in 1985 and 1986) would enable submarines to detect Type 209s at useful ranges.

The British knew that the Argentine submarines were armed with long-range SST-4 electric torpedoes. For protection they relied on towed decoys, the sophisticated Type 182 for warships and the simpler Unifoxer noisemaker for merchant ships pressed into service. Type 182 proved insufficiently robust and had to be modified after the war. Unifoxer noise was sometimes mistaken for the Argentine submarine or its torpedoes. Unaware of the Soviets' wake-following, the British considered Type 182 effective against all known Soviet torpedoes. They also did not know that they had an important advantage: *San Luis'* fire control computer was broken, and the crew was unaware of some of the limitations of the manual technique they had to use instead. For example, the guidance wire would break within a few minutes of firing. The polarity of the connection between the fire control switchboard and the torpedo gyros had been reversed, so torpedoes would not run as set. The torpedoes had been too long between overhauls, and their batteries were time-expired.

The U.S. Navy concluded that given the scale of arms sales to Third World countries, it had to take Western weapons into account as potential threats. ASW tuned to deal with Soviet SSNs might well be irrelevant against diesel-electric submarines operating in shallow water. The U.S. Navy had no cheap urgent attack/classification weapon comparable to the British Mk 11 depth charge or Limbo. It could hardly afford to expend lightweight torpedoes in quantities like those in which the British had expended urgent-attack weapons. Soon after the war the U.S. Navy considered buying a few Mk 11 depth charges. It developed a hydrostatic fuze to convert its standard 500 lb Mk 82 bomb into an air-delivered depth charge. A proposed revived depth charge went nowhere.

Much of the success of the British operation could be traced to how little London interfered with it. Even so, the British suffered from communications overload, which forced them into extensive use of nonsecure HF radio (that also applied to the Argentines). Neither side had enough secure radio links.

The main British government intervention concerned rules of engagement and their application to major Argentine warships. Rules of engagement had to specifically address pre-hostilities action by ships and aircraft suspected of engaging in hostile surveillance. What if they failed to heed warnings? It was vital to define "hostile intent" and "hostile act." Presumably this applied to the Argentine airliner that reconnoitered the British task force prior to the initial Argentine air attack. This issue had already surfaced in U.S. war games. The Soviets relied heavily on preattack surveillance, but they also tracked U.S. formations when they had no intent to attack. When was surveillance hostile? It was an essential part of the Soviet way of attacking using long-range weapons. War games showed that whoever fired first often won. Would U.S. political authority support telling the Soviets during a crisis that the appearance of surveillance aircraft was a hostile act?

The U.S. Navy became interested in improving shallow-water torpedo performance. The British regretted having largely abandoned their best shallow-water weapon, Limbo. The Argentines had retained Hedgehogs on their FRAM II destroyers and had fitted them with single depth-charge racks, but it was not clear that either measure had done them any good.

Submarine IFF was a particular problem for own-force survivability during the early stages of the war.

The British had managed to keep their nuclear submarines at sea much longer than intended; they were designed for 75 days' endurance but remained at sea for 90–115 days during the war.

REPLACING SOSUS

Too much material remains classified to evaluate SOSUS and other sensor performance in the 1980s. Secretary of the Navy John Lehman said at the time that the Navy knew where all the Soviet submarines were all the time, to the extent that if war broke out, all of them (presumably those outside the bastions) could be attacked almost instantly. His later detailed account of the last Soviet submarine operations off the U.S. East Coast presumably drew on knowledge of their movements gathered at the time. The submarines involved were quiet second-generation types.

The SOSUS replacement was the Fixed Distributed System (FDS). Where SOSUS used a few arrays looking across the ocean, FDS would employ numerous upward-looking arrays constituting acoustic fences. Because, like SOSUS, FDS was a completely passive system, a submarine passing over one of these fences would not know that it had been detected. A sequence of two or more fences would establish the submarine's course and speed, which probably would not change very quickly. Thus, like SOSUS, FDS would make it possible to project a submarine's future position. A patrol plane could find it and track it. A U.S. Navy schematic drawing released in 1989 showed lines of upward-looking hydrophones, five in a line. Three such lines, about 50 nm apart, would form a barrier. Individual hydrophones were shown 13 nm apart. This combination would be deployed across a chokepoint.

FDS was announced in 1984; the FDS Navy Development Coordinating Paper was approved 13 May 1986, followed by the DoD Decision Coordinating Paper on 10 May 1989 and the Milestone II decision on 22 September 1989. Full-scale engineering development of the underwater element was approved in January 1990, followed by the shore element in June 1991. The request for proposals (RFP) for the shore station and processor was announced in 1989, at which time the full system was expected to become operational by the second quarter of FY 95.

FDS was conceived as part of a larger Integrated Underwater Surveillance System that would include SOSUS and SURTASS ships, all tied together by the Surveillance Detection System.

After the Soviet Union collapsed, FDS was reduced to technology demonstrator status, a single array being deployed. The plan to have 11 arrays delivered by 2006 was abandoned, but work did continue on 2 test arrays.

The Rapidly Deployable Surveillance System (RDSS) was a kind of localized FDS. It was not a new idea; a long-life surveillance buoy system had been proposed in the 1960s as the MSS, renamed RDSS in FY 77. The project was formally begun in 1978, full-scale engineering development (FSED) being approved in July 1981 and a contract being let (to Sanders) in August of that year. However, the project was cut back to a maintenance level in December 1982, transferred to NAVAIR that year (FY 83), and then canceled in December 1984. It was resurrected in January 1985 and then killed again in February 1986.

SURTASS was upgraded to deal with quieter submarines. A twin-line array was proposed in 1991. The twin-line array offered the same gain as the original long array but was about a fifth as long. With two parallel lines, it could resolve left/right ambiguity without requiring the ship to turn. It consistently outperformed single-line arrays, even in high clutter (including the Adriatic, East China Sea, and Yellow Sea), due in part to its significant directional noise rejection. It could be towed in water as shallow as 180 ft. It was first deployed on USNS *Bold* (T-AGOS 12) in 1994.

As in the past, another counter to quieter submarines was active low-frequency sonar. In this case it was the Low Frequency Adjunct (LFA) for SURTASS ships. Like the twin-line array, the active adjunct came too late for the Cold War. It demonstrated its capability against inherently quiet targets—diesel submarines—even bottomed ones, hence remained valuable in the post–Cold War era. For example, in bistatic mode LFA detected and tracked a diesel submarine in the shallow Ionian basin. In this particular experiment a single LFA system in mid-Mediterranean managed to cover the whole width of that sea, from the northern to the Libyan coasts. Destroyers working with the ship towing the pinger were able to receive echoes.

By 1993, Hughes was producing the first LFA. The array consists of a vertical stack of 18 transducers with its center 400 ft below the surface. It operates at 100–500 Hz to produce an ODT signal that can be steered in the vertical. Source level is 215 dB (actually the level produced by one projector; the projectors are used in sequence). The typical ping includes CW and FM components; it can be 6 to 100 seconds in length, no constant-frequency component lasting more than 10 seconds. Typically time between pings is 6–15 minutes, to allow signals to propagate back from distant targets. Maximum duty cycle is 20 percent (typically 10–15 percent). Typical T-AGOS operating speed with LFA deployed is 3 kts.

LFA was intended for second-generation SWATH ships (T-AGOS 23 class), suspended from a well between the hulls. The operational unit was on board USNS *Impeccable*, the sole second-generation SWATH to be built. Work continued on a smaller, lighter LFA that could be deployed more easily. Twin line and LFA began fleet operations in 1998, but environmentalists have sought to stop LFA operations on the ground that they harm marine mammals.

As before, the point of long-range detection was largely to cue maritime patrol aircraft. Through the 1980s the U.S. Navy, and to a lesser extent RAF

Coastal Command, sought improved sonobuoy/processor performance to overcome silencing. Both the U.S. Orion and the British Nimrod were fully digital, hence relatively easy to upgrade as improved computers became available.[10] The Orion in particular was subject to fleetwide programs called Updates I–IV. Both navies experimented with coherent processing of multiple buoys in the water, forming arrays. During the 1980s such processing required so much computer power that it could be done on only an experimental basis, with the computer filling the airplane. Moore's law guaranteed that the same functions could be translated into operational systems within a few years.

Two new sonobuoys were conceived: a new active buoy (ERAPS, the Expendable Reliable Acoustic Path Sonobuoy, SSQ-75) and an advanced passive buoy (VLAD: Vertical Line Array DIFAR, SSQ-77). Both employed exotic detection paths. ERAPS had been conceived as early as 1961. Looking up, it employed a volumetric array of vertical dipoles. Depth could be set at 60–16,500 ft; typical depths were 15,000 ft in the Atlantic and 6,000 in the Mediterranean. Maximum range was about 15 nm. The first 315 buoys were bought in FY 84–85 for operational tests. Magnavox and Sparton formed a joint company to produce them.

VLAD was intended to benefit from pointing its beam at an angle. It could, for example, look down to avoid surface noise and detect bottom-bounced signals. The signal format was identical with that of the standard DIFAR buoy; the vertical orientation of the buoy was set before it was launched. Pilot production began in May 1984, and full-scale production followed in 1986. Initially buoys were made in alternative bottom-bounce and horizontal-beam versions, but later ones could be set for desired beam elevation/depression angles. A Steered VLA buoy (SSQ-79) did not enter service.

A new generation of sonobuoys was envisaged in the mid-1980s to counter increasing Soviet silencing. By the end of the Cold War, means were wanted to deal with another silencing problem: quiet diesel-electric submarines operating in shallow water. The two fundamental approaches were increased gain (to detect fainter signatures) and the insertion of an active element in a sonobuoy field. One approach to increased gain was individually higher-gain (hence more directional) buoys, exemplified by a program to develop a Horizontal Line Array (HLA). Increasingly compact electronics could provide it with in-buoy data processing (to limit the load on the buoy up-link) and data storage. HLA reached the full-scale engineering development stage in 1986.

An alternative path to higher gain was an array of buoys, coherently processed.[11] Even if the buoys were not used in a coherent array, spreading many more of them would make up for their much-reduced range. The U.S. Navy invested in a low-cost sonobuoy (LCS), which emerged as later versions of DIFAR, such as the current SSQ-53F and SSQ-53G. A follow-on to DICASS (SSQ-90) was in limited production by about 2005. A new low-cost auto-alerting buoy reached the full-scale development stage in FY 83.

There was also interest in moving more of the sonobuoy processing load onto the buoys, exploiting the new microelectronics. It seems not to have reached the operational stage before the end of the Cold War.

A sound source could convert a field of passive buoys into an active system more likely to detect a very quiet submarine. Work on a new active source (adjunct) reached the full-scale engineering development stage in 1987 (it was to reach operational use in 1993). It seems not to have entered service; the active adjunct concept morphed into Enhanced Echo-Ranging (E2R), a return to the much earlier Julie concept in modified form, using a new explosive source, SSQ-110. The receiving buoy was VLAD. An Improved E2R would employ a new receiving buoy (SSQ-101, the Air-Deployed Active Receiver [ADAR]). ADAR could also function as a passive buoy. In the initial SSQ-110 version, charges were held at the water surface and dropped on command. Later there was a coherent sound source, and also a more sophisticated explosive sound source.

A tactical surveillance buoy began full-scale engineering development in FY 86 (for initial operational use in FY 92). It seems to have died at the end of the Cold War.

To deal with quieter submarines, buoys would emphasize broadband (e.g., flow noise) operation. Work to adapt the existing AQA-7 airplane buoy processor to broadband operation reached the full-scale engineering development stage in FY 82.

All of these programs were aside from new approaches to submarine location possible with increasingly powerful computers.

The Soviets lacked anything remotely like the Western, particularly American, computer industry, although they had very competent programmers and could sometimes produce excellent prototypes. As an indication of the size of the problem, when the Soviets' sonar developer began work on their first digital submarine system, it had not only to choose the minicomputer it would use but also to write the machine's operating system. The deep gap between Soviet and Western computer capability seems to have widened appreciably in the 1980s.

Shipboard sonar was similarly to be developed to counter the increasingly quiet Soviet submarines. For surface ships, that meant larger arrays with higher gain. Thus projects in the mid-1980s included development of a conformal or planar hull-receiving array, which would work with a new low-frequency active source, perhaps towed. The current linear towed arrays would eventually give way to multiline towed arrays, which would add gain without becoming too long for maneuvering. Active sources with short, towed-array receivers first appeared in Europe, the most prominent being the French CAPTAS announced in 1996. The concept was later adopted by the Royal Navy (Sonar 2087) and is planned for the U.S. *Constellation*-class frigates.

WEAPONS

After years of research and development, it was finally possible to build some of the weapons first envisaged in the 1970s. Submarines needed a non-nuclear alternative to SUBROC. By 1980 it was the Standoff Weapon (SOW), intended to have the same range (5 nm to the first convergence zone). The project was formally approved in January 1980, four competing concept formulation contracts being awarded in February. Boeing was selected in April 1981. Initially, plans called for a common ASW weapon combining the SOW with a vertical-launch follow-on to ASROC. A major change in direction during FY 82 caused a stretch-out. Like SUBROC, the weapon would be able to reach the first convergence zone. Accuracy was to be better than 1 percent of range, 500 yds. The two programs were split in 1982 but were again merged in 1988. The weapon was named Sea Lance (UUM-125A). To keep launch stealthy, the submarine version was to be launched in a buoyant capsule and would ignite its rocket only when it reached the surface.

In 1986 the program was reoriented. U.S. submarines could no longer passively track the increasingly quiet Soviet SSNs. However, the existence of a SUBROC-like weapon might deter Soviet submarine commanders from running at high noisy speed. The same would be true if they used their active sonars while seeking U.S. submarines. As a quick-reaction weapon, Sea Lance had to be usable without high-level authorization, so it had to be non-nuclear. In December 1986, then, the Mk 50 lightweight torpedo replaced a nuclear depth charge as the primary payload. About 1984 plans called for award of a limited production contract in May 1988, with full production following in May 1989. As of 1988, plans called for 2,400 missiles. Flight trials at White Sands began that summer, the first sea test following on 22 February 1989 (all-up except for a short-burn motor). By this time the program was in trouble; it was canceled late in 1989. Explanations for that seem muddled. There may have been misunderstanding that Sea Lance was not primarily a nuclear weapon.

The approximate shipboard equivalent to Sea Lance was Vertically Launched ASROC (VLA), in effect the successor to the Extended-Range ASROC (ERA) of the 1970s. Compared with ASROC, it had longer range (to exploit SQS-26 direct-path range) and a somewhat heavier payload (a Mk 50 torpedo). As of 1987, the Navy wanted a minimum range of about 5,000 yds and a maximum range of at least

15,000 yds. None was authorized in FY 88 or FY 89, the FY 90 request for 340 being deferred to FY 91. Deferral was due to a 1988 announcement that Sea Lance would be used by both surface ships and submarines, but Sea Lance was canceled in 1989. The destroyer *Hewitt* conducted developmental and operational tests of VLA in December 1988. Full-scale production began in March 1993, the last of 443 being delivered in April 1999. By 2000 they were described as "not operationally effective." The Japanese Maritime Self-Defense Force bought more than 600 VLAs. South Korea developed an analogous missile.

Both the U.S. Navy and the Royal Navy were developing replacement torpedoes—the U.S. Advanced Lightweight Torpedo (Mk 50) and the British Stingray. Both also developed new heavy torpedoes, the U.S. Mk 48 ADCAP (nominally a modified Mk 48, actually a new weapon), and the British Spearfish. See appendix C for details.

Both navies continued to use nuclear depth bombs. The British used both U.S.-supplied and homegrown nuclear depth bombs. The U.S. weapons were called "Lulus," which suggests that initially they had early U.S. nuclear depth bombs. Presumably these weapons were supplied on a dual-key basis, meaning that the United States could veto their use. Later the British seem to have had U.S. B-57s as backups for their own WE 177As, which were sometimes called "600 lb bombs." Reportedly the British weapons had variable yields (0.5 and 10 kT) similar to those of B-57s. The low-yield option was said to be for water shallower than 130 ft. Nimrods and naval helicopters were certified to deliver these weapons. The Royal Navy repeatedly rejected their elimination on the ground that they were the only sure counter to very fast deep-diving Soviet submarines. The United States provided nuclear depth bombs to other NATO allies, to be used only in the event of war. In the 1960s Dutch Neptunes and Italian Trackers were certified to deliver U.S. weapons; in peacetime the weapons were in U.S. custody. Later, one squadron each of Dutch P-3Cs (13 aircraft) and Italian Atlantics (9 aircraft) were reportedly so certified.

THE ASW FORCE

With the election of Ronald Reagan in 1980, the U.S. Navy gained access to resources that enabled it to expand to execute the new Maritime Strategy. The goal of 600 ships, proposed about 1972 by CNO ADM Elmo Zumwalt Jr., was set by the perceptions that ships could be expected to last about 30 years, and that Congress would buy about 20 per year. The strategy set the distribution of types of ships within the desired fleet size. The fleet was built around the battle groups that would carry the fight to the Soviets. Initially that meant 15 active carriers (i.e., not including those refitting) with their necessary escorts, and 100 nuclear attack submarines. With only 12 carriers available when Reagan took office, the new administration sought immediate ways to add more battle groups. At the very least it wanted to complicate Soviet tracking and targeting. The decision was made to recommission the four Iowa-class battleships and arm them with Tomahawk and Harpoon missiles alongside their heavy guns.

The number of battle groups determined the required number of fast battle group–capable escorts. In peacetime, a carrier battle group would consist of a carrier, one or two cruisers, two to three missile destroyers, two to three destroyers, and one to three SSNs in direct support. In wartime, carriers would form battle forces of two to four carriers supported by four to eight cruisers, four to eight missile destroyers, and four SSNs. The substantial number of missile ships reflects the emphasis on the Soviet naval air threat, particularly in waters near the Soviet Union. Ultimately many of the cruisers would be armed with the Aegis combat system. Thus, about 1982, U.S. surface force plans showed that a carrier battle group required two or three Aegis cruisers, two to five destroyers or CGNs, and two to four *Spruance*-class destroyers.

In 1982 plans also envisaged surface action groups (one Aegis cruiser and three Aegis destroyers). The recommissioned battleships would become the cores of surface action groups.

Ships armed with Tomahawk cruise missiles both increased overall naval nuclear firepower and, it was hoped, helped saturate Soviet naval surveillance. The Navy would arm four *Virginia*-class nuclear-powered missile cruisers and nine *Spruance*-class destroyers with Tomahawks in armored launchers. The new *Burke*-class destroyers and refitted *Spruances* were given vertical launchers that could accommodate Tomahawks.

Initial plans also called for seven convoys (each with one *Spruance* and nine frigates), ten underway replenishment groups (each with one Aegis destroyer and three frigates), and a total of ten Aegis destroyers, the four *Kidd*-class destroyers, and eight frigates to cover amphibious groups.

As initially announced, the 600-ship navy would include 45 cruisers, 70 destroyers, and 100–110 frigates. The cruiser force comprised the nuclear cruiser *Long Beach*, 26 large destroyers (DLGs, called frigates before 1975), and the new Aegis cruisers built on *Spruance*-class destroyer hulls. There were 58 destroyers (23 *Charles F. Adams* class, 31 *Spruances*, and 4 *Kidds*).

Through the 1980s the carriers were an increasingly vital means of imposing pressure on the Soviets' submarine bastions, which would be subject to intense air attack, including missile attacks by submarines. These attacks would be mounted without much warning beyond detection of the submarines by towed arrays. The connection between carriers, bastions, and potential Soviet submarine attacks on the SLOCs made the new Aegis air defense system part of the U.S. ASW. The Reagan administration inherited the Aegis program, but it was the administration's choice to expand and accelerate it. When replacing existing destroyers, the Navy Department chose a design (for the *Arleigh Burke* class) armed with Aegis rather than with the slower-reacting and less capable Tartar-D of the *Kidd* class. Aegis offered the kind of rapid reaction required of ships facing the new Soviet missile submarines.

In 1982 none of the missile destroyers or the existing cruisers (except for the Aegis prototype *Ticonderoga*) offered the anti-saturation fleet air defense firepower needed in the Norwegian Sea. Many more Aegis ships were needed, and the administration bought them. By the end of 1989, 15 had been completed, and many of the cruisers had been or were being upgraded with anti-saturation New Threat Upgrade (NTU) air defense systems. The lack of fleet escorts could be blamed partly on a fight between ADM Hyman Rickover and various administrations over his insistence, written into law by Congress, that surface combatants over 8,000 tons be (expensively) nuclear-powered. That was why the initial Aegis ship was based on the *Spruance*-class destroyer hull, which displaced less than 8,000 tons.

Only in the frigate category was the existing (or authorized) force equal to what was needed: 64 SQS-26 frigates of the *Bronstein*, *Garcia*, *Brooke*, and *Knox* classes; and 55 *Oliver Hazard Perrys*. The size of the frigate force Reagan inherited was inevitable given the past emphasis on convoys and their escorts. Thus the fleet included convoy escorts that were not really needed to execute the new strategy. They were a fallback and insurance against the remote possibility that the Soviets really would attack the SLOC. It was not obvious that they would be replaced in comparable numbers as they reached replacement age in the 1990s.

As for submarines, by the end of 1982 the U.S. Navy had 73 ASW-capable fast submarines: 52 *Permit/Sturgeon* class and 21 *Los Angeles*, with many more *Los Angeles*–class building (to a total of 46 by the end of 1989).

Naval air forces were specified to be sufficient for the enlarged fleet. For years OSD had considered land-based patrol aircraft guided by SOSUS and SIGINT the most cost-effective counter to Soviet submarines in the SLOC protection role, which it emphasized. Their numbers were set not by the size or shape of the fleet but by geography, mainly of the North Atlantic. The main impact of the new strategy on this force was that the aircraft would have to hunt farther north, possibly within range of Soviet fighters. For the first time in decades, they might need self-defense. From time to time, it was proposed that they carry Sidewinder missiles as well as defensive IR jammers.

For the fleet-based Maritime Strategy, it was more important that battle fleet ASW aircraft, particularly helicopters, be available in sufficient numbers. The advent of towed arrays made that even more important. In 1980 helicopters were already seriously deficient for the existing fleet. At this time the standard carrier air wing included ten S-3 Vikings and six SH-3H helicopters. Ideally a carrier needed eight SH-3H helicopters, but funding limited the number to six. Thus the required force for 12 carriers was 72 SH-3Hs plus 18 for training, 6 for RDT&E, and 15 for the pipeline (17.2 percent). That added up to 111, but in 1984 only 90 were available. Some carriers were operating non-ASW helicopters. The Reagan administration's goal of 15 active carriers would demand more carrier ASW helicopters. There was no prospect of reopening the SH-3H production line. After a design competition between Kaman and Sikorsky, a modified SH-60 was chosen in March 1985. The first SH-60F flew on 19 March 1987, and the type entered service with HS-10 in March 1989; a total of 92 were made, far below the original goal (about 1995, before deep cuts in the U.S. defense program, the U.S. Navy had 76 SH-60Fs, with another 54 planned). Like the SH-3H, the SH-60F incorporated a dipping sonar, in this case an AQS-13F on a long cable with an integral sonobuoy processor.

The surface ship helicopter situation was also bad. The budget-constrained figure for SH-2 LAMPS helicopters was 85 percent (to allow for ships in refit). At that time 98 ships operated these helicopters (requiring enough for 83 of them). Another 21 helicopters were needed for training, 3 for RDT&E, and 21 for the pipeline, a total of 128, against which only 93 were available. It was unlikely that the SH-2 line would reopen, so the fleet had to wait for the LAMPS III (SH-60B). The objective was two per ship, but with a view to cost the Coast Guard limited it to one per ship. The new SH-60B LAMPS III was one of two competing proposals received by NAVAIR in 1977; it was selected in February 1978 and first flew on 12 December 1979. Deliveries began in 1983, and the helicopter achieved initial operational capability with HSL-41 in 1984; ultimately 283 were delivered.

THE END-GAME

By 1984 the U.S. intelligence community, including the Office of Naval Intelligence (ONI), accepted that in the event of war, the primary drivers of Soviet naval operations would be pre-strategic: the fight to protect Soviet strategic submarines and the fight to destroy Western strategic submarines and warships, particularly carriers, capable of attacking the Soviet Union with nuclear weapons. The U.S. December 1984 NIE on Soviet Naval Strategy and Programs through the 1990s accepted that forces not required for those strategic tasks would be employed to support the advancing Soviet-led army and in a limited SLOC attack. The latter would be intended mainly to divert NATO naval forces. There might also be a mining offensive against NATO European (and possibly American) ports.[12]

Since 1982, when this kind of report had last been prepared, the Soviets had shifted the emphasis in submarine construction from strategic to attack submarines, having agreed to a limit on strategic submarine numbers as part of the SALT process. However, it seemed that even in the 1990s new very quiet submarines would amount to no more than half the total Soviet force, leaving U.S. submarines with clear advantages over the rest. Soviet technical advances would be affected by their continued inability to build a professional navy, as 75 percent of their personnel were draftees who would spend only a few years in the fleet. The problem of inability to fill billets with trained personnel was dramatized when the experimental deep-diving submarine *Komsomolets* was lost in 1989. Her designer later wrote a book placing responsibility on the demand (from above) that she deploy despite lacking trained personnel.

The Soviets prioritized new construction over maintenance. This problem with this strategy was demonstrated when the Yankee-class *K 219* sank off the U.S. coast in October 1986 after suffering a missile tube explosion and subsequent fire. In this and other classes of strategic submarines the liquid-fuel missiles were intended for long-term storage in their tubes, without maintenance, but a tube in this submarine had given trouble on previous patrols.

Overall, only a limited portion of the Soviet submarine force was ever at sea; of the SSBNs, that amounted to 25 percent (about 15 units), although further strategic submarines were considered ready.

By 1984 the Soviets were keeping most of their patrolling strategic submarines at or inside the landward boundaries of their patrol areas, presumably foreshadowing wartime plans to withdraw strategic submarines into defended bastions. Given their readiness problems, it seemed that the "anomalous" patrols closer to the U.S. coast would adversely affect overall readiness from 1987 onward, and that problem quickly became significant.

The intelligence report pointed to frequent Soviet references to the counter-ASW justification for the big surface combatant program: to ensure the "combat stability" (i.e., wartime survival) of submarines, meaning the Soviet SSBN force. In 1984 the Soviets were building their first full-deck aircraft carrier, which would emerge in 1990 as *Kuznetzov*. She was first of a planned class of two, which would have been succeeded by the nuclear-powered *Ulyanovsk*. It was generally assumed that in the event of a general war, fighters on board the carrier would have been used against maritime patrol aircraft and carrier-based bombers launched against the Northern Fleet bastion and its protectors. The ship would also have been useful in limited-war scenarios, for which the Soviets could deploy their substantial surface fleet. The important conclusion of the 1984 report was that *Kuznetzov* and other impressive new surface combatants were intended mainly to ensure the survival of the Soviet strategic submarines (the report did not yet refer to SSBN operating areas as bastions). There was some evidence that the Soviets agreed with the U.S. Navy that new U.S. weapons, such as Aegis, were going to make it much more difficult to destroy U.S. attack carriers during the non-nuclear phase of a war.

The Soviets were developing a greater capacity for a wider variety of possible wars, including protracted ones, but there was no indication that their basic mindset was changing. They would deploy naval forces in the initial stage of a war to tip the nuclear balance, regardless of any expectation that the war would not escalate to nuclear warfare. They had to do so because they considered the nuclear balance ultimately decisive. They would conduct initial operations with an eye to the possibility of later escalation while seeking to preserve their own ability to mount strategic strikes. There was considerable evidence that they expected the NATO nations to surrender if the nuclear balance was favorable enough. That meant both that Soviet SSBNs had to survive, and that NATO SSBNs and other naval nuclear forces had to be destroyed.

The new analysis agreed that the anti-SLOC mission would become more important if, as was hoped, NATO could prolong a war. However, the Soviets almost certainly would not mount a major anti-SLOC offensive until they had tipped the nuclear balance. It seemed likely that through the 1990s the Soviets would not have powerful enough naval forces to conduct a major SLOC campaign while also hunting down NATO nuclear forces and protecting their own SSBNs. The advent of Tomahawk added a lot more NATO nuclear shooters that the Soviets would have to destroy. It seemed to follow that the Soviets would continue to place a low priority on open-ocean SLOC interdiction. The anti-SLOC campaign they did mount would probably be intended mainly to disperse and tie down NATO naval forces and to reduce the efficiency of NATO military shipping. Such a campaign would probably include bombing and mining NATO ports (but the Soviet bomber force might well be otherwise occupied). There was no evidence as to whether they would attack U.S. ports.

Overall, the U.S. intelligence community remained convinced that Soviet naval procurement would be driven primarily by the strategic offensive and defensive tasks.

That good news was tempered by an estimate that a quarter of the large Soviet submarine force might be available for nonstrategic tasks. There was also the estimate that at some point some Soviet nuclear submarines might be released for an anti-SLOC campaign.

The U.S. Navy was already demonstrating its ability to operate in the high Arctic, where it had to go to attack the bastions. It appears that initially the Soviets' obsession with bastions was not well understood or accepted; operations in the far north were designed to attack the concentration of Soviet naval resources there, particularly in the Kola Peninsula. Only later was it understood that this concentration had been created to protect the bastion in the Barents Sea and the White Sea. Smashing it would have opened the door to Western attack submarines attacking the SSBNs, the only really valuable Soviet naval asset, in the Northern Fleet bastion. The 1984 analysis mentioned that the Soviets were assigning Northern Fleet SSBNs to a local defense force in what was later understood as the Northern Fleet bastion.

The U.S. Navy announced its reentry into the high Arctic with the August–September 1981 Ocean Venture '81 exercise built around the Strike Fleet Atlantic, which NATO planners expected to activate in the event of war.[13] A major theme was direct participation in a NATO ground war. The exercise followed the war plan for the Atlantic; its success validated the war plan. The objective was to show that the two-carrier Strike Fleet could operate freely near the Soviet coast. At the outset, the NATO Strike Fleet would head north through the GIUK Gap. The carrier battle force headed for the Kola would also strike the advancing Soviet army. The Strike Fleet demonstrated that it could attack shore targets both in Norway and on the central front in southern Germany while beating off the usual integrated Soviet missile tactics. Although land-based aircraft greatly outnumbered those on board the carriers, it was generally agreed that within a week or so few of those aircraft would still be flying. Given poor Soviet IFF, most would probably be destroyed by friendly fire. Carrier strike aircraft would then have considerable impact—if the carriers could survive that long. Another striking force would head for the Mediterranean to reinforce the Sixth Fleet and attack Soviet bases in southern Russia and in Libya. An ASW barrier would be set up in the GIUK Gap. Sealift would be routed well south to avoid Soviet bombers. The exercise worked even though key technology, such as the F-14/Phoenix combination and Aegis, was not present in the envisaged quantities.

The Strike Fleet, initially built around the carrier *Eisenhower*, simulated an opposed transit to the GIUK Gap, initially engaging the *Forrestal* battle group from the Sixth Fleet, operating in electronic silence. Then the two groups merged. They were joined by the British ASW support group built around the light carrier *Invincible*. The combined force proceeded through the GIUK Gap to the area of the Kola Peninsula, the base of the Soviet Northern Fleet, which protected the nearby SSBN bastion. This was radically different from the usual annual exercise testing NATO forces' ability to protect the transatlantic SLOC. The main exercise included simulated attacks by NATO diesel submarines. It tested towed arrays, which otherwise were considered effective mainly against nuclear submarines.[14]

Among other innovations, Ocean Venture '81 introduced widely dispersed formations not recognizable by satellites. As in UPTIDE, it employed cover and deception tactics and total electronic silence. Given foul weather, for a considerable time the fleet seems to have evaded Soviet surveillance altogether. Apparently the first the Soviets knew about it was when U.S. aircraft appeared near Murmansk, about a thousand miles from the carriers. Once the Strike Fleet revealed itself to the Soviets, they opposed it with numerous air sorties, both for reconnaissance and in simulated missile strikes. More than 20 Soviet surface combatants as well as numerous intelligence-gathering ships (AGIs) and submarines were involved. *Eisenhower* reached a point north of the Kola Peninsula, though not as far east. This exercise was the first to integrate land-based Air Force AWACS aircraft with the fleet, extending its air cover.

Ocean Venture '81 and the associated North Atlantic operation were attuned to earlier concepts of SLOC protection, but they were also attuned to the USSR's obsession with protecting the Barents Sea bastion. They demonstrated that the U.S. Navy and allied navies could destroy the Northern Fleet guarding the Barents Sea bastion, hence that more of the Soviet fleet had to be concentrated there to protect it. Any such concentration detracted from forces that might be available to attack the vital SLOC. NATO's demonstrated ability to strike directly at the bastion and its guardian became more significant for the Soviets. Conversely, the U.S. and NATO fleet coming up from the south and capable of sustained operations in the high north made it difficult for the Soviets to reassign submarines from SSBN protection to an open-ocean assault on the SLOC.

In theory Ocean Venture '81 derived from previous regular NATO exercises, but this time the exercises were considerably extended and merged. Ocean Safari '81, the eastern Atlantic phase, tested shipping protection, including both convoys and a defended lane. Magic Sword was the penetration of the northern Norwegian Sea. Another part of the operation exercised defense of the Danish Straits. French and Spanish battle groups participated. The Spanish group was involved in an opposed amphibious transit of the Strait of Gibraltar. The U.S. Navy sponsored parallel exercises in the Far East and the Mediterranean.

These exercises showed that the strike groups could achieve surprise, as demonstrated by the late appearance of the Soviet reconnaissance aircraft and then missile bombers intended to oppose the carriers. The only exception was a Pacific Fleet exercise apparently betrayed by the Walker spy ring.

Through the 1980s the U.S. Navy continued to develop deception as a means of enabling a strike fleet to survive to strike places like the Kola. It published drawings showing a U.S. fleet spread out in apparently random fashion (to confuse satellite surveillance) over huge areas. That was feasible given the increasing range of submarine detection using, among other things, towed arrays. The new tactics built on the much earlier Haystack concept and on UPTIDE. Deception devices were publicized but never illustrated, one object being to sow distrust of Soviet naval reconnaissance within the Soviet naval staff. At least one exercise tested the ability of a carrier to hide in the Aleutians, an extension of the island tactics tried in the Pacific in 1959. Another placed a carrier in a Norwegian fjord, which could be closed to Soviet submarines by placing Captor mines at its mouth. Examination of the capabilities of Soviet air-launched antiship missiles showed that using their radars bombers could not detect the carrier against the mountainous sides of the fjord before they were inside minimum attack range. In the fjord, too, the carrier could benefit from the assistance of land-based Norwegian fighters.

One object of the exercises was to convince the Soviets that they had to devote all their naval resources to the threats the U.S. Navy was demonstrating. At a post–Cold War conference at Bodo, Soviet general Vladimir Dvorkin told former Secretary of the Navy John Lehman, who had championed the new strategy and a naval buildup to make it work, that he and his staff had concluded from the Ocean Venture '81 exercise that the longest they thought they could keep the Northern Fleet from destruction was a week. After Northern Wedding '86, the Soviet general staff formally notified President Mikhail Gorbachev that successful defense of the northern theater would require trebling of the naval and air force budgets in that sector. This response apparently helped convince Gorbachev that the Soviets could not afford to defend themselves. The Soviets were beginning to feel a visible resource pinch.[15]

The initial Soviet response to initiatives north of the GIUK Gap seems to have been a March 1984 exercise the West called SpringEx-84. The Northern and Baltic fleets moved well out into the approaches to the Soviet Union, such as the Greenland and Norwegian seas, to contest sea control in the area up to 2,000 km from Soviet shores. This exercise was larger than anything attempted in 1983, though still somewhat smaller than Okean 75, the largest Soviet naval exercise. Fifty aircraft were involved, plus 148 surface ships and nearly 50 submarines. Soviet navy Backfire bombers penetrated farther than ever before into the Norwegian Sea. U.S. naval intelligence read the scenario involved as defense of Soviet SSBNs as well as gaining control of the NATO northern flank. In the course of the exercise, about 23 SSBNs were deployed, the most extensive dispersal of such craft yet. Presumably they moved deeper into the defended bastions.

Another big Soviet naval exercise, in the Pacific in the fall of 1985, also demonstrated the Soviets' interest in bastions. This time it included the first Soviet amphibious operation in the Pacific since 1978. This was the last major Soviet out-of-area exercise. After

1986 the Soviets concentrated on defending the bastions.

To defend the bastions, the Soviets were compelled to withdraw their fleet from the world ocean. Soviet worldwide ship deployments peaked in 1984 and then began to decrease. Soviet naval aviation deployments peaked in 1987. Soviet exercise sea-days dropped from 456 in 1985 to 207 in 1986 to 114 in 1987. Probably the last gasp of the Soviet out-of-area fleet was Operation Atrina (March 1987), when five Victor IIIs sortied from the Kola to the Bermuda Triangle and coordinated with Bear-D radar reconnaissance bombers flying from the Kola and from Cuba. There were still some overseas exercises in 1988; for example, a Soviet Mediterranean squadron exercise with the Syrians including the new carrier *Baku*, and also a visit to Cuba by an *Udaloy*-class destroyer, a frigate, and a Tango. However, it was briefer than usual (30 days), involved fewer Bears, and did not include the usual penetration of the Gulf of Mexico.

The notable exception to withdrawal was the establishment and maintenance of the Soviets' only foreign naval base, at Cam Ranh Bay in Vietnam, in November 1983. It can be seen as part of the same Soviet naval strategy as the bastions and the attempt to run down Western SSBNs. The base was near both the U.S. base complex in the Philippines (including Subic Bay) and the likely patrol area of Chinese SSBNs, the South China Sea. The Soviets provided Cam Ranh Bay with defenses in the form of antiaircraft missiles and 14 MiG-23 fighters. Initially the base hosted ten missile-bearing Badger bombers supported by four Bear-D and four naval reconnaissance aircraft, as well as four Bear-F ASW aircraft. The associated naval squadron included four submarines and seven surface combatants. The base survived the end of the Soviet Union and was abandoned only when Vietnam demanded increased rent for occupancy beyond the initial 25-year agreement. By that time the base had been greatly enlarged.

Ultimately the success of the U.S. Maritime Strategy was a triumph of all-arms ASW. Successful trailing of Soviet SSBNs by U.S. and British nuclear submarines led the Soviets to create and defend bastions in which they believed their SSBNs could be protected in wartime. The big NATO exercises, however, showed that carrier strike groups could attack and destroy the Soviet surface forces they thought were essential to guarantee the survival of their strategic submarine force. The main opposition to the carrier strike groups would be mounted by Soviet attack and cruise missile submarines and by Soviet naval aviation. Surface and air ASW would beat off submarine attacks. They would be supported by long-range ASW sensors such as SOSUS. It was essential to destroy not only Soviet submarines but also Soviet naval aviation.

Fortunately, we have no idea of whether any of this would have worked as envisaged, though we certainly do have evidence that the Soviets thought it would have. Many U.S. and British submariners would argue that they could have penetrated the bastions and destroyed Soviet strategic submarines even had the Northern and Pacific surface fleets remained intact. However, what mattered was what the Soviets thought. They believed that the survival of their SSBNs depended on the survival of the covering surface force. That idea may have begun as a way of convincing successive Soviet governments to buy the sort of surface fleet Soviet naval strategists thought they needed for the non-nuclear war they really thought they would have to fight. It did not matter, because the target of the Maritime Strategy was the Soviet Politburo and the Soviet General Staff, both of which did see things very much in nuclear terms.

The attempt to build a modern military, including a big navy, ruined the Soviet Union and brought about its collapse. The Reagan administration seems to have understood better than any others in the West how badly stressed the Soviet Union had become. It saw the opportunity to end the Cold War altogether by increasing the level of stress. The Maritime Strategy was part of that successful effort. It also fit the administration's understanding that the U.S. military had to be able to deal with a wider variety of problems—and therefore with those of the post–Cold War world. The fleet built to execute the Maritime Strategy proved far better adapted for that world than NATO navies carefully adapted to the ASW mission.

AND NOW?

As we look back 30 years after the Cold War, it is important to distinguish special characteristics of Cold War ASW from what we may face in the future. The Soviets' adoption of SSBNs as the vital survivable part of their national deterrent offered uniquely valuable leverage and a special role for Western attack submarines. The futility of Soviet attempts to destroy our own SSBNs absorbed submarines that would otherwise have been quite dangerous to us. Future adversaries may not prioritize SSBNs or indeed national strategic forces in anything like the same way. In that case, future ASW may look a lot more like the ASW envisaged in the 1970s, the question being how to protect vital shipping despite limited resources. That may involve maritime patrol aircraft cued by a future equivalent of SOSUS backing up escorted formations such as convoys. Active sonar, with all its range limitations, may once again be the dominant ASW sensor. Looking back, we can also see the connection between ocean surveillance and offensive submarine warfare, which may make antisatellite weapons an important aspect of future ASW. If anything, we are looking ahead to a world of limited numbers, as in the 1970s. It seems very unlikely, but perhaps we are also looking ahead to a future of successful non-acoustic sensors.

ACKNOWLEDGMENTS

THIS BOOK HAS BEEN a long time developing. Attending a 2018 conference on Sea Control convened by CAPT Peter Haynes, USN (Ret.), I realized that few in the active Navy remembered or knew about the ideas and practices of Cold War ASW. Most officers who had been involved with it had retired. That made this project attractive to me. I am grateful to RADM (Ret.) Jim Stark and to CAPT (Ret.) Pete Swarz for insights into the origins and character of the Maritime Strategy, which was so important in the latter stages of the Cold War. My interpretations are of course my own responsibility. I would like to thank Chris Carlson of the Admiralty Trilogy, Christopher C. Wright of *Warship International*, A. D. Baker III, Dr. Thomas Hone, Paul Marks, and Dr. David Rosenberg for insights and for information. Again, my interpretations and conclusions are my own responsibility. For documentation, I have benefited from the efforts of the staffs of the U.S. National Archives, the U.S. Navy Operational Archives, the British Public Record Office (the National Archive), the Naval War College (via its redoubtable, now retired, archivist Dr. Evelyn Cherpak), the Brass Foundry outstation of the British National Maritime Museum (Jeremy Michell and Andrew Han-Ling Choong), the British Admiralty Library (assisted by its librarian Jenny Wraight), and the French Service Historique de la Marine (now amalgamated with the other service archives as the Service Historique de la Défense). For photographs, I would like to thank the U.S. Naval Institute (photo archivist emeritus Janis Jorgensen and her assistant Emily Hegranes), the U.S. National Archives, the Naval History and Heritage Center, Rick Davis, Dr. Josef Straczek, A. D. Baker III, Chris Wright, Conrad Waters, Paul Marks, and Dr. Raymond Cheung. Above all I thank my wife, Rhea, for her sustained loving support and for her valuable insights.

APPENDIX A
Surface Ship and Aircraft Sonars

INITIAL U.S. POSTWAR PLANS called for mass installation of QHB sonars, to handle fast crossing targets. These sonars did not offer the range needed to detect submarines before they could fire long-range pattern-running or homing torpedoes. The 1947 Undersea Warfare Conference was told what it would cost to increase range from the current 1,700 yds (predicted maximum of 2,150) to slightly less than 10,000 yds. Power level would have to be increased tenfold, and self-noise reduced by a similar factor. Bandwidth would be increased by a factor of 100, and echo recognition ability by a factor of 10—a net improvement of a factor of 100,000 to gain a five-fold increase in range.

The U.S. Naval Research Laboratory tried an alternative, reducing frequency to reduce attenuation in water using an 8 kHz searchlight sonar. NRL's account of the result stunned the 1949 Undersea Warfare Conference. This frequency was initially rejected for an operational sonar because it was thought to require too massive a transducer. The Navy initially chose 14 kHz.

Sangamo Electric built a scaled-up QHB that was tested on board the experimental destroyer *Witek* in 1951. It was designated SQS-4.[1] SQS-4 had a 48-stave transducer, scaled up from QHB dimensions to provide a similar beam width (15 × 15 degrees). The scan rate was reduced from the 1,750 RPM of the earlier sonar to 900 RPM (150 cycles/sec). The operator could select pulse width and had a PPI display similar to that of a QHB. In contrast to QHB, one operator tracked a target in both bearing and range, moving

The massive dome of a prototype SQS-26 sonar is installed on board the large destroyer *Willis A. Lee*, 7 June 1961. The dome was considered comparable in size to a 60-ft personnel boat. USNI

a cursor over the blip on the PPI (QHB had two separate operators). Two parallel beam-formers fed two receiving channels, initially one video and one audio. Operators typically used the video channel to track a contact and the audio channel to classify one selected from the video display. Later both channels were connected to video displays. One could concentrate on a limited sector. The sonar could be linked to a ship's fire control system, which could send it assumed target velocity. The sonar could apply that to predict target position, which could be compared to actual target position, in rate-aided tracking (and feedback to fire control).

Unlike QHB, SQS-4 could depress its receiving beams using special DMCC (dual maintenance of close contact) windings on the staves. An associated SQR-8 mechanism (introduced in 1955) could use this means of vertical beam-switching to measure target depth. Electronic beam depression made it possible to stabilize the sonar.

SQS-4 could also be used passively. A control adjusted for sound velocity, and another adjusted for ship's speed. A manual transmission switch allowed the sonar to transmit Morse code.

The first operational SQS-4 was installed in 1954, and 250 were made for surface ships and submarines. In April 1955 it was scheduled for installation on board all escorts within five years. At that time the low-frequency version was expected to have a range of 6,000–10,000 yds.

Frequency was reduced in further versions, so that ultimately the original 14 kHz version was Mod 4. Mod 1 operated at 8 kHz, the frequency NRL had tried successfully in 1949; Mod 2 at 10 kHz, and Mod 3 at 12 kHz. Mods 1 and 2 used larger transducers than Mods 3 and 4, hence 145 in. rather than 100 in. domes. The frequency difference made it possible for a group of up to four ships to use their sonars at the same time while prosecuting one submarine.

SQS-4 had a source level of 191 dB, and Mods 1 and 2 were credited with a reliable range of 5,500 yds in the layer (4,500 yds for the higher-frequency Mods 3 and 4). The sonar was therefore associated with the 1,500–3,200 yd RAT standoff weapon. In 1956, reported range (in the surface layer/below it, by a 15-kt destroyer) was 4,500/1,500 yds (contact held out to 7,500/2,500 yds). These sonars were later provided with 10,000 and 15,000 yd range scales.

To increase effective power, later versions of SQS-4 replaced the original omnidirectional pulse with focused pulses: first RDT (rotationally directed transmission), the transmitting beam rotating independent of the receiving beam. RDT increased range for a target in the layer by about 50 percent. In 1956 it was estimated that it improved typical range to 7,500 yds in the layer (2,100 yds below). This technique was employed in the SQS-29 series (with suffixes indicating the frequency). The beam was rotated by energizing the staves in sequence. These sonars had a passive capability and a built-in test unit. The next development was a steered pinging beam: SDT (steered directional transmission), in the SQS-39 through SQS-42 series (different designations indicating different frequencies). The SQS-43 series added a variable-depth fish operating off the same

The prototype SQS-23 transducer, in a bow dome, on board the destroyer *Barry*. USNI

A museum display shows SQS-23B consoles from a Royal Australian Navy *Adams*-class missile destroyer. The SQS-23 console is at right, with the corresponding NTDS console next to it and an underwater fire-control console, with its characteristic dials, to the left. The cranks below the CRT enable the operator to move a cursor to measure target range and bearing; above each is a counter. Under the CRT is a bearing counter. The selector switches above the two cursor controls are to control range scale and dwell time. Above the range scale selector on the left is a pulse length switch. On the right is a transmission control switch. The two counters atop the upper plate of the console show sound velocity and ship speed. The much larger CRT in the middle is on a standard NTDS console, which would be in the ship's CIC. The CRT would show the sonar picture, and the operator would use two joysticks to center a cursor above the sonar target indication to input its range and bearing into the ship's tactical system. CHRIS CARLSON

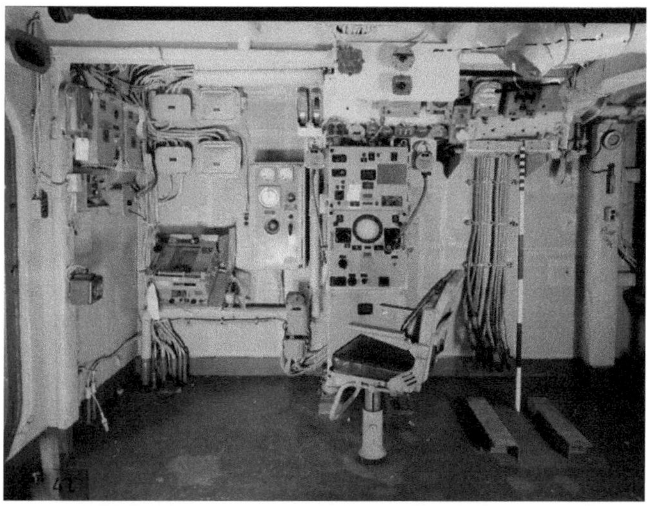

An SQS-23G control compartment shows the printer for its ASPECT feature at left. CHRIS CARLSON

sonar stack, so that a ship could ping from either her hull sonar or her variable-depth fish, but not both at the same time. All versions used the same fish, which could accommodate only a half-size transducer at the two lower frequencies but a full 48-stave transducer at the higher frequencies. The fish and hoist combination was designated SQA-10; it was first tested in 1958–59 on board the frigate *John Willis* (DE 1027). SQA-10 equipped FRAM II destroyers.

SQS-49 through SQS-52 were SQS-4 Mods 1 through 4, respectively, with MARK (maintainability and reliability kit) and SDT.

ASPECT (Acoustic Short-Pulse Echo Classification Technique, approved for service use in 1965) could be installed. Echoes from a burst of short transmissions were printed on a chemical recorder. The scope was blanked out. As its acronym implied, ASPECT could indicate the inclination of the target submarine to the sonar beam. Sangamo developed MCAR (Multi-Channel Aural Reception), in which the sonar listened on three separate beams (7.5 degrees apart) simultaneously, the object being to speed search against fast submarines.

The natural step down in frequency (for greater range) was the 5 kHz SQS-23 (SQS-23A introduced multiple operating frequencies [4.5, 5, and 5.5 kHz]). SQS-23 was generally described as a reliable 10,000 yd sonar, hence was associated with the ASROC ASW weapon. In 1966 rated range was 13,500 yds in the layer and 2,500 yds below it. Later, SQS-23 was credited with a maximum range of 14,000–15,000 yds under ideal conditions. However, this sonar had 1,000, 2,000, 5,000, 10,000, 20,000, and 40,000 yd range scales. SQS-23 was the first sonar to have ASPECT. Source level in an early version was 244 dB. The first SQS-23 was installed in 1958, and 197 were made. In addition to new construction, SQS-23 was fitted to FRAM I destroyers. This sonar was made under license in Italy and Japan.

The main U.S. Navy improvement programs were TRAM (test, reliability, and maintenance, splitting the transmitter into groups each driving eight staves, and offering rotating-beam SDT), MIP (modest improvement program, reducing CRT clutter caused by background reverberation and providing an FM transmission mode), and LORA (long-range addition, largely by self-noise reduction, including a rubber dome, to give a CZ/BB capability in the Mediterranean, for Atlantic Fleet ships only). The last was important because direct-path conditions were terrible there, particularly in the summer. Between May and the end of September surface duct operation was usually impossible, the direct path typically limited to 1,500 yds. Between late fall and early spring, conditions changed, from direct path to convergence zone and then back again to direct path. Range was typically 20,000–50,000 yds.[2]

Many SQS-23 sonars were given additional passive capability as SQQ-23s under the PADLOC (Passive-Active Detection and Localization) and PAIR (Performance and Integration Refit) programs. PADLOC began with trials using an SQS-29 and a submarine-type BQR-2B without its array. The combination had better passive performance than an SQS-29, but worse active performance. In September 1959 the Destroyer Development Group (of the Operational Development Force) began experimenting with a cross-correlation sonar, which had two transducers. By July 1962 the destroyer *Glennon* had used paired transducers to track a 9 kt snorkeler at 15,000 yds while making 27 kts, and at 25,000 yds while making 17 kts. At 6 kts she had tracked a 10 kt snorkeler at 46,000 yds. While increasing speed from 6 to 20 kts she had tracked a submarine on batteries at 8 to 12 kts. At 15 kts, she tracked a 6 kt battery submarine at periscope depth at 16,000 yds. She tracked a shallow nuclear submarine making 5–18 kts at 9,000–12,500 yds while making 10 kts. She also managed to discriminate among multiple targets.

PAIR (SQQ-23) was a solid-state active-passive version of SQS-23 with a second transducer in a separate dome, 60 ft back from the first, for passive tracking. It had rubber rather than steel domes for better sonar transmission. It also added PADLOC features. Displays showed five-ping histories. The system included a BQR-20A narrowband (LOFAR) spectrum analyzer. SQQ-23 was credited with convergence zone capability in the Mediterranean, the Norwegian Sea, and the Sea of Japan above 40 degrees North and below 40 degrees South. SQQ-23A was evaluated in FY 71 and entered service in FY 73. A later SQQ-23 version had a single sonar dome.

While the U.S. Navy pursued greater range through reducing frequency, its laboratories worked to gain range by using the refracting properties of the sea. The Navy sponsored basic sonar propagation research at the Woods Hole Oceanographic Institution (WHOI), the Scripps Institution of Oceanography in California, the Marine Physical Laboratory of the University of California, the Naval Research Laboratory (NRL), and the U.S. Navy Underwater Sound Laboratory at New London.[3] Woods Hole scientists discovered the convergence zone phenomenon.

It turned out that calculation of whether bottom bounce was practicable was complex. It was much simpler to determine whether a convergence zone existed. For any particular ocean basin that depended only on surface temperature and water depth, a sonar with low enough frequency could reach the first convergence zone with a five-degree beam depression. Limiting factors turned out to be noise- and reverberation-making biologics and surface backscattering in high winds. The Naval Underwater Sound Lab (NUSL) conducted an Acoustic, Meteorological, Oceanographic Survey (AMOS) in the North Atlantic in 1949–54, measuring among other things bottom reflection loss at various frequencies. In January 1954 an NUSL engineer was able to estimate the frequency, power, and array size needed to enable bottom-bounce operation at 40,000 yds. In August 1955 NRL reported that a downward-tilted 10 kHz sonar on board the submarine *Guavino* distinguished a target from bottom reverberation. In 2,050 fathoms, echoes could be detected at 17,500 yds.

The Naval Electronics Laboratory (NEL) became interested in the convergence zone. By early 1955, NEL was able to echo-range on submarines in the first convergence zone, 30–35 nm away (an extraordinary performance), using an experimental 2 kHz sonar on board the submarine *Baya* (AGSS-318). Within another year it was receiving echoes from the third convergence zone, 105 nm away. This was 100 times typical performance. NEL's success led directly to work by NUSL on the low-frequency submarine bow arrays of the BQQ-1 and BQQ-2 systems (see appendix B).

Early in 1955, with the SQS-23 program well underway, William A. Downes, head of the NUSL Surface Ship Sonar Department, proposed the logical next step. He seems to have assumed that it would be a bottom-bounce sonar optimized for the maximum possible range, 40,000 yds. It was assumed that really long-range sonar operation would require a specialized ship, typically called a "sonar scout." The scout did not have to be particularly fast, because it would coach killer ships onto a submarine target; 21 kts would be enough. Desired sonar range was set by expected submarine torpedo range, 10,000–20,000 yds. An initial sketch showed a 300 ft, 1,000 ton ship with a large sonar 70 ft abaft the bow, the ship drawing as much as 30 ft. The NUSL bottom-bounce proposal was presented to representatives of BuShips (which would be buying it) and NRL in September 1955. A paper giving basic considerations was presented at the November 1955 Navy Symposium of Underwater Acoustics.

In the spring of 1956 a sonar scout was tentatively included in the FY 59 budget. However, late in 1956 some in OpNav (Office of the Chief of Naval Operations) pressed for an armed prototype, fearing that Congress would reject a purely experimental ship.

Meanwhile BuShips circulated a specification for the associated sonar, which would operate in convergence zone and bottom-bounce modes. It would use Doppler filtering to help it distinguish targets from reverberation. SQS-23 was already demonstrating limited bottom-bounce capability in an area where a negative thermal gradient bent its beam enough for it to strike the bottom and reflect back. It was not large or powerful enough to take full advantage of the bottom-bounce mode.

Initially NUSL proposed a steered flat billboard array (like that of the British Type 177, of which NUSL was apparently unaware). During the spring of 1956 it switched to a fixed cylindrical array, which it argued would be more reliable. BuShips accepted this proposal in July 1956. Sometime later the bureau made the key decision that the sonar should also exploit the convergence zone, which did not depend on bottom reflectivity and the choice of the right depression angle. For the first time a sonar would exploit the properties of the full depth of the ocean.

NUSL showed that the new sonar could not be a scaled-up SQS-23. Scanning loss and waveform were both unacceptable. To overcome scanning loss, the sonar would transmit over a 40 degree sector, stepping that sector rapidly over 120 degrees. This was a form of RDT using a much broader transmitting beam. Twelve receiving beams covered the whole sector. The usual single-frequency CW pulse would perform poorly against a low-Doppler target masked by reverberation. To deal with both high- and low-Doppler targets, NUSL combined the CW pulse with a new FM element. The CW pulse would be effective against high-Doppler targets (6 kts or more; later, against lower speeds as well). The target would be detected by the Doppler shift in the echo, which would be filtered by frequency. The FM part of the pulse could be compressed to make it equivalent to a very short pulse distinguishable from surrounding reverberation. The idea of such a complex pulse (later called a coded pulse) was new.

The new sonar would receive both weak long-range and strong short-range echoes. Existing displays and receivers had limited dynamic ranges. They also lacked persistence: echoes appeared only briefly as flashes on the display (this problem had been identified by the Hartwell study). Without persistence, the operator could not compare new pings with earlier ones. The new sonar would have a storage display associated with each receiving beam that would hold the results of six ping cycles. The necessary display technology had to be developed. In existing sonars, audio search was often more effective than the PPI, since the audio beam did not have the same dynamic problem. The proposed solution was to use four audio beams, using five frequencies (two for each ear, one for both ears).

BuShips issued a formal specification for this SQS-26 in September 1957. It wanted two prototypes:

a conservative XN-1 from EDO, and a more innovative XN-2 from General Electric. EDO would concentrate on early delivery of a bottom-bounce capability. GE would also provide convergence zone capability. Contracts were let in June 1958. In February 1959 the sonar scout morphed into the *Bronstein*-class destroyer escorts after the Navy's Long-Range Objectives Group called for a minimum ocean escort equipped with SQS-26 and a means of prosecuting its contact. Two of these ships were included in the FY 60 program, initially together with an experimental escort (*Glover*, postponed to FY 61 and built to a different design). The two FY 60 ships had the initial production version of the sonar, SQS-26.[4]

The advent of Soviet nuclear submarines made the SQS-26 program urgent: nuclear submarines had to be detected at the greatest possible range. Although experiments had shown that bottom bounce was feasible against at least beam-aspect submarines, it was not known how well the big array could form beams at the envisaged low frequency, or about how clearly the mass of information the big sonar could (or should) amass could be displayed. Nor was anything known about coded waveforms.

The new sonar would impose a considerably greater load on its operator. He had to choose the best propagation path, the optimum depression angle for each, the range window, and the average sound speed to convert elapsed time into range. To do that he had to take into account sound speed down to the bottom, water depth, wind speed (surface scattering), biological scattering, bottom reflectivity, and bottom-scattering strength. He would not have a dedicated computer, only rules of thumb and tabulated data.

In 1960 the prototype XN-1 version of SQS-26 was installed on board the large destroyer *Willis A. Lee* (DL 4). In 1962 the XN-2 prototype followed on board her sister ship USS *Wilkinson* (DL 5).

XN-1 achieved bottom-bounce detection out to 30,000 yds under extremely unfavorable thermal conditions. Later it used bottom bounce to detect a beam-aspect submarine dead astern at 50,000 yds. Reverberation was a problem, and overall reliability

The first ships to carry the huge SQS-26 sonar were the *Bronstein* class; this is USS *McCloy*. Both *Bronstein* and *McCloy* had huge bow domes, with clipper bows to clear them. The bow anchor was adopted because conventional anchors would foul the dome. *McCloy* is shown as built, with a DASH hangar and helipad aft. She and her sister had 3 in./50 guns, but all later SQS-26 destroyer escorts had 5 in. guns, like destroyers. Work on their design began in 1958. Their 24 kt speed was considered inadequate in view of the predicted 30 kt speed of future nuclear submarines, but a 30 kt escort was unaffordable in the desired numbers. NHHC

The *Garcia* class (the first of which is shown off Newport, 18 August 1972) was in effect the tactical version of the *Bronstein* class, with higher speed and 5 in./38 guns. By the time of this photograph, the DASH system initially fitted had been superseded by the manned LAMPS I, hence the landing circle laid out on the helicopter pad right aft. These were still rather austere ships, without ASROC reloads (note the plumb bridge front unmarked by doors for reloads). Note too the cut in the side of the ship for an anchor abaft the bow. NHHC

was poor. In the Pacific, XN-2 was the first operational U.S. warship sonar to demonstrate convergence-zone detection. Again, reliability was poor. In the fall of 1963 the refurbished XN-2 used bottom bounce to track a submarine (at beam aspect) for a protracted time.

Given the urgent need for a long-range sonar, the Navy decided to adopt SQS-26 for new ships, even though it was still in development; it was obviously far better than SQS-23. The key operational problem seems to have been how the operator could adjust sonar operation, given the large number of variables.[5] Twelve of the first production version, SQS-26 (AX), were included in the 1961 program, and another 18 in 1962. Further versions were SQS-26(BX) and SQS-26 (CX). The prototype SQS-26(BX) was tested on board the large destroyer *Wainwright* (DLG 28) in 1966.

A 1964 estimate of SQS-26 sonar performance on two likely Atlantic convoy routes showed how bottom-bounce, convergence zone, and direct-path operating modes complemented each other. In summer, direct path and convergence zone would be good enough, though bottom-bounce performance would be poor. In winter the surface layer would be as much as 1,000 ft deep, enabling exceptional direct-path performance.

A high-level June 1964 meeting endorsed continued SQS-26 development: planned FY 64 and FY 65 procurement of SQS-26 (CX) would go ahead. The XN-2 prototype on board *Wilkinson* was modified to serve as prototype for the CX production version. An early CX on board the destroyer escort *W. S. Sims* (DE 1059) was operationally evaluated. The Operational Test and Evaluation Force (OpDevFor) tested SQS-26CX extensively.[6] Performance was sensitive to sea state, submarine depth, and relative closing speed, in that order. Test results were very dependent on operator alertness, detection ratios varying between 1 detection for 5 opportunities and 13 for 15. Sonars generally failed to reach the predicted 50 percent detection ratio for a particular range, but on the other hand, ranges generally corresponded to the 50 percent prediction based on the sonar equation. Ships were consistently able to hold contact with very evasive submarines at convergence zone range, maneuvering to keep the submarine within the annulus of the convergence zone. Probability of correct classification (as a submarine) was 95 percent, and there were only 8 false contacts per 100 hours. In low sea states submarines at periscope depth were detected at high rates (100 percent for low relative speeds, 71 percent for high). However, submarines operating below the surface layer were much more difficult to detect, with no detections at all for high relative speeds. Few test runs were made in high sea states (sea states 3–6).

Early experience with bottom-bounce detection was unfavorable. Of 103 runs, with a reflective bottom and 3,000-fathom depth, the submarine was detected by bottom bounce on only 12 of them. Submarines were not detected at all on the 29 occasions when aspect was less than 65 degrees; that is, when the submarine was not nearly beam-on. When the submarine was at the best depth to avoid detection, it was detected only once (4 percent). Probability of classification was 63 percent, and there were 11 false contacts per 100 hours. In theory, a ship had a 50 percent or better chance of detecting a submarine this way if bottom reflectivity was good, water depth was less than 3,000 fathoms, and there was no concentrated biological activity, the ship moving at less than 20 kts in moderate seas, with a wind speed of less than 15 kts. Estimated range would be about 15–25 nm for a depth of 3,000 fathoms and depression angles of 15–42 degrees. Detection was sensitive to target aspect, to target depth, and to relative speed, in that order. Favorable conditions existed about 20 percent of the time in the Northern Hemisphere. Not surprisingly, the U.S. Navy emphasized convergence zone detection over bottom bounce.

Sonar operators found their task extremely complex because they had to choose many different system parameters without computer assistance. Ultimately, a special computer using in situ data would be needed.

Even so, SQS-26 performed much better than its predecessors during a September 1965 NATO convoy exercise that included USS *McCloy* (DE 1038), one of the two SQS-26 ships. Conditions precluded convergence zone detection, but *McCloy* managed four direct-path detections at ranges between 10,000 and 22,000 yds, conforming to theoretical expectation. *Brooke* (DEG 1), which had an SQS-26(AX), made the first documented free-play (i.e., unscripted) convergence zone detections (30 nm, off San Diego) in July 1966. In January 1967 she detected a target at 26 nm in the same area, again in free play. These successes were despite the fact that the ship's sonar had not been modified to reflect lessons of ongoing work with the XN-2 prototype. For example, she had a paper recorder rather than a CRT.

SQS-26(CX) added a preformed-beam (DIMUS) passive element operating at a frequency other than the active frequency, hence usable while the sonar was pinging. As such it promised much longer detection ranges. However, destroyer crews were not used to continuously monitoring a passive sonar, so in active and passive exercises against unalerted operators, the submarine was detected passively on only 11 of 370 runs—3 percent. Probability of correct classification was only 43 percent. Predicted direct-path passive range (50 percent probability of detection) on a second-generation Soviet SSN (CVY) operating at 15 kts, the frigate also making 15 kts, was about 20,000 yds. In an exercise with alerted operators, quiet (3–5 kts) submarines at depths of 150–200 ft were detected in the convergence zone 29–32 nm away.

Convergence zone performance was successfully demonstrated in the Pacific (28 nm) and in the Mediterranean (20 nm).

SQS-26CX was also tested in the direct-path mode. Predicted range with a 200 ft layer was 23,000 yds for a 10-kt ship and 20,000 yds for a ship at 16 kts. With a layer at 400 ft, these figures increased to 29,000 and 25,000 yds, respectively. These ranges were for a 120-degree sector search; when the sonar was in omnidirectional mode, range decreased by about 30 percent. OpDevFor tests showed that performance was worst when there was no surface layer at all (average of 5,400 yds) and best for a periscope depth target in the layer (average 13,600 yds). Submarines being tracked at periscope depth could dive under the layer to evade.

Off Hawai'i on 1 October 1974, USS *Rathburne* (*Knox* class) shows the sloped bridge front that covers ASROC reloads. Her single Mk 42 5 in./54 gun mounting was intended to fire twice as fast as the earlier hand-loaded 5 in./38. In theory she had a gun battery equivalent to that of the two-gun *Bronstein*s, but the automated gun was prone to jam. By this time *Rathburne* had been refitted with a manned helicopter and a Basic Point Defense Missile System (whose box launcher is shown right aft). Note that she has an anchor on the side of her hull, rather than on her bow itself. The *Knox* class was the largest postwar escort program, amounting to 46 ships. Ten more (DE 1098–1107) were authorized in the FY 68 program, but they were dropped, six to shift funds to the follow-on DX/DXG (*Spruance* class) and the others to cover overruns in the nuclear submarine program. In the fall of 1966, a rate of 12 per year was proposed, cut to 10 by OSD. At that time construction was to have continued at least through FY 72, leading to construction of at least 96 ships rather than the 46 actually built. *Knox*-class production was capped in FY 68 (and then in FY 67) when the entire escort program was reviewed, ships being divided into those equipped with area air defense missiles (DXG) and ASW escorts (DX). Initially DX would simply have been more *Knox*-class escorts, but the *Spruance* class was built instead. NHHC

In the North Atlantic SQS-26 often failed to detect targets at convergence zone range (35–40 nm). The problem was losses in the steel sonar dome due to corrosion of its paint covering. In 1963 the B. F. Goodrich company received a contract to develop a rubber sound window in the steel sonar dome. It was first tested on board *Willis A. Lee* between July 1965 and April 1966; August 1966 measurements showed a 6 dB improvement below 12 kts and 3 dB above 20 kts. A second rubber dome was installed on board

Bradley (DE 1041) between 17 November 1971 and 20 March 1972. This time the improvement was 15 dB over the speed range between 12 and 28 kts. The greater improvement was attributed to the better quieting of the destroyer escort, whose Prairie-Masker shielded the sonar from propeller and machinery noises.[7] Late in 1972 *Knox* (DE 1052, with a rubber dome) was compared with *Kirk* (DE 1087) with the usual steel window, the submarine *Guitarro* acting as the sonar target. At higher speed (22 vs. 12–15 kts) *Knox* performed much better: her effective convergence zone was four times as wide, and her ratio of echoes to pings was 80 rather than 50 percent. The improvement due to the rubber dome was far greater than those due to all of the electronic modifications to the sonar.

In a September 1973 exercise *Belknap* (rubber dome) gained contact 41 percent of the time, as opposed to 17 and 14 percent, respectively, for the steel-dome frigates *Montgomery* and *Vreeland*. Passive performance was also much better. *Belknap* held contact with a slow SSN at twice the range (7,000–9,000 yds vs. 3,000–4,000 yds) achieved by the steel-dome *Hewes*. Not long afterward the Sixth Fleet was tracking Soviet submarines in the Mediterranean during the Yom Kippur War. SQS-26 significantly outperformed SQS-23. A rubber-dome ship achieved the longest ranges recorded at that time: *McDonnell* held passive contact at 17,500 yds and direct path active contact at 14,000 yds. A few weeks later *McDonnell*, the towed-array frigate *Hewes*, and patrol aircraft tracked a Victor-class submarine. *McDonnell* again held confirmed passive contact out to 14,000 yds and, in several unconfirmed instances, to 40,000 yds. A rubber-window ship managed to hold contact for a total of 30 hours, not far from the 36 attributed to the towed array. According to the Naval War College course material on sonar, this experience greatly encouraged installation of both towed arrays and rubber domes.

A 1973–74 study compared rubber and steel domes in a notional convoy run from Norfolk to Gibraltar. The formation took advantage of the expected 35 nm convergence zone detection range, with four escorts spaced 50 nm apart. The formation was adjusted for conditions in which convergence zone detection was impossible. In that case assumed mean detection range in shallow water would be 25,000 yds, in bottom-bounce mode 25 kts, and in the surface duct 26,000 yds. Calculations were made for 30 points along the convoy route, 21 of which would support convergence zone detection (86,000 yds). The calculation showed that convergence zone was by far the most reliable long-range path in deep water, offering the longest detection range. Bottom-bounce tracking was used about half the time either to close the target (i.e., to overcome the ambiguity due to the width of the convergence zone) or to maintain contact. In 82 percent of the deep-water points, the surface duct could be used either as primary search mode or as a backup to either convergence zone or bottom-bounce search. Often bottom-bounce search could maintain contact on a target detected in the surface duct.

It also appeared that the rubber window would dramatically widen the usable convergence zone. A steel-dome SQS-26CX could not detect deep submarines (1,200 and 1,600 ft) at all. For a submarine at 50 ft depth, the zone widened from 500 to 12,900 yds. For a submarine at 400 ft depth it widened from 500 to 9,800 yds. Maximum bottom-bounce search range increased from 36,000 to 46,000 yds, and maximum tracking range from 50,000 to 60,000 yds. Perhaps most important, the percentage of the sea bottom that could be used in sea state 3 increased from 5 to 55 percent, making bottom bounce a viable sonar mode. Mean detection range for a periscope depth direct path target increased from 20,000 to 30,000 yds.

SQS-53A, deliveries of which began in October 1972, was SQS-26CX with a digital interface, using the same transducers but with better signal processing. The Surface Ship Sonar Modification Program (SSMP) began in 1974 to counter Soviet anechoic coatings. It also addressed problems inherent in analog systems and their inability to integrate with other systems. Phase I of this program was a modified SQS-53A, engineering development of which began in 1979. It added a passive narrowband processor, the type used in submarine sonars. An interface unit made it possible to connect the sonar to a digital display, of the new standard UYQ-21 type.

SQS-53B was a fully digital sonar, a preformed-beam version of SQS-53A, hence offered multiple-target tracking (on a TWS basis only) and substantially better passive (including narrowband) performance. SQS-53B used preformed beams. In addition to an active mode, it had passive narrowband and broadband modes. Contacts could be fed into a digital computer to create a track-while-scan combination. The sonar could correlate active and passive contacts, and it could track multiple contacts. The arc over which the sonar could search was greatly widened. The same computer was fed by the ship's towed array (SQR-19) and the ship's SQQ-28 sonobuoy processor and tracker, working with her LAMPS helicopter. Tracks created by SQQ-89 were fed into the ship's Mk 116 fire control computer, which conducted submarine-like target motion analysis (TMA).

Perry-class frigates lacked either a long-range sonar or a destroyer-type fire control system. Their version of SQQ-89 used data from an SQR-19 towed array and the SQQ-28 LAMPS processor, track analysis being conducted by a weapons alternative processor (WAP). Reliability and maintainability were much improved, and time between failure was increased to 500 hours, and manning was cut by 60 percent.

Ship installations of SQQ-89 began in 1985 with the *Spruance*-class destroyer *Moosebrugger* (which also had the first SQR-19 towed array) and the *Perry*-class frigate *Curts*. Three destroyers were fitted in 1987, five more in 1988. Two frigates were fitted in 1986, and then five each in 1987 and 1988. From *San Jacinto* (CG 56) on, new *Ticonderoga*-class cruisers had SQQ-89, as did all *Burke*-class destroyers.

SQS-53C (*Arleigh Burke* class, Phase II of the SQS-53 improvement program) enjoyed better performance due in part to higher power and new transducers. The analog transmitter beamformer was replaced by a more powerful broader-band one that electronically controlled beam shape and pointing. As in a phased-array radar, multiple targets could be tracked automatically by quickly creating beams in different directions. The system's Westinghouse-developed Active Adjunct Processor detected and tracked targets, handling up to 100 active and passive target tracks. Reaction time was also much improved. During trials in USS *Stump* in August 1986 through 1989, the EDM model of SQS-53C regularly demonstrated second convergence zone capability. The ship regularly ownship speeds detected targets at what were formerly considered transit speeds. *Stump* detected targets at extended ranges in shallow water, using the sonar's improved waveforms (to overcome reverberation) and frequency agility. Earlier analog systems tend to suffer from alignment problems and offer many fewer waveform options. Reportedly SQS-53C had a stealthy (presumably spread-spectrum) signal, hence the significance of the wider bandwidth waveform.

The other significant late U.S. Cold War active sonar was SQS-56, which equipped the *Perry*-class frigates. This class was conceived specifically for large-scale production at minimum cost, the idea being that they would provide air defense for convoys whose ASW defense was provided by frigates equipped with SQS-26. The hull sonar was considered mainly a means of self-defense, the frigate being equipped with lightweight torpedoes to drive off submarines at relatively short range. Initially the preferred sonar was SQQ-23, but the Canadian SQS-505 was preferred because its performance was not much worse and it cost much less. Raytheon's private-venture sonar roughly matched the Canadian one in size, weight, and performance. It was entirely digital in both active and passive modes, operating at 5.6, 7.5, and 8.4 kHz. Raytheon produced an export version as its DE-1160.

The British and Preformed Beams

A Staff Requirement for Type 177, the first British preformed-beam sonar, was issued in 1949.[8] It became the standard British surface ship sonar of the late 1950s. Beams were formed using delay lines connected to all the sonar elements, the differences in delays imposed on their outputs corresponding to beam-forming phases. Because connecting each element to several different delay lines was cumbersome, this type of preformed beam creation was cumbersome, and only a few beams could be formed. Because delay lines were cumbersome, the British opted for a planar array with only five transmitting beams (each 10 degrees wide) and four receiving beams (covering a 40-degree sector). The array could be turned to cover different sectors. To cover different range bands, the British opted for a complex transmission: a sequence of three sub-pulses, one of 150 msec, one of 20 msec, and one of 5 msec, staggered so that pulses of a given length were sent out in quick succession. The full five-pulse transmission lasted 875 msec. The long pulse was intended to reach out to long range, and the others to shorter ranges. They were staggered above and below the main transmitting frequency. A bust of 5 msec pulses was intended to indicate target aspect. The 20 msec pulse was for target tracking and for water conditions in which it would be better than the longer pulse. Bandwidth sufficed to give target speed (relative to the sonar) within ±1 kt on the long pulse, ± 5 kts on 20 msec pulses, and ± 20 kts on the 5 msec pulses. The long pulses were passed through Doppler filters; they could handle relative speeds of up to ± 22 kts. There were 4 paper range recorders per pulse length, 3 per beam, plus 1 for a reverberation-suppressed pulse, a total of 16 paper displays. The paper range recorders gave the results of four successive transmissions, enabling the operator to estimate the rate of approach for an attack. There was also a sector display for the two middle beams, which used phase comparison to locate a target more precisely (it used all three pulses). Audio was available for all pulses. The Doppler displays showed approximate target bearing. To avoid mutual interference among ships, the sonar could operate at three nominal frequencies: 6, 7.5, and 9 kHz.

Passive signals were presented either aurally or on a special display using the Doppler display CRT.

Source level was 138 dB, which might be compared with 110.4 dB for the late-war Type 164 searchlight set operating at higher frequency (14–22 kHz). It was much lower than that of the U.S. QHB and its successors, because there was no scanning loss. Average effective range was 8,000 yds for a target above the layer and 3,000 yds for a target below the layer. Range scales were initially 5,000, 10,000, and 20,000 yds, a 30,000 yd scale being added later.

This extremely complex sonar embodied British wartime experience, which favored paper recorders over the PPIs the U.S. Navy used, and emphasized the value of the operator's ears. In 1952 the Royal Navy director of torpedo, A/S, and mine warfare thought Type 177 was at very nearly the limit of sonar performance.[9] Its main perceived limitation was its sector coverage. That was not too serious a problem as long as the targets were diesel-electric submarines.[10] A British scientific officer recalled an afternoon on board a Type 81 (Tribal-class) frigate in the early 1970s keeping contact continuously with a U.S. diesel submarine up and down the AUTEC range. Sector coverage was acceptable as long as there were numerous ASW ships.

The other great underwater problem was long-range torpedoes.[11] At the least, the British needed a sonar that could warn high-value warships so that they could evade in time. Type 177 would not be good enough, because it did not stare in all directions. Work on a torpedo detection sonar began in 1955. It was intended to detect a 45 kt torpedo at 4,000 yds, the ship moving at 18 kts (later it was credited with a 6,000 yd range under those conditions). The sonar had a secondary role: contributing to the ASW potential of the force as a whole (the first installation was planned for the carrier *Eagle*). In addition to all-round active search, it would search a trainable sector 60 degrees or more wide at longer range. The short-range search would be presented on a PPI, the longer-range search (comparable to Type 177) on a range recorder and a Doppler display. The trainable sector seems to have been conceived as an adapted but simplified Type 177. The all-round display would offer about two-thirds of Type 177 range.

The situation changed with the advent of very fast submarines. Exercises with the fast submarine HMS *Explorer* and USS *Nautilus* when she visited in 1958 convinced the British that Type 177 could not cope with their high crossing rates. The British needed all-around coverage. Their solution was to adapt the practices embodied in Type 177 and concepts developed for the all-around torpedo detector in a new Type 184 sonar.[12] It would make maximum use of the components of the earlier set: 30 percent were retained, and another 30 percent were modified. Type 184 did not match Type 177's range, because it did not revisit any sector as often, but it searched a much wider area, hence could catch a much faster target.

The new sonar would operate either at Type 177's frequency or at a lower frequency with a larger array. As in earlier British sonars, the size of the array was limited by a requirement that it be retractable through the hull. The British chose a 32-stave cylindrical array that produced 8 preformed receiving beams, each 45 degrees wide. There were two versions, one with a smaller transducer operating at either 7.5 or 9 kHz, and one with a larger transducer operating at 6 or 7.5 kHz, probably with diameters of 3 and 4 ft.

Each of three separate receiving functions had its own display: a PPI (a departure in British practice) for active all-around search, a searchlight (a focused beam over a 90-degree sector) function with a Doppler display; and a passive searchlight. Two targets could be marked and tracked simultaneously on the PPI. There was a separate torpedo warning display. The operator could listen to pings from any target, selected by a joystick. Typically Type 184 had three electronic displays side by side: a PPI, a Doppler display, and a passive (hydrophone effect) display. Like Type 177, Type 184 was considerably more complex than the contemporary U.S. SQS-23.

Like Type 177, Type 184 had a kind of RDT, in this case employing groups of 12 staves at a time. The operator could focus on a target by splitting the beam (five staves for each sub-beam, the middle staves shared by the two sub-beams). Type 184 was required to handle targets making up to 40 kts, which corresponded to ± 200 Hz at the frequencies used.

The design team assumed that nuclear submarines would soon be able to make 30 kts, which would give a very high crossing rate. Minimum echo detection range of a 30-kt nuclear submarine was set at 6,000 yds.

Fast targets required the sonar to handle large Doppler shifts, that is to accept pulses over a broader frequency range. Unfortunately, the wider the bandwidth, the less sensitive the receiver. Type 184 solved the problem by using numerous narrowband filters (10–40 Hz) and displaying their outputs separately. The Doppler channels made possible detection at greater ranges than were possible with the PPI. Initially Type 184 used the same combination of pulses of different widths as Type 177, but by 1961 it had been reduced to a single compromise pulse width (45 msec). The sonar transmitted a sequence of 32 pulses (total 1.6 sec). Under bad conditions, when maximum range was short, the sonar had an ODT mode in which it transmitted a single 45 msec pulse (blind range 210 yds).

Type 184 suffered badly from reverberation. Like Type 177, it was difficult to maintain. There were problems with vacuum tube electronics, with high-speed commutator wear, and with analog Doppler filters working at the limits of their technology, as well as internal electrical noise. Type 184 entered service in 1963 and by 1968 was fitted to eight *Leander*-class frigates as well as to five missile destroyers and two aircraft carriers. It was projected for all future surface ships. Although by 1968 Type 184 was not deemed cost-effective to modernize, so many were in service that efforts had to be made to keep it effective. Type 184M was a solid-state upgrade. In 1987 the Royal Navy began a Type 184P modernization program to improve reliability and to smooth tracking.

Type 184 and Type 177 were based on 1949 technology. The British claimed that they were the best of their type in any navy, but they would probably become more and more difficult to maintain as that technology became more and more dated.[13] The prototype of a replacement sonar would be needed about 1974, production beginning in 1977.

Type 184 seemed to perform poorly in free-play exercises, typically reaching about half its theoretical range or the range reached in sea acceptance trials, and often giving inaccurate bearings. It relied on a human operator watching a PPI. As the U.S. Navy discovered at about this time, a fleeting blip was easy to miss. The range loss could be attributed to human factors (both in the sonar room and in the operations room), to inability to keep the sonar at peak performance, and to mutual interference between ships. Computer assistance could significantly reduce the impact of both operator inattention and high data rates. Computers would rapidly compute fire control solutions to attack fleeting targets. Computer-aided performance monitoring and maintenance could help keep up performance. An ability to change frequency easily could greatly reduce mutual interference between ships. Other desired improvements were better classification and fire control accuracy and improved passive capability, particularly against torpedoes and diesel submarines. Personnel were becoming more and more expensive (and hence scarcer), so it was hoped that manning per watch could be reduced to two (possibly even one) control rating rather than the eight needed by the current combination of 177/170/176.

Transducer size, hence gain, was set by a requirement that the new sonar be able to replace existing sets in current ships. That in turn was thought to determine effective range, because transducer diameter determined gain. At best, transducer size might be doubled in a new hull outfit. The current Type 184M (3 ft diameter) could reach out to 7,500 yds. With new signal processing and data handling, that might be increased to 8,500 yds. Doubling transducer size (about the maximum for existing frigate/destroyer designs) and using a new hull outfit would increase potential range to 10,000 yds.

The British were intensely interested in shallow-water performance, which was limited by interference due to bottom reverberation. The current Type 184M was effective in 65 percent of shallow UK coastal waters (100 fathoms or less); the object was to increase that to 90 percent. In the Thames estuary almost any sonar would be good enough. In Portland and the Southwest Approaches a 3 ft diameter transducer "might be quite reasonable." However, in the North Western Approaches even a 6 ft diameter transducer would be marginal at best. A survey of British bottom conditions indicated that Type 184M would cover about 65 percent of the shallow-water area (100 fathoms or less). With new signal processing that might rise to 90 percent. Doubling transducer diameter would increase that to 95 percent and offer some chance of success in the North Western Approaches.

As of 1970, the British planned to optimize the new sonar for 20,000 yd range. A speaker at the 1973 ASW symposium referred to convergence zone capability, making contacts at 13–15 nm (26,000–30,000 yds).[14] This was Mediterranean performance. Another speaker referred to a three- to fourfold increase in the range at which a submarine had a 50 percent probability of being detected, compared with the 2,500 yd free-play range of Type 184. That would mean a reliable 10,000 yd range. The new sonar would offer automatic torpedo warning against non-acoustic torpedoes and some of the noisier homing torpedoes.

The Staff Requirement for the new Type 2016 sonar was approved in November 1970. At that time it was hoped that a pre-prototype would be on board the sonar test ship HMS *Matapan* for sea trials in 1972–73, a prototype would be ready for fitting to an operational ship in 1975, and production sets by 1977. In fact, trials on board *Matapan* lasted through 1976. The operational prototype was tested on board HMS *Broadsword* in 1978. The trials seem to have shown that the 6 ft diameter transducer was necessary.

The cylindrical transducer had 64 staves as well as an auxiliary vertical array that could rotate around the cylinder to monitor its performance, an important requirement. The array was roll-stabilized, initially hydraulically but later electronically. Like earlier British hull sonars, it could retract into a ship's hull. The transducer produced two alternative sets of beams, one nearly horizontal and one tipped down (7-degree angle) to penetrate the surface layer and for convergence zone operation in areas like the Mediterranean. As with Type 184, there were three side-by-side consoles, each with a circular CRT: one on the left for classification and tracking; one for surveillance (PPI); and one on the right for passive tracking and torpedo warning. The lack of an audio channel proved a considerable drawback in the unfamiliar waters of the South Atlantic in 1982: operators could have used it to classify contacts. The Falklands experience also

The British Type 22 frigates were conceived as *Leander* replacements with the new Type 2016 sonar. The Batch II *Sheffield*, shown, replaced the destroyer of that name sunk in the Falklands. These ships had a towed array, which they could stream from the transom. Their active sonar transducers (for Type 2016) were moved up to the bow, hence the clipper bow with anchors well forward to clear them. Like some converted *Leander*s, she was equipped with the Seawolf point defense missile system and with Exocet antiship missiles visible forward of them. The ship's mainmast carries antennas for the Anglo-American Classic Outboard communications intelligence system. They may have been lengthened to provide a long enough baseline for its hull antennas. USNI

showed that operators were confused by the rectangular displays that had replaced the circular CRTs on earlier sonars. Type 2016 operated at 5.5, 6.5, or 7.5 kHz, using 1 second FM pulses (i.e., 1,000 msec) in ODT and SDT modes.

In 1970 a total of 58 sets was envisaged, but in the end 34 were made as the British escort fleet shrank. These sonars were installed on board Type 22 and Batch 3 *Leander*-class frigates, later Type 42 class destroyers, and the *Invincible*-class carriers (replacing Type 184s).

Trials of the follow-on Type 2050 begun on board HMS *Jupiter* in 1986 were followed by sea trials on HMS *Scylla* in 1987. At that time 17 were on order. The first production unit was delivered in 1987. This set apparently uses the same transducer but has very different electronics. Reportedly it operates at 4.5–7.5 kHz. It has additional types of pulses: a long constant-frequency (CW) pulse to exploit Doppler processing (and for better performance in areas of high reverberation), as well as long and short FM pulses (with the same bandwidth) and long and short CW pulses, combined in five operating modes.[15] Type 2050 has been credited with an excellent software-based ability to reject bottom echoes. The three consoles of the earlier sonar are replaced by a single workstation with two side-by-side displays. Beam-forming and processing use the Curtis architecture developed for towed arrays.

After the Cold War the Royal Navy fitted some surviving Type 23 frigates with Type 2087, an integrated sonar that combined Type 2050 with a low-frequency adjunct to the ships' towed arrays. This choice presumably reflected the new situation, in which the threat shifted to diesel submarines that would not produce much of a passive signature.

Beginning in 1968, AUWE argued for a very different sonar for the 1980s or 1990s that would fully exploit bottom-bounce propagation. It would need a huge transducer (in 1970 a minimum diameter of 70 ft was cited to enter the "bottom-bounce club"). Unlike convergence zone, bottom bounce offered continuous coverage over a wide zone. Ranges might be 10–25 nm, depending on water depth and bottom conditions, far beyond duct (i.e., direct-path) sonars. Based on depth estimates, half the area of the ocean would support bottom bounce (later it was claimed that a combination of bottom bounce and convergence zone would cover 80 percent of the ocean). A bottom-bounce ship would have two flank arrays, leaving 20-degree blank arcs around her bow and stern. AUWE called this the high-grade sonar. It could operate in convergence-zone mode but did not have to.

The project was likely to be quite expensive, so the British looked for partners. Only the U.S. Navy was interested. The French wanted to concentrate on the Reliable Acoustic Path optimized for the Mediterranean and the eastern Atlantic but marginal elsewhere because the body involved would have to be towed very deep, down to 10,000 ft or more. The other potential partner, Canada, was also concentrating on RAP because it did not want to duplicate U.S. and UK work on bottom bounce.

Asked how the proposed system would compare with SQS-26, in 1970 AUWE's Dr. A. P. Gordon said that SQS-26 had only a very restricted bottom-bounce capability: "The Americans say that it is not good enough, which is why they are putting money into our project." He considered the 16 ft SQS-26 transducer too small. Gordon explained that the Royal Navy was not buying SQS-26 because it was too large to fit in British ships, too expensive (£1.5 million, compared to a desired price for Type 2016 of £300,000 and an expected price of £650,000), and lacked the desired computer data-handling facility.

AUWE envisaged an array 150 ft long, projecting a beam 140 degrees wide. The test ship was the destroyer *Matapan*, completed at the end of the war but never commissioned. The U.S. Navy was interested in conformal sonar partly because its SQS-26 suffered from quenching as a pitching ship lifted a bow transducer out of the water. Replacing the big bow sonar would make it possible to optimize a hull for speed and seakeeping. The Americans built the transmitter, the British the receiver. For a time the joint UK-U.S. program included a second test ship, to be converted from the U.S. light cruiser *Spokane* (SCB 191), but she presumably fell victim to the cost of the Vietnam War. Initially AUWE expected that a prototype would go to sea in 1974, followed by a true prototype, with an operational system in service on board large destroyers in the 1980s. By 1973 the time scale had stretched out, bottom bounce being seen as an option for the 1990s rather than the 1980s. Tests were completed in 1977, but the big ASW destroyers were never built.

VDS

The surface layer limited the range of any hull sonar. It seemed that a dipping sonar (variable depth sonar, or VDS) was the way to defeat submarines that could dive below the duct and thus reduce detection range to 1.5–2.5 nm. However, VDS sonars generally operated at lower source levels. Because their towed bodies were relatively small, they generally operated at higher frequencies and lost more of their energy to attenuation.

The Royal Canadian Navy seems to have been the first Western navy to try a VDS. The navy wanted to exploit a sound channel lying between the warm surface layer and the warm bottom layer on the Scotian Shelf off the Canadian east coast. Conventional sonar was often useless there. The Canadians seem to have begun VDS experiments in 1947, followed by their "Dunker" project in April 1949. Trials with an adapted British Type 144 on board the minesweeper *New Liskeard* showed that a sonar in the axis of the channel could detect a submarine there at 9,400 yds, both being stopped. Even with both at the upper edge of the channel, range was 4,200 yds, far beyond what a hull-mounted Type 144 could achieve. Through the early 1950s the Canadians worked to produce a viable towed sonar body, testing their CAST 1/X on board HMCS *Crusader*. In 1958 CAST 1/X was tested off Bermuda, off New London, on the Scotian Shelf, and in the Gulf of St. Lawrence. It outperformed a hull-mounted Type 144 by a factor of five.

HMCS *St. Laurent* shows her SQS-504 variable-depth sonar. RCN

CAST 1/X was reengineered as SQS-504: its sonar was a modified version of the British Type 170 towed in a body based on the British 100-in. sonar dome. Work began in 1956, the first full prototype being completed in 1961. SQS-504 operated at fairly high frequency (9.5, 11, 12.5, or 14 kHz), hence had limited range (it averaged 6,000 yds although it had 5,000, 10,000, and 20,000 yd range scales).[16] Production sets were planned for the midlife modernization of the *St. Laurent*–class frigates. Their only drawback was their handling gear. Lessons learned in the get-well program were applied to the SQS-505 VDS element, which used the same scanning transducer as the SQS-505 hull sonar. It was claimed that the VDS offered 20 times the overall search capability of

HMCS *Gatineau* shows the big fish of her SQS-505 variable-depth sonar, August 1988. USNI

SQS-504; in early trials it achieved a range of 27,500 yds, compared with 5,000–7,000 yds for SQS-504.

The British tested a VDS roughly equivalent to Type 177 under the designation Beta (Type 192). They found that SQS-504 (which they designated Type 199) outperformed Type 192 despite being smaller. Even so, the British were unhappy with Type 199 after adopting it. They considered modernizing it with a cylindrical transducer in an improved towed body. When that project was abandoned, they began looking at alternatives, the only attractive one being the U.S. SQS-35, which was too expensive. A project for an entirely new VDS was abortive.

The U.S. Navy became interested in variable-depth sonars in the late 1940s. Work was suspended during the Korean War, then revived later. In 1955 one Atlantic Fleet destroyer escort and three Atlantic Fleet DDEs were fitted with the SQA-5 hoist for a 25 kHz transducer. It was unsuccessful, but interest in a VDS adjunct to the SQS-4 series led to two alternative hoists, SQA-8 and SQA-10. The latter was widely installed in FRAM II destroyers in the 1960s. It operated off the same electronic "stack" as the hull sonar. In the early 1960s the U.S. Navy became interested in a fully integrated sonar provisionally designated SQQ-20, but never fully defined. It would have employed a VDS fish the size of an SQS-23 transducer to fill the shadow region beneath its hull and to penetrate the surface layer. It seems to have died about 1964.

Twenty-five *Knox*-class frigates (DE 1052, later FF 1052 class) were fitted with independent VDS (IVDS), meaning that they operated independently of the hull sonar. Work on the sonar experimental ship *Glover* was pressed ahead partly to learn to what extent the IVDS would interfere with the ship's hull sonar. Apparently, all ships of the class had been designed for such installation, with their torpedo tube spaces right aft adapted for it (the torpedo tubes were not fitted). The VDS was EDO's SQS-35, derived from earlier hull sonars. EDO later sold variable-depth sonars to

The SQA-10 variable depth sonar fish on board the FRAM II destroyer *John W. Thomason*, the lead ship of the class, 4 January 1960. NHHC

The fish of the SQS-35 variable-depth sonar on its hoist on board a *Knox*-class frigate. This sonar was developed by the EDO company under a 1966 contract, based on its earlier SQS-17. It operated at 13 kHz, average detection range being 5,000 yds. EDO went on to develop variable-depth sonars that, it pointed out, would not be quenched by the motion of a small ship in rough weather. USNI

many navies. It argued that, because it did not share hull motion, a VDS "fish" offered superior sonar performance to a small ship. Developed in 1966, SQS-35 may have been the first U.S. preformed-beam sonar, with 12 beams. It operated at 13 kHz. Range scales were 1,000, 2,000, 4,000, 8,000, and 16,000 yds. For 50 percent probability of detection of a submarine at optimum detection depth, with sea state 3, wind speed 13 kts, normal biological scattering, mild below-layer gradient, and average target strength, predicted range was 7,200 yds at a ship speed of 10 kts (6,200 yds at 20 kts) with a layer depth of 100 ft, rising to 9,000/6,500 yds with a 300 ft layer and 9,600/7,000 yds with a 500 ft layer. These ranges were for the ship operating her Prairie-Masker silencer; without it ranges would be somewhat longer (and less affected by ship speed) because Prairie-Masker created noise abaft the ship.[17] The SQS-35 fish later became the depressor for U.S. SQR-18 passive towed arrays.

Helicopter Sonars

Despite their limited range, helicopter sonars were initially attractive as a way of extending a sonar screen well beyond ships. They also had some of the attributes of VDS sonars. The major Western Cold War developers were the United States and the United Kingdom.

After rejecting its first helicopter dipping sonar (AQS-1) as too heavy, the U.S. Navy adopted the AQS-4 searchlight sonar, with a 30 degree beam, its transducer in a rubber ball on a 70 ft cable. It operated in the same high-frequency band as contemporary shipboard sonars (20–22 kHz). To search, the transducer automatically made 10 degree steps. Average time for a complete 360-degree sweep was 6 minutes. Once a target was detected, the transducer was hand-trained. Power was only a fifth of that intended, so range was typically 1,900 yds or less rather than the planned 4,700 yds. The display was the usual paper range recorder. Typically the helicopter hovered 10 ft above water, with the transducer 20–60 ft below. An initial contract for 152 units was increased by 600 under the third supplemental FY 51 program, part of the Korean War mobilization. AQS-5 was an AQS-4 with a much longer (200 ft) cable and an underwater communications feature. Royal Navy representatives who participated in the U.S. AQS-4 trials in 1951 thought U.S. acceptance of AQS-4 was premature, that the Americans were "trying to run before they can walk."[18]

The successor AQS-10 (1955) was a 10 kHz scanning sonar offering two to six times the area search rate of AQS-4, its dipping body a cylinder rather than the ball of the earlier sonar. It had a 210 ft cable. Average range was 7,500/4,000 yds. AQS-10 became operational in 1959 on HSS-1N and HSS-2 (later designated SH-3A and SH-3B) helicopters. In 1957 NRL proposed a drastic frequency reduction (to 2 kHz) in its proposed LOMASS-2 dipping or blimp-towed sonar, promising detection at 12,000 yds in the layer or 3,200 below it (15,000/3,500 yds if towed by an airship). This performance would be achieved at an average speed of advance of 35 kts, compared with 11 for AQS-4. Nothing came of the idea. Instead, AQS-10

An SH-60F carrier inner-zone defense helicopter dips its AQS-13 sonar. This version of the SH-60 replaced the SH-3 Sea King.
USNI

was developed into AQS-13 (1960), the standard late Cold War helicopter dipper.

AQS-13 had a longer cable (500 ft) and was reeled faster (13 vs. 5 ft/sec), for a faster dipping cycle. The initial version operated at 9.2–10.75 kHz; range scales were 1,000, 3,000, 5,000, 8,000, 12,000, and 20,000 yds. The sonar scanned on reception, emitting a pulse 180 degrees wide. The dip cycle was 2–3 minutes, the helicopter hovering at 50 ft and the transducer 450 ft below the surface. Typical range was 4,000 yds (range scales were 1,000, 3000, 5,000, 8,000, 12,000, and 20,000 yds). AQS-13B (1969) had 16 preformed beams. AQS-13F, which had a sonobuoy interface, was adopted for the SH-60F carrier ASW helicopter. Commanders initially disliked AQS-13; in 1968 ASWFORLANT reported that the combination of design errors in the sonar and navigational errors by the helicopters made for only marginal capability at the upper end of the 8,000 yd scale. The report noted, "Operational commanders cannot rely upon the accuracy of helicopter generated contacts beyond 6 kyds from the reporting helicopter." That made it impossible to rely on the helicopters for screening.

Under MDAP the British received (presumably having requested) a squadron of HO4S-3 (S-55) Whirlwind helicopters equipped with AQS-4. The British experimented with them for 16 months in the Mediterranean and then in home waters. Like its U.S. counterpart, the Whirlwind was initially a search-and-rescue helicopter. The prototype British license-produced HAS 7 first flew on 17 October 1956. Whirlwinds were initially equipped with transferred AQS-4s, which became less and less serviceable as spares ran out. The British Anglicized AQS-4 as their Type 194, which entered service in 1959. It was credited with an effective range of 2,500–3,000 yds.[19]

The Whirlwind could operate effectively in winds up to 45 kts and in sea heights of up to 25 ft, but it could not safely hover at night. It could carry either the sonar or a Mk 30 torpedo, but not both, and changing from one to the other took two to three hours. Given its very short sonar range, it had to jump frequently from dip to dip, making navigation (usually by dead reckoning) difficult at best.

As of 1958, plans called for replacing the Whirlwind with a British version of the larger and more powerful S-58 (the Wessex, powered by a gas turbine rather than the piston engine of the U.S. S-58/HSS-1), which would carry a new Type 195 dipping sonar with a range of 6,000 yds. The new helicopter would also carry a U.S. lightweight torpedo, initially Mk 43-3 and later Mk 44. It would be capable of flying in all weather and at night. Wessex HAS 1 entered service in 1961, well before the new dipping sonar was available. It had no radar and no onboard plotting facility. Although in theory it could carry the Mk 44 torpedo, the loss of endurance involved was deemed unacceptable.

A Staff Requirement for a replacement Mk 3 version formulated in 1960 called for the ability to carry the airborne ASW systems then under development, such as a new dipping sonar, and two Mk 44 torpedoes. This reengined Mk 1 would enter service in 1964/65 as an interim ASW helicopter, to be replaced in 1972 by a new one. It was considered unacceptable for the 1970s. Development was protracted, so that Mk 3 did not enter service until 1967. It was not canceled because a contract for 40 aircraft had already been let.

The delay may have been partly due to lagging Type 195 development. In 1958 it was expected in 1963, and apparently the schedule for the follow-on version of the Wessex was keyed to that. Type 195 was introduced in 1966 (presumably approved that year); the first helicopter to carry it, Wessex HAS 3, entered service in 1967.

Mk 3 had an integrated ASW system developed concurrently with the Type 195 dipping sonar: a lightweight radar, an automatic plotting board, and the Blue Orchid Doppler navigator. The British considered this combination a revolutionary advance, the world's first automatic radar/navigation plot. It was integrated with the sonar and ground-stabilized by Doppler. It enabled the helicopter to maintain its correct position in a screen, to navigate more accurately and safely, and to control other units. The helicopter now had a system comparable with the combat direction system on board an escort, on which the earlier ASW helicopters had relied for much of their direction and control.

By 1968 Type 195 was demonstrating a reliable range of 6,000–8,000 yds, and double that in good conditions: a threefold or more factor of improvement over Type 194, not counting the advantage that it was a scanning (preformed beam) sonar rather than a searchlight. As in Type 177, the staves were excited in turn to sweep the beam across the sector. Three receiving beams were created by delay-line networks. The 90-degree scan was combined with better displays and greatly increased sonar search depth. It offered a search and detection capability comparable to that of a surface ship with the sector-scan Type 177, as well as the advantage of a VDS. Type 195 was designed for a 6,000 yd range under average conditions (3,000 yds against a target beneath the layer, but 8,000 yds if both transducer and submarine were beneath the layer). In 1967 the British thought that the only comparable helicopter sonar in the world was the U.S. AQS-13, which lagged behind the British in development, had not yet been proven, and at best was unlikely to have any marked advantage (this was part of a justification for the British project). Type 195 production ended in 1985 with the Type 195M version. Type 2069 was an upgraded Type 195M operating at greater depth (1,000 ft) and integrated with an AQS-902 sonobuoy processor.

HAS Mk 3 was considered no more than an interim helicopter, because it lacked performance. The British decided that the U.S. SH-3D Sea King offered the most advantages within a reasonable cost and timetable. As a twin-engine helicopter, it could survive an engine failure under most conditions. It could carry four weapons; it was estimated that with four torpedoes it had 1.85 times the kill probability of a two-torpedo Wessex 3. Its higher speed reduced time late at a datum, increasing its detection probability from 15 to 27 percent at a 40,000 yd datum. At this time the Royal Navy was experimenting with SSNs in direct support, offering sonar ranges as great as 19 nm. Helicopters were ideally suited to running down such contacts. The faster the helicopter, the shorter the time getting there. The greater weapon load was significant when distance from surface units precluded rapid reinforcement. In 1968 the weapon load was four Mk 44 torpedoes or two or three NASR 7511 (Stingray) or two Mk 44 or NASR 7511s and one WE 177 nuclear depth bomb.

Such considerations were again significant when the Royal Navy introduced a towed array in the Type 23 frigate; it needed an integral means of running down a contact. The Type 23 project was therefore linked directly to a project to buy a long-range helicopter, ultimately the EH-101 Merlin. This helicopter uses a new lower-frequency FLASH dipping sonar, a version of which also equips the U.S. SH-60R.

Sonobuoys

Sonobuoys were the equivalent of sonars for fixed-wing aircraft and, often, for helicopters. After World War II both the U.S. Navy and the Royal Navy used remaining stocks of wartime U.S.-built sonobuoys, which were retrieved and repaired after use. These buoys operated at sonic (audible) frequencies, which limited their effective range. Operators' ears were the processors. The operator compared the loudness of sound transmitted by different buoys to decide the submarine's position and heading. The monitoring airplane had to fly directly over a buoy to register its location, using an "on-top" indicator. As noted in the 1950 Hartwell study these techniques were hardly likely to be entirely effective. Some graphic way of indicating that a sonobuoy had picked up usable sound was needed.

By 1950 the wartime buoys had worn out, so both navies sought replacements. The United States produced a version of the wartime directional buoy (CRT-4) as SSQ-1. A new omnidirectional buoy was developed as SSQ-2. The first acceptable SSQ-2 was delivered in 1951, SSQ-2B becoming the first U.S. mass-production sonobuoy.[20] By 1956 an improved SSQ-2 could report on 16 radio channels, meaning that in theory an airplane could operate as many as 16 simultaneously. Development of SSQ-1 proved difficult, so in 1954–56 the U.S. Navy bought 3,500 of an Americanized version of the British T-1946 as

Sonobuoys are loaded onto an S2F Tracker. The Tracker had a very unusual form of buoy stowage: in its nacelles to minimize fuselage space required. The objects atop the buoys are rotochutes. The rotochute would deploy at launching and rotate to slow the buoy's descent. Holes in the sides of the buoy covered switches that could set parameters such as hydrophone depth. All of the buoys are the standard "A" size. The Tracker could also carry four larger "B" size buoys of British type, but by the time the aircraft were in service the U.S. Navy was no longer using adapted British buoys. USNI

SSQ-20. In theory a submarine would be detected initially by omnidirectional buoys, then localized using the more expensive directional ones.

Some means of ranging was needed. The first solution was the SSQ-15 active pinger, which appeared in 1952. It was succeeded by explosive echo-ranging (Julie). Given problems with Julie, the range-only SSQ-15 was revived and was in active service by early 1961. Because the SSQ-15 was omnidirectional, the submarine position indicted by ranges from two buoys was ambiguous; a third was needed. This buoy was command-activated but not turned off by command, so it had a short operating life.

By 1961 an AQA-1 airborne buoy processor was in service. It demodulated the VHF radio outputs of two buoys and displayed the results on a screen, which was switchable between SSQ-1D, SSQ-15, and SSQ-20 inputs. When an operator placed a cursor on the indicated output, it was translated onto a CRT functioning as a plotter. For Julie, two SSQ-1D buoys were used, the operator distinguishing between the initial direct sound of the explosion, the echo off the bottom, and the echo off the target submarine. Data from two buoys could, in theory, place the submarine at one of two positions at which range signals from the two buoys intersected. The tracking airplane could resolve that ambiguity by dropping a third buoy.

Airborne LOFAR (Jezebel) required a new processor (AQA-3) but not new buoys. The display was a chemical recorder, whose stylus darkened the paper in proportion to the strength of the signal at various frequencies. As the paper unrolled, the operator could see areas corresponding to broadband sound and lines corresponding to narrowband frequencies. The sweep rate of the stylus across the paper corresponded to the rate at which the signal was analyzed. This was not simple to arrange, because the filters could not be applied in sequence; it would have been impossible to follow changing signals. Instead, the incoming signal was recorded and transferred to an auxiliary analog memory which could be rapidly scanned. AQA-3 and its successors could also apply CODAR analysis. AQA-3 could handle up to four buoys: four LOFAR or two CODAR; AQA-3A could handle two LOFAR and two CODAR. It employed analog time compression for frequency analysis, at 10–200 Hz. AQA-4 was a similar analog processor, on board some P-3s and the S-2E. Like AQA-3A, it could handle two LOFAR and two CODAR or four LOFAR channels. It was similar to AQA-3 but used a single internal band to handle both broadband and narrowband analysis. It could process data from DIFAR buoys only as omnidirectional. Buoys capable of detecting low-frequency sound were called Julie-Jezebel, because their outputs could be processed for LOFAR.

AQA-3 could not be accommodated on board carrier-based S2F Trackers. Instead they were given a relay system that sent sonobuoy data to the parent aircraft carrier (CVS) for processing.[21] The enlarged S2F-3 (S-2D) could accommodate the processor, though carriers had the capacity to receive and process their sonobuoy data.

Remarkably, despite the need to protect the LOFAR secret embodied in SOSUS, LOFAR processors were released to NATO by about 1964–65. The related SOSUS technology was not. It is not clear whether Neptunes in NATO service received LOFAR processors. They certainly did appear on board NATO P-3B Orions, and on board next-generation NATO ASW aircraft.

AQA-5 employed a DELTIC (delay-line time compression) solid-state compression technique, which offered a higher compression ratio and hence better frequency resolution—that is, a greater capacity to pick a submarine out of surrounding noise. It could handle twice as many buoys. AQA-5 was introduced in the 111th U.S. P-3A and then in U.S. P-3Bs (and the export versions) as well as the British Nimrod, the French Atlantic, and the Japanese PS-1 and PS-2. Modified versions were made for Australia, Britain, and Germany. The frequency band was 80–2,400 Hz. AQA-5 could provide graphic displays of up to four buoys, and audible analysis for eight. The Atlantic had two AQA-5s, so it could print out data from up to eight buoys.

The big break with the past was the digital AQA-7, which could process DIFAR buoys. It fit the same space as the earlier AQA-5 (but AQA-5 could not be modified into it). The specification demanded an accuracy of 15 degrees, but in 1967 prototypes were showing an estimated accuracy of 2–5 degrees. AQA-7 could handle eight LOFAR analyses (four on paper, four on CRTs) or six DIFAR (two on CRT, four on paper) or two CASS (CRT only) or two DICASS (CRT only). The frequency band was 10–2,400 Hz.

In connection with their own decision to adopt the directional Barra buoy, the British later claimed that as late as 1973 the U.S. Navy was fully occupied in solving DIFAR problems of insufficient sensitivity and directional accuracy. U.S. documents of this period do not seem to show any dissatisfaction with DIFAR.[22]

A P-3C shows the ports of a mass of preloaded sonobuoy chutes abaft its wing. The airplane's computer system (including a flight director) could shape a path over the desired field location and automatically drop buoys as required. In all previous large ASW aircraft, crewmembers loaded a few chutes from inside the airplane. It relied on an "on top" indicator to know when it was above any particular buoy, so that buoy position could be plotted. The P-3C carried additional buoys and had chutes that could be loaded from inside, but the preloaded chutes were primary. Note the Maverick surface-attack missile on the port pylon, outboard of the outer turboprop; this was a post–Cold War fitting. The small turret under the trailing edge of the wing was for electro-optical sensors used in Afghanistan. U.S. NAVY

Fully electronic processing was implemented in the S-3 Viking carrier ASW airplane, which had an OL-82 system. Its tactical suite had been conceived as a smaller version of that of the P-3C, benefiting to some extent by better automation and a faster computer. Given more automated sonobuoy processing, the two acoustic operators of a P-3C could be replaced by one. The airplane retained the capacity to monitor 16 buoys simultaneously. Targets were detected primarily by automatic line integration (ALI), which was not quite the same as visual integration of a lofargram. In some cases such integration would not bring up a discontinuous or changing line that might be evident from visual examination. Signatures were automatically identified using a library. This technique was probably not altogether successful (mainly automatic detection has not generally been adopted in other ASW systems), but it was unavoidable in a small airplane.

The U.S. Navy continued to develop active buoys. SSQ-47 was a deep-operating version of the command-activated SSQ-15. SSQ-50 could be set even deeper (1,500 rather than 800 ft). The active buoys could be used, for example, to refine LOFAR data. The next step was a directional command-activated buoy, DICASS (SSQ-62), which could use an FM waveform to determine a submarine's speed. It was developed after a 1968 feasibility contract was let to Sparton. Much more elaborate buoys were developed late in the Cold War, such as ERAPS and VLAD.

Unlike the Americans, the British assumed that the Soviets would deploy the silencing techniques the Royal Navy was developing. That left them with propeller cavitation, which could be detected only at relatively short range. Acoustic performance had to be good enough for a single operator to classify the sound and monitor a 12-buoy barrier quickly enough to be sure of hearing a submarine passing through the barrier. The initial British nondirectional buoy (T.1945) was considerably larger than its U.S. counterpart (diameter 10 in. and length 50 in.), compared with the U.S. A-size (4 ⅞ in. × 36 in.).

The companion directional buoy (T.1946) used a rotating searchlight, the rate of rotation being limited by the operator's ability to sense and classify the sound (1.5–2 sec). A full rotation would take from 18 to 32 seconds, the hydrophone having an effective beam width of 30 degrees. Given sonobuoy dimensions, maximum hydrophone diameter was 12 in. (folded up in the buoy). Better range (due to lower frequency) was traded off against higher gain (due to shorter wavelength), the British selecting 12 kHz as a good compromise. Hydrophone depth was set at 40 ft to place it in the layer. The hydrophone (and the entire buoy) was turned by a small motor. Initial technical trials were carried out in August 1949. The buoy could detect a 3 kt submarine at 5,000 yds and an 8 kt submarine at 8,000. In practice (1954) a 4 kt submarine could typically be detected at 5,000 yds, and sometimes at 9,000. The limiting factor in tracking in higher sea states was the marine marker, which became difficult to keep in sight in sea states 4 to 5. The U.S. Navy bought T.1946 as an interim directional buoy.

In 1949 the Royal Navy operated the Firefly AS Mk 5 because the projected Gannet carrier ASW airplane was not yet available. It could monitor only the American nondirectional buoy. A new Firefly AS Mk 6 was needed to monitor the Mk 1 version of the British directional buoy. It had one sonobuoy receiver and indicator; the follow-on Firefly AS Mk 7 had two sonobuoy receivers (it had a crew of three rather than two). The Gannet similarly had two receivers. Unfortunately, the Firefly was not considered suitable for night deck landing, a problem solved when the Royal Navy received Grumman Avengers under MDAP.

By 1954 the standard British directional sonobuoy receiver used a CRT to provide both a visual and an aural indication. It was so effective that the emphasis in training shifted from nondirectional to directional buoys. Nondirectional buoys were to be relegated to barriers, containing circles, and to investigate contacts and unconfirmed sightings.

In 1951 there was an ongoing project to develop an Anglo-American active buoy for prototype trials in 1955. With it, Sonics Mk 1 would become Mk 1B. Owing to delays in the active buoy program, it was decided to keep using Mk 1. A new Mk 1C would use both passive and active buoys. In 1955 it was expected to begin to enter service in 1959. In May 1955 the prototype TX 1901 active buoy ran successful trials in the Mediterranean, achieving a maximum range of 3,600 yds and an average of 1,900 yds. In 1958, looking ahead to introduction of Sonics Mk 1C in 1959, expected range under average conditions was only 1,500 yds. At that time the best hope of greater range seemed to be either an improved active buoy or explosive echo-ranging.

RAF Coastal Command used the same buoys. In 1949 the new Shackleton maritime patrol aircraft was expected to accommodate two Pentane torpedoes and 30 sonobuoys plus marine markers and radar transponder buoys within a 6,000 lb payload. Coastal Command wanted to be able to locate a submarine within a radius of 1,000 yds. To use directional buoys (i.e., to use cross-bearings) the airplane needed a large chart table with seats for three operators. The buoy load planned for the Shackleton was based on a 20-minute run against a 12 kt submarine. Five different buoy patterns were developed, each requiring six to eight directional buoys plus the same number of marine markers, since the pilot had to know where the buoys were. Hence the bomb bay was to house eight buoys for each Pentane homing torpedo. For each such pattern an additional nondirectional buoy would be provided. Extra nondirectional buoys might be carried outside the bomb bay for searches.[23] Early trials showed the importance of automatic plotting to help visualize the complex tactical situation presented by multiple buoys; an automatic plotter was successfully tested in a Shackleton in 1954. The plotter was a glass table with two projectors underneath, synchronized with the buoy receivers. It did away with manual plotting of sonobuoy directions. It was effective enough that the attack team in the airplane could be reduced to one operator to monitor the two indicators and one tactical navigator. Mk 1C entered service in 1960.

By 1968 the passive component of the Mk 1C system was credited with a range of 9,000 yds, but in practice the maximum was about 1,000 yds for every knot the submarine made above silent speed. In theory the active buoy had a 3,000 yd range, but on average range was 1,500–2,500 yds. The active element of the system was tested in 1968 at the AUTEC range, in near-perfect conditions in deep (6,000 ft) water, with specially trained operators. It confirmed that active range was 3,000 yds at best. The probability of acquiring a target was best at a range of 600 yds (0.49), and was only 0.1 at 3,000 yds. The data rate was only one fix in 50 sec. Buoy reliability was described as indifferent. Mk 1C demanded a high level of operator expertise. It also needed close interaction between the sonics operator, the tactical navigator, and the pilot. Buoys were dropped near a contact datum. The probability of relocating a diesel submarine from a 5,000 yd datum was at best about 0.27, and less against an SSN exceeding 15 kts. The probability of placing a buoy within acquisition range of a target was poor, and the data rate was low. Overall performance was marginal against a submarine making over 10 kts, and submarines already in service had silent speeds of about that. Even against a noisy target, the probability of kill declined rapidly for a submarine making 15 kts or more.

Nimrod Mk 1, the Shackleton successor, had Jezebel in addition to the Sonics Mk 1C of the earlier airplane. The British had been informed of the project at the outset and had discussed its potential. For the first time Jezebel provided long-range search and classification, but not localization, hence retention of the earlier system. In 1968 the Air Staff pointed out that it was suited mainly to deep water, its performance falling off in shallower water and in areas of high shipping density, both of which were the case in large areas around the British Isles, Scandinavia, Western Europe, and the Mediterranean. The RAF wanted either a better passive system or a better active one, or preferably both, for the coming Nimrod Mk 2. In January 1968 an RAE (Farnborough) working party concluded that there was a clear case for a dual-mode active/passive buoy that could fix and attack both quiet slow submarines and fast noisy ones, including SSNs.[24]

The Nimrod Mk 1 acoustic system was a compromise because the desired airplane could not be developed in time to meet the full requirement; the Mk 2 project was an attempt to cure some of its deficiencies.[25] By 1971 the British considered that they had to adopt an off-the-shelf directional buoy and a new processor.[26] They hoped that a passive directional system could locate quieter submarines 10 dB below current capability, enabling a Nimrod with new homing torpedoes (Mk 46 or NASR 7511, which became Stingray) to contain and attack them despite speeds up to 30 kts, compared with the current 15. The new long-range buoy should be accompanied by a new attack sonobuoy system needed to deal with modern diesel submarines running on battery. The analog AQA-5 of the initial Nimrod was to be replaced by a fully flexible digital system that could provide an efficient information flow to a tactical navigation system (which Nimrod already had). It should provide enough automatic processing capacity to fully exploit future buoys. AQA-5 was limited to 10–300 Hz, but it was known that SSNs emitted detectable noise up to at least 2,400 Hz. Extending the ability to do line analysis in the broader band would enhance Nimrod performance against current Soviet submarines, though not against future quieted ones. The British also wanted higher gain. In 1973 they envisaged a buoy operating over 10–2,000 Hz, adapted to a modern digital processor, which might offer a system gain of 2 dB in the 10–300 Hz band. CODAR was increasingly seen as ineffective, so the British wanted a directional LOFAR buoy.[27] They rejected the U.S. DIFAR buoy as offering insufficient gain.

The Australians and Canadians were considering buying more advanced avionics, but both had budget problems. The Canadians were interested in developing an acoustic processor on a shared-cost basis. In 1973 the British apparently discounted AQA-7 in favor of something like the future U.S. Advanced Processing System (UYS-1), which they thought was well behind schedule because effort was being concentrated on curing DIFAR problems. Of the two U.S. projects for active attack buoys, CASS had failed its operational evaluation and DICASS was still under development, with problems similar to those of DIFAR. The U.S. Navy had a future array buoy, MAPS, in early development. It is impossible to say whether these evaluations were simply intended to support the decision, already taken, to buy Barra.

The British wanted an unambiguous bearing accuracy of 5 degrees. Their projected active buoy was to operate in sea states up to 5 to have a probability of detection of at least 50 percent on a 15 dB return from a submarine at 5,000 yds range. The buoy might have to operate as much as 100 nm from the monitoring aircraft (at 8,000 ft). The British decided to invest in (and adopt) the ongoing Australian Barra program. They chose a home-grown GEC Avionics AQS-901 processor over the Canadian UYS-503; the Australians followed suit. Two such computers on an airplane could process eight Barra buoys. Barra employed an array of 25 hydrophones in a horizontal array to create 60 beams. The Australian version was SSQ-801; buoys made by Plessey in the UK were SSQ-981.

The candidate active attack buoys were CASS, DICASS, the French DSTA 3, and the British CAMBS. Compared with the U.S. DICASS, CAMBS was expected to be less expensive (£1,000 vs. £1,400; CASS cost £1,260). Because CASS and DSTA 3 were omnidirectional, three were needed to fix a target. Only CASS used FM radio. Low power output generally precluded Doppler resolution below 12.5 Hz, with corresponding reduction of target localization capability. The U.S. CASS was better technically, but it offered a relatively low data rate and a delay while the three buoys were planted. There was uncertainty that it would be produced. The directional DICASS and CAMBS offered better detection and discrimination of slow targets despite reverberation and buoy drift. In deep water, where performance would be limited by ambient sea noise rather than reverberation, they were likely to perform similarly, the higher output of DICASS being offset by the better directionality of CAMBS. Although both were reverberation-limited and affected by drift, the narrower beam of CAMBS was expected to offer a 5 dB advantage. CAMBS was therefore expected to be better against a quiet target operating in the relatively noisy shallow waters around the UK and in focal areas in the Mediterranean. It also offered a higher data rate.

British Towed Arrays

Unlike the U.S. Navy, when the Royal Navy developed surface ship arrays, it sought to combine the surveillance and tactical roles. In November 1972 the Royal Navy's long-range shipbuilding plan showed three towed-array surveillance ships, the first to be ordered in February 1979. Presumably they would have been analogous to the U.S. T-AGOS ships. However, it was soon evident that tactical arrays were a very real possibility. Royal Navy tactical towed arrays appeared in plans for the first time in 1975. In 1978 they were central to the destroyer/frigate building plan. The prototype Type 2031X was tested on board the test ship Matapan and then on board the frigate Lowestoft. Initial plans called for fitting 4 more Rothesays, 5 Batch III Leanders, and 13 Type 22 frigates. The British Fleet Requirements Committee agreed that although Type 22s were the best available vessels, a study of a light towed array ship should be set in hand. It became Type 23 (Duke class).

The first two of the planned *Rothesay* installations, *Berwick* and *Rhyl*, fell afoul of industrial problems. Scheduling probably made it easier to install the first production array on board HMS *Cleopatra*; that effort produced a design applicable to Batch IIB *Leanders*. Her sisters *Sirius*, *Phoebe*, and *Argonaut* were converted. Plans to convert five Batch III *Leanders* were canceled as part of the Nott cuts in British naval spending, but in 1985 *Arethusa* was converted using the refurbished prototype, now designated 2031Y (that year *Lowestoft* was withdrawn). All of these arrays had hardwired processors. The long five-octave 663 m (2174 ft) array (with its 1,830 m [6,000 ft] tow rope) severely strained *Leander* hulls, causing stress cracking, splits, and structural failures.

The frigate HMS *Phoebe* shows her towed array reel aft. The tail boom of her helicopter is visible abaft her hangar, presumably full of towed-array data-processing equipment. Unlike some other ships of her type, the big Type 965 air search radar on *Phoebe*'s stub mainmast was removed as part of the installation. USNI

The reel and other mechanism of the Type 2031 towed array on board HMS *Phoebe*, 28 August 1989. AUTHOR'S COLLECTION

In November 1978 the Royal Navy decided to split the program into an interim (formerly Phase 1) Type 2031 program for the *Rothesays* and a Type 2038 (formerly Phase 2) to full naval standards for later ships, which at that point amounted to 5 Batch III *Leanders* and 16 Type 22s. The Type 2038 program was expected to reach project definition stage in June 1979, to be reviewed after *Lowestoft* had completed trials with the "thin" (2.5 in.) array. The projected building program had been rebalanced to favor Type 22s over Type 42 missile destroyers. Now towed arrays were to be installed on board the first two Type 22s at their major refit, in one Batch 2 (hull 05) at a special refit, five Batch 2s on build, and eight Batch 3s (hulls 11–18) on build. The refit of the third Batch 1 was outside the time period involved. Studies of the light towed array ship, either a Royal Navy version of the commercial Type 24 or a later version of Type 22 (after hull 10), were underway. A proposal to fit up to 12 "expendable" offshore patrol vessels (OPVs) was dropped because their surveillance and damage-absorption roles (in war) were not compatible. Too, fitting OPVs would divert resources from the Type 2031 program. Too, no OPV program could be justified until much more was known about towed array performance in the shallow water where OPVs might well operate.

Initial tests used a 3.5 in. array covering a limited frequency range not considered suitable for the fleet. An advanced-technology 2.5 in. array (capable of detecting frequencies down to 10 Hz) was wanted, because it was expected to remain against Soviet submarines for many years despite silencing. When the new requirement was endorsed in July 1975 it was hoped to fit production 2.5 in. arrays beginning in 1981, as the Royal Navy wanted to get arrays to sea as soon as possible. During 1977, however, it became clear that the thin-line array could not be available before 1983, which was unacceptable. The Royal Navy therefore decided to fit an interim array based on the *Lowestoft* experiments to get operational experience as soon as possible. These arrays could be fitted in 1981–83, followed by a standard Type 2031 program from 1983 on, and then by a more advanced array for follow-on ASW ships in the 1990s. At this time the Royal Navy wanted at least 17 towed array frigates. In 1975 the Naval Staff proposed fitting 29 frigates by 1980; this was then reduced to the 26 Batch III *Leanders*, Type 21s, and Type 2s, and was then reduced again with the 1978 long-term costing. Then the plan was dismissed as impractical: Type 2031 procurement would coincide with too many other ASW programs, including the new Type 2016 surface ship duct sonar. The plan was therefore cut to 18 towed array frigates by 1988: 8 Type 21s and 10 Batch III *Leanders*. The Type 22s were wanted for the ultimate system, though they were also candidates for the developed Type 2031. Initially the *Rothesays* were dropped altogether, but they were brought back when the Navy pointed out that it would be unwise to fit the interim (initial) system to long-life, extensively modernized ships. The Type 21s were dropped because their commercial design was too tight; they would sacrifice too much in weapons for the array. *Leanders* were preferable. *Rothesays* and Type 22s were firm candidates.

Steam-powered ships were likely to be quieter than gas turbine ships in the 5–7 kt range at which arrays would be towed, because the gas turbine ships increased their radiated noise when they reduced propeller pitch to go slowly. How serious this might be was not yet known.

In 1981 it became apparent that the ultimate array (Type 2038) would be too expensive, so a review called for a production-standard array that would have much the same capability but would require less development. It would have been functionally similar to Type 2031, so the new project was canceled in favor of a variant of Type 2031.

Meanwhile Tom Curtis at AUWE proposed a new signal-processing technique that would drastically reduce the cost and volume of signal processing associated with the array. It was generally described as a shortcut that avoided calculations but would not much affect the result. Probably that entailed first analyzing the output of the array elements for frequency, and only then forming beams. Doing the frequency analysis first would simplify calculation, since the high-frequency part of what was received would be discounted. The Curtis technique was first tried in October 1981. The success of the trials and the prospect of considerable savings reoriented the British towed array program. Work on the four *Leanders* continued because they were already under contract, but further production of this version was cut. Four sets of experimental Curtis processors were bought from Waverley Electronics, three for Type 22s (ships 05 through 07) and one as a reference unit for the development of tougher equipment for future ships. The version of the array using Curtis processing was Type 2031Z. Compared with 2031I, it needed one rather than four electronic racks. Type 2031Z was also incorporated in the later Type 22s and in the Type 23 (Duke class). The array is reportedly 500 m (1,640 ft) long, twice the length of SQR-19; including the tow cable and the rope tail it is reportedly about 2 km (6,560 ft).

The array was credited with multiple convergence zone capability in the Norwegian Sea against noisy Soviet submarines. Typically, British towed array ships were cued by SOSUS and then handed targets over to submarines or to maritime patrol aircraft They could estimate target speed from blade rate, and could resolve bearing ambiguities by changing heading. They used submarine-style target motion analysis.

The British Type 23 frigate was conceived primarily to tow an array. HMS *Northumberland* shows the array aperture in her transom, as well as the large hangar for her Merlin (EH-101) helicopter, intended to exploit second (or even third) convergence zone contacts. Also prominent is the after director for her vertically launched Seawolf self-defense system. The objects abaft it are radar absorbers. USNI

APPENDIX B
Submarine Sonars

IN 1945 THE GERMANS surprised the Allies with their extensive use of low-frequency (sonic, as opposed to supersonic) sonar systems, generally passive array sonars. The U.S. Navy was impressed by the performance of the passive array sonar (GHG) of the captured cruiser *Prinz Eugen*, and also by the circular *Balkon* (Balcony) sonars of captured U-boats, particularly the two Type XXIs the U.S. Navy operated postwar. At 15 kts, a Type XXI using *Balkon* could detect targets at 4 km (4,360 yds). *Prinz Eugen* had a much larger set that used 240 hydrophones and was said to have tracked HMS *Hood* over (actually under) the horizon before her battle with *Bismarck*. Postwar the U.S. Navy tested it on board the U.S. submarine *Flying Fish*. Its performance may have inspired the U.S. SSK program. The *Balkon* from the seized Type XXI U-boat *U 2513* was installed on board the Guppy submarine *Cochino*. She was testing it in northern waters when she sank in 1949 following a battery explosion.

Balkon was the basis of the U.S. BQR-2, a 68 in. circular array of elements forming an 18 degree beam, listening at frequencies between 500 Hz and 15 kHz. Because the array was circular, it could be scanned using a commutator. The listening beam could be used in either searchlight or scanning mode. In 1956 it was claimed that a quiet submarine could detect a snorkeler or cavitating target in the layer at 160,000 yds and at 8,000 yds below it. About 1960 BQR-2B was credited with a range of 15–20 nm against a snorkeling Guppy in the deep Atlantic. The later BQR-21

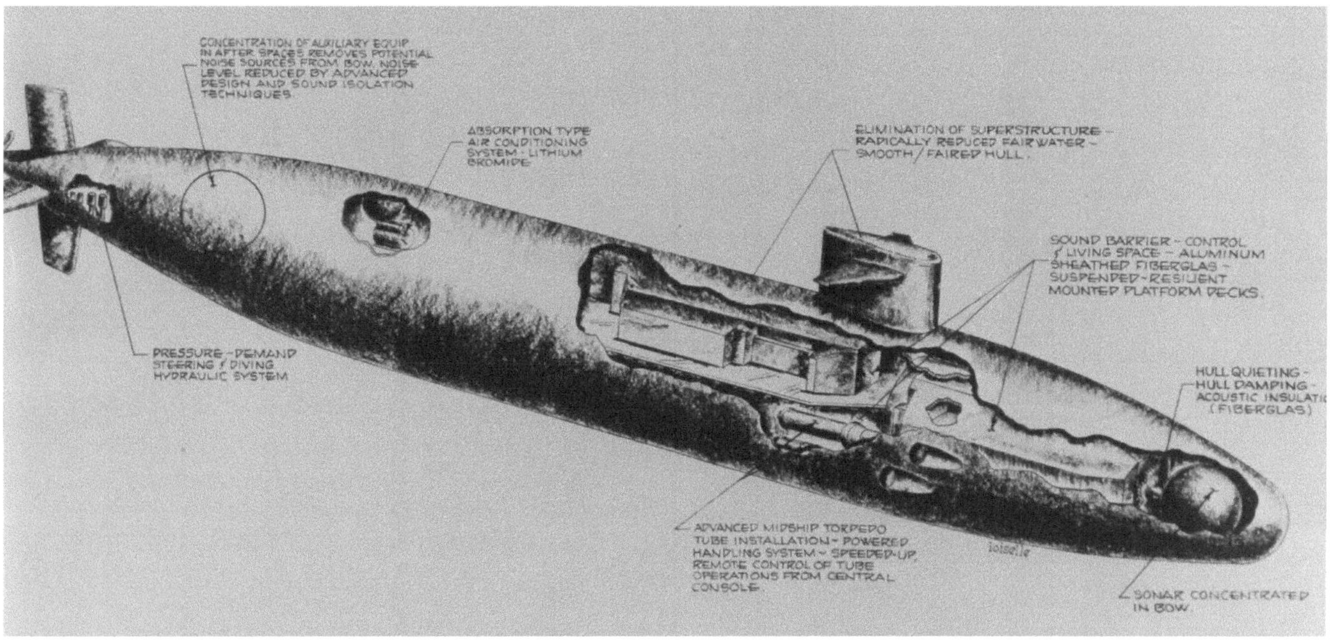

USS *Thresher* was the first fast U.S. submarine conceived specifically to hunt down enemy submarines. This official sketch shows her bow filled with the spherical transducer of a BQS-2 active sonar. Not shown is the passive BQR-7 array wrapped around it. Filling the bow with sonar equipment required that the torpedo tubes be moved well back; the number of tubes was reduced from the usual six. To enable the submarine to fire more rapidly (to compensate for the reduced number of tubes), the tubes were power-loaded. The sail was drastically reduced compared with that of previous U.S. nuclear submarines, and the hull radically faired. The "advanced . . . sound isolation techniques" mentioned in the caption refers to the rafting introduced in this design. NHHC

USS *Flying Fish* in Long Island Sound, 21 April 1952. The big fairing around her bridge covers an ex-German GHG passive array, probably from the cruiser *Prinz Eugen*. NHHC

USS *Grouper* shows an experimental active/passive billboard array (14 × 14 elements, 0.5–1.2 kHz) abaft her sail and what appears to be a single PUFFS sonar dome right aft. She had been converted into an ASW submarine (SSK) and was reclassified as an auxiliary submarine (AGSS) in 1958. USNI

(appeared in 1974) was a digital preformed-beam (DIMUS) replacement, using a hull-mounted conformal array; it could track five targets simultaneously. BQR-21 was a DIMUS version of BQR-2. There was never an operational DIMUS version of BQR-7. BQS-4 and BQS-4B used an omnidirectional 6.4 kHz pinger inside the BQR-2C array, the array being used as a receiver (BQS-4D employed BQR-2D). BQS-4 replaced SQS-4 sonars on board submarines. Typical range was 6,000–8,000 yds.

The German GHG (*Gruppenhorchgerät*, "group listening device") used an irregularly shaped array, a directional beam being created by what the Germans called "electric water," an array of delay lines. The U.S. Navy standardized its version of GHG as BQR-4. The full horseshoe-shaped array of 58 hydrophones was 10 ft wide and 20 ft long, operating between 150 and 5,000 Hz. A quiet submarine could detect a snorkeler in the layer at 160,000 yds (30,000 yds for a target below the layer). In 1956 it was estimated that a quiet 13 kt submarine could detect a snorkeler in the layer at 20,000 yds (3,000 for a target below the layer). U.S. ASW submarines (SSKs) were typically equipped with both BQR-2 and BQR-4, which were usually supplemented by an active scanning sonar such as SQS-4 (and later, BQS-4).

With the appearance of nuclear submarines, something considerably better was wanted. The Underwater Sound Laboratory at New London began feasibility studies in January 1956 on an integrated sonar for the new *Skipjack* class. As formally proposed in August, the sonar was presented as the outcome of laboratory programs to improve submarine sonar active and passive detection, attack information, and communications. The laboratory envisaged a pair of hemispherical 8 ft active transducers, one

Grouper showing additional arrays as a test submarine, presumably for the Bottom Reflected Active Sonar System (BRASS) program to evaluate bottom-bounce possibilities: a long flat receiving array and the characteristic three sonar domes of a PUFFS passive-ranging system. The flat array was presumably the 85-ft BRASS II receiver (for bottom-bounce experiments), which had two 25-degree preformed beams. A 144 dB 4.5 kHz transmitting array was in her bow. BRASS III employed a more powerful parabolic bow transducer (2.15 kHz, 150 dB). USNI

above and one below the torpedo tubes. They would be surrounded by a conformal array of spot hydrophones extending back 36 ft and supplemented by a communications transducer well back on the underside of the submarine. The half spheres were considered better than the usual cylindrical transducers at high speed. They would also be used for coded omnidirectional messages. They would operate at 5 kHz (SQS-23 frequency), meaning that they would function in direct-path mode. Since the sonar beam could elevate or depress, they might also be usable in bottom-bounce mode. They could measure the vertical angle to a target for passive ranging in water of known depth. The sphere would cover 280 degrees in bearing. The laboratory estimated that at low speed in directional active mode the sonar would detect targets at 20,000–57,000 yds (16,000 yds at high speed). In scanning mode, range would be 12,500 yds (5,500 at high speed). At low speed, the conformal array would detect and track targets at 185,000 yds (14,000 yds at high speed). With automatic target following, effective range would be reduced to 35,000 yds. BQR-2B was credited with a range of 105,000 yds (11,000 yds at high speed).

A final specification was written at NUSL between March and October 1957; in May 1958 Raytheon received a contract for two systems.[1] Given interest in the lower-frequency SQS-26, the laboratory reduced active frequency to 3.5 kHz. The active element had to grow (to 15 ft diameter) to maintain the desired beam width; the two hemispheres were merged. It had 1,245 elements. Transmission was omni-directed, rotationally directed, or tri-beam. Torpedo tubes were moved aft to clear the big bow sphere. One of the sonar developers later wrote that for the first time sonar was the driving force in submarine design; the advent of a new sonar became the occasion for a new submarine design.

The spherical active sonar was designated BQS-6. Like other contemporary U.S. active sonars, it used commutators to form its beams, the commutators consisting of contacts crawling over spherical models of the sonars to connect the appropriate elements with the appropriate phasing. The beam was steered in 6-degree steps in elevation between +15 and −45 degrees. Typically there were three commutators, one for active operation, one for passive operation, and one for fire control. Because there were so many possible beam positions, scanning the space around

Baya in her second conversion (1958–59) as an experimental submarine, showing an enlarged bow for a transducer and a folded-back boom for receiving transducers. The forward torpedo tubes have been removed and the blunt sonar bow installed; 23 feet was added amidships. She had a 1.5 kHz billboard projector forward and two 40-ft linear receivers on the extendable booms at the bow. She also had a 4.3 kHz active/passive transducer forward of the conning tower. USNI

USS *Tigrone*, a former submarine radar picket, was also employed in the BRASS tests. She shows a large, apparently cylindrical, free-flooding bow sonar dome. Later she had a big billboard abaft her bridge structure. USNI

a submarine was a lengthy process (typically 45 minutes for a full scan).[2]

Submarines typically employed an associated two-dimensional passive sonar (BQR-7) to search for targets, using both LOFAR and broad-band techniques. It had 156 hydrophones (3 × 52). It turned out that LOFAR processing could give the array considerable range. In 1956 an experimental model was considered capable of detecting a snorkeling submarine at 30–100 nm (a cavitating submarine at 10–50 nm). Once a target was being tracked, the sphere could focus a beam on it and obtain a single-ping range and bearing for a SUBROC shot. In *Thresher* and her successors, the bow sphere was associated with a digital fire control computer system (Mk 113) intended to perform target motion analysis.

The combination of BQS-6 and BQR-7 (without the digital LOFAR processor [BQQ-3] and the planned PUFFS passive ranger) was designated BQQ-1. It was designed to incorporate a secure underwater communications system (SESCO) to promote group operations, but that seems not to have been successful. The first BQQ-1 went on board the ASW submarine *Tullibee*. The second was installed on USS *Thresher*, the first U.S. attack submarine designed for ASW in accord with the Nobska recommendation. BQQ-1 was apparently compromised to the Soviets in 1964 along with the associated SUBROC missile system. It was thought (probably incorrectly) that the sonar on Victor-class submarines was the result.

BQQ-2 was the follow-on production system, incorporating the PUFFS passive ranging sonar and the LOFAR processor. Technical evaluation was completed in 1963, delayed by the loss of *Thresher* on 10 April 1963. Operational evaluation had to be delayed to 1964. The sonar was formally introduced into the fleet in November 1964.

In deep water (Bermuda) a 10 kt snorkeler could be detected passively at low speed by the conformal array at 250,000 yds and by the sphere at 210,000 yds. It could be tracked automatically by the conformal array at 37,000 yds. By way of comparison, BQR-4 could detect the snorkeler at 105,000 yds and track it automatically at 12,000. The LOFAR processor proved crucial, because it offered the submarine its longest passive detection range, using the BQR-7 array. The spherical sonar typically was not used for search, because it took so long to sample all of its beams one by one, using commutators, allowing sufficient integration time on each beam to pick up faint signatures. Simply because it had only two-dimensional beams, BQR-7 could be scanned far more rapidly.

In July 1964 a requirement for a new fully integrated sonar, which eventually became BQQ-6, was also formulated. It was to have twice the passive range of BQQ-2, faster target acquisition, and simultaneous reception over all azimuths covered by its passive array (BQR-7 or a successor). It would combine eight separate arrays, its spherical sonar (BQS-6) providing an output compatible with the spectrum analyzer (BQQ-3) for classification. The associated BQR-7 array would use preformed beams. The sonar would use vertical angle of arrival for passive ranging, the vertical angle being converted into range and automatically displayed to the sonar operator. In August an Advanced Development Objective for a completely new sonar was issued. As yet there was no program to develop a successor to BQQ-2 (which was just entering service).

The new sonar was wanted to deal with future fast Soviet submarines, so it had to detect them quickly. Active search rate had to be faster, perhaps using omnidirectional pings or tri-beam. All propagation paths, not just direct path, had to be used. A spherical sonar might exploit the vertical angle of the target for localization. Self-noise limited passive operation, because it required longer integration times to overcome quiet targets. The sonar should enable a single operator to track 6 targets passively (12 with two operators). Accuracy should suffice for the Mk 113 fire control system of the *Permit* class, and the sonar was wanted in FY 69. BQQ-2 was credited with a passive range of 4,000 yds with 20 minutes of integrating time on a 10 kt target, at 22 kts own-ship speed. The new sonar should be able to extend that to 50,000 yds with 5 minutes of integration time. It should track 6 targets passively rather than the current 3. Active range (omnidirectional transmission) should be doubled to 20,000 yds.

Given considerable development, particularly digital, the next sonar (BQQ-5) was considerably more capable than desired. Concept formulation began in January 1969, and elements of the system were tested on board *Pogy* (SSN 647). Preproduction versions were tested on board *Guitarro* and *Pintado* in 1972, and a production version was tested on *Bergall* in December 1972; production was authorized in February 1973. The prototype system was installed on *Philadelphia* (SSN 690). BQQ-5 was approved for service use in April 1979; an improvement program began in FY 80. While the sonar was under development, it was decided (approved September 1970) to augment the hull sonar by adding a BQH-4 type towed array; the BQH-4 low-frequency receiver and the array were integrated into the BQQ-5 design. BQQ-5 therefore had three beam-formers: one active and one passive for the sphere, and one passive for either the chin array or the towed array, forming 25 port and 25 starboard beams and 2 noise-nulling beams for the chin array. The active beamformer formed a total of 600 beams (10 elevation/depression beams between +19 and –53 degrees for each of 60 6-degree azimuth beams. Incorporation of a towed array was approved in 1973.

The 15-ft spherical active element (BQS-13) incorporated 1,241 hydrophones operating actively at about 3.5 kHz and passively at 0.5–5 kHz. Parallel beamformers for active and passive reception each formed a total of 600 beams (each 6 × 7.2 deg) over 60 azimuth positions at each of 10 elevation angles (+19/–53 deg). The passive beamformer fed a spectrum analyzer, a feature absent from earlier U.S. spherical bow sonars. A new DIMUS chin array replaced the BQR-7 of the earlier BQQ-2 system. It formed 24 beams plus 2 noise beams intended to reject self-noise. When a towed array was added, the chin array beamformer could be switched to form its beams. Associated with the main arrays (and sharing their displays) were a BQS-14 mine avoidance sonar and a WLR-9 intercept receiver.

BQQ-5A added a towed-array broadband processor and more computing power, plus new active and passive signal processors. BQQ-5B incorporated SHAB, a steerable hull-array beam-former for the chin classification array. BQQ-5C, which passed its operational evaluation in June 1984, expanded DIFAR reception. BQQ-5D introduced a multi-array signal beamformer and conditioner as well as new broadband and narrowband processors. It was designed to work with the TB-23 thin-line array. BQQ-5 completed Operational Evaluation in November 1988 and was certified for service in May 1989. The next version, BQQ-5E, underwent sea trials in 1993.

BQQ-6 was a parallel passive-only system for Trident submarines, with about 73 percent of its components in common with BQQ-5.

The U.S. Navy also became interested in passive range-finding, developing a system it called PUFFS (Passive Underwater Fire-Control Feasibility Study). It employed three transducers—forward, amidships, and aft. The pairs forward and amidships and amidships and aft measured directions to the target, and it could be triangulated to provide target range and bearing. Actual processing was manual, each of the pairs being processed to give a corellogram on a CRT. In diesel submarine, PUFFS (BQG-2 and BQG-4) employed three prominent sonar domes spaced along the ship's length. Typically such a submarine would have 3 equally spaced 6 ft vertical hydrophones along a 250 ft baseline, operating at 200 Hz–8 kHz, which were expected to give a range accurate to within 2 percent against a snorkeling submarine at 10,000–15,000 yds. In 1964, BQG-2 and BQG-4 were considered PUFFS I, with a maximum range of 20,000 yds. A new PUFFS II was expected to double the range and to be able to exploit transients or noise bursts rather than relying on continuous noise. It seems never to have materialized, perhaps having been cut during the Vietnam War.

In the *Permit* class, these domes were replaced by steel plates on which were mounted line hydrophones.[3] Because there was no topside space for arrays, they were mounted in pairs off the hull on both sides. This version of PUFFS could track a target. Claimed bearing accuracy was a tenth of a degree, and range error was about 2 percent between 7,000 and 9,000 yds.

In diesel submarines PUFFS (BQG-4) was associated with the Mk 45 nuclear torpedo. PUFFS was similar in principle to the French DUUX system, which was widely installed on board Type 209 export submarines, and also on early French nuclear submarines.[4] The French mounted all their arrays forward of the sail to avoid self-noise in the array. British observers noted that the French made no effort to mount their sensors resiliently.

A much more massive Wide Aperture Array (WAA) was developed later in the Cold War. The concept was tested as early as 1971; by 1980 WAA was considered an essential element of any post–*Los Angeles*–type submarine because it was capable of passive ranging out to Mk 48 range. It employed three large conformal arrays on each side of a submarine. It was valued not only for ranging but as the most precise means of measuring bearing and bearing rate, for quicker target motion analysis without any own-ship maneuver and at longer ranges. It was sometimes called RAPLOC (rapid passive localization), the "rapid" differentiating it from the much slower target motion analysis. Unfortunately, it was quite massive, having been conceived for the *Seawolf* class; it employed 404 hydrophones per subarray (6 in all), compared with 10 per subarray in the SSN version of PUFFS, BQG-2A. USS *Augusta* (SSN 710) tested an EDM model beginning in July 1987.

British Submarine Sonars

The British did not adopt German passive sonars, perhaps because they did not experiment with them after the war. When submarines were assigned the ASW role, initially they were to have had a Type 718 passive set in a big dome right forward, operating at 2.5 kHz at low submarine speed and at 10 kHz at high submarine speed. Its big split-beam transducer (5 ft by 2 ft) would have scanned mechanically at 4 RPM. Attached to the transducer would have been a four-square (Type 171, the submarine version of Type 170) for precision tracking and mine detection. The submarine would switch from 2.5 to 10 kHz as she neared her target; at about 10,000 yds Type 171 would also gain contact. At 5,000 yds the Type 171 operator would use a sharp ping to determine range. Type 171 was also the only way to determine target depth relative to the attacking submarine. These sonars were tested on board HMS *Thermopylae*, and about 1951 it was hoped that they could be operational in 1959.

A British diesel submarine in drydock shows her Type 187 transducer without the usual sonar dome. It belonged to the same preformed-beam generation as the surface ship Type 177. The submarine also had omnidirectional hydrophones (Type 186) on each side, wired together to create a single beam. The submarine scanned it by turning. Type 186 functioned as the searcher, Type 187 being the associated attack sonar. It operated actively at 10.5 kHz and passively at 2.5 or 10 kHz. A submarine typically used it as a high-definition passive sensor, with a range of 13,000–30,000 yds at 2.5 kHz and a bearing accuracy of 0.5 degree at that frequency. ALAN RAVEN

In fact, Type 171 was canceled about 1952 after initial trials, but Type 718 survived as the Type 187 attack sonar. It worked with the Type 186 "searcher" sonar, a fixed array of 12 hydrophones on each side of the submarine. The hydrophones formed a single beam on each side of the submarine, which scanned it by circling or by following a sinuous path.

Type 186 was the survivor of a British alternative to SOSUS called Corsair. Because the British knew that they could drastically reduce the narrowband sound that SOSUS used, they assumed the Soviets could and would do the same. They therefore focused on broadband low-frequency noise, integrated over a substantial time interval. For example, this technique could exploit the low-frequency part of the flow noise around a submarine. An experimental fixed Corsair array detected a quiet *Porpoise*-class submarine at a range of about 70 nm, which was better than twice the SOSUS range on such a submarine. The fixed version of Corsair (Type 191) was abandoned after the 1957 Defence Review because it was not relevant to the out-of-area operations the Royal Navy then contemplated. Type 186 was the submarine equivalent of Corsair (KNOUT), tested on board HMS *Thule* in 1954. Its single broadband beam detected submarines at tens of miles. Type 186 operated at about 1 kHz.

In 1957 the then–Underwater Detection Establishment (UDE) issued a design study of a new low-frequency active/passive Type 2001 asdic, intended specifically for the prototype nuclear attack submarine HMS *Dreadnought*. *Dreadnought* seems to have been conceived largely for close support, working with a surface link ship. The concept was operationally evaluated in December 1963 using *Dreadnought*.

Like Type 177 and Type 184, Type 2001 used preformed beams created by delay lines. It was far more massive, and far more complex. Each beam was stabilized. The fixed horseshoe-shaped array, 40 ft long, stretched around the submarine's bow over a 240 degree arc. The array was 7 ft high, tilted back 20 degrees from the vertical in HMS *Dreadnought*. The transmission mode was correspondingly complex: a ripple—one pulse immediately following another—of 23 pulses (each 100 msec long at 3.25 kHz followed by 23 20 msec pulses at 2.95 kHz), each 10 degrees wide, scanning from 110 degrees red (port) to 110 degrees green (starboard). The follow-on ripple of short pulses was intended to pick up short-range targets. A second mode was Sector or FM: 612 msec, swept from 3.5 to 3.8 kHz, over a selected 20-degree arc. A third was broadcast: a single transmission over a 180-degree sector centered on the ahead bearing. It employed a 100 msec, 20 msec, or FM pulse. Reception was directional. This mode was normally employed for fire control. Transmission interval could be set to give maximum range of 10,000, 20,000, 40,000, or 80,000 yds. The usual scale was 40,000 yds, corresponding to an interval of about one minute. There was one display in the control room, and three displays in a rather cramped sound room. In addition to the active mode, the sonar had a passive mode using two alternative bands, LF (1.6–2.3 kHz) and HF (5.5–6.5 kHz).

It seems to have been clear from the outset that operators would find it difficult to handle the vast amount of data. As others had found, moreover, flashes of light on a CRT often vanished before operators could see them. Computer technology to store information did not exist. The British solution was complex and somewhat awkward. The CRT was photographed during the whole reception period. The film frame was then rapidly processed, dried, and rear-projected onto a large translucent screen at the front of the main sonar console. This image remained during the next reception period, enabling operators to see the difference between the two images. The screen was blank for about nine seconds between images. Each image was called a frame, since it corresponded to one frame of film. Two operators marked up all contacts on this initial detection display (IDD), reporting all that seemed valid given their persistence and their Doppler and other characteristics. All dark spots on each frame were called contacts. An operator marked and numbered each contact on a frame, repeating the process for five frames and then starting over. Data from the IDD sector display, PPI, and passive displays went to the ARL plotting table, to the target bearing indicator, to the torpedo control calculator, and to the torpedo guidance control unit (for wire-guided torpedoes). This baroque arrangement lasted into the 1980s.

Detections were compared with contacts already held on the local operations plot from other sources, such as surface ships from the link ship working with the submarine and from other sensors. In this way known surface ships could be filtered out, leaving possible submarines to the sector display for closer investigation and accurate tracking. Command could classify contacts using the echo characteristics on the sector display together with correlation of its Doppler and apparent motion. A contact classified as a possible submarine (or more positively) was tracked using the sector display and reported to the link ship for relocation and prosecution by air or surface forces.

The IDD, the sector display, and a passive display were in the cramped sound room; there was also a PPI built into the PCO console in the control room to support tactical decision-making.

Evaluation of HMS *Dreadnought* in the direct support role revealed problems. She reported far too many contacts because she lacked any means of classifying them. If she turned through more than ten degrees, the IDD display had to be recentered, breaking continuity. Continuity could also be broken if the IDD ran out of photographic chemicals. Manning was insufficient, with only three operators for the sonar console (IDD, sector display, and passive display). Maximum operating speed using long-range sonar was 15 kts, but even when zigzagging to launch aircraft, carriers sometimes achieved a net speed of 18 kts. A close-support SSN would have to sprint and drift. On the other hand, in trials a half-array passively detected USS *Nautilus*, running at high speed, 200 nm away in the Mediterranean. In active mode, it detected the submarine *Cachalot* at 34,000 yds in the North Sea.

Type 2001 equipped most Royal Navy nuclear attack submarines. Improvements included relocation of the array to a "chinstrap" position lower in the bow, deeper-diving transducers, and an interface for a digital data-handling system in the *Swiftsure* class. As relocated, the sonar could exploit convergence-zone and bottom-bounce modes of operation. The transducers, initially mounted on the outside of the hull, were covered by a dome to reduce excessive flow noise. The modified sonar was Type 2020, the first installed on board HMS *Trafalgar*. The sonar was now considered primarily passive.

When towed arrays were adopted, Type 2020 was modified to provide forward-looking very-low-frequency coverage (the towed array did not look dead ahead). Type 2032 used 15 detuned bow array transducers for LOFAR passive detection; it was equivalent to the U.S. linear array wrapped around submarines' bows. It was first delivered in March 1989, and all 13 had been delivered by March 1991. It was generally considered part of the long-range towed array project. A Type 2076 upgrade was competed in 1989–90 and so falls outside the Cold War.

Resolution class ballistic missile submarines were given a version of the Type 186 flank array. It was considered essential because it offered twice the passive range of Type 2001 (in 1967 it was credited with a range of 100 nm).

Like the U.S. Navy, the Royal Navy gathered acoustic intelligence on Soviet submarines using spectrum analyzers called AILEEN. They seem to have been adopted after the British became aware of the U.S. TUBA program.

The Royal Navy decided to build its ballistic missile submarines at about the same time the *Thresher/Permit* group began to enter service, with their BQQ-3 frequency analyzers. The Royal Navy bought BQQ-3 (which it designated Type 2006) for both Type 2001 and Type 186, initially for the new ballistic missile submarines. They were intended to classify submarines detected at long range. The Royal Navy knew that the U.S. Navy was finding that flow noise blanked out the lowest-frequency (hence longest-range) parts of the spectrum. U.S. submarines relied on the higher-frequency part of the spectrum, including noise that could be processed (demodulated) to reveal signature data, using a process called DEMON.

By 1964 the Royal Navy was interested in buying BQQ-3 (Type 2006) for the coming Improved *Valiant*-class (*Swiftsure*-class) attack submarines. BQQ-3 could not fit the earlier SSNs or the diesel-electric submarines, so the British looked at alternatives. Initially it seemed that the only available one was the U.S. Navy's BQH-2 intelligence-gathering analyzer. For a time the Royal Navy wanted to buy four sets of BQH-2Cs for evaluation on board "P" and "O" class diesel-electric submarines and on board early SSNs. With its inbuilt high precision, BQH-2 was expected to be expensive, so the Royal Navy sought a British signal analyzer (Type 2012). The U.S. AQA-3 and AQA-5, used by patrol aircraft, had already been rejected. A breadboard version of Type 2012 was tested at sea on board HMS *Walrus* in May 1967.

After the British became aware of the U.S. submarine towed array project a Staff Requirement for a British equivalent was approved on 3 February 1965. A British mission to the United States learned of the

U.S. "hindsight" project, and of efforts to adapt a towed array to attack submarines. The British group recommended buying U.S. towed arrays for British SSBNs and for conventional submarines with more than ten years of remaining life. In 1965 the U.S. Navy planned to test an operational towed array on board an SSBN in 1967 and to begin production at that time. The British version of the short U.S. towed array (BQR-15) was Type 2023, using a Plessey-developed Type 2012 frequency analyzer. Type 2023 was authorized in 1973 and entered service in 1977. By the 1970s British SSBNs used U.S. BQR-22 and BQR-23 processors for their BQR-15 arrays, under the designations Type 2030 and Type 2035.

Type 2024, the parallel attack submarine towed array, was substantially longer than the array of Type 2023. The first contract was let in 1970, the prototype appearing in 1974, and the first array entering service in 1975. The project was accelerated by using a U.S. processor and a British array. The processors were Types 2017 (DEMON and LOFAR) and 2018 (DEMON only). By accepting a simplified system, the Royal Navy got a towed array into service three years earlier than would have been the case with something more elaborate. To exploit its full potential it was given more U.S. analyzers (Types 2030 and 2035). The operational date slipped from 1978 to 1980. It was installed on eight SSNs (instead of ten as planned), two SSBNs (rather than one as planned), and one SSK. HMS *Conqueror* used this clip-on array to track the Argentine cruiser *Belgrano* from beyond the horizon, keeping station for two days.

A new all-British Type 2026 towed array was planned; in the interim, mating a wider-frequency version of the Type 2024 array with the new Type 2026 processor. Type 2026 exploited new technology developed for the Type 2031(I) surface ship system. It could be switched to classify signals from the intercept array and from the Type 2008 underwater telephone. The R & D contract was let in 1977–78, the system was tested in 1983–85, and it entered British service in 1987. Type 2026 equipped the first five *Trafalgar* class. Because the array was all-British, it was possible to make Type 2026 an Anglo-Dutch project, which helped finance it. The Dutch version used a Marconi beam-former and a Signal display processor. HMS *Turbulent* tested a reelable version, the prototype for the reelable array planned for the *Vanguard* class; its winch was housed in a hump visible aft, above the waterline. The array itself was presumably stowed in the narrow space between the pressure and outer hulls. Other submarines used Type 2026 with a clip-on array.

The next British submarine towed array was Type 2046, developed in parallel with the mature version of the surface ship Type 2031. The designation applied both to a towed array (using fiber optics to transmit data back to the submarine) and to its three-octave Curtis architecture core processor/display. Compared with Type 2024, Type 2046 reportedly provided an improved narrowband search capability. The first of 27 arrays was delivered in 1987. Type 2046 was installed on board the *Upholder* class SSKs as well as the SSNs. Reportedly at least some submarines (such as HMS *Triumph* and HMS *Turbulent*) have carried two Type 2046s each.

Type 2052 was an interim clip-on towed array for nuclear attack submarines, derived from Type 2024.

Type 2007 was a passive SSN flank array on board the *Swiftsures* and the first five *Trafalgar*-class SSNs. It seems to have superseded a proposal for Type 2005 (1965–66), which would have been a passive ranging array comparable to PUFFS.

Soviet Submarine Sonars

Western knowledge of Soviet submarine sonars and related systems during the Cold War was very limited, because they operated largely passively. Much more became known after the end of the Cold War.[5]

The Soviets ended World War II with a German passive array (their Mars) with transducers flush against the submarine hull, and hence subject to considerable flow noise. Given its very limited height, the array offered little vertical directivity, which also made it more vulnerable to noise. Bearing accuracy was considered inadequate. Mars equipped first-generation postwar Soviet diesel submarines (some later Whiskeys had the successor Feniks). The last Mars was installed on board the first November-class SSN; others in the class had Feniks.

Second-generation diesel submarines also had an active searchlight sonar (Plutoniy), which was accepted for service in 1958 (it was developed in 1948–56). It was used for echo-ranging, for mine detection, and for directional two-way Morse communication. It could operate passively instead of actively, but only in its narrow frequency band (i.e., it lacked any auxiliary passive transducer). This sonar was on board Romeo- (Project 633), Whiskey Long Bin– (Project 665), and Juliett- (Project 651) class submarines.

Requirements for new sonars were set in 1947. Submarine sonars were to operate at up to 20 kts and at depths to 250 m. The Soviets sought an "objective" direction-finding technique, much as the U.S. Navy did, to replace reliance on operator earphones. The low-frequency Feniks was developed to meet this requirement. Based on the wartime German *Balkon*, which the Soviets had captured in 1945, it was developed between 1948 and 1951, and accepted for service in 1952. Feniks was comparable to the contemporary U.S. BQR-2, which was also based on *Balkon*. It was the first Soviet system to use a cylindrical array and a sound-transparent dome; for years it was the main Soviet submarine sonar. The design of the magnetostrictive transducers was based on that of the rotatable JT line hydrophone of U.S. submarines. Feniks introduced U.S.-style commutator scanning to Soviet practice. Scan time was 1–3 minutes, and there were 132 channels (corresponding to 132 transducers). At 15 kts a submarine using Feniks could detect a destroyer making 15–18 kts at a range of 7 km. Feniks was later upgraded to incorporate an automatic target-tracking mode; data from the tracker was transferred to the Leningrad torpedo FCS. The improved version was Feniks-M (MG-10). The Soviet designers considered Feniks to be well ahead of U.S. technology (probably meaning U.S. wartime technology).

Feniks could be recognized by its large stainless-steel wraparound window on the submarine's bow. The window in turn was a major feature distinguishing second-generation postwar Soviet diesel-electric submarines, helping to distinguish Foxtrots (Project 641) from Zulus (Project 611) and Romeos (Project 633) from Whiskeys (Project 613). These ships also had Arktika fire control sonars—which used some Feniks components.

The first major upgrade added a target-tracking mode. A second upgrade (1957–58) provided correlation signal processing, enabling detection below the nominal noise level, apparently for the first time in Soviet practice. So modified, Feniks became MG-10 (Feniks-M). It was tested on board two Whiskeys in 1958, and was deployed on board Foxtrots, Romeos, and Novembers (Project 627A). With an integrated sonar intercept receiver, it became MG-10M. This version, developed in the 1960s, offered more precise torpedo fire–control data, and its passive range in scanning mode was about 30 percent greater. The power requirement was better than halved. Effectiveness was improved by using the same control panel and display for both the scanning and automatic target tracking modes.

Feniks-M had two NATO nicknames, Shark Teeth for nuclear submarines and Trout Cheek for diesel submarines, because it was passive and therefore produced no telltale signatures distinguishing one from the other. Apparently, NATO analysts assumed that the same array was part of the second-generation nuclear submarine system, since Charlie and Victor I were credited with Shark Teeth plus a Pike Jaw (Arktika) successor called Shark Fin, an MF attack sonar. These assumptions were entirely incorrect; they reflect the virtual impossibility of determining what sort of passive sonars a submarine uses.

An attempt to extend performance further by placing an omnidirectional sound source inside the array, as in the U.S. BQS-4 (as the Tuloma sonar), failed.

The companion active searchlight set was Arktika (MG-200, NATO Pike Jaw). It was the sensor of the Leningrad fire control system, which for the first time made it possible for the Soviet submarines to fire torpedoes without exposing a periscope. The Soviets described it as an integrated sonar, because it combined passive search and tracking with active ranging. Arktika was first tested in 1955. The modified

Victor and later Soviet attack submarines had large chin sonars, with their torpedo tubes above them. This Victor, which had collided with a U.S. ship, shows the covers of her two upper torpedo tubes and some of the staves of her big bow sonar. USNI

Arktika-M was first installed in 1957. Arktika and Arktika-M equipped Foxtrot and Romeo diesel submarines and first-generation nuclear submarines.

The first really big Soviet submarine sonar, the cylindrical Rubin (MGK-300; NATO Shark Teeth/Shark Fin), was proposed in 1958, and work began in 1959. It was designed for the Papa (Project 661) class submarine, but was also on the Victor (Project 671) class. A project completed in 1958 had shown that large arrays would passively detect targets at much longer ranges. Because Project 661 was expected to be so fast, Rubin had to operate at high (35–40 kts) as well as low (8–10 kts) speeds. The designers planned to use bottom-bounce and convergence-zone propagation. A Russian historian later remarked that, in effect, the units for measuring sonar range were changed from cables (tenths of a nautical mile) to kilometers, an increase by roughly a factor of 5. Rubin was conceived from the first as an integrated system, controlled from a single console, for detection, underwater communication, and echo-sounding. Its arrays had to be multipurpose, since there would not be enough space in the bow of the submarine (the quietest place) for multiple large ones. In turn, the designers of the Project 661 submarine allocated an enormous volume, by contemporary standards, to the bow sonar. The missile-firing Project 661 and its new sonar received high priority, Rubin itself being the subject of a Council of Ministers decree. Hardware volume was more than 60 times that of earlier systems. The entire project was nonetheless completed in six years (1958–65).

By the 1970s Western intelligence was aware that Soviet SSN sonars incorporated both underwater communications systems and distance-measuring equipment (DME; the sonar sent out a ping and another sonar replied, the timing of the reply giving distance). Presumably the combination showed that the Soviets planned to operate their submarines in groups. Observation of Soviet training exercises apparently showed much the same thing. Group operation made it possible for a Charlie to participate in a group that together could localize a target. Submarines without integrated underwater communications systems employed underwater telephones, which could be keyed for Morse transmission. Later submarines had frequency-coded (tonal) communication systems at low enough frequencies to exploit the convergence zone (out to second zone), with an information rate estimated as five to ten times greater than that of the earlier system. The variety of known Soviet underwater communications systems testified to their intense interest in such devices, and, by implication, to their interest in group tactics.

A 19 November 1971 U.S. exercise using the voice low-frequency mode of the standard WQC-2 underwater telephone demonstrated that even without much training, submarines could pass enough information to develop a missile fire-control solution using triangulation. Solutions became more accurate as the submariners gained experience. Worse, nearby surface ships failed to intercept the low-frequency signals, despite their high power (the exception was a ship using SQS-26 sonar, which could intercept signals in the 3 kHz range). The U.S. submarines managed without the Soviet DME capability. They relied on consort submarine grid location, passive sonar bearing, and target motion analysis techniques. Compared to DME, their technique was time-consuming and complex if the target maneuvered. Using DME, Soviet submarines could gain an instant fire-control solution not subject to navigational error, and also a solution from astern (which would make the submarine less vulnerable to detection). The only inputs would be single passive sonar bearings.

Two Charlie-class submarines participated in a 16 August 1971 anticarrier exercise, with Soviet surface ships simulating a carrier task force. They were seen within 15 nm of each other and within 20 nm astern of the simulated task force. An underwater communication signal (Dabbler) was heard from a Charlie trailing the carrier *Forrestal* (5 October 1972), indicating that a second submarine had to be nearby. This signal was intercepted by an SH-3 using dipping sonar. Dabbler used a sequence of tones transmitted in ascending and descending steps at 2.5–3.2 kHz.[6] Transmissions were known as Musical Tones. Brazenface was a higher-frequency system (4.8–8.2 kHz) known as Shifting Tones. It was known to have DME/IFF capability, which was only suspected for Dabbler. There was also an AM voice system, Windrow; and a Morse code signal, Dachshund. Breadfruit (5–8 kHz) was a lower single-sideband voice and Morse system, and Enactment was a voice, Morse, and tone system operating at higher frequency (12–15 kHz). By the 1970s, when this information was compiled, the U.S. Navy had tried and failed to develop reliable long-range underwater communications, mainly because of how sound behaves in water. It was generally assumed that the Soviets had succeeded because they were willing to limit themselves to a few stereotyped messages that could survive considerable distortion in transmission. The postwar Russian sonar history is silent on this topic, except that it seems that new generations of sonars had new generations of underwater communications systems.

The perceived keys to much greater range were lower-frequency, higher-array gain (which, given the lower frequency, meant a much larger array); better passive signal processing (for another kind of gain); and, in active mode, greater power. The big cylindrical array was used for passive and active operation, for LF acoustic intercepts, and for communications. Communication included range measurement between submarines. Longer higher-powered pulses were used. The beam could be elevated or depressed electronically. The sonar had a single electrically steered transmitting beam. Preformed receiving beams were used over a limited sector. Data from a set of preformed beams was stored together, the slope of lines produced on a paper recorder indicating target bearings and other results. For the first time the Soviets examined the potential of, and the requirements imposed by, target motion analysis (TMA). The most important result of this study (1963–66) was to justify investing in much better direction-finding accuracy.

A smaller cylindrical array in the sail, similar in concept to the main array, was used for HF acoustic interception. There was also an omnidirectional (in the horizontal plane) communications radiator.

The Soviets considered Rubin equivalent to the U.S. BQQ-2. It became the basis for further development, the next sonar being Yenisey, the first Soviet sonar with full preformed beam-forming (with one elevation angle). Yenisey abandoned the use of the same transducers for reception and transmission, the lesson having been learned that it was difficult to match the two functions in one unit.

Western naval intelligence considered Rubin similar to the British Type 2001, presumably because Soviet spies at Portland were known to have obtained information on that sonar (British submarine officers were told that Rubin and Kerch were equivalent to Type 2001 and that their knowledge of Type 2001 should guide them when confronting its Soviet equivalents). In this context, Type 2001 was described as an active/passive search and fire control system operating mainly in direct path but with a limited bottom-bounce capability, with a source level of 127 or 137 dB depending on amplifier type; under good sonar conditions it would have an active detection range of 5–10 nm. It might reach out to the first convergence zone. It was considered likely that Charlie and Echo II–class submarines had conformal-array sonars; the passive sonar on a Charlie might be similar to Type 2001. Estimated passive sonar range against a carrier at high speed under optimum conditions was about 150 nm, though most operational ranges would be less, and tracking range would always be significantly less.

Kerch MGK-100 (NATO Shark Teeth/Shark Fin) was a parallel development, probably initially for Charlie (Project 670) missile submarines. It was also on board Echo II (Project 675), Yankee (Project 667A), and Delta I and II–class submarines (Projects 667B and 667BD). The project assignment for the Kerch R&D project was approved in the last quarter of 1959. Development was completed in 1964–65, and two prototypes were built. The first, on board *K 57* (Echo II class), passed state tests in the Pacific Fleet and was commissioned in February–April 1966. The second, on board a Charlie, was tested in the Northern Fleet (in the White Sea, Barents Sea, and Norwegian Sea) and provided the targeting for the first Ametist (SS-N-7) cruise missiles. The developers claimed that Kerch was comparable to the U.S. BQQ-2. The project was urgent enough that production of the first 8 sonars (of an eventual total of 110) began in 1964–65, using drawings that had to be updated based on initial prototype tests. The first order for batch production was given in April 1963 (5 systems), followed that fall by a second order for 8 more.

Given the different sizes of the submarines involved, three different arrays would have been needed. Instead, the designers broke the single array into parts with different functions; different versions used different sets of arrays. They were a cylindrical passive array (4 m diameter, 2.4 m high), a rotating active planar array (operating at 3.5 and 7 kHz, covering the 135 degree forward sector), a cylindrical sonar intercept array, an echo-ranging array, two mine detection arrays (receiving and transmitting), and an underwater communications array. The tilting active transducer was used for bottom-bounce ranging. The two passive arrays were based on Feniks technology but had more sensitive transducers, for twice the overall sensitivity. In addition to these arrays, at least the Charlie-class version had a pair of flank arrays (up to 33 × 3 m) mounted between their inner and outer hulls operating at very low frequency (0.25 kHz to 2 kHz). They could detect surface ships at two convergence zones (ranges of 150–200 km). These arrays were tested on board a Pacific Fleet Whiskey-class (Project 613) submarine in 1960–61, initially in the Sea of Japan, in the area of Kuril-Kamchatka Depression with its known convergence zones; surface ships were detected at ranges of more than 250 km. The sheer complexity of the system drove the developers to adopt semiconductor amplifiers (Rubin used vacuum tubes).

The associated fire control center (FCS) was Brest, which could position-keep and develop attack solutions for two targets simultaneously (rather than on a single target, as in Leningrad). Torpedoes were spindle-set (without any electrical connection through the breech of the torpedo tube). An American visitor to a Delta I (which had Brest) saw a single PPI display on the fire control panel. Parameters were adjusted by push-button. There was no provision, at least in the control room, for manual TMA.

Balaklava was an associated upgrade system on which work began in 1967. It was particularly intended to improve performance against quiet targets. An active mode was intended to achieve a mean effective range of 120 km to support missile firing (presumably for the P-120 Malakhit missile). Alternatively, the system would support underwater communication between two submarines to triangulate a target passively at up to 200 km.

The first integrated Soviet sonar system was Okean (MGK-1000, NATO Squid Arm) designed for the Alfa (Project 705) class. The integrated sonar element was Yenisey. Okean would include a navigational system which, among other things, had to measure the absolute speed of the submarine, and also to guarantee navigational precision and safety of surfacing under ice; and a mine detection system. The sonar was considered so integral to the success of the submarine that its chief designer was appointed deputy chief designer of the submarine. Okean was controlled by its own digital computer, which was separate from that of the submarine combat direction system. All data processed by the Okean system were fed into the Akkord computer, which controlled the submarine and its weapons. The system computer added new functions, such as automatic range prediction based on the oceanographic environment. The Okean design cycle lasted about a decade. The system embodied many innovations that appeared in later Soviet submarine sonars, such as the first mass use of semiconductor elements—more than 50,000 of them—in a Soviet sonar.

Yenisey was the first Soviet submarine sonar intended primarily to detect quiet submarines. By way of contrast, the nearly contemporary Rubin and Kerch systems were intended primarily to locate noisy surface ships. In contrast to Arktika and Rubin, which searched passively and used active pinging only for ranging, Okean searched at long range in both passive and active modes. Both active and passive signals were automatically detected. In the passive mode, detected targets were automatically passed to the tracking mode, and bearing angles sent to the Akkord combat direction system. Targets detected actively were automatically passed to the combat direction system.

To minimize submarine size, sonar array size and inboard electronics volume were limited. The number of sonar operators had to be cut, the watch cut from the usual two or three to only one. Reliability was ensured by standardization and duplication. Besides its main functions, Yenisey was responsible for acoustic communication (including distance measurement) and for sonar interception. The submarine's designers saw the acoustic communication channel as a data link between the combat systems of submarines operating together. The communications equipment was therefore combined with the submarine's central computer complex. When in code communication mode, submarines could exchange data between their combat control systems without operator intervention, using a special code to transmit up to 60 decimal characters in a string. Signal transmission and reception were separated to make reception more reliable under noisy conditions.

Initially the designers wanted to use a single spherical array for both active and passive operation, but it could not be developed in time; they were compelled to use separate passive and active elements, partly because the same transducers could not be optimized simultaneously for active and passive operation. This was the first Soviet sonar split this way. Separation made it possible to adopt different forms of automatic processing for the active and passive parts of the system. In a departure from previous practice, the sonar used beam preforming for all functions (active, passive, mine detection, and acoustic intercept). Earlier Soviet array sonars had been scanned sequentially, so they spent much less time listening in any one direction. The designers claimed that they had reduced by orders of magnitude the time required to detect a signal. The passive array was a cylinder, the active array a trainable elevatable slab hung from it.

Two prototypes were tested, one for the prototype Alfa and one to be installed (1969–70) on board a Pacific Fleet Zulu-class diesel submarine specially modified for accelerated trials. Trials more than met expectations.

Rubikon (MGK-400, NATO Shark Gill) was conceived in 1964 to replace sonars in first-generation nuclear submarines and as new equipment for some diesel submarines. It was to achieve efficiency approaching that of Rubin in the space occupied by existing sets (MG-10, MG-200, MG-13, and MG-15). Inboard space had to be equivalent to 7.5–8.5 equipment racks, compared with 40–60 for Rubin or Kerch. Weight was also quite limited, as was AC power. It was therefore impossible simply to modify existing Rubin and Yenisey components, though Yenisey transducers were used. Many navy requests, such as integrating a mine detection function, had to be rejected; the project went through eight versions of the technical and tactical requirements before one was agreed. Because development work was proceeding meanwhile, the preliminary design was completed in December 1966, the contract design in May 1968. The system name, a reference to the decisiveness (and perils) of Caesar's crossing of the Rubicon, seems to have been chosen by the designer to reflect the problems involved.

To maximize array size, the designers adopted an unusual upside-down truncated cone form that nearly filled the space available in the bows of existing submarines. The cylinder was used only for reception, both for echo-ranging and for passive detection. Active sonar pulses were emitted by flat transducer arrays in the sail and on the flanks (in the spaces between the inner and outer hulls. To reduce electronics volume inside the pressure hull, for the first time in a Soviet sonar the designers placed some of the sonar electronics in a pressure capsule inside the array. Submarines generally lacked sufficient power output for the active pulses, so their pulse generator was fed directly from the submarine's batteries. Submarine designers were less than enthusiastic, as the battery was the submarine's only emergency power source. Estimated performance was 6 nm active, 9 nm passive. This sonar could automatically track two targets. Operators could manually track another three targets. Like the earlier Brest, Rubikon could engage two targets simultaneously, in this case using wire-guided torpedoes (whose paths appeared on the system's CRT).

Between late 1973 and early 1974 the first prototype was successfully tested on board a Northern Fleet nuclear submarine. The second prototype was tested in July 1973 on a new Black Sea diesel submarine (a Project 641B Tango). Preparation for series production had begun as early as 1969, and the first two production sonars were completed in 1973. Rubikon was officially adopted in 1976. It is now widely used as the Kilo-class sonar. Rubikon was also used on board Soviet Tango-class diesel submarines (Project 641BUKI); modernized November, Echo II, and Charlie-class submarines; Yankee Mod (Project 667AT), and Delta III (Project 667BDR).

Skat (Skate: MGK-500, NATO Shark Gill) was the last operational Soviet Cold War submarine sonar. It was conceived in 1964 as a third-generation sonar for all submarine classes. The codename Skat was formally applied in 1969. Plans called for two versions: large for new submarines and small for retrofit to second-generation nuclear submarines. The planned new-submarine platform was Project 649, which in the end did not materialize.

The main objective was to detect quiet submarines and surface ships at long range (e.g., at multiple convergence zones). That in turn would entail low-frequency passive detection of both the target's narrowband spectrum (i.e., LOFAR) and its low-frequency flow noise (DEMON). For this purpose the sonar used a towed array. Efforts were also made to improve the efficiency of active sonar search against a variety of targets at different speeds (hence different Dopplers). The system could classify the objects it detected. Secondary data processing (e.g., TMA) and decision-making were automated as far as possible. Thus, a planned system element was a target classifier, a new type of equipment. Man-machine interaction was optimized, and the operator relieved of secondary functions. For the first time, psychologists were brought in to help develop the sonar displays. As in other Soviet submarine sonars, Skat incorporated a communications system, stealthier than past ones yet more reliable. It was more reliable than in the past, with automatic test equipment (equivalent to BITE). Early experiments were conducted on a Pacific Fleet Zulu (Project 611) class submarine.

Skat was a hybrid sonar. Signal processing was still analog. However, secondary signal processing, such as track management, and system management were digital. Signal classification was also digital. Skat-3 (see below) represented the rest of the move toward fully digital operation.

For long-range active detection the largest possible transducer was needed. Initially a sphere was wanted, but its diameter would have been limited because submarine designers were unwilling to move torpedo tubes abaft it. Quite late in the project, then, a traditional cylinder was adopted. The designers found that its simplicity enabled them to eliminate 12 equipment racks. They were also able to mix transmitting and receiving transducers on a single array structure. Skat was the first Soviet sonar to use complex pulses (with intrapulse modulation) and coherent processing. Coding signals greatly improved noise resistance and also improved range resolution. The sonar had an extensive library of signals.

Skat development was delayed by the need to develop new components and even to choose an appropriate computer (both candidates were still under development).

Users were said to have been particularly impressed by the quality of the displays and the simplicity of control and clarity of data display. The system had the first Soviet flicker-free CRT display of passive and active sonar and sonar intercepts. The displays were finally bright enough that the system could operate under normal lighting. Skat also did not flicker, as other Soviet electronic display systems did. A single screen with panoramic sweep simultaneously showed information from outputs of the detection circuits in all frequency ranges, information on the positions of targets tracked by the auto target-tracking system, indications that passive targets had been detected, the parameters of the last three sonar signals detected by the acoustic intercept receiver, and the outputs of the sonar target classifier. The console also had a display showing self-noise at the arrays, apparently an important new feature.

The system was described as automated, which implies both digital beam-forming and a higher degree of integration. In a Typhoon-class SSBN, Skat is described as a digital sonar capable of tracking 10 to 12 targets. The passive part of the system includes a magnetic recorder.

For trials, the question was whether to use an existing submarine or wait for a new one. A proposal to fit a Charlie with the small version of Skat was rejected (this class was modernized instead with Rubikon). The Malakhit design bureau, responsible for the Victor class, proposed that the class be modernized and equipped with Skat (including the towed array), a new navigational system, and a new combat direction system (Omnibus). This radical proposal made it possible to test several new developments very quickly. In fact, it required virtual redesign of the submarine (as Project 671RTM, Victor III). Deliveries of the arrays to the builder (the Admiralty Yard) began in 1976, and the planned completion date of the ship was December 1977. The system ran trials in the Norwegian Sea, the Barents Sea, and in the Arctic Ocean, followed by Pacific Ocean and Sea of Japan trials in 1980. Tests were completed in 1982. By 1978, however, production Skats were being installed on board Typhoon (Project 941) and Oscar (Project 949) class submarines.

Apparently, only in the Victor III class were the transducers of the cylinder wrapped around a sphere housing the processing elements; this was not a spherical sonar. Victor III lacked internal space which in other designs was used for the sonar processor. The version in Typhoon and Oscar class missile submarines is Skat-KS (MGK-501). Charlie II had Skat-M. Known derivatives of the basic Skat were Skat-Plavnik (for the Mike-class submarine, Project 685 Plavnik), Skat-BDRM or Skat-2 (for Delta III/IV-class ballistic missile submarines), and Skat-2M.

Skat-3 (MGK-540) is the next-generation version, using fully digital signal processing, developed specifically for the Akula (Project 971) and Sierra (Project 945/945A) class submarines. Development was driven by the perception that sonar array development was unlikely to yield the required greater level of sensitivity necessary to deal with quieter U.S. submarines. Only better (which meant digital) signal processing could do that. Outside specialists had claimed that introducing Skat-3 would give a threefold increase in range. It was hoped that simultaneous automation of data processing and decision-making procedures would increase the sensitivity of the system by almost an order of magnitude. Not only was it necessary to improve the system's ability to distinguish weak sources in ambient noise, but the emphasis had to be on classifying the detected signals. Development was urgent, given the fear that Soviet submarines would be inferior to U.S. in duel situations.

The developers considered Skat-3 quite comparable to the U.S. BQQ-5/6, and they were proud of how quickly—in six years—they had completed development of so complex a system. In Skat they already had limited digital systems for predisplay processing and classification. Now they had to develop a real-time operating system to handle spatial, frequency, and time processing. Software developed specially for Skat-3 amounted to more than half a million words, using a total computer power of 200 GOPS. The developers considered that their experience was not too different from that involved in the development of the U.S. *Los Angeles*–class system (presumably BSY-1), and that their own effort involved more complex problems. They claimed that the sheer size of their effort was convincingly demonstrated in comparison with descriptions of space, aviation, ground, and naval information system development programs. Presumably the key problem was that there was no background of other Soviet digital signal-processing efforts that would have provided basics such as the necessary real-time operating system.

Skat-3 functions matched those of Skat, except for the demand for much greater sensitivity. Target acquisition and tracking and acoustic intercept were all automatic. Narrowband and broadband signal processing applied both to detected targets and to intercepted sonar signals. As in other Soviet sonars, the system included an acoustic communications channel by means of which information and identification could be exchanged among submarines. For the first time, Skat-3 included a system to measure the vertical sound velocity profile directly. That profile in turn was used for performance prediction. A network of digital computers performed primary and secondary data processing.

The array differs from that of Skat, in that the bow cylinder is passive only. Active signals are emitted by flat arrays on the sail and between the hulls.

The towed array pod of a Victor III–class submarine, two of her propeller blades visible to the left. She had fouled a U.S. towed array cable; the crewman is cutting it away so that she can escape. The submarine's own array was wound up inside the pod, with its tow line. Larger Soviet submarines carried their arrays internally and deployed them through tubes atop their vertical fins. USNI

The version of Skat used in Victor III submarines (Project 671RTM) employed a towed array. It evolved from a 1964–67 project for a very low frequency (subsonic: 100–200 Hz and below) passive system that would concentrate on submarine blade rate, detected either directly or by demodulating flow noise (DEMON). Initially the project examined not only towed arrays but also hull arrays (bow, sides, rails) and fixed bottom arrays. The submarine trials platform was the Pacific Fleet Zulu-class submarine *B-68*. Her towed array successfully detected the blade rates of nuclear submarines (30 Hz and below). This and work on a bottom array (Liman-M) led directly in 1970 to an experimental project to develop the Avrora towed array in three versions: for submarines (Avrora-P), for surface ships (Avrora-N), and for installation in Arctic drift ice (Avrora-D). Since first-generation Soviet nuclear submarines had strong blade rates, the arrays were to be used on a sprint-and-drift basis, the submarine stopping while listening. The array would be retracted once the submarine moved at (loud) speed. Hence the characteristic array stowage nacelle introduced in the Victor III class. In order to be reelable within a relatively small nacelle, the array had to be quite thin.

The Soviets' experience in detecting their own blade rates probably led to efforts to reduce that signature. Once that had been done, in second-generation nuclear submarines, they found that submarines were still detectable by discrete spectral components at somewhat higher frequencies (LOFAR signatures, as in Western practice). These components became the priority for towed array detection. Now the array could be substantially shortened. The submarine did not have to stop to listen, and the array therefore did not have to be retracted periodically. The new type of array, a modified Avrora-P, passed its tests both as an independent sonar and as part of Skat in the late 1970s and early 1980s. Work on the towed arrays was conducted in 1978–85). The stand-alone Pelmida, conceived for strategic submarines, included a pair of linear flank arrays. All of the towed arrays were nicknamed Shark Tail by NATO.

The major new sonar development of the 1970s, intended for fourth-generation submarines such as Project 885, was Amfora. It employed the first spherical bow array in a Soviet submarine, so large that the torpedo tubes were moved back (as in U.S. submarines) to fire around it.

The Kit (MG-20) passive ranging sonar was apparently unknown in the West. It was analogous to DUUX 1 and PUFFS. The first prototype was tested on board a Romeo-class (Project 633) submarine in 1966, the second on board a November (Project 627A). Production began in 1967.

The Soviets tried non-acoustics of various types as alternatives to sonar. This Hotel-class nuclear submarine was apparently converted into a testbed for a variety of probes. At the end of the Cold War, the U.S. Navy and the Royal Navy gained access to this Soviet work, and both seem to have tried analogous systems without success. USNI

A Victor-class submarine shows non-acoustic probes at deck level forward of her sail. The object forward of her forward periscope collected water for analysis, the goal of the system being to enable the submarine to trail a silenced Western submarine. The technique seems not to have been successful. USNI

APPENDIX C
Torpedoes

HOMING GENERALLY TURNS into a stern chase; hence an ASW torpedo needs a speed advantage. The rule of thumb is that a homing torpedo should be 50 percent faster than its target. In 1968, in connection with a torpedo project, the British made numerous computer runs simulating homing torpedoes attacks. They concluded that the torpedo needed a 30 percent speed advantage, so that a 45 kt homing torpedo could, in theory, successfully attack a 34.6 kt target. The faster the torpedo, the worse its self-noise and flow noise over its sonar.

Passive homing generally gives the bearing of a target but not its elevation in the water. Some torpedo developers argued that snorkel submarines would be too quiet to detect except when snorkeling, in which case their depth would be obvious. Too, a passive torpedo with relatively low speed could be effective because submarines generally could not snorkel at speeds over about 10 kts.

The first U.S. ASW torpedo was the wartime Mk 24 "mine" (Fido) developed by NDRC to meet a BuOrd requirement. The project moved very quickly. At a 10 December 1941 meeting at the Harvard Underwater Sound Laboratory, a feasibility study of a homing antisubmarine torpedo was organized. General characteristics were outlined at a 24 December meeting at Bell Telephone Laboratories. The first experimental run was made on 10 May 1942. Fido received top development priority in August. On 7 December an air-dropped prototype had its first successful acoustic test. By that time production was already planned. Fido was service-tested on 1 February 1943; the first was delivered to the Navy in March 1943, and by May 500 had been delivered.

Fido homed at supersonic (i.e., nonaudible) frequency (24 kHz), hence would not respond to audible mechanical whines or to propeller beats that sonobuoys could hear at ranges beyond that of cavitation swish. This frequency was considered the lowest that would be unaffected by the mine's self-noise. Fido responded to cavitation swish at about the same range at which sonobuoys could distinguish it. Because the torpedo homed passively, its detection range depended on the speed and depth of the target. Average listening range varied from 1,350 yds on a 6 kt submarine at periscope depth to 120 yds on a 3 kt submarine at 250 ft depth. Torpedo speed

A U.S. FRAM II destroyer shows her tubes for both long and short torpedoes. The long tubes were for relatively long-range Mk 37s, which gave her a capability intermediate between that of her DASH drone and that of her lightweight tubes firing Mk 44 or Mk 46 torpedoes. It proved difficult to fire wire-guided torpedoes from the long tubes shown here. USS *O'Brien* (DD 725) is shown as refitted, 26 August 1961, at Mare Island. NHHC

was limited by the noise of water flowing over the seeker. A friendly ship within 3 nm of the dropping point would mask submarine noise. Thus Mk 24 was not to be dropped when within 3 nm of a ship or if it seemed likely that a ship would enter the area within 15 minutes. It had to be dropped at low altitude. For example, at a ground speed of 150 kts, it should be dropped between altitudes of 200 and 300 ft. The torpedo searched passively to a depth of 125 ft, with a 30 ft ceiling. Rising above 30 ft disrupted depth control, so it sank back to 125 ft. Speed was 12 kts. Endurance with a full-strength battery was 12–15 minutes with a battery at full strength (3 nm). The explosive charge was 95 lbs of Torpex. Mk 24 weighed 680 lbs. Diameter was 19 in. and length 83 in.

Fido was too slow to attack the Type XXI submarine. Knowing Type XXI's capability, in October 1944 the U.S. Navy established a requirement for a fast successor, which became Mk 35.[1] Initially it was a passive homer, usable by aircraft, surface ships, and submarines. Speed would be 20 kts (for 20 minutes), and it would dive to 750 ft. It would weigh 1,400 lbs (250 lb warhead). An August 1945 OpNav specification omitted submarines, so the standard air-launched caliber, 22.4 in., was adopted. Speed increased to 30 kts and maximum depth to 1,000 ft. Homing range would increase to 2,500 ft. Charge increased to 400 lbs. Then submarines became prime ASW platforms, so in June 1946 caliber changed again back to 21 in.

By the spring of 1947 the torpedo was assigned an active (not passive) seeker; passive seekers were considered too vulnerable to noise-making decoys. This Mod 1 version was expected to reach 30 kts and a range of 15,000 yds. Maximum depth was 1,000 ft. The torpedo could be launched by both submarines and surface ships. It could be set to attack surface/snorkel targets only (floor 30 ft), submerged targets only (ceiling 50 ft), or targets anywhere in the whole depth of the water. Mk 35 was intended to home on a 175 × 20 ft target, with at least 100 ft between keel and sea bottom. Enabling range could be set at 300–10,000 yds. Mk 35 was to be 161 in. long and to weigh 2,500 lbs (in fact it was 162 in. long and weighed 1,770 lbs). Powered by a seawater battery,

The first U.S. postwar fleet destroyers, the *Forrest Sherman* class, had four fixed 21 in. torpedo tubes. USS *Barry* is shown in 1960 with her new clipper bow, which keeps her bow anchor clear of the SQS-23 sonar transducer in her forefoot. Her two port torpedo tubes, mounted at an angle to her centerline, are visible at the after end of the superstructure block which includes her forefunnel. Abaft them on the same level, just forward of her after funnel, is a Mk 32 triple tube for lightweight torpedoes. Note the 01-level Hedgehog abeam her bridge. The shield on her fantail protects the winch and spare bodies of her Fanfare T Mk 6 torpedo decoy. NHHC

it was credited with a range of 15,000 yds at 27 kts. Presumably the original 30 kt requirement was associated with a failed project to replace the battery with an internal combustion engine.

Prototypes were being tested at Key West by the fall of 1948. Formal evaluation of the Mod 2 production version began in March 1953. There was also a production Mod 3. Before it was superseded by Mk 37, about 400 Mk 35s were made in 1949–52. In addition to submarines, Mk 35 was planned for surface ships; it featured in the specifications for early postwar ASW ships, such as DDEs.

Mk 35 was too long to share the bomb bay of the new standard patrol bomber (the P2V Neptune) with an anti-snorkel APS-20 radar. Since it was undesirable to operate separate radar and attack airplanes, Mk 35 was shortened to become Mk 41 (120 in., 1,327 vs. 1,770 lbs). About 200 were made.[2]

When it became apparent that Mk 35 would not be ready as soon as expected, CNO directed BuOrd to reengineer the earlier Mk 27 Mod 4 for possible interim use by surface ships and submarines.[3] As of mid-1948, BuOrd expected that development would be complete by 1 April 1949 and planned to produce a lot of 200 for training. Mod 4 was generally described as a submarine anti-escort weapon. However, its ability to home passively enabled it to attack snorkeling or surfaced submarines. It was replaced by Mk 37 Mod 1 in 1960.

The interim surface ship torpedo was Mk 32. It had the first U.S. active (echo-ranging) torpedo seeker. Mk 32 could be launched by surface ships or aircraft; as a development of Fido, it had much the same performance (initially 6,000 yds at 12 kts, but 9,600 yds at 12 kts in the postwar Mod 2). About ten were made during World War II for evaluation, but under the postwar (presumably Korean War) program about 3,300 were made, serving until about 1955, when they were replaced by the lighter Mk 43.

A Mk 32, the interim surface ship ASW torpedo, on its launcher on board the preserved destroyer *Cassin Young*, 13 June 2012. RICK DAVIS

Launching a Mk 32. USNI

As it was 19 in. in diameter, it was typically launched by being thrown over the side.

CNO also directed that the earlier Mk 34 Mod 1 (initially Mine Mk 44) be engineered as a possible interim air-launched ASW torpedo. It had been developed between March 1944 and March 1945 by the U.S. Navy Mine Warfare Test Station at Solomons, Maryland, to test the feasibility of a faster homing torpedo. Initially it was wanted as an anti-escort torpedo faster than Mk 27, so its fins and rudders were cut down to fit a standard torpedo tube, from which it would swim out. Like Mk 27, it had two tandem batteries. Work was revived when Type XXI appeared, the control surfaces being enlarged again. The torpedo's longer body made it possible to drop at higher aircraft speeds. The batteries could be connected in parallel or in series for two-speed operation: 12,000 yds at 11 kts or 3,600 at 17 kts, or a 30-minute search at 11 kts followed by 6–8 minutes of higher-speed attack at 17 kts. Corresponding running times were 10 minutes for Mk 24 and 30 minutes at low speed for Mk 34 (6–8 minutes at attack speed). Mk 34 also had a heavier warhead (116 vs. 92 lbs) and was considerably heavier (1,150 vs. 680 lbs) and longer (125 vs. 84 in.). Diameter was the same as Mk 24, 19 in. Fin span was 26.4 in. Mk 34 used magnetostrictive rather than crystal hydrophones. Production of 300 for evaluation tests was completed in July 1945. A batch of 200 was assembled and placed in storage.

As of 1948, it was expected that CNO would soon direct production of 100 new Mk 34 Mod 1 torpedoes. Mk 34 was released for service use in July 1952 as an interim ASW weapon. The first were 25 of the original Mk 34s, converted by Naval Ordnance Plant Forrest Park. Maximum speed of 17 kts was marginal against a Type XXI, but it was much better than the 12 kts of Mk 24. In 1947 the experimental ASW squadron (VX-1) reported that Mk 34 had been found very effective against a World War II target. Against a 13 kt target, if dropped within 300 ft, it would make 8 to 10 attacks within 3 minutes. Against a 15 kt target, if dropped within 200 ft, it would make one or two attacks and then lose the target. It was also susceptible to decoys

The Mk 37 was the most important heavy U.S. torpedo of the 1960s.It was a swim-out torpedo; to provide sufficient space for water to flow back as it moved down the torpedo tube, its diameter was 19 rather than 21 in., with studs (visible here) to keep it from colliding with the tube walls. This one was not wire-guided, hence does not have a wire dispenser under one of the vertical fins. USNI

but was considered useful if it could be dropped close enough to its target. Ultimately 4,050 Mk 34s were made.

For U.S. submarines, the important ASW torpedo of the 1950s and 1960s was Mk 37, which had been conceived as an anti-escort complement to Mk 35 and antiship homing torpedoes.

While Mk 37 was in the early stage of development, work was proceeding on an active homing panel developed from 1942 onward by the wartime Harvard University Sound Laboratory. It was Doppler-enabled to avoid homing on static targets and on reverberations. Conical scanning enabled it to use a single amplifier and receiver, which sampled the inputs of its four quadrant receivers in sequence to generate steering commands.[4] In 1946 Westinghouse and the Ordnance Research Lab at Penn State (ORL) combined this active homing system with a passive one and a new body and new propulsion to create Mk 37. Adoption of quadrant homing, enabling the torpedo to find a target above or below it, suggests that Mk 37 was already considered an ASW weapon. As with other U.S. homing torpedoes of this period, the active seeker worked at 60 kHz, its high frequency (hence short wavelength) giving good directional definition even in a small torpedo head. The primary battery fed a two-speed electric motor. Initially range was set at 3,000 yds (later much exceeded). Mk 37 was designed to swim out of its tube, both for quietness and to enable launch down to 1,000 ft. Its reduced diameter (19 in.) left room for water to flow back over it as it powered itself out of its 21-in. tube. It had visible strakes to support it against the sides of the tube.

It took about a decade (and $90 million) to show that the faster Mk 35 was never going to work. Mk 37 became the only large-caliber U.S. ASW torpedo, being approved for service use in 1959. Given its limited speed, it was always a single-purpose (ASW) torpedo. In developed form it had a range of 10,000 yds at 26 kts or 23,500 yds at 17 kts. The warhead weighed 330 lbs. Mod 0 and Mod 3 (a refurbished Mod 0) became standard submarine weapons.

Initial production was very slow. As of April 1960, 99 U.S. submarines had array-type sonars making them suitable for ASW missions, but there were only enough modern torpedoes (Mk 37 and the antiship Mk 16) for 15 of them; the others had obsolescent World War II weapons (Mk 27 and the antiship Mk 14). About 50 Mk 37s were being delivered each month; each submarine needed 17–25 of them. Nearly 1,500 Mk 37-0 torpedoes had been programmed through the FY 60 budget, amounting to about 40 percent of the initial requirement.

Some U.S. FRAM II destroyers had fixed torpedo tubes enabling them to fire Mk 37s. As a surface ship weapon, Mk 37 had a minimum range of 300 yds, minimum depth 4 ft, and maximum depth 1,000 ft. It could home actively or passively or use a combination of the two. Range was 8,000 yds at high speed and 18,000 yds at low speed. The torpedo followed a preset horizontal search pattern, circling left or right or snaking. It could search vertically up from the ordered running depth within preset limits.

To protect the firing submarine, Mk 37 had an initial straight enabling run. Unfortunately, that might last as much as 15 minutes, giving a submarine that heard it time to evade. The solution was wire guidance, which enabled the firing submarine or surface ship to maneuver the torpedo as the launching ship tracked the submarine target. The U.S. wire guidance system was tested using modified Mk 27 Mod 4 torpedoes designated Mk 39 (120 were made). They were employed mainly by the seven fleet submarines converted into SSKs.

Wire guidance added 25 in. and 260 lbs to the Mod 1 version of Mk 37. Operating modes were predicted intercept, corrected intercept (midcourse corrected run), and bearing-rider (wire-guided, the torpedo pointing continuously toward the target, as in a pursuit-course interception). Speeds were 24 kts (range 8,000 yds) and 16 kts (18,000 yds).

Mod 2 was a refurbished Mod 1 with new sonar transducers, which increased acquisition range to about 1,000 yds. The first was delivered in March 1967. Active acquisition range was over 900 yds at 900 ft depth. The torpedo could follow a helical snake search pattern when water depth was greater than 150 ft (2,700 ft diameter, 1,000 ft maximum depth). Otherwise it could be set to run at constant depth while homing, or to search vertically.

While developing Mk 37, BuOrd also worked on a quiet passive wire-guided torpedo developed from Mk 27 (though modification was so extensive that very little of the earlier torpedo remained). By 1956 it was designated EX-4 in the bureau's experimental series. It was intended to attack snorkeling submarines, so running depth was only 50 ft. Range was 13,000 yds at 15 kts. This project seems to have been abandoned in favor of the EX-10, which became Mk 48. In the fall of 1956 BuOrd wanted to develop a 35 kt, 20,000 yd wire-guided torpedo, the main obstacle to the desired range being signal attenuation in so long a wire.

While these torpedoes were being developed, the wartime Mk 32, the first U.S. active-homing torpedo, was resurrected for aircraft and destroyers. Fifty had been made on an experimental basis in 1945. As Mod 2 (1951), 3,300 were made for destroyers. The torpedo weighed 805 lbs (19 in. diameter, 93 in. long, 107-lb warhead) and made 9,600 yds (24 minutes) at 12 kts using a secondary (rechargeable) battery. They searched in a helical pattern. These torpedoes employed Mk 2 over-the-side launchers. Mk 32 was withdrawn once lightweight Mk 43 torpedoes entered service about 1955.

Given the manifest problems of early ASW torpedo development, unguided weapons remained important. The converted *Fletcher*-class destroyer *Murray* shows her Weapon A in its vertical (loading) position. She also shows one of the two Low Report fixed Hedgehogs, on her 01 level, with ready-use ammunition boxes. The muzzles of two fixed torpedo tubes are visible in line with her HF/DF mainmast. USNI

Alongside the large-caliber Mk 35 and Mk 37, BuOrd developed lightweight torpedoes, initially for helicopters. The first such U.S. lightweight (Boswell) was developed by the Advanced Weapons Unit (AWU) of the Key West Operational Development Force. It was derived from the abortive wartime Mk 30 small-diameter (10-in.) homing torpedo demonstrated in 1943, which had been developed by the Brush Company. The 10-in. AWU weapon was 109 in. long and weighed only 196 lbs as a practice shot. At this time the BuAer Armament Division was proposing the much heavier Mk 41.

Successful development of Boswell seems to have inspired a requirement for a lightweight weapon. BuOrd requested a study in February 1950, and desired characteristics were set out in March 1950, but this was an unfunded study at the time.[5] This Mk 43 was intended to use a Mk 35–type active acoustic homing system, albeit on a smaller scale. It began its search by diving at 30 degrees, making a flat turn at the initial search depth, then sinking at 12.5 ft/sec while turning to port, in a helical pattern. Acquisition zone, defined by the cylinder within which a target would be detected (not acquisition range) was 500 yds. It was acquisition range plus the radius of the torpedo turning circle (140 ft). Initial search depth was typically set at 150 ft but could be set at 75–500 ft. Depth limits of Mod 1 were 75 and 650 ft.

Mod 2 was conceived as an acoustic equivalent of the passive-homing Mk 34, passive range being considered greater than active against a snorkeler (to achieve sufficient range, homing frequency was cut from 45 to 30 kHz). The passive control panel was interchangeable with the active one of Mod 1. Technical evaluation was completed in April 1955.

Mod 3 increased speed to 21 kts with the same 4,500-yd range. Its ring tail came within its 10-in. overall diameter. It could dive to 1,000 ft (minimum search depth 50 ft); it could be set to begin searching at 50–850 ft. Its search circle radius was 240 ft, and it sank at 17.5 ft/sec. Total production of all versions of Mk 43 was about 5,000, between 1951 and 1959.

A modified Mk 43 Mod 0, designated EX-5, was used in tests of the RAT standoff weapon.

Mk 43 could not approach the requirements set by the Hartwell Committee, so in 1952 BuOrd let competitive development contracts under the designation

The Mk 43 Mod 3 torpedo had a ring tail like that of its successor Mk 44, but only a single propeller. Mk 43 was the first lightweight ASW torpedo a helicopter could launch. The initial Mod 0, with a 12.75 in. diameter, had conventional tail fins, as did the 10 in. diameter Mod 1. Both versions were electric and carried out helical searches to find their targets. Mod 1 was selected for further development and production. Its Mod 3 production version had the ring tail shown here and was much faster (21 vs. 15 kts). NHHC

EX-2: the Naval Ordnance Test Station (Pasadena) EX-2A and the General Electric EX-2B.[6] In 1956, when development was complete, the required speed of 30 kts exceeded that of any operational U.S. homing torpedo. It was initially considered enough to kill a 30 kt submarine (but later was considered sufficient against a 17 kt target). Required range was 6,000 or 7,000 yds (6 minutes) using passive/active homing. The warhead weighed 75 lbs. Both competitors used 60 kHz active seekers, although by 1956 NOTS was interested in a sophisticated new electronic panel using FM pulses and doubling and quadrupling the pulse rate as the torpedo closed in on the target, frequency being reduced to 28 kHz. The torpedo would search at a series of depths, at each beginning with a passive search and then switching to active before going to the next depth.

The winning EX-2B, which became Mk 44, was powered by a seawater battery fed water by a scoop. It introduced the standard 12.75-in. diameter of U.S. and other Western lightweight torpedoes (length 100 in.). It weighed 425 lbs, with a 75 lb warhead (Mod 1 was 101.3 in. long, 433 lbs, with a 73 lb warhead). A characteristic feature was a circular ring around its tail, a device introduced on wartime U.S. air-launched torpedoes to improve their in-water performance. Mk 44 was a pure active homer that searched in a helical pattern. Average acquisition range was 650 yds (maximum 1,000), compared with 800 yds for Mk 43. The torpedo could search between 50 and 1,000 ft, search depth being preset between 50 and 900 ft. It circled as it sank at 25 ft/sec, the search circle having a 440 ft diameter. Initial dive angle was 30 degrees, but as it searched the torpedo rose or sank though a helix at a set dive or climb angle of 4.5, 6, or 7 degrees, executing a flat turn at the bottom and then climbing back up to continue searching until its battery was exhausted. In an alternative snake search the torpedo ran at a set search depth. Maximum depth was 1,000 ft (minimum 50 ft). Estimated kill probability for a two-shot salvo was 28 percent.

The air-dropped Mod 0 version became active as soon as it entered the water. To make the torpedo

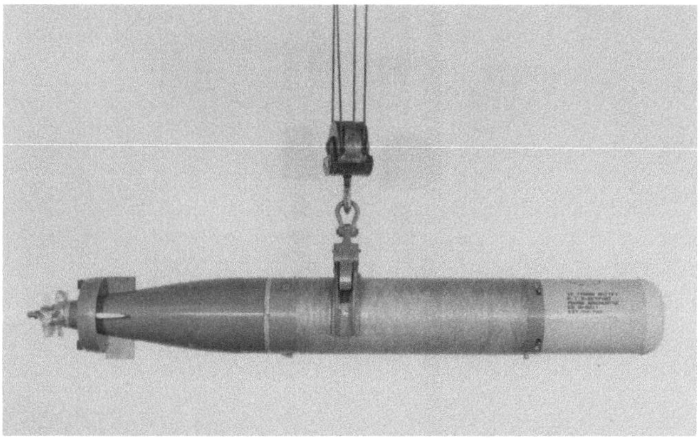

A Mk 44 torpedo, showing its contra props abaft its ring tail. Unlike previous homing torpedoes, it had two alternative ping rates, one for search and the other to close the target. General Electric, which developed it, initially planned to use a gas turbine burning propyl nitrate monopropellant. An accident halted its use, and GE had to switch to electric propulsion using a seawater battery and a converted jet engine starter motor. In theory the shroud ring surrounding the control surfaces would make them more effective, but test runs showed that was not true; the desired "end plate effect" was not forthcoming because the shroud was not flat. In a compromise redesign, the span of the control surfaces was reduced and the chord lengthened so that much of the control surfaces was outside the shroud. The production version was Mod 1. USNI

usable from surface ships Mods 1–4 could be set with a 1,000-yd fore-run before becoming active.

Production was initially slow, only 1,000 (less than 20 percent of the initial requirement) being programmed through FY 60. In all, 10,583 Mk 44s were made for the U.S. Navy under the FY 59–65 programs, the last being delivered in December 1967. Mk 44 was employed on board aircraft (including helicopters) and surface ships, and it was the initial payload of the ASROC system. From 1967 on the U.S. Navy replaced it with Mk 46.

Mk 44 became a standard NATO weapon, manufactured in Canada, France, Italy, Japan, and the United Kingdom. The Italian A244, conceived as a replacement for Mk 44 Mod 2, is sometimes described as an improved Mk 44 Mod 2.

The 1956 Nobska study recommended a new generation of torpedoes to deal with nuclear submarines, whose speed was expected to be about 30 kts; the corresponding minimum torpedo speed was 45 kts. Given problems in designing a fast homing torpedo, Nobska recommended an interim unguided nuclear torpedo: Mk 45 (ASTOR, Anti-Submarine Torpedo). It was associated with the passive PUFFS fire control system. For stealth, Mk 45 was a quiet electric torpedo that quietly swam out of its tube. The warhead was command-detonated via its guidance wire. To swim out, the torpedo had a diameter of 19 in.; length of Mod 0 was 225 in., and weight was 2,330 lbs. Range was 11,000–15,000 yds at 40 kts (later versions gained range by using seawater batteries), the minimum burst range being 2,600 yds. The torpedo ran out at a preset depth of 100–1,000 ft. Its shock wave could destroy submarines at considerably greater depths, a unique capability at the time. In Mods 1 and 2 enabling range was 2,050 yds, presumably somewhat beyond lethal radius. Development was completed in FY 60, and Mk 45 was approved for service use in FY 61; deliveries began in FY 63. Planned procurement was 528 torpedoes, of which 80 were Mod 0s and 342 were Mod 1s. As of 1969, 106 Mod 2s were being bought to complete the program. Mk 45 was withdrawn in 1976, when the conventional Mk 48 finally demonstrated the required capability. There was some concern that the shock effect of an exploding Mk 45 might damage the firing submarine, and there was also interest in reducing the number of naval tactical nuclear weapons. As Mk 45 was withdrawn from service, torpedo bodies were converted into "Freedom Torpedoes" for sale abroad (apparently unsuccessfully).

The two non-nuclear Nobska torpedoes were developed under a RETORC (Research TORpedo Configuration) program begun in November 1956. Initially RETORC was a general torpedo development program that would produce test vehicles to which promising concepts could be adapted. The development of a new torpedo, based more or less closely on a RETORC vehicle, would be started from time to time. Orientation of the program toward the two Nobska torpedoes departed from this model. RETORC I was the replacement for Mk 44, and RETORC II the replacement for Mk 37. In the fall of 1956 BuOrd was pursuing development of a wire-guided passive/

active torpedo capable of 27 kts, with a goal of 45 kts and an acquisition range of 2,000 yds.[7]

Torpedo development was affected by Nobska's recommendation to develop a submarine that could dive to 4,000 ft. *Thresher* nearly doubled the diving depths of previous U.S. submarines in one step, and it seemed reasonable that much greater depths could soon be achieved. A March 1962 submarine ASW program envisaged progressive increases in diving depth, to be achieved by weight-saving in the basic *Thresher* design. Diving depth might increase to 2,000 ft in FY 65, and a new mass-production submarine begun then might dive to 3,000 ft. A 4,000-ft-depth submarine might be begun in FY 66, and by 1975 a general purpose submarine might dive to 10,000 ft. None of these diving depths was approached.

The perception that future submarines would dive much deeper seems to have shaped U.S. long-term torpedo planners' thinking in the early 1960s.

A Mk 45 torpedo is loaded onto a torpedo retriever boat. Note the contra props and the small tail fins. Studs along the sides of the torpedo held it in place in its larger-diameter tube, providing space for water to flow around it as it swam out. The dispenser for the guidance wire is visible atop the tail of the torpedo. USNI

While the Mk 46 lightweight torpedo (below) was being developed, in May 1962 the U.S. Navy issued a Specific Operational Requirement for a deeper-diving Mk 44 replacement, consistent with its interest in deeper-diving submarines. It was called the Advanced Lightweight Torpedo (EX-11), but it bore no relation to the later ALWT. It would function down to 2,500 or 3,000 ft, searching a cylinder at least 4,000 yds in diameter through all layers, acquiring a target at 2,000 yds or more. The attack run at 45 kts would last at least 5 minutes. The warhead was to be powerful enough to rupture a submarine with a diving depth of 3,000 ft, and the torpedo was to be compatible with ASROC and DASH. The project presumably died when the Soviets did not field the envisaged deep-divers and the Vietnam War cut R & D money.

By 1960 the Mk 46 (RETORC I) program was well underway.[8] A final development characteristic issued in June 1956 called for a torpedo capable of attacking a 30 kt (i.e., nuclear) submarine, with dimensions of about 15 × 98 in. and operating depths of 50–1,000 ft. A feasibility study was completed in December 1956; Aerojet received a development contract in May 1958. The operational requirement, issued in November 1960, presumably embodied Aerojet's design: acquisition range of 1,200 yds and a maximum speed of 45–50 kts. Long acquisition range could make up for sonar errors, blind time, and target evasion. Mk 46 could search at a set depth, typically 750 ft. Maximum depth was 1,500 ft. In a technical evaluation (September 1962–June 1963), Mod 0 showed 70 percent reliability and the ability to attack 20 kt submarines. Computer simulation showed that it would be effective against targets with speeds up to 33 kts. Operational evaluation (April–November 1964) showed that in deep water, without countermeasures, acquisition probability was 60 percent; hit probability against a submarine after acquisition was 77 percent (88 percent against a mobile target). The torpedo could be dropped at airspeeds of up to 500 kts. When air-launched, torpedoes could be set with attack ceilings of 50 or even 20 ft, making it possible for them to attack periscope-depth targets.

Mod 0 (initially EX-8) was powered by Aerojet's solid-fuel rocket engine, its gas exhaust driving a turbine. When operationally evaluated, Mk 46 Mod 0 was 68 percent reliable.[9] Although the Mod 0 had not achieved the desired 90 percent, the U.S. Navy was confident that it would, to the extent that it planned no "onboard" testing of air-launched and ASROC torpedoes, and only a simple go/no go check of the run-out gyro actuating squib for tube-launched torpedoes. Torpedoes would be issued to the fleet for periods of 12 months without further maintenance.

The single homing frequency was considered acceptable because it seemed that torpedoes would be fired only rarely in salvo. Whether that was true was still unresolved. Apart from mutual pinging interference, torpedoes could interfere with each other by their radiated noise.

Data on shallow-water performance, which British observers considered crucial, was scanty. The original initial depth setting of 75 ft in Mod 0 was discarded due to excessive self-noise at that depth. The operational evaluation was run off Key West in depths of 300–500 ft. Some subsequent West Coast trials were run in 300 ft, but no trials were run in shallower water. The torpedoes avoided bottom or surface capture thanks to the complex logic of the homing panel, which incorporated a 3 kt Doppler gate and a reverberation filter. Except when launched from a surface ship tube, the torpedo normally overshot its initial search depth by about 90 ft. Minimum depth (except for surface launch) was taken as 250 ft. British observers (trying to decide whether to buy Mk 46 for the Royal Navy) were told that shallow-water capability had never been specified by the U.S. Navy, and that no steps were being taken to enable Mk 46 to run at shallow depths.

Mk 46 programmed logic was revised because it showed poor countermeasures resistance during op eval. The British observers doubted that measure would be effective against anything more sophisticated than a simple static decoy. They thought that R & D would be concentrated on better countermeasures resistance; improved shallow-water performance; and increased depth capacity, first to 2,500 and then perhaps to 4,000 ft. A more powerful warhead would be desirable. There seemed to be no U.S. interest in increased speed and endurance.

A Mk 46 lightweight torpedo being fired from its triple launcher. Despite the presence of the personnel around the launcher, it was automatic, controlled from the CIC. Note the characteristic swept-back fins. USNI

The first production Mod 0 torpedoes were expected to reach the fleet about February 1966.

An alternative Mod 1 program employed a swashplate piston engine using Otto liquid fuel. The first production contract was placed in 1965, the first production torpedoes being expected in January 1967. Mod 1 was the first major production version. Evaluation (November–December 1965) demonstrated an 86 percent probability of acquiring, attacking, and hitting a submarine. The new power plant made for better reliability and maintainability and less expensive production. The movable shroud ring used in Mod 0 (and in Mk 44) was replaced by four movable fins.

Operation was either active or passive/active. Maximum active homing range was 1,600 yds (1,300 yds average) using a 30 kHz sonar. The torpedo was initialized by setting its autopilot, which turned on the onboard fixed-program computer once the torpedo had passed a preset enable distance. Presets were homing mode, initial search and ceiling depths, search pattern, and course heading; some versions also had a search floor to avoid hitting a shallow bottom. The two main search modes were snake and circle, both flat, unlike the helix modes of earlier torpedoes. Course headings were programmed only for snake search (5 for a surface tube but 12 for a helicopter, so the latter could attack in any direction from the hover). Operating depth was 20–1,500 ft. Comparison with later torpedoes shows that the computer did not track several targets simultaneously to decide which was worth attacking (and which was a decoy), and it used preset tactics. The difference was presumably entirely a function of the much greater power of newer computers. A stronger hull (for greater depth) and better counter-countermeasures were introduced in 1968. Computer logic changes (1969) improved performance against slow and/or periscope-depth targets (masked by reverberation).

Speed was 45 kts, with an endurance of 6 minutes (9,000 yds). Maximum depth was 1,500 ft for acquisition, but the torpedo could survive down to 2,500 ft when in hot pursuit. Minimum depth was 226 ft. In 1975 the British credited Mk 46 with an acquisition range of 1,480 m (1,613 yds), compared with 550 m (600 yds) for the Mk 44 Mod 1 in British service. When searching actively, the torpedo turned at 10 deg/sec, but maximum turn rate was 50 deg/sec. These figures compared with 13 deg/sec for Mk 44 search and 23 deg/sec maximum. Compared with Mk 44, which had virtually no countermeasures resistance, Mk 46 Mod 2 was considered effective against a noisemaker, with some capability against an echo repeater. It could operate effectively in water 100 m deep (Mk 44 had no shallow-water capability and was prone to bottom capture).

Mod 1 began to search immediately upon hitting the water, hence could not be used by a helicopter with dipping sonar that detected a distant contact. Mod 2 (1971) solved this problem; for single-helicopter attack (HATS, the helicopter attack tactical system), it provided a 1,000-yd run-out (modes were AIR/ASROC, without a run-out, and HATS). Without HATS, two helicopters were needed for an attack, one vectoring the other. HATS used a modified autopilot that offered a wider set of initial courses (95, 120, 150, and 180 degrees in addition to the earlier 0, 35, and 70 degrees), so it could be launched while its helicopter hovered into the wind, regardless of target bearing. This version introduced PBXN-103 explosive (Mk 103 warhead), 27 percent more powerful than its predecessor.

Many U.S. ships were rebuilt with ASROCs carrying lightweight Mk 44 and then Mk 46 torpedoes and nuclear depth bombs. The large destroyer *John S. McCain* approaches the carrier *Constellation* to replenish, 14 April 1977. In effect she exchanged her forward Weapon A and her forward twin 3 in./70 for the ASROC box; aft, she exchanged the gun for a Tartar launcher and its radars. As modified, she was redesignated a large missile destroyer (DLG rather than DL). NHHC

Mod 3, which incorporated control improvements, was canceled, presumably in favor of NEARTIP.

Mod 4 was the payload of the Captor (Mk 60) mine. Work on this version began in 1973. Mk 46 procurement for Captor began in 1976. Captor procurement was stopped in FY 85, freeing about 800 Mk 46 torpedoes for the fleet. Procurement was not resumed, perhaps because it seemed less and less likely that Soviet submarines would be surged into the Atlantic through potential Captor minefields.

A major NEARTIP (near-term improvement program) upgrade (Mod 5) was prompted by a combination of Soviet silencing, anechoic materials (to reduce echoes from high-frequency torpedo sonars), and more sophisticated countermeasures (echo repeaters, pulse transmissions, and FM sweeps). Submarines were operating at greater depths. Air and ASROC launches were currently limited due to deep and erratic initial plunges. There was also interest in reducing propulsion noise and increasing warhead effectiveness. The goal was to restore initial detection range (1,600 yds) despite a 5 dB reduction in target strength. The initial plunge was to be reduced to less than 200 ft, a better autopilot pulling the torpedo out of its initial dive more quickly. Initial search depth cover would be increased to 2,500 ft (probably initially 2,000 ft). Improvement to the sonar transducer would reduce depth sensitivity, and logic (e.g., search patterns) would be improved. The seeker would operate at lower frequency (reportedly 24 kHz) at higher power. Propulsion would be silenced.[10] The program was intended to cover the period from 1979 through the late 1980s, when the follow-on ALWT (Mk 50, see below) would enter service.

Work on the upgrade began in October 1972; it became NEARTIP in July 1974. The upgrade was intended to apply to all existing Mk 46 torpedoes (at that time, about 10,000). The modified torpedo was designated Mk 46 Mod 5. It underwent technical and operational evaluation in 1977 and was approved for service use in February 1978. It was released for production in May 1979, the initial contract being awarded in July (for 570 torpedoes). The last U.S. order was awarded in February 1987. Mitsubishi in Japan received a license to coproduce Mod 5 torpedoes (coproduction was approved in July 1982).

NEARTIP ended up with a largely new seeker, guidance, and propulsion. Conversion of a Mod 2, for example, involved replacing the transmitter and receiver sections, replacing the control section, replacing the existing fuel pump (with some wiring changes), and replacing the forward cap over the sonar transducers.

In Mod 5 all operation was digital, with improved electronic stability and precision and an improved digital computer. Improved search and reattack patterns enhanced countermeasures resistance (discrimination) and target acquisition and reacquisition. Sonar improvements were intended to deal with Soviet anechoic coatings, which were expected to reduce target strength by as much as 13 dB, so that a submarine target with a strength of 8 dB would have that reduced to –5 dB. The primary object of the program was to maintain the same performance against the –5 dB target.[11] To do that, the sonar emitted a long single-frequency pulse supplemented by a short pulse for terminal homing (a narrow vertical beam intended to improve shallow-water performance was omitted). The new sonar waveform enabled better use of target Doppler for signal processing. The correlation channel used to determine whether sounds the torpedo sonar received really were echoes of its pings was improved to make up for weakening of echoes due to the new coatings. Better signal processing could better reject false targets. Complex new signal processing and counter-countermeasures logic enabled the torpedo to classify several kinds of countermeasure and to conduct more complex search patterns. Reverberation effects in shallow water were reduced, so self-noise could better be measured and disregarded. The fuze was modified to function even if the torpedo struck the target at a shallow angle.

The torpedo could begin its search at a maximum depth of 1,500 ft rather than at a maximum of 1,250 ft. NEARTIP could shift from active to passive homing to attack a submarine that surfaced to evade attack (earlier active homing torpedoes would be unable to attack a surfaced submarine). This mode was publicized by photos of surfaced submarines being sunk by NEARTIPs. The torpedo was modified to pull out more quickly in shallow water (to enable it to be launched in depths as shallow as 200 ft). The motor and fuel pump were changed to provide an alternative lower quieter speed (36 kts), which increased endurance (30–50 percent in terms of time). The torpedo could now search at low speed, accelerating to maximum speed (43 rather than 45 kts) to attack.

During technical evaluation NEARTIP achieved a success rate of 0.8, which was expected at that time to reach 0.9 after further wiring modifications. During technical evaluation the torpedo demonstrated satisfactory acquisition against low-strength targets, both coated submarines and simulated targets. There was some evidence (as expected) that in shallow rocky areas the torpedo would have a poor acquisition performance.

The U.S. Navy introduced further modifications in 1984 to improve performance against slow submarines at periscope depth in both deep and shallow water. Such targets provide little Doppler (which the torpedo uses to distinguish them from background noise), and the torpedo may be confused by reflection (both of its pulses and of its self-noise) from the surface. The torpedo was given a more coherent transmitter and better steering logic. This version was successfully tested in Eckernförde Bay in 1987, confirming that it met the NATO 40 m (131 ft) depth requirement. Honeywell (now Raytheon), the manufacturer, developed a further shallow-water modification, Mod 5A(S), beginning about September 1989 (it entered production about 1990).

Total Mk 46 production for FY 80 and earlier years was 11,646 torpedoes. That included 1,252 Mod 0s and about 9,200 Mod 1s. Mk 46 exports, which kept production open, began in FY 67. U.S. Mod 5 production amounted to 5,881 new-build torpedoes, not counting conversion kits. Reported total Mk 46 production was about 26,000 torpedoes. The first NEARTIP (Mod 5 conversion) kit contract, for 600 kits ($18.5 million), was awarded in February 1978. Contracts for another 5,140 kits were let in 1979–83 (total 5,740). Additional Mod 5s were made for the Dutch and British navies. In mid-1995, the U.S. Navy had about 14,000 lightweight torpedoes on hand (about 300 Mk 50s and 9,500 Mk 46s active plus 4,000 Mk 46s being cannibalized or in deep storage).

Once Mk 46 was well underway, RETORC I turned to developing a much more efficient power plant offering much greater range (12,000–19,000 yds) and operating depth (3,000–4,000 ft).[12] By 1965, the objective had shifted to a combination of better shallow-water and deep (to 4,500 ft) water capability. A new homing system using coherent processing would have a high search rate. Good range and Doppler resolution would protect it against countermeasures. It would search at low speed and attack at high speed. The propulsion goal was lowered to 40 kts (range about 7,000 yds) because in shallow water submarines would operate at fairly low speeds. It was also believed that submarine speed would not rise above 30–35 kts for a long time. Electric propulsion was favored, but thermal had not been ruled out. The planned vehicle was 119 in. long, with enough space for a 250-lb warhead, rather heavy for a lightweight torpedo.

Meanwhile work proceeded on the heavy RETORC II (EX-10) torpedo, which became Mk 48. The feasibility study was completed in September 1957, and the operational requirement issued in November 1960. Compared with Mk 37, EX-10 was to offer 400 percent of the range, 150 percent of the diving depth, twice the speed, and 17 percent greater acquisition range. The existing Mk 37 was credited with only 10 percent effectiveness against a 20 kt submarine. The proposed Mk 48 should be 40 percent effective against a 35 kt submarine. Mk 48 was later officially described as the only U.S. torpedo capable of reaching the Alfa's operating depth, reported (incorrectly) as 800 m (2,624 ft).

This program was much more focused than RETORC I. Its Phase Alfa produced the Mk 48 torpedo, which in 1962 was expected to complete development by the end of 1965. Work was accelerated after system tests on board the submarine *Permit* in the Pacific in 1964. A Beta program was intended to extend the torpedo to much greater depths. It was not pursued urgently because those involved were fully occupied in developing the Alfa torpedo and also an (ultimately abortive) EX-13 antiship torpedo.

BuOrd's novel solution to the problem of self-noise at high torpedo speed was to split an attack into

The pump-jet propulsor of a Mk 48 torpedo at the DSEI 2009 show. The tab above the torpedo, forward of the pump-jet, is the guidance wire dispenser. AUTHOR'S COLLECTION

Some surface ships were adapted to fire Mk 48s. *Brooke* (DEG 1, later FFG 1) in the Pacific on 15 June 1969 shows the two long torpedo tubes in her transom, intended to fire Mk 48 torpedoes. The torpedoes would have been fired dead astern to simplify deployment of their guidance wires. Similar arrangements were designed into the parallel *Garcia* class and into the follow-on *Knox* class. Given limited Mk 48 production (due to the cost of the war in Vietnam), it was decided that these weapons would be carried only by submarines. In many *Knox*-class frigates, the spaces were used instead for the machinery to deploy SQS-35 variable-depth sonars. NHHC

transport (runout) and attack phases. During the transport phase the torpedo would be controlled by the firing submarine, so it did not have to limit self-noise that might deafen its sonar. Near the target the torpedo would slow to search. Once it was near the target, target echoes would be stronger than flow noise. It could accelerate before attacking. The British characterized this as the "fast and noisy" approach to torpedo design. It could be argued, moreover, that if the target submarine heard Mk 48 coming, it would probably start a high-speed run of its own, making more noise and thus promoting homing by the Mk 48.

About 1962, planned transport speed was 50 kts and search speed was 25 kts. The transport speed somewhat exceeded the Nobska goal of 45 kts, but the expected range of 20,000 yds was short of Nobska's recommended 35,000. By 1961 RETORC II was using a pump-jet to minimize self-noise. By 1965 BuOrd announced that the test vehicle approached the noise level of the quietest torpedo then in service (Mk 37) while running at more than twice its speed (54 kts). At this time the Nobska target acquisition goal (2,000 yds, twice that of Mk 37) had been increased by 50 percent, so the acquisition cone was more than 20 times the area of the one associated with Mk 37. Maximum depth was 2,500 ft.

The combination of desired speed and range required an internal-combustion engine. Initially BuOrd planned to use 90 percent hydrogen peroxide plus diesel oil, but in 1963 it switched to a new mono-propellant, Otto II (named after the bureau's Dr. Otto W. Reitlinger; it is a nitrogen ester with oxidant). Tests began in July 1963. Otto fuel proved far less temperamental than peroxide, and it was formally adopted for the new torpedo in February 1964.

Mk 48 introduced digital wire guidance that could support complex commands. Too, digital signals tended to remain intelligible at greater ranges. The U.S. Navy was introducing digital fire control systems (later versions of Mk 113 and then Mk 117). The all-digital Mk 117 in particular (on board the *Los Angeles* class) had no provision for analog links to torpedoes, hence could not support the Mk 14 antiship torpedoes they carried because existing homing torpedoes were ineffective against surface ships. Hence the need to produce a dual-purpose form of Mk 48.

In 1961 Westinghouse received a project definition contract, and Clevite a contract for an alternative acoustic system. Westinghouse received a contract to develop the turbine-powered Mod 0, which was limited to ASW.[13] A production contract was placed in July 1964, trials of the first test vehicle (non-acoustic) being expected to begin in November 1965, and the first acoustic vehicle to be in the water in the spring of 1966. At this time operational depth was 10–2,500 ft. Speed was continuously variable between 25 and 55 kts. Range was 25,000 yds at 55 kts at 100 ft, or 13,000 yds at 55 kts at 2,500 ft. Acoustic homing range was 4,000 yds. The guidance wire was 20,000 yds long. Countermeasures resistance was a high priority. As of 1964, operational evaluation was expected to begin in March 1968. Mk 48 Mod 0 entered service in 1972, apparently delayed by limited funding during the Vietnam War era.

Clevite received the Mod 1 contract for a torpedo powered by a swash-plate engine. It was louder than the turbine, but it turned out that flow noise over the torpedo was far stronger than engine noise. BuOrd therefore chose the less expensive if noisier swashplate, which maintained its power level against the increased back-pressure at greater depths, though it used more fuel and therefore lost endurance. Turbines were quieter but lost power (hence speed) at depth. The first Mod 1s were delivered in 1972.

In 1967–68 contracts were awarded to modify the basic torpedo for antiship as well as antisubmarine use, the single-purpose Mk 47 antiship torpedo project having been abandoned. This requirement presumably accounts for the torpedo's non-acoustic mode, presumably a straight or pattern-running shot at a surface target, the only intelligent element of the torpedo being its fuze. Numerous published photographs showing Mk 48s blowing substantial old warships in half strongly suggest that the torpedo uses a standoff (proximity) fuze for under-the-keel explosion. Part of the difference from Mod 0 may have been the weight of the warhead, the 250 lbs of Mod 0 being insufficient. The Mk 48 Mod 1 warhead was 645 lbs of PBXN-103. The ADCAP version has a 650 lb PBXN-105 warhead. Westinghouse redesigned Mod 0 as Mod 2 but lost out in a renewed competition.

In 1977, the British considered buying Mk 48 Mod 1 instead of developing a new torpedo (which became Spearfish).[14] The U.S. Navy told them that overall "mission success" rate was 66 percent. They estimated that reliability was 89 percent, excluding problems with the wire dispenser. Rated Mk 48 endurance at 2,000 ft depth at 55 kts was 17,000 yds. At 500 ft depth it was 21,000 yds. Normal runout speeds were 40 and 55 kts. Total wire length was 30,000 yds, of which 20,000 was in the torpedo and 10,000 in the firing submarine. The British considered Mk 48 quite noisy when running out at high speed, so much so that the firing submarine probably would be unable to track it after firing. It would surely alert a target, which would maneuver away and deploy decoys. However, within a critical range the target submarine could not escape: 8,000 yds for a fast submarine, 12,000 for a slow one. The British considered the 55 kt attack speed near-optimum for chasing fleeing future targets.

In deep water, against a low-Doppler target in the Norwegian Sea in winter at 40 kts, active acquisition range was 2,600 yds. Against a high-Doppler target it was 3,500 yds. Under subtropical conditions, at 40 kts against the low-Doppler target it was 3,500 yds, and against a high-Doppler target it was 5,100 yds. The torpedo would do somewhat better at its 28 kt search speed—for example, 4,000 yds against a high-Doppler target in the Norwegian Sea in winter. In shallow water (700 ft), reverberation would limit acquisition range to less than 2,000 yds against a high-Doppler target. The torpedo would be unable to home on a low-Doppler target. In a sea state greater than 3, the torpedo was reverberation-limited; otherwise it was limited by self-noise. When reverberation-limited the torpedo might have difficulty acquiring submarines with anechoic coatings (the British thought this was why the U.S. Navy was "gravely" concerned with the development of Soviet anechoic coatings). In shallow water detection range was generally about 1,500–2,000 m (about 1,300–2,020 yds).

Under good conditions homing beam coverage against high-Doppler targets would extend from the surface down to 2,000 or even 2,500 ft in low sea states. That was cut to 1,200 ft under reverberation-limited conditions against a low-Doppler target in sea state 3 or worse. In good deep water the swept volume of the active homing system would enable a Mk 48 fired from within 8,000 yds to acquire a submarine despite evasive maneuvers (without countermeasures). In shallow-water high-Doppler conditions that would be reduced to 5,000 yds. Homing performance would degrade in continental shelf areas with "hard" bottoms and high sea states.

The U.S. Navy did not use passive homing against submarines, but the British evaluating Mk 48 estimated that passive homing range was 1,150 yds for a "90 dB" radiated noise target (a 10 kt Y- or D-class submarine).

Mod 3 (1977) added a two-way wire link, enabling the torpedo operator to exploit the torpedo's sonar data.

Mod 4 resulted from a directive by the Secretary of the Navy for a short-term fix of the Mod 1–Mod 3 inventory to counter the Soviet Alfa class, opening the Doppler window for higher target speeds and increasing depth. The first modification kits were bought in FY 82.

A more radical ADCAP (advanced capability) redesign was inspired by tests that showed that the basic Mk 48 structure could survive at the depths credited to the new Soviet Alfa-class (Project 705) submarine, and that targets could be recognized at the necessarily high torpedo speeds involved. The first step was Mod 4, with improved ADCAP electronics (a digital processor and an improved sonar with greater source level). It added a fire-and-forget mode.

ADCAP would have merited a new Mk 49 designation. Instead, the initial ADCAP became Mk 48 Mod 5. Presumably using the Mk 48 designation simplified the programmatic side of the development process. The Operational Requirement was issued in 1975, and the formal request for proposals in November 1978. Hughes received the contract to develop the torpedo's digital guidance and control electronics in August 1979. The program was accelerated in September 1979 in view of the demonstrated characteristics of the Alfa-class submarine. ADCAP reached full production in 1989.

Unlike the analog Mk 48, ADCAP was a fully digital weapon. It had greater speed, endurance, and depth capability as well as better electronics; a quieter acoustic array and a new antiship under-the-keel exploder; and a new seeker. The new nose (transducer, electronically steered sonar with beam-former, and new signal processor) improved acquisition range, including range at high torpedo speed. A wider sonar field of view reduced the need for the torpedo to maneuver (and thus to lose net forward speed) as it approached a target. The fuel rate to the swash-plate engine was increased to add power (and more allowance was made for speed variation); more fuel was carried, in space liberated by a more compact computer. In 1980 the British claimed that about half of the torpedo had to be replaced. In 1996 the U.S. Navy credited ADCAP with a speed of 65 kts at a range of 20 km (22,000 yds).

ADCAP was first tested on the Canadian Nanoose range early in 1982. Technical evaluation began in August 1986. The first war shot (which sank a decommissioned destroyer) was reported in July 1988. It had already been decided (in 1985) to accelerate production. ADCAP was operationally evaluated in August 1988. It was considered effective against some but not all threats. Reportedly there was some question as to whether it was sufficiently lethal against large surface combatants. Congress limited production because of the evaluation. The Navy ordered modifications, including reliability upgrades and improved fuzing. Upgrades continued after the end of the Cold War, and newer versions have been developed.

In 1985 the reported requirement was 4,659 Mk 48 torpedoes. A total of 3,887 had been authorized through FY 85. As of mid-1995, the U.S. Navy had about 4,000 Mk 48/ADCAPs on hand, of which 3,000 were active (about 1,000 ADCAPs and 2,000 Mk 48s [mostly Mod 4]); but with the decline in submarine numbers only about 2,000 were needed.

NEARTIP was only an interim solution to the new-generation deep-diving Soviet submarines. A technical assessment of a replacement ALWT began at about the same time as NEARTIP. As of 1977, ALWT was envisaged as a 55 kt weapon with a range of 15,000 yds and 3,000-yd acquisition range, effective in both very deep and shallow water. The speed was wanted to deal with coming Soviet submarines. Assuming that the torpedo should be 50 percent faster than the target, ALWT could deal with 36.7 kt submarines. Probably the most challenging requirement was that the torpedo maintain its speed and range when pursuing future fast deep-diving submarines. That precluded the sort of open-cycle engines used in Mk 46 and Mk 48. Two torpedoes were developed in parallel, the chemically powered EX-50 and the silver-oxide battery EX-51 (which had a side-looking array sonar). EX-50 was selected in 1981 to become the Mk 50 (Barracuda) torpedo. Full-scale engineering development began in 1983. The SCEPS (stored chemical energy power system) power plant contains a block of solid lithium and a tank of sulfur hexafluoride. When the torpedo is fired, electric squibs melt the lithium. It

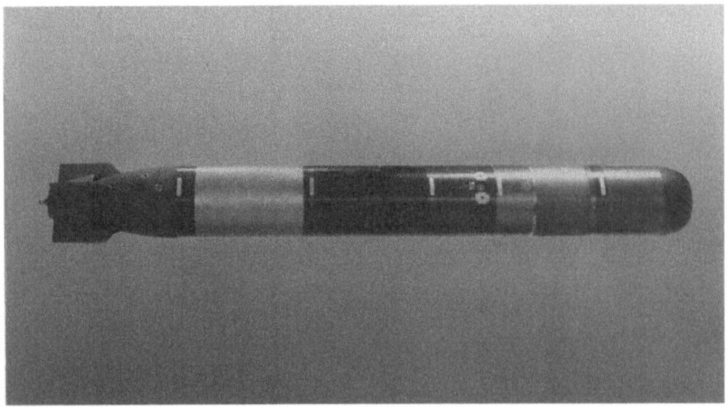

A Mk 50 Advanced Lightweight Torpedo, intended to solve the deep-diving target problem. Note its pump-jet, for quieter propulsion. AUTHOR'S COLLECTION

Mines were always included in the ASW arsenal but received only limited attention. Here they are being loaded on board a *Permit*-class nuclear submarine. These are unpowered ground mines, probably moored magnetic Mk 57s. The postwar Undersea Warfare Conferences stressed the importance of mines as a way of dealing with otherwise elusive Soviet submarines, and through the 1950s the White Oak Naval Ordnance Laboratory developed the Mk 50 mine series. There were also Captor rising mines and standoff (mobile) mines converted from torpedoes, which could travel a distance before sinking to the bottom. USNI

reacts with the sulfur hexafluoride to create heat (to generate propulsion steam) and a solid ash, which has less volume than the reactants, so does not have to be dumped outside the torpedo during its run. Garrett, which developed the system, said that it is easier to produce than the Otto-fuel power plant of a Mk 46. Reportedly Mk 50 actively searches (reportedly at 25 kHz) at reduced speed and then attacks at higher speed. Reported maximum operating depth is 1,100 m (about 3,600 ft), but the torpedo can operate in less than 10 m depth. The warhead is a shaped charge. This torpedo is much heavier than Mk 46, about 771 lbs rather than 508 lbs for Mk 46 Mod 1.

As of 1988, the reported production objective was 7,743 torpedoes plus 160 for research and testing. At the end of the Cold War, the Mk 50 program was drastically cut, production being completed in July 1996 with a total of 1,063 torpedoes.

Despite the exotic power plant, the ALWT's speed was considered barely adequate; in theory it would take a 70 kt torpedo to counter a 45 kt Soviet submarine. A new version of SCEPS designed to drive the torpedo at 70 kts with a range of 18,000 yds was overtaken by the end of the Cold War.

The Mk 50 front end was later combined with the engine and body of a NEARTIP to create a better lightweight torpedo (Mk 54, initially Mk 46 Mod 8) suitable to attack the diesel submarines the U.S. Navy considered the main post–Cold War threat.

British Torpedoes

The British developed, but could not deploy, an equivalent to Fido during the war: Dealer, sponsored largely by the RAF (for Coastal Command).[15] It was 72 in. long with a diameter of 18 in., carried a 90-lb warhead, and had a speed of 12–15 kts. Unusually, it had two propellers but no control fins. Each propeller was connected to one hydrophone angled to a side. Power was set by what the hydrophone heard, the difference in power steering the torpedo. The batteries could be shifted back and forth to move the center of gravity, to make the torpedo rise or dive as required.

The Royal Navy Torpedo Factory produced a pilot batch of 100 Dealers in 1945, but the war ended before they entered service. At 15 kts Dealer was too slow to attack a Type XXI; it was considered effective only against submarines making 2–12 kts.[16] Homing trials against a live submarine (1947) showed that Dealer would not always home on a submarine moving slower than 3.5 kts, even at a range as close as 70 yds. It would home and hit a 5 kt submarine from about 300 yds. It would home on, but not actually hit, a submarine at 7 kts and at a range of about 350 yds.[17] Trials conducted in 1949 finally demonstrated the value of a passive homing torpedo against a submarine running above silent speed, given a small enough turning circle.

Bidder A was a surface-launched torpedo with the same control system.

The postwar Dealer B, unrelated to the wartime Dealer, was an interim passive weapon with a small turning circle. It circled at 12.5 kts while listening, accelerating to 19.5 kts when it heard a target, with a 2,000-yd run at that speed. The attack speed was chosen because above 20 kts self-noise would swamp the seeker. Conversely, a seeker operating at low speed could be highly sensitive. To make up for its low speed, the torpedo attacked on a collision course (constant lead angle, about 20 degrees) rather than in a tail chase. The British claimed in 1950 that Dealer B outperformed contemporary U.S. homing torpedoes, presumably including Mk 35.[18] Two torpedoes sent to the United States for evaluation homed on a 7 kt submarine at 300 ft from 1,450 yds.

Dealer B was redesignated Mk 30 for service use by both the Royal Navy and RAF Coastal Command.[19] By 1953 it was being license-produced in Canada, and the U.S. Navy had just asked for ten more for evaluation after successful trials at Key West. In 1955, already in service with Coastal Command, Mk 30 was due to be issued to carriers in mid-1956. Mod 1, with an improved homer, was provisionally released in June 1954 and was expected to enter service in 1958.[20] Mk 30 survived in British service to 1971 It could detect a 3.5 kt submarine at over 1,000 yds, although against a 2.5 kt submarine that was reduced to 100 yds.[21] Superintendent of Torpedo Experiment and Design wrote that the new Dealer B passive seeker was undervalued. American passive seekers, which were improved versions of that of the wartime Mk 24 mine, were not proving very satisfactory. British seekers, which would be less expensive, outclassed their American counterparts.

However, by 1956 the Royal Navy was planning to abandon fixed-wing ASW aircraft in favor of helicopters. It bought 50 U.S. Mk 43s and contracted with Plessey to anglicize the weapon for British production. The Mk 43s were bought with money freed by cancellation of the Dealer B Mod 1. Because funds were limited, British projects for lightweight torpedoes, including a 14-in. active/passive 25 kt helicopter-borne one, were abandoned at this time. As the Gannet was retired, Royal Navy Mk 30 torpedoes were given to the RAF (for Coastal Command) and offered to West Germany, which operated land-based Gannets.

A roughly contemporary "Pentane" program sought to develop a more satisfactory air-launched weapon, target performance being a speed of 30 kts and a run of about 6,000 yds.[22] These characteristics were comparable to those of the later U.S. Mk 44, but Pentane was nearly five times as heavy, about 2,000 lbs (Dealer B weighed about 660). Pentane would normally home actively, but it could home passively on a noisy target. Pentane was already enjoying "supreme priority" in 1949. Like Dealer B, it would circle until it detected a target. Pentane was running satisfactory trials by 1951. A final Pentane (now designated Mk 21) design was chosen in 1954. In 1955, it was expected to enter service in 1961. Its developers said that its active seeker was far more sophisticated than the very simple ones of its U.S. counterparts, hence far better adapted for future fast submarine targets. By 1958 Pentane was credited with a range of 12,000 yds. It was sometimes promoted as a possible surface antisubmarine torpedo, since it could be fired from standard 21-in. tubes. Cancellation at that time was justified on the ground that its search rate was too low. Experience gained in the Pentane program with both silver-zinc batteries and thermal propulsion fed into later torpedo projects.

By 1951 the only carrier airplane that could lift Pentane was the new Gannet, and then only by reducing its sonobuoy load. Pentane was large and heavy because it had to find its target from a considerable distance, based on ambiguous data from a nondirectional sonobuoy. However, once it was mature, the British had directional buoys that made it possible to drop a homing weapon 500–1,000 yds from the submarine, in a favorable sector for homing. A patrolling airplane carrying more weapons could attack multiple targets, and it could afford to attack ambiguous ones. Lightweight torpedoes were also wanted because British mobilization plans envisaged converting merchant ships into escort carriers that would operate the lightweight Short Seamew airplane.

The Dealer B homing technique was embodied in the Bidder (Mk 20) passive large-caliber homing torpedo.[23] The requirement (TASW 118) issued in 1949 covered both submarine (Mk 20S or Bidder S) and escort (Mk 20E) versions. Mk 20S had a single speed: 20 kts (range 12,000 yds). In 1958 estimated acquisition range on a noisy target was 800 yds. The acoustic head looked only down, so the torpedo had to be set to run shallower than its target. If the torpedo lost its target, it would rise back to its set depth and attempt to redetect it. Bidder could be set for a straight run out

Type 16 frigates retained their after quadruple torpedo tubes so that they too could fire the new homing weapons. HMS *Petard* is shown in 1956. Note her considerable close-in air defense battery of single and twin Bofors guns. She shows a pair of Squids on the shelter deck aft, pointing forward. CONRAD WATERS

The Type 15 frigate HMS *Relentless* shows a torpedo tube for long torpedoes abaft her funnel. Given problems with British homing torpedoes, these tubes did not last. Once they were gone, British frigates lacked long-range ASW weapons. MATCH and Ikara were attempts to recover that capability, which was badly wanted as the Royal Navy deployed longer-range sonars. USNI

to 250 yds or more, up to its full range, after which it would circle while listening. Operating depth was 200 to 800 ft. Homing frequency was 45 kHz, the center frequency of cavitating Soviet W class submarines. Typical acquisition range was 400 yds. Total weight was 1,800 lbs, including a 200-lb warhead.

Initially Mk 20E was welcomed for surface escorts as the long-range complement to the much-shorter-range Limbo mortar, much as the U.S. Navy saw its Mk 35 as a complement to Weapon Alfa. Thus initially the Type 12 frigate (*Whitby* class) would have had 4 fixed torpedo tubes and 12 torpedoes, later changed to 12 tubes, 8 of them fixed, and 2 trainable twins (there was insufficient deck space for all the preferred fixed tubes). When Staff Requirements for the Type 12 frigate were reviewed in June 1952, the requirement was changed to salvo firing (three-torpedo salvoes). All six Type 12s except *Whitby* and *Torquay* eventually received torpedo tubes, but all were removed by 1959. The follow-on *Rothesays* all had torpedo tubes. The companion Type 14 was assigned a pair of twin rotatable tubes, but only five ships had them (*Blackwood, Duncan, Exmouth, Malcolm,* and *Palliser*). Type 15 was changed from four fixed tubes on each side to six lightweight tubes on each side. This armament was canceled in 1953, only *Ulster* being fitted (with eight fixed tubes). The associated underwater fire control system was delayed. By 1956 many captains had no use for Bidder. The projected but abortive successor was a surface-launched version of the later Ongar torpedo (see below). The helicopter-delivered lightweight torpedo (MATCH system) provided the long-range firepower that had been hoped for with Mk 20E.

The British submarine force had no alternative to Mk 20S. In his 1969 critique of British submarine weapons, the chief British submariner (FOSM) described Mk 20 as obsolescent and far from what British nuclear attack submarines needed. It was still designed to home on the cavitation produced by the obsolescent Whiskey (and, presumably, Zulu), which did not match signatures of later Soviet submarines. Bidder remained in submarine service through at least 1970 because its planned successor was badly delayed. The British submarine service looked to the United States for wartime torpedoes. The need for a fire control system flexible enough to accommodate both British and U.S. weapons led to early British interest in computerized submarine command/fire control systems.

The British sought to implement wire guidance, which the Germans had invented during the war, under the codename Mackle (Mk 22). The idea was initially attractive because the firing submarine could avoid the danger inherent in a self-guided torpedo. Only later was it clear that the submarine could use her own low-frequency sonar to guide the torpedo well beyond the distance the torpedo's own higher-frequency sonar could reach. At least initially the torpedo was intended to run out at a relatively shallow depth, looking down passively and diving into the target when it detected sufficient noise. Fire control would be far simpler than for Bidder, requiring only a single operator. Because the target would be tracked continuously, Mackle could counter evasive maneuvers such as zigzagging. Much of the guidance equipment (for lead angle control) embodied in Bidder could be on board the launching submarine. In 1952 the Mackle working party concluded that a wire-guided Bidder might suffice against the current submarine target, essentially equivalent to Type XXI. Moreover, a wire-guided version of Bidder would be available earlier than any more sophisticated weapon.

Mk 22 died in favor of Mk 23 (Grog), a wire-guided Bidder lengthened to accommodate a wire drum. Although the intention was to use active/passive homing near the target, Mk 23 was limited to the passive homing used by Bidder. By 1959, when Mk 23 began trials, it was considered an interim weapon pending development of the next torpedo, Mk 24. Speed was initially given as 28 kts with a range of 8,750 yds. However, the torpedo could not listen at that speed. As delivered it had a range of 12,000 yds at 19 kts, with maximum running depth of 800 ft. The contact-fuzed (as in Mk 20) warhead was 200 lbs of Torpex. Like Bidder, Grog was designed to home on the 45 kHz frequency radiated by cavitating Soviet W-class submarines. Because its seeker was hard-wired, it could not home passively on Soviet nuclear submarines. The Royal Navy had to develop direct support tactics in which its nuclear submarines used their long-range sonars to direct attacks by an associated surface force.

Trials began in 1959, and Mk 23 entered service in 1966. As of 1963, plans called for producing 600 by April 1967. Given delays in the Mk 24 program, in 1965 plans called for stopping Mk 23 production at the end of 1967–68 with a total of 750.

By this time the first British nuclear submarines, with long-range Type 2001 sonars, were entering service. The Royal Navy wanted a weapon to match. The next torpedo (Ongar, Mk 24) was intended to meet the 1970 threat, a nuclear submarine running at depths as great as 1,000 ft and at speeds of up to 30 kts. The codename Ongar seems to have been chosen because at the time Ongar was the farthest stop on the London Central Line; this was to be the ultimate

A Tigerfish torpedo on display at the Royal Navy Submarine Museum, Gosport. Not visible are the retractable stub wings that enabled it to maintain depth at reduced listening speed. JOHN GOURLEY

British torpedo. The Staff Requirement was written in 1959, with the first British nuclear submarine, *Dreadnought*, under construction. In 1963 plans called for producing 770 Ongars by 1972, already down from an earlier target of 1,130. In 1965 plans called for a total of 1,000 Ongars. The first order for 100 was placed in 1967 with the Royal Navy Torpedo Factory.

Ongar roughly paralleled the U.S. Mk 48 project. Like Mk 48, it seems to have been conceived as a two-speed torpedo with a maximum speed of 55 kts, which would have been needed to engage a fast target. However, the British did not have a powerful enough motor, so Ongar was considerably slower. Ongar had retractable wings to provide lift at slow homing speed and improve roll control.

The development program seems to have been planned in two stages, the first concentrating on the guidance problem. The guidance wire had to pay out properly at longer range and higher speed than with Grog. Probably a second phase would have introduced a high-performance power plant; in 1958 the Admiralty was financing work on new internal combustion techniques that seem comparable to that adopted for the U.S. Mk 48. Any second phase was buried by the need to solve problems with the initial Ongar design. In effect the later Spearfish was the second-phase torpedo.

As of 1969, Ongar was a two-speed electric torpedo with wire guidance and a terminal seeker. Maximum guidance range, set by the length of the guidance wire, was 15,000 yds. Running range was 32,000 yds at 24 kts and 15,000 yds at 35 kts. Maximum running depth was 1,150 ft, but maximum discharge depth was estimated at 1,000 ft. Minimum running depth was 50 ft. The fuze was either active magnetic or inertial (contact). The warhead was 300 lbs of Torpex, compared with 805 in the old straight-running Mk 8. As of 1969, Mk 24 was expected to home passively at 33 kHz. At 35 kts it could home actively at 1,000 yds and passively (against a noisy target) at 2,000, at a depth greater than 300 ft. It was considerably more sensitive than Mk 20/23, but it had to overcome self-noise at higher speed. In 1969 FOSM observed that although Mk 24 could catch a Soviet nuclear submarine, its limited operating depth gave its targets a safe haven.

Mk 24 had only limited antiship capability, because it was difficult to control at depths less than 50 ft. The 15 ft range of its influence fuze severely limited its ability to attack any ship with a draft less than 35 ft, which in 1968 meant the entire Soviet surface fleet.

The program ran into serious delays. As of early 1969 it was supposed to enter fleet service in January 1970, but the 1969 acceptance trials were so unsuccessful that a special Torpedo Acceptance Executive was established, the contract being transferred to GEC. Marconi was asked to design a new narrow-beam seeker and a more powerful onboard computer that could take over more of the guidance role.

The Royal Navy particularly wanted surface ship attack capability and tried to join the U.S. antiship torpedo program. When that program collapsed in favor of Mk 48 Mod 1, the British solution was to modify Mk 24, one rationale being that the torpedo had to be able to attack a submarine that might either dive or surface during an attack. The Admiralty also wanted deeper diving. Some Soviet submarines were credited with an operating depth of 1,500 ft; emergency diving depth might be as great as 2,000 ft. It appeared that by 1975 at least half of Soviet operational submarine fleet might be able to exceed the 1,150 ft floor of the Mk 24 torpedo.

GEC made a feasibility study of an all-passive dual-purpose Mk 24 under a 1965 contract. This Mod 1 project was approved in 1968. Maximum running depth was 1,600 ft, and minimum depth was reduced to 25 ft. This torpedo would attack targets drawing only 10 ft (such as surfaced submarines) at speeds of up to 26 kts. It would run out at greater depth, then climb to attack its ship target. The Admiralty Board approved the project on 27 June 1968. At that time it was expected to enter service in 1974 (the original torpedo, now designated Mod 0, was scheduled to enter service in 1969). However, in August 1970 Mod 0 was not expected to enter service until late 1973.

The British had estimated that for the next few years Soviet nuclear submarines would not much exceed 22 kts, but as the time scale of Mk 24 stretched, the British looked toward 1975, when they expected that Soviet nuclear submarines would generally have maximum speeds of 30 kts (in fact, they already did). Like their American counterparts, they overestimated Soviet diving depths; they thought that many Soviet submarines could dive to 1,700 ft. Collapse depth would be more than 2,000 ft. By 1985 a quarter of Soviet submarines would be able either to proceed at 30 kts and dive, on occasion, to about 2,500 ft or proceed at 35 kts and dive, on occasion, to about 2,000 ft. Unfortunately, the crush depth of the Mk 24 hull was 1,400 ft (it had been given a search floor of 1,000 ft to overcome decoying). A Navy working party suggested an extreme solution, Mk 24N—a Mk 24 with a nuclear warhead that could crush submarines running at greater depths. It would be wire-guided. It is not clear whether Mk 24N was ever fielded. In 1971, when it was being seriously considered, British carriers had nuclear antiship bombs on board, and the British had nuclear depth bombs.

The time scales of the original version of Mk 24 and the dual-purpose version converged. It would be impossible to significantly change basic propulsion, homing, and fuzing (which made it impossible to attack surface targets with drafts of less than 8 ft). The torpedo would be unable to attack targets faster than 26 kts. FOSM pointed out that the warhead was only 40 percent as heavy as that of the old straight-running Mk 8; it might be unable to sink surface ships. A further study showed that by using the full warshot battery power the torpedo could reach 37.5 kts, but that did not solve the 30 kt submarine problem. Redesign could reduce radiated noise by 10 dB at 24 kts, to reduce the range at which a target would hear the torpedo from 4 to 2 km. That would give the target less time to evade.

Trials of Mod 0 in 1973–74 showed that although the torpedo depended heavily on data provided by the firing submarine, it tended to break its guidance wire. Reliability in different trials varied wildly between 20 and 80 percent, with a 40 percent average. Guidance, for which Marconi was responsible, was clearly the main issue, so the company received the overall Mk 24 contract in 1974. Mod 0 failed provisional fleet acceptance trials in 1979 but was issued to the fleet anyway, beginning in August 1980.

The Mod 0 project badly delayed the dual-purpose Mod 1. As of 1972, conversions of Mod 0 to Mod 1 torpedoes were expected to begin in 1977. Mod 1 passed acceptance trials in 1978. It still had problems: when three Mod 1s were fired into the burned-out hulk of the LSL *Sir Galahad* during the Falklands War, two of them did not hit because their batteries failed. Deficiencies in Mk 24 explain why HMS *Conqueror* fired Mk 8 torpedoes at the Argentine cruiser *Belgrano* in the Falklands.

Marconi received a get-well contract in 1983, and tests in 1985 finally showed 80 percent reliability. The company's improved Mod 2 (Tigerfish Mod 1) incorporated some technology from its lightweight Stingray torpedo. By 1987 all 600 war-stock torpedoes had been modified to Mod 2 standard. Total production was about 750.

It was already time to develop an entirely new torpedo, the current Spearfish. In 1968 the Royal Navy defined a notional Mk 25 powered by a monofuel (Otto II, as in Mk 48) engine powering a pump-jet propulsor. It might eject long-chain polymers to smooth boundary flow and reduce drag (and noise). Desired range exceeded that of the U.S. Mk 48: 39,000 yds at 55 kts at 150 ft depth. It could be fired at any depth from 1,250 ft up, and could run as deep as 2,500 ft. Terminal homing would be passive/active acoustic, with a 500-lb influence-fuzed warhead. Dimensions would be 21 × 254 in. (all-up weight 3,600 lbs), and its warhead would weigh 500 lbs. At a depth of 2,500 ft, range at 55 kts would be 26,000 yds. These characteristics were presumably inspired by those of the U.S. Mk 48, then in the early test stage.

In 1974–75 the Mk 24 Way Ahead Working Party examined options for a successor torpedo. The Naval Staff rejected the U.S. Mk 48 as a viable alternative on the grounds that it was not compatible with existing British fire control systems or practices. It did not transmit its data back down its guidance wire to be exploited by the firing submarine, as Mk 24 did.

A new requirement (NST 7525) was formulated for an interim replacement torpedo, on the assumption that either an evolved version or a new torpedo would be needed in the 1990s. Targets might be able to sprint at up to 38 kts (35 kts sustained) and dive to 760 m (2,493 ft). The torpedo would be dual-purpose, although submarines would be the primary targets. For faster development, NST 7525 would use currently available technology, including some already-proven major components. A prefeasibility study was completed in December 1976. At that time a yearlong feasibility study was expected to begin in May 1977, followed by project definition (June 1979–June 1980); the weapon would enter service in 1986–87. Length was set by the current NATO standard (5.85 m, 230.3 in.) rather than Mk 24 (6.464 m, 254.5 in.), presumably in hopes of exports. The submarine fire control computer had to be modified to deal with much higher data rates, and the guidance wire had to be dispensed at much higher speed. Like Mk 24 and the U.S. Mk 48, the new torpedo would have two speeds (at the time, 35 and 55 kts; later a search speed of 30 kts was chosen) and, as in Mk 48, a pump-jet. Instead of the piston engine of Mk 48, the torpedo would be powered by a Sundstrand turbine, the power plant the U.S. Navy had used (and dropped) in Mk 48 Mod 0. Electric propulsion did not offer the desired maximum speed. To maintain high speed at depth, Spearfish was given a very high speed when running shallow, reportedly as much as 81 kts. The reported depth requirement was 3,000 ft.

By this time microcomputers were advanced enough that the new torpedo could have a very adaptable programmable seeker driving a multifrequency sonar that could generate various beam patterns as required. By discriminating in range, direction, Doppler, and signal amplitude it could distinguish real from decoy targets, and it could filter out reverberation. It could vary its frequency according to the phase of the attack. Too, the tactics of the torpedo could be modified by changing its software, to avoid obsolescence. A warhead whose size was set by the antisubmarine mission might weigh about 300 lbs, but that would not be enough to sink a large warship. Reported warhead weight was 300 kg (661 lbs).

Problems with Mk 24 made it necessary to accelerate the program, using existing components: the Sundstrand turbine and a digital homing head that Marconi had already developed for the lightweight Stingray torpedo. Probably the main entirely new element of the program was the two-way guidance wire. The new torpedo used HAP (hydrogen ammonium perchlorate) Otto fuel, which produced only water as its combustion product, hence was trackless. It has also been reported that Marconi adopted this fuel (early in 1980) to increase torpedo speed (to deal with the Alfa class) by increasing power from 550 to more than 1,000 HP.

Marconi won the contract, against stiff opposition from Gould in the United States, with its Mk 48 ADCAP, both technically and on the basis of its fixed-price package proposal. It appears that Marconi offered considerable savings if the Stingray and Spearfish programs were combined into a single British torpedo program. The U.S. Defense Department offered offsets if the Royal Navy adopted ADCAP; it seems to have dangled the possibility that the U.S. Navy would adopt Stingray. As of mid-1981 it was not altogether clear that the British could afford the Marconi weapon. Advocates of the British torpedo claimed that the future of ADCAP was uncertain; the Royal Navy could not be sure of what would happen during the torpedo's 20-year life. Thus, according to an internal British memo explaining the decision to Prime Minister Thatcher, Spearfish embodied the latest technology being proven in Stingray, and could be improved to meet an evolving threat, but Mk 48 had been in service for ten years and was now undergoing its last update (the U.S. torpedo has since shown quite a bit of life, Mk 48 ADCAP being virtually a new torpedo). The technical advantage offered for Spearfish seems to have been a digital seeker based on that of Stingray, with considerable potential for stretch as digital technology advanced. It was also suggested that buying ADCAP would damage the prospect of selling Stingray in a market currently dominated by the United States and Germany.

Marconi's Spearfish heavy torpedo on display at the Royal Navy Submarine Museum, Gosport. Note the pump-jet and the control surface actuators on the ends of the tail fins. The visible tube abaft the fins is presumably the dispenser for the guidance wire. JOHN GOURLEY

The initial contract with Marconi was the largest fixed-price contract that had ever been placed, up to that time, by the Ministry of Defence. The development and initial production contract was let in December 1981. The new torpedo was named Spearfish on 4 March 1982.

No Spearfish were delivered before the end of the Cold War. The original planned in-service date was December 1987; the actual date (using production torpedoes) was March 1994. The delay was assessed as due to propulsion (9 months) and reliability assurance (62 months). The main production order, reportedly for 350 torpedoes, was given in February 1994. Part of the delay in production may have been due to an extensive reliability program (current reliability as published is better than 90 percent). First deliveries of the main production batch were in June 1999, and fleet weapon acceptance came only in January 2001. The first submarine to deploy with a full load of Spearfish was *Tireless* (May 1999). The last production torpedo was handed over on 24 November 2003. Tigerfish was replaced altogether in 2004. An official British audit reported that the program had slipped 75 months.

The British lightweight torpedo story was more complex. Having abandoned development of these weapons, they bought the U.S. Mk 44 and anglicized it as Mk 44 Mod 1 (not a U.S. designation), adopting it in August 1964. They were not pleased with the result. Anglicization took longer than expected. Mk 44 Mod 1 proved unreliable partly because it used an earlier generation of electronic components. Too few of the early ones passed proof test. The extent to which other NATO countries encountered similar problems as they manufactured Mk 44 under license is not clear. The Italians seem to have been unique in developing the torpedo further as their A244 and A244/S.

The Royal Navy was oriented largely toward limited-war operations like the recently concluded Confrontation with Indonesia, so its assessment for 1970–80 emphasized the threat of diesel submarines. Opponents such as Egypt or Indonesia might deploy up to eight such submarines in one theater; up to four might be concentrated against a seaborne force. Toward the end of the period, Indonesia and Egypt might possess one or two nuclear submarines. The Soviets might intervene covertly in limited war, perhaps with up to four nuclear submarines. They actually did so with the Echo IIs near the Tonkin Gulf in 1972.

Given their interest in wars East of Suez, the British observed that in many sea areas (China, Java, and Arafura Seas, the Persian Gulf, the North Sea, and UK approaches were cited) submarines might run near the seabed, so it was vital for a torpedo to distinguish them from the seabed. Mk 44 Mod 1 could not do that. It was liable to be captured by the bottom when within 300 ft of it. At best it could operate only to within 150 ft of the bottom. The British wanted a passive capability (which Mk 44 Mod 1 lacked) for situations in which active search was limited by seabed or surface reflection (capture) effects or by other reverberation. Mk 44 Mod 1 searched too slowly and was considered unreliable. It could not attack surface ships or snorkeling submarines because (to avoid surface capture) it could not run shallower than 50 ft. Due to transducer losses, acquisition range decreased with increasing depth below 500 ft: an effective acquisition range of 750 yds at 100 ft fell to about 500 yds at 1,000 ft. Because they had only one operating frequency, torpedoes dropped in pairs would interfere with each other. The time-consuming helical search pattern used up much of the six-minute battery life and therefore reduced the probability of acquisition and kill.

The British Mk 44 Mod 1 (equivalent to the U.S. Mod 0) could not be fired by ships because it had no set runout (the feature that made the torpedo suitable for shipboard use by precluding an attack on the launching ship). That feature had been eliminated as an unnecessary complication. Unfortunately, having accepted Mk 44 Mod 1, the Royal Navy wanted to replace Limbo mortars on board *Leander*-class frigates with lightweight torpedo tubes when they were equipped with ASW helicopters.

The anticipation of shallow-water operations prompted the British to pursue a redesign, initially called Mk 44 Mod 2. The British wanted the torpedo to operate in shallow water about 200 ft deep, with an optional attack capability (selected before flight) against surfaced or snorkeling submarines. It should be suitable for dropping in pairs or more; with an improved search pattern, improved endurance, speed, and stability; and a larger warhead (if that could be achieved without increasing torpedo size and weight).[24] This was far more than a limited modification, so the torpedo was designated Mk 31 (in the British series).[25]

At this time the British thought that Soviet nuclear submarines would not exceed 23 kts in the near term. Mk 31 was expected to be effective against targets not exceeding 28 kts and not diving deeper than 1,200 ft (corresponding limits for the U.S. Mk 46 were 35 kts and 1,500 ft). Neither Mk 31 nor Mk 46 could attack targets with less than 5 kts of Doppler. In the 1970s Soviet nuclear submarines might greatly exceed these figures, with speeds of up to 35 kts and operating depths up to 2,000 ft. That justified a new NAST 7511 lightweight, due in 1975, which would be the first available torpedo capable of meeting this threat. In fact, the November was already capable of 30 kts, but apart from the Alfa, the 35 kt SSN did not materialize during the Cold War.

The British were already developing the lightweight torpedo-sized Deep Mobile Target, based on Mk 31, to support the Mk 24 program.[26] The Admiralty approved a staff target for the Mk 31 torpedo on 18 May 1965, immediately after the Treasury approved development of the Deep Mobile Target.

Mk 31 (NASR 7709) was seen as an interim weapon embodying technology already developed for other programs: homing from the Mk 24 heavy torpedo, much of the electronics and propulsion from the Deep Mobile Target, and some work already done for NAST 7511. It would have an improved seeker to reduce susceptibility to surface capture, hence make the torpedo effective against submarines at periscope depth. A bottom sensor would protect it from striking the seabed. A new transducer would give constant performance at all depths. A new search beam that was much broader vertically (as in the U.S. Mk 46) would render helical search unnecessary. The air-dropped torpedo would circle to search. Fired by a ship, it would snake search. It would have a passive homing mode, selectable (when airborne) or automatically selected when active operation became reverberation limited. Noise would be reduced. Countermeasures resistance would be improved, particularly against multiple-pulse decoy signals, broadband pulse jamming, and echo repeaters. A new contra-rotating motor lighter than that of Mk 44 would eliminate gearing noise (for better sonar performance). Weight saved in the motor would go into a heavier warhead. A short-range influence/contact pistol would preclude nonactuation due to a near miss or a glancing blow. To improve capability against a retiring target, the torpedo would have a burst speed of 35 kts in the final stages of homing.[27]

Feasibility studies were conducted in September 1965 and January 1966. A formal Staff Requirement (NASR 7709) was approved by the Admiralty Board on 15 March 1966 and endorsed by the secretary of state (Royal Navy) on 24 May 1966. Initially it seemed that enough of the development work had already been done that the new torpedo could enter service by the end of 1969. The British R & D Board considered Mk 31 worthwhile even though it thought development would take a year more than expected.

The Mk 31 project was attractive because the British knew that other Mk 44 users were dissatisfied, particularly with the shallow-water performance of Mk 44. The United States tried to protect foreign sales by prohibiting the British from selling their anglicized version, but now they were no longer making Mk 44s. Mk 31 was also quite different internally. The considerable potential market helped justify the Mk 31 program. Assumed total production, including torpedoes for RAF Coastal Command and foreign sales, was 2,000, an initial order for 200 torpedoes being expected in January 1968. Initially the British thought that a Mk 44 could be converted into a Mk 31 relatively inexpensively.

The initial Staff Requirement was amended (approved 14 March 1968) to require increased countermeasures resistance, an increased maximum diving depth (1,500 rather than 1,200 ft), a heavier warhead (90 rather than 75 lbs, using RDX explosive equivalent to 130 lbs of TNT), and a variable forerun (500–5,000 yds) for surface ship operation. The increased diving depth was associated with the supposedly known capacity of Soviet submarines to operate down to at least 1,500 ft. In fact, operating depth of Soviet first-generation nuclear submarines was 300 or 320 m (984 or 1,229 ft); maximum depth, not be exceeded, was 400 m (1,312 ft). The stated figure may reflect British estimates based on assumed structure of these submarines and also British safety factors. The torpedo would operate in 20 fathoms and within 20 fathoms of the bottom in deeper water. It is not clear where this figure originated. Demands for better reliability and more complex operation would be met within the existing torpedo body thanks to rapid advances in electronics miniaturization.

By February 1969 Mk 31 had reached the prototype stage. Homing was primarily active, but the system would respond to any noise within its frequency range. Active homing range was 1,200 yds and rated passive homing range was 3,000 yds. Surface ships were safeguarded by limiting the torpedo to depths greater than 50 ft. Homing frequency could be set at either 44.6 or 33.6 kHz, making it possible for two torpedoes to be fired together without acoustic interference. In deep water the torpedo would begin its search at a set depth, then search at a series of greater depths. For example, if it was preset for an initial depth of 400 ft, it circled at that depth, then circled twice at 600, 1,000, and then 1,400 ft. In shallow water its sensor measuring height above seabed could override the dive program. When attacking, the torpedo accelerated to 35 kts for the last 150 yds; its rudder rate increased to enable it to turn more tightly to follow the target. The torpedo could reattack if it lost contact with the target. Operation was largely digital, using integrated circuits. Search was controlled by a digital search pattern generator.

The obvious alternative to Mk 31 was the U.S. Mk 46, already in service. Although much slower than Mk 46, Mk 31 was fast enough to deal with the expected targets. However, the longer run of a Mk 31 would give a target more warning time in which to launch decoys. In deep water, Mk 31 would have a similar search range to Mk 46, but greater coverage within it due to its broader vertical search beam. In water depths of 300–600 ft Mk 46 would have complete coverage out to about 1,000 yds against high-Doppler targets (10 kts or more), even those very nearly bow- or stern-on. However, it would not detect low-Doppler targets except at very short range. Mk 31 would enjoy full coverage in similar conditions, regardless of target speed, out to 700–1,000 yds except when the target was nearly end-on. Both torpedoes would be about as susceptible to decoys, but there was time to improve Mk 31 in that respect. Because the British were interested in shallow-water operation, Mk 31 was expected to have a quicker response after entering the water, hence could be used in depths as shallow as 120 ft (Mk 46 required 250 ft for its initial dive). Mk 31 had a dedicated bottom sensor to protect it against bottom capture. Overall, in deep water the two torpedoes would probably be equally effective against conventional targets. For nuclear submarines running faster than 20 kts, the requirement noted, "the higher speed of the Mk 46 begins to assume greater importance, though the Mk 31 should retain some capability against a new generation of Soviet submarines."[28]

In 1971 Mk 31 was canceled to free resources needed to develop the next-generation NSR 7511, which eventually emerged as Stingray. The British bought Mk 46 despite its perceived weaknesses. NSR 7511 survived to protect the British foothold in the world torpedo market. This point was made by the British Operational Requirements Committee in May 1972. Cancellation would be particularly embarrassing given the decision 3 years earlier to build up a British design capacity starting with the Mk 24 "get well" program and the decision 18 months earlier to concentrate on NSR 7511 instead of Mk 31. As recently as 20 April the Operational Requirements Committee had endorsed an Operational Requirement for Mk 24 Mod 1. That program had been assigned to Marconi Defence and Space Systems, which was also likely to develop the new lightweight torpedo.

Probably the strongest argument against buying Mk 46 was that it would cost scarce dollars and thus weaken the British pound.[29] With no prospect of producing it in the UK, the fuel would add to the cost of deploying Mk 46. In water 300–600 ft deep Mk 46 was ineffective against slow (low-Doppler) targets. Its Otto fuel left a gaseous wake, on which torpedoes had a marked tendency to home when the Doppler feature was disabled (at this time the U.S. Navy was considering using the Doppler feature at all depths, despite reduced effectiveness against slow [less than 5 kt] targets). The Otto fuel tended to decompose at temperatures above 50° C (144° F). At lower temperatures it could generate a toxic gas.

NASR 7511 was then expected to appear in the late 1970s, but it was officially accepted in 1986. Initially it was expected to achieve over 50 kts, a 10,000-yd run, and a maximum depth of 2,000 ft. It would be quieter than Mk 46 and nearly wakeless, using internal combustion (perhaps HTP). Shallow-water performance would be better than that of Mk 46 Mod 1, with improved homing logic derived from the Mk 31 program.

Stingray was powered by a magnesium/silver-chloride seawater battery driving a pump-jet. Reported speed was 45 kts (reported endurance 8 minutes, meaning 12,000 yds). Weight was 588 lbs, compared with 425 lbs for Mk 44 and 508 lbs for Mk 46 Mod 1. In March 1981 Stingray was credited with depth performance equal to that of an Alfa-class submarine, then (wrongly) thought to be about 800 m (2,624 ft).

Stingray on display at Euronaval 2006, showing its pump-jet propulsor and its characteristic bodies, presumably actuators, at the ends of its tail fins. AUTHOR'S COLLECTION

Marconi claimed that Stingray was the first lightweight torpedo to incorporate a programmable general-purpose computer rather than a hardwired special-purpose device. Its software could be modified for different water environments and for different targets. The computer selected a search mode (for deep or shallow water, for example, in either case a snake rather than helix) and a ping mode (CW or FM). In shallow water, Stingray ran at a fixed height above the bottom. Observed target data could be compared with stored characteristics; the computer calculated an interception path that took into account the target's potential speed and maneuverability. At short range, Stingray used highlights in the target's sonar image to measure target speed and aspect, so that it could make a terminal maneuver (a sharp turn) to hit at right angles for maximum shaped-charge effect.

The torpedo was negatively buoyant: to avoid crashing into the bottom, it had to accelerate quickly (reportedly it pulled out of an air-drop dive in less than 4–5 seconds, to avoid hitting a shallow bottom). Minimum depth was about 20 m.

Stingray production began in April 1981, with orders for 280 and then for 550. Some early-production torpedoes went to the South Atlantic for the Falklands War in 1982, but they were not used in action. On 19 October 1985, in its first live-warhead test, a Stingray dropped from a Nimrod aircraft destroyed a double-hulled submarine, HMS *Porpoise*, moored at a depth of 65 m in the Mediterranean. The torpedo was expected to cut a small-diameter (albeit fatal) hole in both hulls, after which the submarine would have been raised for examination. Instead, the vessel was completely destroyed and could not be raised. This result may indicate the power of all current lightweight torpedoes equipped with warheads made of modern explosives.

When Stingray was offered to it in 1981, the U.S. Navy found its speed unacceptable. Gould offered an alternative electric variable-speed power plant with a maximum speed of 62 kts. That figure may indicate the sort of performance achieved by the U.S. Mk 50.

Soviet Torpedoes

Torpedoes were a serious gap in Western naval intelligence. In 1953–54 German specialists who had worked in the Soviet Union were allowed to return home. Interviews with them provided the first (and, for many years the only) insights into the Soviet R & D program. Western naval intelligence became aware of the designations of some Soviet torpedoes, typically in the form caliber-year, as in 53–57 for a 53 cm (21 in.) torpedo deployed in 1957. Electric torpedoes were designated in a separate ET series. For many years NATO applied Soviet-style designations to torpedoes the Soviets were thought to have.

The Germans said that the Soviets were showing interest in long-range trackless torpedoes, which were not yet in service. They were pursuing alternative power plants, including HTP, oxygen, and the closed-cycle Junkers Jumo diesel. According to a March 1955 report on Soviet war potential, the Soviets had built 20 HTP prototypes designated 53/47D, which were undergoing sea trials.[30] Estimated performance was 9,000–10,000 yds at 55 kts or 16,000 at 48 kts. An improved production version was expected by about 1960. According to a 1956 report on Soviet war potential, German returnees had clarified the position on Soviet torpedo development: the Soviets were intensively pursuing both HTP and closed-cycle petrol engines, and they had made 30 prototype HTP torpedoes.[31] According to the 1956 periodic intelligence summary, the interim HTP torpedo might have been in production since 1955, so it might already be in service.[32] It was unlikely to be a pattern-runner, and probably used only a contact pistol (exploder). An improved version was probably in hand. Pattern-running would probably cost 2–3 kts. The first such torpedo in Soviet service was 53–57. German returnees said that the Soviets had shown considerable interest in wartime German active and passive homing; it seemed that they could produce a passive homer in 1954–55 and an active one in 1957–58. They were known to be interested in wire guidance, which would make a very-long-range torpedo more effective.

The known (or imagined) "zoo" of Soviet torpedoes remained classified until a list was published in the British magazine *Maritime Defence* in 1987. The article announced the shocking reality that all current Soviet antiship torpedoes employed wake-following, a type of homing extremely difficult to counter. The torpedo detects first one side and then the other of the ship's wake as it crosses. When it emerges from the wake, it turns back across it at an angle that takes it closer to its target. It keeps crossing back and forth until it reaches the target, typically using a proximity fuze to explode under the target's bottom. There are now also edge-riders that follow the edge of the wake.

The Germans had invented (but not perfected) this technique, and Western navies experimented with it after the war. A 1967 U.S. analysis summarized what was learned: the technique worked very well and was very difficult to counter. The U.S. Navy would be well advised not to introduce it (but, unknown to the writer, the Soviets were already using it). The first

Western wake-follower seems to have been the U.S. Mk 45F Freedom Torpedo, a non-nuclear use of the Mk 45 body and motor. It was bought by Turkey, but deliveries were embargoed due to the 1974 Cyprus War between Greece and Turkey. No other buyer seems to have emerged. Wake-following is now virtually universal.

In 1984 the U.S. Navy learned that the Soviets had 65 cm torpedoes, with better than twice the range of their U.S. counterparts and four times the warhead. In fact, the first such torpedo had been introduced in 1973 (as 65-73). Its successor 65-76 carried a 500 kg (1,100 lb) warhead and could reach 50,000 m at 50 kts. Initially this torpedo was thought (wrongly) to have introduced wake-following in Soviet torpedoes. The 65-73 was conceived as a straight-running underwater nuclear missile; it was designed for the Victor II class. The follow-on 65-76 was a wake-follower that could carry either a conventional or a nuclear warhead. The 65-76 was powered by hydrogen peroxide, which was later blamed for the loss of the submarine *Kursk*. It seems likely that the 65 cm torpedoes were designed because 65 cm tubes had already been adopted to fire the SS-N-16 ASW missile.

Like most of their Western counterparts, most Soviet submarine torpedoes were 53 cm (21 in.) in diameter. The Soviets adopted German-type pattern-running, probably initially in the 53-39PM version of their standard steam torpedo, which entered service in 1949. Range was 8,000 m at 34.5 kts or 10,000 m at 30.5 kts. The export 53-56V (1964) was based on 53-39PM. The first postwar Soviet torpedo, 53-51, was a straight- or pattern-runner with a reported range of 10,000 yds at 39 kts.

The successor torpedo was 53-57 (NATO 53-56), the long-awaited Soviet HTP torpedo. This modified 53-51 could be fired by both submarines and surface ships. Its new power plant offered much greater range: 9,000 m at 51 kts and 18,000 m at 45 kts. This performance would enable a submarine to destroy a convoy from beyond contemporary sonar range. The 53-61 was an improved version, with wake-following seeker (Andromeda) torpedoes designated 53-57M and 53-61M. Andromeda was formally adopted on 20 July 1964 after tests on board the Whiskey submarine *S 380*. It could detect the wake of a 10 kt to 12 kt target (over 100 m long) from 600 m astern; of an 18 kt target, 1,000 to 1,500 m astern; and of a 24 kt target, 2,000 m astern. Torpedo running depth was 3–4 m. T-5 was a nuclear version of 53-57, armed with a 15 kT RDS-9 warhead. It became operational in 1959 or 1960. Many 53-57 and 53-61 torpedoes were converted to 53-65K.

The 53-65 (NATO 53-68) wake-follower, which superseded 53-61, became operational in 1968. Like 53-61, it used HTP, reportedly in a semi-closed cycle, using oxygen to enrich reused exhaust gas. Range was 12,000 m at 68 kts or 22,000 m at 44 kts. It introduced wake-homing into Soviet practice. In 1969, 53-65 was upgraded to 53-65M (NATO 53-68), with an improved seeker. Production torpedoes were derated from the full 1,350 kW (1,800 HP) for better run performance. A nuclear version (NATO 53-68N) that carried a 20 kT warhead became operational in 1970. The export version was 53-65KE.

It appears that submarine torpedoes used HTP, and their surface ship equivalents used oxygen-kerosene engines, which the Germans had seen as an alternative to HTP. The first weapon so powered was the 53-56 surface ship torpedo; 53-65K was a 53-61 upgraded to 53-65 standards, using oxygen rather than HTP as its oxidizer.

The Soviets exported a more conventional 53-56VA acoustic homing torpedo, which was formally accepted in 1964 and entered service in 1966. Speed was 29 kts. The homing sensor could detect a 12–18 kt ship at 600 m. This torpedo was standard in Soviet client state navies, on board surface torpedo craft and on submarines through the Romeo and Foxtrot classes.

The first Soviet antiship homing torpedo was SAET-50 (NATO ET-80[50]), based on the wartime German Gnat (T-5). SET is the acronym for *somonavodyashchayasya electricheskaya torpeda*, or "self-guiding electric torpedo." Details became known in the West after the German engineers who helped develop it got home. According to one report, experience with

SET-53 and SET-53M demonstrated that purely passive homing was insufficient to deal with submarines.

The follow-on antiship homing torpedo, SAET-60, entered service in 1965–66 (SAET-60M followed in 1969). Using passive 25 kHz homing, the torpedo could home on a 16–24 kt target at 600–800 m.

SET-53 or—SAET 53 (NATO ET-80A[60])—was the first Soviet ASW homing torpedo, the standard weapon of the 1960s. Work began in 1950. It was intended to engage targets moving at not less than 9 kts and at 20–200 m depth. It was tested in Lake Ladoga in 1954–55, with formal state trials between August and October 1957; it was accepted into service in 1958 as SET-53 (Izdeliye 238). The formal designation was presumably SET-53-58. According to Western sources, ET-80A could acquire a cavitating submarine (21 dB at 24.8 kHz) at about 700 yds. SET-53M was an improved SET-53 adopted in 1964. Performance improved to 29 kts/14 km. This version may have incorporated pirated British Mk 20 active/passive technology. This was reportedly the most widely produced of all Soviet torpedoes. It was supplied to users of all the first-generation postwar submarines (Whiskey, Romeo, and Foxtrot) and thus was in service in China and in North Korea.

The next major Soviet ASW homing torpedo, SET-65 (NATO ET-80[67]) was developed as part of the massive anti-Polaris program. It was authorized by the 13 October 1960 anti-Polaris decree. Compared to SET-53, it had 1.7 times the speed and twice the range (15,000 m at 40 kts, also given as 10,000 m at 35 kts). It used an active/passive seeker. Active seeker range was 800 m. Maximum depth was 400 m. SET-65 was formally accepted into service in 1965 and became operational in 1967. Standard practice was to fire pairs, at ranges up to 30–35 cables (3–3.5 nm) and depths to 400 m. SET-65KE is credited with a target detection range of 1,500 m and with depth limits of 27 and 400 m. There is also a SET-65M (NATO ET-80A[76]). SET-65A is an electrically set version for nuclear submarines. Modernization, probably in the 1980s, including installation of a new Sapfir unified (presumably surface and submarine) homing system.

A version of SET-65 was probably the first Soviet wire-guided torpedo, TEST-68 (TEST is the acronym for *teleupravlyayannaya torpeda*, or "teleguided torpedo"). Research on wire-guided torpedoes had been conducted in 1961–63.

TEST-71 (NATO 53-71) was the standard Soviet wire-guided torpedo, developed from SET-65 for surface ships and submarines. The modernized version was TEST-71M. Diving depth is 400 m. TEST-71M searches at 24 kts and attacks at 40; maximum range is 15–20 km. Homing range against a submarine is 1,000 m. Homing can be shut down and the torpedo re-aimed to overcome decoys. Modernization included installation of a unified Sapfir homing system, probably in the 1980s. Probably all buyers of Kilo-class submarines received TEST-71ME as standard equipment. NATO's TEST-83 was probably TEST-71M. UTEST-80 was a dual-purpose follow-on that may not have entered service. The U.S. Mk 48 was compromised in the 1980s, and the Soviets seem to have copied its monofuel engine for a TEST torpedo.

Also unknown to the West was Shkval, a 200 kt rocket torpedo that runs inside a supercavitating bubble created by a gas generator. It was developed under the 13 October 1960 anti-Polaris decree, under which SET-65, Vikhr (RPK- 1, NATO SUW-N-1), Vyoga (RPK-2, NATO SS-N-15), the air-launched AT-1 and AT-2 torpedoes, and an abortive Purga rocket torpedo were also developed. Shkval was sold as an anti-Polaris weapon that could destroy an SSBN before it finished firing its missiles. Design requirements were a range of 15–20 km at a speed of 100 m/sec (194 kts). Apparently Shkval was intended for the Alfa-class (Project 705) submarine. The draft design was accepted in 1963; the first experimental vehicles were fired in 1964. It was test-fired in May–June 1966 near Feodosia from the Whiskey submarine (*S 65*), which had previously tested the Vyuga standoff ASW missile (SS-N-15). Shkval was accepted into service in November 1977 as VA-111. It was apparently entirely unknown to the West until the Russians disclosed it about 1992. At least some U.S. submariners saw it as a counter–Mk 48 weapon, to be fired down the guidance wire of the U.S. torpedo to force the firing submarine to break off. It is not clear whether the Soviets saw it that way.

Reportedly the Soviets' interest in a high-speed underwater rocket forming a cavity (the shell of the rocket met the water, hence caused friction, only at nose and tail) in which to run dates back to the early 1950s, the original object being to counterattack ASW ships attacking with depth charges. A diesel submarine fitted with a modified Katyusha rocket launcher hit a target at 500 m range. The project was abandoned as potential enemy ships were fitted with ASW rocket launchers that outranged the underwater rocket. However, sufficient interest remained to inspire development of a new type of hydromotor. This technology matured at the end of the 1950s, just as reports were received of a fast new U.S. EX-8 torpedo, presumably the nuclear Mk 45. Reportedly the Soviet government then approved development of an underwater rocket capable of more than 100 m/sec, to be used for counterattack.

Shkval was a straight-runner, designed originally to carry a nuclear warhead, with two burst settings. It was considered effective against submarines moving at up to 50 kts at depths as great as 400 m.

NOTES

CHAPTER 1. THE COLD WAR AND ASW

1. U.S. rationales from a CIA paper, Staff Memorandum 52-57 of the Office of National Estimates, dated 22 November 1957 and declassified in August 2000, "The NATO Defense Problem," prepared as background for President Eisenhower, who was attending a December 1957 NATO Council meeting. Remarkably, the paper makes no direct reference to the Korean War.
2. The conference was not permitted to include this resolution in its formal recommendations, but no ships of this type were transferred before the 1960s.
3. According to the 1957 CIA paper, European governments emphasized the corrosive effect of internal weaknesses and the threat of subversion over the military threat. They tended to regard their military efforts as payment for U.S. protection, which in turn would restore sufficient confidence to resist Soviet pressure. They also saw this effort as a quid pro quo for retention of substantial American ground forces to guarantee U.S. involvement in any European conflict. Given the disparate European and U.S. views, not surprisingly the gap between goals set by NATO authorities (strongly backed by the United States) and European military contributions widened.
4. According to the April 1950 Low Report. At this time, it was assumed that a mothballed ship could be made mobile within a month and brought to operational standard in two more. Low asked for test recommissionings. Another document of this period mentions a total of 600 escorts, including those needed for Pacific convoy routes. The difference between this and U.S. figures may have been the assumed Allied contribution.
5. French records of the Conseil Supérieure de la Marine (in SHD, Vincennes) give an idea of the gap between what was wanted and reality. Desired French naval forces to support NATO (as listed in NATO MC 26/1 in 1952) included 31 first-class escorts, 52 second-class escorts, and 20 coastal escorts (some other NATO lists called for many more). In 1952 France had 17 first-class and 43 second-class escorts plus 18 coastal escorts; but given French national requirements, the totals required were 18 first-class and 54 second-class escorts and 31 coastal escorts. Many of the ships involved could not meet a NATO requirement for 4,500-nm endurance at 12 kts to cross the Atlantic without refueling: the French lacked an efficient high-temperature high-pressure steam plant comparable with those in the U.S. Navy and the Royal Navy. The only foreign source of escorts was the United States, which was supplying six slow destroyer escorts to supplement the eight supplied in wartime. In addition, the United States bought ocean escorts and coastal escorts for France under its OSP program: the first 4 postwar French frigates (E 50 type) and 7 of the 14 follow-on E 52 class under the U.S. FY 52 budget, as DE 1007–1013 and DE 1016–1019 (E 50 type), plus six *Le Fougeux*–class coastal escorts (PC 1610–1614 and PC 1618). OSP also paid for frigates and coastal escorts for other NATO countries. In theory, OSP would finance reconstruction of French defense industry, which would fill French government orders. However, through the 1950s France was in terrible financial trouble, due partly to the stress of the wars in Indochina and Algeria.
6. The exercise was conducted between 13 and 25 September 1952 to exercise SACEUR and SACLANT forces in the defense of the northern flank of NATO. Volume III of the British Cabinet report is ADM 116/6437; the report on operations by the carrier *Midway* is ADM 116/6438.
7. Isabel Campbell, "Canadian Insights into NATO Maritime Strategy, 1949–70: The Role of National and Service Interests," *Northern Mariner* 25, no. 3 (July 2015), summarizes the evolution of this argument, the Royal Canadian Navy standing between U.S. and British views. Campbell thought the U.S. Navy's strong advocacy of early nuclear strikes by carriers to hit Soviet submarines at source partly reflected the dominance of carrier aviators in the U.S. Navy. The carrier-poor Royal Navy was much more interested in convoy ASW (arguing that U.S. carriers would be quite vulnerable to Soviet submarines).
8. This account of the evolution of SACLANT strategy is based on Campbell, who made extensive use of both Canadian internal files and British official material. She mentions forward barriers such as the Northern Chain and surveillance of the GIUK Gap, presumably meaning HF/DF surveillance and air surveillance (SOSUS did not yet cover these areas). British comments on SACLANT's 1952 Emergency Defense Plan (in DEFE 6/22) suggest that the barrier patrols would employ aircraft using snorkel-detecting radar.

9. As described in a 25 August 1955 paper by the British Joint Planning Staff, in DEFE 6/31.
10. At the 1947 U.S. Navy Undersea Warfare Conference LCDR Edward L. Beach Jr., who later wrote the novel *Run Silent, Run Deep*, raised the question of defense against such missiles. Although the possible threat of submarine missiles was decried at the 1949 meeting, by that time the U.S. Navy was actively developing its own submarine-launched cruise missiles, Regulus and Triton.
11. As discussed in a 7 January 1954 letter from British Defense Attaché ADM C. F. Hallett to First Sea Lord Admiral of the Fleet Rhoderick R. McGrigor, file 8C in 1954 First Sea Lord papers, ADM 205/102. Hallett wrote that "because they are working with experimental submarines firing guided missiles themselves, they are scared stiff of the political effect of a Russian submarine fitted with guided missiles with atomic warheads appearing off their East Coast and firing a few shots into their very vulnerable East Coast ports and towns." Hallett thought the U.S. submarine force was largely responsible for the shift toward home defense against missile submarines. The idea of an SSK barrier was being revived, this time to stretch between Halifax and Bermuda. There were rumors that hunter/killer groups and a number of A/S vessels might be allocated to the national eastern frontier commanders (i.e., non-NATO). Although the idea might not be widely held, it affirmed that "uppermost in some U.S. naval minds is that the defense of the continental United States is the primary role of its Naval Forces. I am uncertain how widely these unbalanced views are held." Any such shift would reduce the dominant U.S. contribution of ocean escorts in a future war. Hallett and others in the British mission were "trying to place the threat in its true perspective in our discussions with the United States Navy." However, the U.S. Navy told him that it could not afford to have the East Coast attacked, hence would devote "a totally disproportionate amount of Naval and Air Forces and scientific effort in guarding their Eastern sea-board." Long-range VLF detection devices (SOSUS) at Bermuda, Sandy Hook, and near Cape Sable in conjunction with the Canadians would be backed by intensive air patrols looking for snorkels. ASW submarines would back up this line. Behind it would be hunter-killer groups. Hallett argued that the places to catch Soviet submarines were in their bases (attack at source), coming out of their bases (mines), and by a Corsair/Jezebel (SOSUS) chain in the Orkneys/Shetlands/Faroes/Iceland line to stop as many as possible from getting into the Atlantic. He wrote that his American contacts were inclined to agree, but "in the meantime it was their duty to ensure the inviolability of the coasts of the United States." Hallett wrote that if only the Russians knew the impact of the threat, surely they would build some missile submarines and publicize them.
12. As quoted in ADM 1/24533, to justify cuts in current air ASW programs.
13. This point is made by Cross Associates, *Sea-Based Antisubmarine Warfare 1940–1977* (2nd ed., 17 February 1978), report prepared for the Office of Naval Research and later declassified from Secret (henceforth the Cross Report), I, 152–53. Continued interest in convoy operations is shown by the creation of a third experimental ASW task group (Charlie) in October 1958 specifically to develop convoy tactics; it seems to have pioneered drone operations (DASH) and the use of variable-depth sonar. The first experimental task group (Alfa) was established on 1 April 1958, built around the ASW carrier *Valley Forge*, with a destroyer squadron, two Guppy submarines, and aircraft, including shore-based Neptunes. This was also the first Atlantic Fleet task group intended to remain continuously at sea to investigate submarine contacts, meaning to prosecute submarines detected by SOSUS and by the HF/DF net. Task Group Bravo was formed in October 1958 alongside Charlie specifically to modernize HUK group area search and hold-down and regain contact procedures.
14. Cross Report I, 152, citing a 1959 ONI report, claiming general concurrence with the idea that the Soviets would use their submarines this way.
15. Under the doctrine of "flexible response," NATO would initially fight conventionally. If that failed it would mount a limited nuclear attack to convince the Soviets to stop. If that failed, NATO would employ tactical nuclear weapons on a larger scale. If those failed to stop the Soviets, the alliance would use strategic nuclear weapons. Skeptics observed that this last option would be as suicidal for NATO as it would be devastating for the Soviets, hence it was less than convincing. Gloomy U.S. studies in the 1970s showed that NATO lacked the forces to defeat a Soviet invasion, and that if NATO used nuclear weapons its defeat would be worse.
16. POMCUS was conceived in October 1961 by CINCEUR and the JCS as a means of speeding crisis response. To an extent it was part of the Kennedy administration's attempt to provide a credible non-nuclear capability in Europe. It is not clear to what extent the POMCUS concept was initially tied to skepticism about the time required to gain control over Soviet submarines, but even in the 1960s it was assumed that would take some time. OSD skepticism about sea control grew over time, making POMCUS more important in the 1970s.
17. A British Admiralty paper written in June 1963 (in DEFE 7/2031) for the British Chiefs of Staff argued that current NATO naval force requirements were based on an

outdated strategy. The requirements placed undue emphasis on operations after the nuclear exchange, whereas greater priority should be given to operations up to and including the exchange; NATO needed maritime shield forces whose role would be analogous to that currently accorded land and air shield forces. NATO land forces were intended to (1) deter aggression; (2) meet it, if it occurred, with sufficient force to force the aggressor to choose between escalation and withdrawal; and (3) make escalation increasingly difficult for the Soviets. Given the Admiralty's shift in the late 1950s, and its acceptance by the then minister of defence, it is remarkable that the Royal Navy was first arguing this case in NATO in 1963. NATO maritime strategy was still based on the assumption that there would be little or no warning of attack; that nuclear weapons would be used by both sides from the outset; and that a war would consist of an intense nuclear exchange followed by a subsequent phase of broken-backed war of indeterminate length and lesser intensity.

18. The British Chiefs of Staff approved this concept in 1969. The one-week figure appears in contemporary documents but also in John Nott's memoir of his time as defence minister (1981–82), justifying his rejection of Royal Navy arguments that its surface fleet was needed to cover convoys.
19. Admiral Kidd's 1977 letter encouraging a British study of shipping defense is in DEFE 69/208. This was new for Kidd, who had been in office since 1975. NATO had not issued a convoy study since August 1973 (and that one had not been officially published). The most recent relevant NATO effort was a revision of the convoy portions of the standard Allied Tactical Publication ATP-1B called "Attrition at Sea"; it discussed high-speed convoys. Kidd's new strategy made protection of reinforcement shipping—on which, he wrote, the fate of a European land battle—a major part of the NATO deterrent—would depend. By the late fall of 1977 SACLANT was planning a standing Seahawk ASW conference, the first scheduled for 7–9 February 1978. One of its main topics would be "An Operational Concept for the ASW Protection of Trans-Atlantic Reinforcement Shipping," which was described late in 1977 as a "hot potato on the NATO scene." There was increasing interest in employing the NATO Strike Fleet to cover the SLOCs, to the point that the British director of plans asked whether the SLOCs could be protected at all without support by the Strike Fleet. At this time the Strike Fleet was expected to deploy into the Norwegian Sea to prevent the Soviet fleet from breaking out.
20. For example, the CIA released "The Soviet Attack Submarine Force and Western Sea Lines of Communication: An Intelligence Assessment" in April 1979, using information available as of September 1978 (the declassified version was released on 14 June 2017). Director of Central Intelligence VADM Standfield Turner wrote in the foreword that although planners in both NATO and the Warsaw Pact expected a short war, planners in both world wars had also wrongly expected short wars; "I therefore believe there is a high probability that if a war between the Warsaw Pact and NATO became an extended conventional conflict, the naval side of the war would turn into a protracted campaign of sea denial versus sea control in the North Atlantic." Should the West concentrate on what the Soviets were saying or on what all those submarines might do in wartime? Naval forces are inherently flexible. Before December 1941, for example, U.S. submariners were trained mainly to scout for the fleet and to attack Japanese capital ships. On 7 December, however, they were ordered to execute unrestricted submarine warfare against Japan—to sink Japanese merchant ships on sight. They proved hugely effective, particularly after early torpedo problems were solved. Only some other more urgent submarine role would preclude a Soviet antishipping campaign.
21. DEFE 69/208, the file on the 1978 Maritime Tactical School study of seaborne reinforcement of Europe.
22. It appears that Mikhail Gorbachev was made general secretary (leader) of the Soviet Communist Party on the strength of his pledge to solve the computer problem ("medium machine building," in Soviet-speak). His efforts ultimately proved fatal to the USSR.
23. The Soviet account of the meeting mentions a Soviet naval staff study of the naval lessons of the war, but unfortunately no copy of this study has ever surfaced in the West.
24. According to Juergen Rowher and Mikhail S. Monakov, *Stalin's Ocean-Going Fleet* (London: Frank Cass, 2001), 93, in August 1939 the Soviets had in commission (and under construction) 27 (19) large submarines, 77 (38) medium submarines, and 59 (17) small submarines, a total of 163 submarines in commission and another 74 laid down. At that time the planned 1947 fleet would have included 88 large submarines, 229 medium submarines, and 123 small submarines (plus 2 experimental boats), a total of 440 submarines. Stalin rejected proposals for various specialized submarines in 1945, and the 10-year program he approved on 27 November 1945 included 40 large, 204 medium, and 123 small submarines (Rowher and Monakov, 185). The latter were mostly updated M class plus 30 closed-cycle Quebec (Project 615) class. By 1955, the Soviets had actually completed 6 large, 151 medium, and 66 small submarines. According to Natalia I. Yegorova in "Stalin's Conception of Maritime Power: Revelations from the Russian Archives," *Journal*

of Strategic Studies 28, no. 2 (April 2005), submarine priority was increased in 1952 (she cites an April 1952 resolution by the Council of Ministers), the number of large and medium submarines envisaged in the Ten-Year Plan being increased from the 179 envisaged in a 5 February resolution to 277. She states that the submarine fleet was given the greatest attention at this time. The amphibious ship program was apparently cut to compensate. Yegorova, who worked at the Center for Cold War Studies of the Institute of General History of the Russian Academy of Sciences, based her account on documents from the Russian State Archive of the Economy, which included holdings of the State Planning Committee and the Soviet Ministry of Finance.

25. It appears that the M class was conceived to provide the maximum number of torpedo tubes at a fixed cost; boats were built in sections that could be transported by rail.
26. According to V. P. Kuzin and V. I. Nikol'skiy, *The Soviet Navy 1945–1991* (St. Petersburg: Historical Naval Society), Stalin was unimpressed by the potential of submarine attacks on shipping. He opposed large submarines and attempts to increase their numbers. He seems to have envisaged two stages of naval development: a first ten-year program to develop a fleet to dominate waters close to the Soviet Union, and a second ten-year program to develop a truly oceanic navy. Stalin talked about structuring the initial program to train cadres for the oceanic navy. Soviet building plans are described in A. N. Sokolov, "Our Ambitious Plans: Soviet Shipbuilding Plans of the Post-war Decades," *Warship International* no. 3 (2010): 229–39, and no. 4. According to Sokolov, Stalin wanted to be ready for an attack by the capitalist West in a decade, which mirrors the British "year of maximum danger" idea. The March 1945 plan was unrealistically large; on 27 September 1945 Stalin chaired a conference intended to cut it to manageable proportions. The amended program was ratified by a 27 November 1945 decree. According to Sokolov's table, which probably reflects the original approved plan, it provided for 40 Project 611s (later adjusted to 30), 204 Project 613s (not listed as adjusted), and 123 small Project 612 submarines, later dropped in favor of fewer Q class (Project 615). These figures did not include a prototype closed-cycle submarine (Project 95, a prewar project) or the completion of 57 M-class coastal submarines, many of them begun before the war. The new designs for medium (Project 613 Whiskey) and large (Project 611 Zulu) submarines apparently took longer than expected to complete. No large submarines were completed during the first plan period, and only six in the second. Later adjustment seems to have replaced some of the large submarines with medium ones.
27. According to Kuzin and Nikol'skiy, projected totals were 350 Whiskeys, 160 Foxtrots, and 560 Romeos. These figures for planned submarine production roughly match the 1,000-submarine figure. Kuzin and Nikol'skiy are credible because they were former naval officers, presumably with access to planning documents. It is not clear why their figures diverge from Sokolov's, but presumably ideas changed drastically from 1957 onward. Of the types they list, Foxtrots (Project 641) were successors to Zulus, and Romeos (Project 633) were updated Whiskeys. Note, however, that the mass-production Romeo had twice the rated stores endurance of a Whiskey—60 rather than 30 days—although rated range was similar. They criticize Zhukov as impractical and ignorant of naval practicalities. To sink the necessary more than 2 million tons per month, "by the most modest calculation" the Soviets would have required about 1,000 submarines and monthly production of 15,000 torpedoes. The Soviets abandoned production of short-range submarines altogether in the late 1950s, selling the Romeo tooling to the Chinese (who were also given Whiskeys). The Chinese Type 033 is their version of the Project 633 Romeo. By 1971 the U.S. Navy accepted that Whiskey was essentially a coast-defense submarine. A declassified collection of documents used for a Naval War College ASW course in 1972–73 included a September 1971 CIA paper on the evolution and operations of the Soviet attack submarine force (SR-IM 71-11-8), presumably incorporating U.S. Navy input. It credited a Whiskey with 20 days' endurance on station per patrol at a radius of 1,200 nm, compared with a similar period on patrol for a Zulu at a distance of 2,400 nm. Under optimum conditions a Whiskey could spend about 10 days on a patrol station 1,800 nm from its base. In the CIA view, a Whiskey was better able to defend the sea approaches to the USSR and interdict naval vessels in the Norwegian Sea. Of submarines built during the initial Soviet building program, only the Zulus were suited to reach U.S. SLOCs to Europe and conduct lengthy operations. The CIA account mentions the 1,000-submarine goal of the 1950s. According to this paper, in the late 1950s the Soviets recognized a more serious threat in the form of U.S. carriers capable of launching nuclear strikes against the Soviet Union. In the CIA's view, the Soviet Union developed cruise missile submarines to counter the carriers, while torpedo submarines were intended for interdiction. However, Soviet writings about attacks on carriers envisaged coordinated missile and torpedo attacks.
28. Sergei Chernyavskii, "The Era of Gorshkov: Triumph and Contradictions," *Journal of Strategic Studies* 28, no. 2 (April 2005), special issue *The Cold War at Sea*. Chernyavskii saw Zhukov, with Khrushchev's support, reacting against the attempt by Kuznetzov to build a powerful balanced

fleet. According to Chernyavskii, Kuznetzov proposed a 10-year program (1955–64) to build 9 aircraft carriers, 21 cruisers, 118 destroyers, and 324 large submarines, to form 3 or 4 surface squadrons. Kuznetzov was not invited to a government meeting held at Sevastopol in fall 1955 to determine the future of the Soviet navy. Chernyavskii worked for the Central Naval Museum in St. Petersburg, in effect the Soviet naval archive. He doubted that Gorshkov had any particular long-range plan. Initially he charmed Khrushchev (and thereby replaced Kuznetzov) by acceding to his view that surface ships were obsolete. He then seized an opportunity presented by Brezhnev to build his desired fleet. Historians of the Soviet Union have pointed out that Soviet-era politics was personal. Views expressed were inevitably tactical. Those espousing them could easily shift to views that had proved fatal for their failed opponents. One might see Gorshkov's replacement of Kuznetzov in this light.

29. Andreyev was quoted at the 1959 U.S. Undersea Warfare Conference: "The United States is the smithy [foundry] of war. . . . Overseas transport of weapons to allies and providing the overseas requirements of the U.S. Army and Air Force will be of paramount importance. If Allied sea routes are severed or disturbed, it will unavoidably lead to the critical weakening of the military participants of the aggressive North Atlantic bloc." No context for Andreyev's remarks was given.
30. Nikolai Simonov, *Voenno-promyshlenniy complex SSSR v 1920–1950-e gody* (Moscow: Rosspen, 1996), translated as *Military-Industrial Complex of the USSR 1920–1950s* by Steve Zaloga. The naval shipbuilding plan is tabulated on page 299. Other warships were 12 light cruisers, 10 missile ships, 22 destroyers, 4 air warfare ships, 10 radar pickets, and 94 ASW ships. None of these figures was approached. Another table shows projected production to meet the goals, including 18 reactors in 1958 and 134 in 1968. Sokolov gives much smaller numbers in *Warship International* no. 3 (2010): for the 1959–65 program he lists 20 Novembers (13 were actually built), the single liquid-metal version of that class (Project 645), 24 Romeos (20 built), 32 improved Romeos (Project 654, none built), 22 Project 629M (SSBN, none built), an unknown number of Project 658s (SSBN; he does not give numbers built), 10 Whiskey Long Bin (conversions: 6 actually done), 6 Echo I (5 built), 30 Echo II (29 built), 48 Juliett (16 built), 1 P627A (November version carrying long-range cruise missile, unbuilt), 4 Project 653 (nuclear powered submarines with two long-range cruise missiles each, unbuilt), 12 transport submarines (Project 664, none built), 7 nuclear transport submarines (Project 681, not built), 20 nuclear submarine tankers (Project 681, not built), and 30 A-658 (Project 658 with D-7 missile, not built). This would amount to 240 submarines. Sokolov also lists some studies without planned building quantities: Project 654 (a multipurpose submarine), Project 663 (a 1959 midget submarine armed with torpedoes), Project 673 (a study with an *Albacore* hull form), Project 683 (a modified Juliett), and Project 688 (a nuclear submarine carrying P-120 or P-500 missiles). Sokolov does not list any of the next-generation submarines, such as the mass-production Project 670 or the very fast Project 661 (Papa), presumably because they would have been completed after 1965.
31. Brezhnev apparently saw Khrushchev's fall (which he had engineered) as proof that it was far too dangerous to tamper with the core military element of the Soviet economy. He therefore accepted continued and even uncontrolled growth of the military sector—including the naval sector—to the point that it nearly collapsed in the 1980s, forcing project cancellations before the Soviet Union itself collapsed. Among other things, after Khrushchev's fall the Soviet army was rebuilt.
32. Chernyavskii, "The Era of Gorshkov," 281–308.
33. Planning included not only construction but also planned refits. Later Soviet practice was to plan a midlife refit (to bring it back to zero-hours condition) for a ship or submarine, and also to provide spares for onboard repair. This planning naturally omitted major changes during the midlife refit, which could not have been planned in advance.
34. According to the declassified NIE 11-15-82, "Soviet Naval Strategy and Programs through the 1990s," based on information available as of 19 October 1982, the primary Soviet navy tasks would remain to protect Soviet ballistic missile submarines and to attack Western carriers and ballistic missile submarines. "Accomplishment of these tasks would entail attempts to control all or portions of the Kara, Barents, and northern Norwegian and Greenland Seas, the Seas of Japan and of Okhotsk, and the Northwest Pacific Basin, and to conduct sea denial operations beyond those areas to about 2,000 km from Soviet territory." The Soviet navy would also support ground operations in peripheral areas such as Norway, and it might conduct some SLOC interdiction. Soviet naval strategy would be unchanged, but the need to counter Western units armed with the new Tomahawk cruise missile would drive the Soviets to expand sea control operations, possibly out to 3,000 km from Soviet territory.

CHAPTER 2. FINDING SUBMARINES

1. The translation of asdic into an acronym for an Allied Submarine Detection Committee was created in 1939 by Winston Churchill when he was First Lord of the Admiralty. In Parliament he was asked what "asdic" meant, and he invented the story on the spot.

2. Prewar British research showed that this kind of attenuation was proportional to the square of the frequency. In 1948 it was discovered that absorption was due to ionic relaxation of a trace compound in sea water, magnesium sulphate. According to Willem Hackmann, *Seek & Strike: Sonar, Anti-Submarine Warfare, and the Royal Navy 1914–54* (London: HMSO, 1982), 145, only in 1933 did the British separate attenuation from spreading from attenuation due to absorption of sound by the water. Within a shallow duct, sound waves spread out cylindrically, so their intensity is reduced roughly as the inverse of range. However, if the duct is deep enough, the sound spreads spherically, falling off as the square of range. There was also attenuation due to scattering in the water (reverberation).
3. Purchases of bulk supplies of quartz were stopped in 1929, then resumed in 1935. At that time the existing supply was enough for 650 more transducers (70 of which were included in the 1935 program). In 1935 the expected emergency production rate was ten per week. After 1935 between one and three hundredweights (112 lbs. each) were bought each year. Hackmann, *Seek & Strike*, 153. The British did not adopt magnetostriction because they wanted to avoid major production changes, such as new sonar domes.
4. NRL began developing sonars in 1924 after ordering ultrasonic gear from France and being told that would take two years to produce. NRL claimed that its own gear proved superior when the French gear finally arrived. The QA series equipped only the initial experimental ships. QBA was the initial submarine set. QBF was a surface ship sonar for ships up to frigate size (300 built from 1943 on) operating at 22–28 kHz. The main destroyer sets were in the QC series; QCB, a 1938 destroyer sonar, operated at 24 kHz. QCD was a depth-determining sonar equivalent to the British Type 147, which it was intended to replace. QG-series sonars were console versions of the QC series, with better operator facilities. QJs were console versions of the QB series. QHs were scanning sonars, important after the war. Submarines had J-series passive equipment. There was also a W series of combined ranging and sounding devices.
5. This procedure is from Admiralty handbook CB 4127(43), *Asdic Operating Procedure for Surface Craft,* issued 1943 as amended. RG 38 Foreign Documents collection, NARA II.
6. Details of chemical range recorder as used by the U.S. Navy are from 1944 handbook, *Operation of the Sonar Range Recorder*, issued by HQ, U.S. Fleet, in NARA II collection, former NHHC collection.
7. ADM 259/384 includes a description of Type 144 written in 1943.
8. According to Hackmann, 280, the first confirmed report of a U-boat diving deeper than 700 ft (June 1943) led to rapid modification of depth charge and recorder settings. Late in 1943 a Staff Target assumed that U-boats could dive to 1,200 ft, and in 1944 that was increased to 1,300 ft. According to Hackmann, the greatest U-boat depth measured by the Royal Navy during the war was 780 ft.
9. U.S. Fleet *Operation of the Anti-Submarine Depth Finder (147B),* February 1945, NARA II, former NHHC collection.
10. These efforts are described in the NDRC volume on *Echo-Ranging Systems* (Underwater Sound Equipment II), volume 15 in the series of NDRC Summary Reports.
11. The wartime U.S. program is described in an NDRC volume *Underwater Sound Equipment III: Scanning Sonar Systems* (Washington, 1946). After some false starts, a commutator sonar was first demonstrated in November 1943. In February 1944 Sangamo Electric Co. received a prime contract to produce three prototype QH scanning sonars. This sonar included both a motor-driven commutator for scanning and a hand-trained listening commutator, to identify targets and noise. Raytheon received a contract for an electronically scanned alternative, which became QHA.
12. Hackmann, 274. The first British approach to rapid scanning, Type 149, used a single rapidly spinning directional magnetostriction hydrophone feeding a PPI, showing the bearings of all sound sources. It was installed in four cruisers by the summer of 1944.
13. Hackmann, 139 and 142–43, dates British awareness of variation with depth to trials in 1922 using the destroyer *Rocket* and the submarine *H 22*, operating in the Mediterranean, the Sea of Marmora, and the Black Sea. The first formal measurements of temperature and salinity at different depths, which were converted into sound velocity gradients, were made in 1930. These gradients indicated how sound would be refracted (bent). In 1933 the Admiralty accepted that further oceanographic research was necessary for effective use of asdic.
14. The British experienced the afternoon effect, but the Americans at Woods Hole discovered its cause to be the very shallow surface duct. In 1937 Columbus Iselin of the Woods Hole Oceanographic Institution and I. Batchelder of the Submarine Signal Company (which made U.S. sonars), operating off Guantánamo, found out why the afternoon effect occurred. Initially they thought gas bubbles were involved, but when they took temperatures at unusually short depth intervals, 4 m rather than 25 m, they discovered the cause. The same explanation was advanced soon afterward by the West Coast Sound School. According to the NDRC volume on *The Application of Oceanography to Subsurface Warfare* (volume 6A of the Summary Technical Report of Division 6 of NDRC), both off Guantánamo and at San Diego (where the West Coast School was), echo-ranging was typically conducted in

deep water in places where the effect of downward refraction was particularly striking. Many earlier tests had been made in places where refraction effects were not so frequent.

15. Because the Atlantic has deeper warmer water (due to the Gulf Stream and the Sargasso Sea), the deep channel is deeper in the Atlantic than in the Pacific. Typical convergence zone ranges in nautical miles for summer/winter are: North Atlantic 35/30, GIUK Gap 27/26, Northern Pacific 27/32, Tyrrhenian Sea 20/none, and Bermuda 33/33. A convergence zone is most prevalent in the Mediterranean during the summer, when high temperatures evaporate enough water to raise salinity, which in turn raises the deep sound channel to a depth between 200 and 1,400 ft (in the Atlantic and Pacific it is typically 4,000 ft deep). This shallow deep channel traps surface noise, including shipping noise, which can drown out sonar exploiting the deep channel. In northern latitudes convergence zones can exist at 20 nm range in water at least 700 fathoms deep.
16. The first description of the convergence zone seems to have been by Maurice Ewing in his Interim Report No. 1, Woods Hole Oceanographic Institution, 25 August 1945; it was also described in M. Ewing and J. I. Worzel, "Long-Range Sound Transmission," *Geological Society of America Memoir* 27 (October 1948). According to Gary E. Weir, *An Ocean in Common: American Naval Officers, Scientists, and the Ocean Environment* (College Station: Texas A&M University Press, 2001), 172, Maurice Ewing conceived the deep sound channel during 1937 experiments with signal bombs. He found that their sound reflected seven times from the bottom and the surface before dying out to the extent that he could no longer hear it unaided. He was impressed that sound could travel that far, and speculated that a deep sound channel existed, and that with proper amplifiers, sound in it might be heard over thousands of miles. Ewing resumed work on a possible deep channel in 1942, proposing it as a communication medium in a July 1943 report to the U.S. Bureau of Ships. BuShips was less than enthusiastic due to the possibility that the channel would be subject to interception and jamming. However, the bureau's Underwater Sound Design Division took up sponsorship, allocating the oceanographic research vessel *Saluda* and three ASW units, including the destroyer escort *Buckley*. They set off 1 lb bombs at 80 ft depth and 4 lb bombs at 4,000 ft depth at 100 mile intervals. *Saluda* lowered a hydrophone to the axis of the Atlantic sound channel, 3,500 ft down. The bomb could be heard as far away as instructions allowed, 900 nm. A seafloor array emplaced at Eleuthera was able to hear bombs exploded across the Atlantic at Dakar, 2,300 nm away. The deep channel was not mentioned in the 1946 NDRC report on oceanography, which carried a Restricted (i.e., official use only) classification, the lowest one.
17. Sonar conditions described in declassified ASW Reference Manual (AD 522643) for the Mediterranean, issued August 1972.

CHAPTER 3. LEGACY

1. GCHQ history, 128–33.
2. For an account, see W. J. Holmes, *Double-Edged Secrets* (Annapolis: Naval Institute Press, 1978).
3. According to Hackmann, *Seek & Strike*, 294, the Allies did not really appreciate the capabilities of German passive arrays until after the war, even though the British had captured an example on board *U 110* and the Americans had taken one on board *U 505*.
4. The anonymous authors of the postwar British study on German use of special intelligence, *The German Navy's Use of Special Intelligence* (in the NSA Historical Documents file, RG 457, in NARA II), pointed out that in 1931 the World War I commander of U-boats, ADM Hermann Bauer, published *Das Unterseeboot*, in which he argued that the only suitable counters to convoy were a combination of reconnaissance and assembling the maximum possible number of U-boats for attacks. He referred to but discounted radio intelligence. He suggested that U-cruisers lying in wait outside ports of departure could shadow convoys. The document is a carbon copy, presumably of a document written to contribute to the Government Code and Cipher School (GCCS) book on the Battle of the Atlantic, unsigned, undated, without security markings, and paginated only in pencil. It appears to have been compiled immediately after the war by the British codebreaking organization (whose histories were often cosigned by U.S. Naval Reservists who had worked at Bletchley Park or in the U.S. codebreaking organization). The authors worked from copies of the daily and weekly reports of the German codebreakers and from the logs kept by the admiral commanding U-boats (BdU). It was not, however, always possible to say whether the U-boat command was using current decodes or correlations and information from other material. When the Germans lost the ability to break the convoy codes, they relied on estimates of convoy routes based on earlier experience. The U-boats relied heavily on the Anglo-U.S. Combined Naval Cipher, to the point where the staff responsible for breaking it was doubled after the cipher was changed on 15 December 1942.
5. Hackmann, 256.
6. According to OEG 51, the lethal range of a 300 lb depth charge was 21 ft, compared with 6 ft for a 30 lb Hedgehog projectile. During the war, TNT was replaced by the

more lethal Torpex, so the lethal range of a depth charge increased. Too, 600 lb depth charges were introduced.

7. There were two U.S. versions: Projectors Mk 10 and Mk 11. Mk 10 fired an elliptical pattern, 120 by 140 ft. Mk 11 fired a circular pattern with a diameter of about 200 ft. Hedgehog was a gun, so it had a gunlike deck reaction. The U.S. Navy introduced a rocket equivalent (i.e., without recoil), Mousetrap (rocket launchers Mks 20 and 22), that could be installed on board sub-chasers (later it was installed on board destroyers assigned to Mediterranean convoy duty).
8. Blackett's May 1941 proposal (in ADM 1/15194) was a directional buoy that could be dropped from a standard bomb rack. The microphone would be rotated by a paddle or screw, and the buoy would have a magnetic compass that would enable it to indicate bearing. Reference from Paul Marks, "Sonobuoys," in the e-Defence *Electronics Newsletter*, No. 91 (February 2020, pt. II).
9. My summary of the course of the Battle of the Atlantic is largely based on OSRD, *A Summary of Antisubmarine Warfare Operations in World War II* (summary technical report of Division 6; NDRC: Washington, DC, 1946), originally issued as a classified report. It is often referred to as OEG (Operational Evaluation Group) 51. The editors who signed the introduction were C. M. Sternhell and A. M. Thorndike. That account omits coding and codebreaking as factors. To reintroduce them I have relied on an early postwar account of Germans' use of codebreaking and on the GCHQ volume on the Battle of the Atlantic, Naval History, vol. XVIII, *The Battle of the Atlantic*, by LCDR R. J. Goodman RNVR, LCDR K. W. McMahan USNR, LT E. J. Carpenter USNR, and others, edited by LCDR L. A. Griffiths RNVR. My copy is in NARA II.
10. OEG 51 adds that the light aircraft helped the British develop an accurate plot of U-boat positions.
11. The first escort radar (Type 286) was derived from an RAF air-to-surface vessel (ASV II) set. It could detect a surfaced U-boat about 15 nm away from an airplane at 2,500 ft. Installations of Type 286 began about November 1940; by April 1941 it was on board about 40 Western Approaches Command destroyers. It had a fixed transmitting aerial and typically used two receiving aerials. Together they covered an arc of about 50 degrees each side of the bow. The transmitting antenna had large back lobes, so the radar could inadvertently detect ships astern of the transmitting ship. An improved Type 290 was less subject to back echoes. In either set, range over water was limited by the wavelength, the radar suffering from multipath interference off the sea.
12. According to the GCHQ history (pp. 112–13) COMINT became a regular source at this time, previously having been sporadic and tardy. All U-boat signals of February 1941 were read at various dates after 10 March as a result of captured documents. March messages were not read, but April messages became available by cryptography between 22 April and 10 May. Messages of 21 days of May were read with a delay of 3 to 7 days. June and July were read nearly currently with the help of captured documents, though a change in the indicator tables on 15 June increased delays. With the exception of December, when the average delay rose sharply to 80 hours as a result of another change in indicator tables, almost all U-boat messages until the end of January 1942 were read with an average delay of 36 hours.
13. Coastal Command began equipping its aircraft with depth bombs as well as conventional bombs; on 16 August 1941 a Sunderland using the new weapon severely damaged a U-boat. After spotting the U-boat, it dropped a charge set for 100 ft about 20 yds ahead of the U-boat, which had spotted the airplane and was crash-diving. It was forced to the surface, and the Sunderland dropped a second charge, set for 150 ft. The U-boat was again forced to surface and was then observed to sink sideways. The airplane dropped a stick of its four 250 lb bombs on the submerged U-boat. Coastal Command decided that all of its aircraft should be armed with depth charges. In two cases during this period, ending in December 1941, Coastal Command attacks seem to have killed U-boats. In May 1941 Coastal Command made the first (unsuccessful) attempt to hunt a U-boat to exhaustion using a radar-equipped airplane. Operational analysis showed that attacks should be concentrated on submarines caught on or near the surface that crash-dived when they sighted an airplane. In June 1941 Coastal Command aircrew were told to set all depth charges to 50 ft depth, space them 60 ft apart, and drop all of them in a single stick. Crews were to be trained to drop their bombs within a 20 ft error circle. The proportion of attacks resulting in at least some damage to a U-boat rose from 10 to 25 percent. By the end of 1941 it had been realized that attacks could be made much more lethal by setting depth charges to 25 ft. Coastal Command also began to pay more attention to U-boats transiting the Bay of Biscay en route to or from their French bases. By November 1941 U-boats were being attacked on one of every three cruises, so they transited the bay submerged during the day. By December Coastal Command aircraft operating over the bay had radar.
14. GCHQ history, 123.
15. According to his war diary, Doenitz decided to withdraw the last two U-boats operating off the U.S. East Coast on 19 July 1942. At this time U-boats could still operate in the Caribbean and in the Gulf of Mexico. GCHQ history, 231.
16. GCHQ history, 241. The Germans attributed their failures against convoys during the operations on the U.S.

East Coast partly to the inexperience of the U-boat commanders involved. They regarded the campaign partly as a training exercise. Through the summer, the few experienced U-boat commanders accounted for nearly all sinkings.

17. GCHQ history, 143.
18. GCHQ history, 126–28, 161–62. According to the GCHQ history, the failure of the U-boats in the Atlantic (due to convoy evasion) was directly connected with the redistribution of U-boats, as it became more and more difficult to justify concentrating U-boats in the Atlantic without any great successes.
19. GCHQ history, 269.
20. GCHQ history, 321.
21. History of German wartime naval codebreaking. The reference is to page 110, the first of chapter VI, "Decline and Fall" (of the *B-dienst*). This is apparently the final version of the draft in NARA II.
22. GCHQ history, 351.
23. GCHQ history, 360. Doenitz wrote in his war diary: "Once more it is evident that the main problem in U-boat warfare is a problem of *finding*, and that a final solution of this problem is possible only through constant use of *long-range* aircraft of our own."
24. Summaries of NID reports in a U.S. Navy attaché's report on "underwater internal combustion propulsion in U-Boats" dated 10 April 1944, RG 38, NARA II.
25. In March 1945 the British summarized their experiences fighting snorkel submarines in a memo for the U.S. Navy, the expectation being that the snorkel U-boats would soon be operating off the U.S. coast (ADM 1/17595).
26. GCHQ history, 402.
27. According to the GCHQ history, 408, they were actually a last attempt to revive convoy warfare in the open Atlantic. Early in April 1945 these boats were ordered to form a reconnaissance line sweeping west-southwest along the U.S.–UK convoy line. They were ordered to stand on 12 successive lines, but never sighted a convoy. The group's only success was to sink the destroyer escort *Frederick C. Davis.*
28. Examples from Cross Associates, *Sea-Based Antisubmarine Warfare 1940–1977* (17 February 1978, 2nd ed.), report prepared for the Office of Naval Research and later declassified from Secret, I, 48–50.

CHAPTER 4. NEW KINDS OF SUBMARINE

1. ADM 1/16495, Director of Anti-Submarine Warfare of Naval Staff, 24 March 1944. The enclosed paper is undated; the cover paper refers to the minutes of ACNS(W)'s meeting held on 28 February 1944, which called for a panel to consider the implications of the fast and/or deep-diving U-boat. The main paper was circulated under a 28 March 1944 memo.
2. NID 413 (5 January 1944) described German work on underwater internal combustion propulsion in U-boats, based on information received up to 31 December 1943. The Germans were known to be experimenting with internal combustion turbines or diesels using compressed or liquid oxygen and some kind of fuel. They were probably in the advanced trial stage, but there was no evidence of when they might be in operational use. Nor was there evidence of how many such boats were being tested or of their propulsion program. According to U.S. report x-103 in series P-10-j, 12142 in the RG 38 wartime naval attaché series, NARA II, "most reports refer to boats of about 500 tons or larger; small boats of various types are also reported. Both kinds may be rubber-covered." The U.S. report mentioned that the Italians were known to have built a 14-ton closed-cycle submarine that ran on alcohol and air (closed cycle) when submerged.
3. ADM 259/465, W. E. Dawson, "The Asdic and Its Associated Weapons," January 1947, Report No. 30 of the Underwater Detection Establishment, Portland. It covers both world wars.
4. DEFE 10/483, proceedings of the Operational Requirements Committee for 1966. The Canadians had tested the fuze in 1964–65. The Swedish ZAMBO active acoustic fuze was rejected because it did not meet British safety standards and because it would be more expensive than the Canadian NC 20 fuze.
5. *Progress in Underwater Warfare 1946* (ADM 239/420). I have been unable to find Staff Requirements for new asdics framed in 1944–45.
6. Account of Type 160X from *Progress in Underwater Warfare 1947* (ADM 239/421). It did not mention Type 170. Limbo first appears in the 1947 edition, with a note that trials of a single-barrel version are underway. Limbo and its fire control appear in the 1948 edition (ADM 239/422).
7. This plan responded to a COMINCH/CNO secret letter (FF1/S1Ser 00772) dated 6 March 1944, presumably based on the report of the Admiralty committee. The joint reply by the chiefs of the Bureaus of Ordnance and Ships, BuShips (900–940) 001909, is undated, but the Cominch/CNO letter required a response by 1 September 1944. Requirements were modified in July, and the time limit was extended to the spring of 1945. References in the report include breakdowns of Squid attack errors (presumably based on trials in the United States) dated 15 February and 13 March 1945, as well as reports on scanning sonar development.
8. Design figures for Type XXI, according to the postwar U.S. Portsmouth Navy Yard report on the Type XXI as turned over to the U.S. Navy, Report 2G-21, were 17–18 kts for 1 hour, 11 hours at 10 kts, and 21 hours 15 min at 8 kts. These were the same as figures given by the

Japanese naval attaché in August 1944. However, he gave an endurance of 150 hours at 3 kts, whereas the design figures quoted in 1946 was 95 hours at 3 kts. The authors of the U.S. report stressed that these submarines never achieved 18 kts submerged, but they did not give trial figures. At the U.S. Third Undersea Symposium, Type XXI was credited with a maximum submerged speed of 15.2 kts, an operational approach speed of 10 kts, and a maximum distance of submerged escape of 360 nm, compared with 72 nm for Type VIIC. Average attack evasion speed was 12 rather than 6 kts. These figures were given in a British summary of remarks at the symposium, ADM 213/908. Postwar statements of Type XXI performance varied, as it turned out that the Germans had not achieved what they had planned. In June 1948, for example, Director of Torpedo and A/S Warfare issued a summary of past, current, and likely future trends in submarine tactics (TASW 329/47, ADM 1/27214) that credited Type XXI with 15 kts on batteries (endurance 1.5 hours), with an endurance of 4 hours at 12 kts, 8 hours at 10 kts, 10 hours at 8 kts, and over 4 days at 3 kts. Snorting speed was 10 kts, compared to 6 for converted conventional submarines. Type XXI had a silent electric motor that could propel it at 0.5 to 1 kt. Portsmouth Navy Yard report courtesy of Chris Carlson.

9. Test depth is the "designer's depth" quoted in the July 1946 Portsmouth Navy Yard report. The report warns that not all designed characteristics were met; it is possible that actual test depth was 120 m (394 ft), as in Type VIIC/42. The report gives a designed safety factor of 2.5, rather than the usual 2.0.

10. SRH-025 in the NARA II SIGINT series on the Battle of the Atlantic. This traffic was broken in the late spring of 1944. The attaché's December 1943 message, translated on 20 June 1944, accurately described Type XXI, with its figure-eight pressure hulls and its fishlike overall shape. Dimensions were given as 76 × 6.5 m, and displacement as 1,600 tons. It was correctly credited with six bow torpedo tubes and no stern tubes. Submerged speed was given as 16 kts. The description of the smaller U-boats was more confused, mixing features of Type XVIIB (Walter closed-cycle type), Type XXIII, and midget submarines. The attaché remarked that "considering the fact that our nation is a step ahead as regards this type of submarine . . . we ought to conduct prompt experimentation on construction." The "step ahead" would seem to refer to Japanese experiments with submarines with high underwater speed apparently unknown to the Allies. He mentioned the Germans' amazement at the "great capacity of our storage batteries. They were also astonished at their shape, which permits of their being arranged in banks and ventilates the . . . liquid in the side units." He estimated that based on the number of new U-boats then in early stages of construction, they should begin completions about May of the following year (1944).

11. Malcolm Llewelyn-Jones, *The Royal Navy and Anti-Submarine Warfare, 1917–1949* (London: Frank Cass, 2006), 69. The British moved extremely fast. Modifications to *Seraph* were begun on 16 June 1944, one month after the launch of the first Type XXI, *U 2501*; *Seraph* was running trials seven months before the first operational cruise by a Type XXI, *U 2511* (Llewellyn-Jones, 70). Representatives of the U.S. Navy in Europe were involved in the trials; the U.S. Navy streamlined the old submarine *R 1* but made no structural changes and then used it for training.

12. Llewellyn-Jones, 71. The asdic recorder held her only to 750 yds stern-on, but the asdic operator could hear her at 2,700 yds. She could not be heard (hydrophone effect) beyond 500 yds at speeds below 6 kts, but was easily audible above 6 kts. The intensity of the hydrophone effect peaked at 9 kts and could be detected out to 5,000 yds. However, as escorts' speed increased so did their self-noise, so that their sonar and hydrophone ranges were drastically reduced. The British estimated that above 16 kts asdic range might be reduced by 45 percent. *Seraph* could not dive below 200 ft, but it was estimated that a submarine's cavitation noise would decrease at depth, since pressure would reduce cavitation. Hydrophone effect was particularly important because an asdic could sweep about ten times more quickly for it than for echoes (since with echoes the operator had to wait for a reflected ping before searching in another direction). A fast submarine might rapidly move outside the asdic beam, but it could not easily evade a listening asdic.

13. Llewellyn-Jones, *The Royal Navy and Anti-Submarine Warfare*, 72.

14. For example, the Hermann Goering Werke at Straßburg was captured about October 1944, with about 5,000 relevant documents. ADM 1/17568.

15. *Report of Second Visit to Germany*, by Constr. Capt. A. J. Sims (the senior British submarine designer), ADM 281/143, dated July–August 1945. Sims learned that the Germans were disappointed in the speed of the submarine, attributing the unexpected resistance to the flooding holes. A compromise between diving time and speed made it possible to increase the latter to 16.4 kts at the cost of increasing diving time from 16 to 25 sec.

16. Llewelyn-Jones, *The Royal Navy and Anti-Submarine Warfare*, 64. He does not give a specific reference, so mid-1944 is my assumption. He dates identification of the Type XXI as the new U-boat to the spring of 1944, pointing out that through 1943–44 the Naval Intelligence Division (NID) had suspected that the Germans were developing a new U-boat type. He dates a brief provisional assessment of

the Type XXI to the September 1944 Monthly Anti-Submarine Report, based on a more detailed analysis by the Admiralty Department of Naval Operational Research. CAB /1536, a file on the U-boat threat during 1945, includes the 31 January 1945 paper by the British Combined Chiefs of Staff.

17. Llewellyn-Jones, *The Royal Navy and Anti-Submarine Warfare*, 89, points out that the British may have been willing to abandon efforts to make a Type XXI operational because their Type XXIII (*U 2326*) failed to make her advertised underwater speed (this turned out to be quite wrong in the case of Type XXI). They were also unwilling to expend much effort to make *U 3017* safe enough to test, given the urgent need to refit existing British submarines. The Germans had failed to make most of their Type XXIs operational. British Type XXI trials were formally canceled on 14 October 1945.
18. Transcript of 1947 Undersea Warfare Conference.
19. Llewellen-Jones, *The Royal Navy and Anti-Submarine Warfare*, 78.
20. NARA II, British documents collection formerly in NHHC, RG 38. It was issued 19 December 1945 and was classified Secret.
21. U.S. tactical trials quoted in ADM 292/214 gave a tactical diameter of 479 yds with 25 degrees rudder, or 398 yds with 32 degrees rudder. Acceleration was 6 to 13 kts in 55 seconds, and 6 to 15 kts in 2 minutes 30 seconds. Deceleration was 15.3 to 6 kts in 45 seconds.
22. ADM 1/27214 notes that "only one occasion has been reported in the war of a submarine being heard, and that was because she transmitted directly at an escort with full power at short range."
23. In 1946 the British loaned their one usable Type XXI (*U 2518*) and one of their two Type XXIIIs (*U 2326*) to France; *U 2326*, which was used for snorkel trials, was lost on 6 December 1946. The British conventional submarines were one Type XVIID (*U 1407*) and five Type VIIC (*U 1171*, *U 1108*, *U 975*, *U 953*, and *U 712*). Later *U 190* (Type IXC) was substituted for *U 975*. *U 1105*, which had a rubber covering, was allocated to the United States, but the British wanted to test her in comparison with the unmodified *U 1171*. Both arrived for trials on 29 June 1945; the trials lasted to almost the end of the year. Non–Type XXIs received by the U.S. Navy were one Type VIIC (*U 1105*), five Type IXC (*U 530*, *U 805*, *U 858*, *U 889*, and *U 1228*), one Type IXD (*U 873*), and one Type XB (*U 234*). Two U-boats handed over in Argentina after the end of the war (*U 530* and *U 977*) apparently did not count in the Potsdam totals. In addition, France retained four U-boats that had been seized in her Atlantic ports: two Type VIIC (*U 477* and *U 766*), one Type IXB (*U 123*), and one Type IXC (*U 510*). The Royal Navy transferred three Type VIICs to Norway in 1950 (*U 926*, *U 995*, and *U 1202*). Note that none of the transfers to Norway were among the Type VIICs allocated to the Royal Navy by the Potsdam commission.
24. Juergen Rowher and Mikhail S. Monakov, *Stalin's Ocean-Going Fleet* (London: Frank Cass, 2001), 205.
25. Data from table, p. 21, in *British Progress in Underwater Warfare*, 1947. "Guppy" was an acronym for Greater Underwater Propulsive Power. The conversion included drastic streamlining, addition of batteries, and installation of more powerful electric motors.
26. According to the British account of the occupation of the Walterwerke between 5 May and 25 November 1945 (ADM 199/2434) by CDR(E) I. C. Aylen, DSC, RN. An 18 May 1944 U.S. Navy intelligence report referring to the British NID 486 of 15 May summarized evidence mainly from late 1943 that there was a W-boat.
27. This account is based on Walter's postwar account, dated 24 April 1946 (ADM 213/247), and on the British 12 December 1945 report on the Walterwerke (ADM 199/2434), which in turn is based partly on Walter's interrogation. At his initial interrogation, Walter said that his company had seven branches; the Russians had seized four and the others were in British hands. They included torpedoes and catapults (for the V-1 missile, important later as a step toward the steam catapult). The Soviet-occupied plants were mainly for production and, Walter said, for air work. The British account ends with the comment that altogether Walter spent about £5 million to develop his various weapons projects, most of it on the submarine and the HTP torpedoes, noting, "this figure is very large in comparison with our own meagre allowances."
28. Message dated 13 February 1945, translated 20 February, crediting Type XXVI with 7 hours' endurance submerged at 22 kts. It gave principal measurements and performance data. A second message (13 February, translated 21 February) provided more details of the engine room arrangement. A third (16 February, translated 24 February) explained how the power plant worked. These messages are all in RG 457, entry 9013, NARA II, covering the period May 1940–November 1945.
29. Llewellyn-Jones, *The Royal Navy and Anti-Submarine Warfare*, 84, describes extraordinary measures to keep the HTP submarine out of Soviet hands. British members of the Tripartite Naval Commission were told specifically not to discuss HTP with the Soviets. Instead of refitting *U 1407* in Kiel, as initially planned, the boat was taken to Barrow-in-Furness.
30. This was a closed-cycle diesel using liquid oxygen. Work began before the war, but the two small submarines involved seem never to have operated. The project was

revived after the war. The prototype *M 254* was completed on 30 May 1953, the project having begun in December 1946. In 1956, *M 257* suffered a fire and *M 259* an explosion. Reportedly, Soviet naval personnel referred to these boats as "cigarette lighters." The Soviets developed an alternative oxidant, glass beads that could be made to emit oxygen, but the appearance of nuclear submarines made the project unnecessary. Plans for larger closed-cycle submarines were abandoned. These boats were powered by three diesels and were rated at 15 kts submerged, capable of spending 100 hours submerged. Western naval intelligence considered these boats closed cycle but seems to have assumed they were Walter submarines. The Russians publicly described the project after the end of the Cold War.

31. Admiralty Technical Staff Monograph, *German Torpedoes and Development of German Torpedo Control,* 1952 (BR 1972), ADM 234/955. G indicated torpedo diameter, 533 mm (21 in.).
32. *German Torpedoes and Development of German Torpedo Control*, 30. The receiver in the torpedo was modified so that its characteristics could change during the torpedo run. Initially it did not respond to sounds inside 15 degrees on each side, hence it could home on a destroyer once it closed in, the destroyer now being abaft 15 degrees off its side. Once it detected the target, the torpedo would turn through a semicircle with its acoustics turned off, then switch to ordinary homing, with a very narrow dead angle.
33. The British 1952 account of German torpedoes devotes very little space to sonar-only firing. Type VII seems to have been the first U-boat able to angle torpedoes (90 degrees either side); once that problem had been solved, a U-boat torpedo director similar to that used by surface ships was devised. The underwater dead-reckoner introduced in Type IX was intended to work with the sound gear introduced in that class. The text suggests that at least initially the idea was to enable the U-boat to fire entirely submerged after the target had been sighted once or twice by periscope.
34. Torpedo tubes limited firing depth to about 70 ft, but the torpedoes themselves could function at 250 ft. ADM 1/17602, "Enemy Torpedoes," a British document compiled in 1945 after the end of the war in Europe but not in the Pacific (as evidenced by descriptions of Japanese torpedoes).
35. Shipboard HF/DF systems used during World War II varied greatly. The British FH4 was a "wide-open" detector that broke an incoming signal into its two components and used them to drive the plates of a CRT. Its display showed a signal as a line pointing toward the emitter. The U.S. wartime HF/DF system used a rapidly spinning detector to achieve much the same result. Presumably shore-based systems had to be much more sensitive to operate at vast distances, so they could not employ these technologies.
36. M. E. Coles, GCHQ History of German Naval Communications, ed. LCDR L. A. Griffiths, RNVR, 256. M. E. Coles, vol. VI of the Government Code and Cipher School Naval History, *The German Navy: Communications*, ed. LCDR L. A. Griffiths, RNVR, NARA.

CHAPTER 5. FACING THE POSTWAR NIGHTMARE

1. The transcript of the 1946 high-level conference is in NHHC, as are transcripts of the 1947–49 secret conferences (the 1948 transcript is partial). There was apparently no 1950 conference, presumably due to the outbreak of the Korean War. Individual papers from the 1951 conference have survived, but NHHC has no transcript. The next U.S. conference was not held until 1959 (NHHC has a transcript).
2. Transcript of 1947 Undersea Warfare Conference, 209.
3. Changing air strength reflected the impact of the 1948 cuts. According to an article on the naval aeronautical organization in the April 1951 issue of the *Naval Aviation Confidential Bulletin*, in the summer of 1949 the U.S. Navy had 14 carrier air groups, 8 ASW squadrons (VS), and 30 patrol squadrons (VP). The secretary of defense ordered a cut down to 6 carrier air groups, 7 VS, and 20 VP (the less drastic cuts in VS and VP reflected an attempt to force the Navy to specialize in ASW. It took time to make the cuts, so in the summer of 1950 there were 9 carrier air groups, but the ASW cuts had taken place. In the winter of 1950–51, Korean War mobilization increased the force to 12 carrier air groups, 10 VS, and 27 VP. In the summer of 1951 and in 1952 NATO plans called for 14 carrier air groups, 15 VS, and 34 VP. These figures could be contrasted with the plans made in 1948–49, before the cuts: 24 carrier air groups, 7 VS, and 34 VP.
4. For example, in 1946 OEG was down to about a third of its wartime strength; only 8 of the 25 staff members were veterans of the wartime Anti-Submarine Warfare Operations Research Group (ASWORG).
5. As expressed at that year's Undersea Warfare Conference.
6. On 5 March 1946 CNO ordered conversion of two submarines to fire the Loon (modified V-1) missile. *Cusk*, scheduled for conversion that November, entered Mare Island Navy Yard on 8 January 1947 and emerged on 6 February; she fired her first Loon (unsuccessfully) on 18 February. By that time the planned FY 49 guided missile development program included a 500 nm missile carrying a 3,000 lb payload; it emerged as Regulus, which entered service. Design studies sponsored by the Bureau

of Aeronautics (BuAer) had begun in October 1945. Regulus studies began in June 1946. Los Alamos began work on the (relatively) lightweight warhead in May 1948. The Navy formally proposed a nuclear warhead for Regulus in June 1949, and that September the Atomic Energy Commission formally proposed four missiles as candidates for a new lightweight (3,300 lb) warhead (the others were the Navy's Rigel, the Army's Matador, and the Air Force's air-launched Rascal). In January 1950 the secretary of defense approved using the W-5 warhead in the Regulus missile. The bomb version (Mk 5) was tested in 1951.

7. Transcript of 1949 Undersea Warfare Conference, 84.
8. Transcript of 1948 Undersea Warfare Conference, 90.
9. In his 1948 address to the Naval War College, Vice Admiral Low offered an HF/DF net as a possible basis for hunter-killer operations. The wartime U.S. HF/DF net covered the area from Iceland to the South Atlantic. Low said that the refinements of the wartime net "extended to the point where we could identify individual boats by number, due to the characteristics of the radio set and also, a considerable number of individual operators as a result of personal peculiarities of keying." As examples of HF/DF-supported successes, he cited the destruction of a cargo submarine carrying both material and Japanese experts to Germany in June 1944, the capture of *U 505* in the Caribbean, and the destruction of the U-tankers in 1943. To some extent these successes were due to codebreaking, which Low could not mention. However, he alluded to the need to keep sources of intelligence secret lest the enemy stop using communications that could be exploited. Low warned that *Kurier* might destroy much of the value of HF/DF in a future war. The NSA historical collection (RG 457 at College Park) contains several papers written in 1945–47 on possible solutions to the *Kurier* problem. A summary of progress written in November 1947 concluded that "we are fundamentally no more able to cope with the *Kurier* problem now than then" (i.e., in September 1945).
10. As described, for example, in a 31 March 1949 lecture on U.S. ASW given by RADM T. H. Robbins Jr. (in the Naval War College archives). Robbins claimed that World War II hunter-killer operations had broken the German U-boat force. Much of his lecture described HUK group operations he had commanded, with an emphasis on air operations.
11. Figures from a 1 November 1948 speech to the Naval War College by VADM F. S. Low, who had been Tenth Fleet chief of staff during the war. Despite the demise of the *United States*, the U.S. Navy gained sufficient nuclear air firepower to make attack at source viable. A study by the British Joint Planning Staff (24 September 1953, in DEFE 6/23) evaluated the potential of attack at source in a war beginning in December 1954. It was assumed that attacks would be nuclear. Unfortunately, most large Soviet submarines would already be at sea, and others would be dispersed. Priority would go to destroying Baltic bases (the Soviets had not yet shifted their submarines to the Northern Fleet) and to mining the exits from the Baltic. Also in the priority category would be Baltic naval air bases and naval bases in the north. It would not be worthwhile to attack the Soviet submarine industry, since that would have little impact until well into a war. High priority was accorded operations in the north to deny the Soviets the ability to overrun Norway and to operate submarines and surface ships in the northern approaches to British waters. An interesting feature of the report is the large number of sorties required to guarantee delivery of limited numbers of nuclear bombs, for example to destroy 10 Baltic naval bases, between 15 and 20 atomic bombs—and 90 to 120 sorties. It was assumed that U.S. carriers would operate in the Norwegian Sea, the Arctic Sea, and the Far East. U.S. carrier aircraft listed in the report were the A2J-1 (radius of action 1,025 nm), the AJ-1 Savage (790 nm), the F10F (745 nm), and the F2H Banshee (540 nm), of which the A2J and the F10F never entered service. Assumed dive-bombing accuracy was a CEP of 31 yds (95 percent within 125 yds), figures consistent with Korean War visual bombing. The study assumed that nuclear attacks (using 10 and 30 kT bombs) would be made with a CEP of 800 yds (95 percent within 1,600 yds). The planners admitted that it was difficult to evaluate the effect of nuclear weapons, and that estimation by comparison to conventional attack was likely to be flawed. Damage from bombs falling within 800 yds would be much greater than assumed, and for those falling farther away it would be much less. The total number of sorties took into account the need to provide additional sorties for diversion or escort, and also to allow for abortive sorties. An accompanying map showed areas that could be covered by various British bombers and by U.S. carrier-based fighters and bombers.
12. World War II special naval operations, such as the midget submarine attack on the *Tirpitz*, were a form of attack at source. An attack at source working party in April 1948 (DEFE 2/1660) concluded that the Royal Navy might usefully attack enemy bases. Midget submarines were of interest because neither harbor defense nor attack was receiving much attention in the United States; a British or Commonwealth (Australian/Canadian) project might therefore complement U.S. work. The study omitted air attack, which the Royal Navy could not mount. The Royal Navy's interest in nuclear attack by a midget submarine evaporated about 1956 as the strategic focus shifted away from a central war.

13. Transcript of 1948 Undersea Warfare Conference, 200, remarks by CDR A. W. F. Sutton, RN, 29 September 1948.
14. Transcript of 1949 Undersea Warfare Conference, 87.
15. As described by Commander Roach of the 2nd Task Fleet for the 1949 Undersea Warfare Conference.
16. Conference held 17 June 1946, transcript in NHHC Operational Archives. This conference was called to implement a 7 June 1946 directive from CNO. The speaker was Captain Phelan of the Office of Naval Intelligence (ONI). The opening of the 1946 conference coincided with the creation of an Office of the Coordinator of Under-Sea Warfare.
17. In an estimate of the world situation in 1957 (JIC (47) 42, 12 June 1948) in CAB 158/1.
18. JIC (47) 7/B (6 August 1947, CAB 158/1) credited the Soviets with 230 submarines.
19. A report on Soviet capabilities and intentions issued on 18 June 1949 (JIC (49) 55 in CAB 158/7 Pt 2).
20. ADM 213/908, notes by British participants, a 15 April 1948 report by the British Joint Services Mission. Unfortunately, the transcript in NHHC is fragmentary. British reports issued by the Joint Intelligence Committee give other figures.
21. Transcript of 1949 Undersea Warfare Conference, 17.
22. According to (JIC (50) 7 of 13 March 1950 in CAB 158/9 Pt 1), "the effects of [adaption of German technology] are likely to become apparent about 1954, and from then onwards the offensive capacity of the Soviet Fleet will be very greatly increased." According to an 18 July 1950 estimate of the threat to UK maritime communications in the event war broke out in 1954 (JIC (50) 65 of 18 July 1950 in CAB 158/11), the Soviet Baltic and Northern fleets would deploy 65 large submarines (35 percent fast electric or Walter), 32 slow medium submarines (Shch class), 36 coastal, and numerous midgets in the Atlantic. By 1954 submarines would be able to use Soviet inland waters to move between the Black Sea, the Baltic, and the Arctic. It was estimated that the Black Sea Fleet included about 10 medium slow submarines and about 35 coastal submarines, plus numerous midgets. CAB 158/11 includes a joint Anglo-American paper on Soviet intentions and capabilities for 1950–54 dated 22 September 1950. In 1950 the British estimated that the Soviets had 70–90 large, 75 medium, and 120–130 coastal submarines. The U.S. estimate for that year was 88 large, 75 medium, and 121 coastal submarines. The British did not project submarine numbers beyond 1950, but the U.S. estimate for 1954 was 144–164 large oceangoing submarines, 90 medium range, and 150 coastal. Both navies knew that the Soviets had midget submarines, but no numbers were given. Current disposition was 27–29 large and 2–3 medium submarines in the Northern Fleet, 32–38 large and 25–27 medium in the Baltic, 7–10 large and 7–8 medium in the Black Sea, and 13–17 large and 37–41 medium submarines in the Far East.
23. A report on the scale of attack against Allied sea communications (JIC (50) 103 of 9 March 1951) in CAB 158/11.
24. JIC (52)1/9 of 12 September 1951 in CAB 158/14 Pt 3.
25. JIC (51) 2/7 in CAB 158/13.
26. JIC (53) 16 in CAB 158/14 Pt 1.
27. The first mention of the two nicknames in JIC papers is in JIC (54) 1/9, the 1 September 1954 periodic intelligence summary for NATO commands (CAB 158/17).
28. JIC (54) in CAB 158/17, a 15 February 1954 report on Soviet War Potential 1954–58.
29. JIC (55) 1/9 of 20 September 1955 in CAB 158/19, the 1955 periodic intelligence survey.
30. For example, JIC (55) 3 of 24 March 1955, a report on Soviet war potential 1955–59, claimed that construction of a new midget 75–100 ft long was underway in at least one yard.
31. Material from interviews with returning Germans are in the British national archives under the title "Dragon Returns." A report on German contributions to Soviet torpedo development is DEFE 44/10.
32. The first mention of likely Soviet adoption of the new German torpedo technology in the JIC series seems to have been in a report on Russian R & D issued on 27 February 1952 (JIC (53)) 16 in CAB 158/14 Pt. II.
33. 1947 Undersea Warfare Conference, 131, report of OpDevFor.
34. Transcript of 1949 Undersea Warfare Conference, 77.
35. Synchros are analog means of transmitting information, causing a remote device to follow the movement of another device, such as a fire control output.
36. More elaborate (but abortive) trainable Hedgehogs were planned. Mk 16 would have fired 36 charges. Mk 17, based on a 5 in./38 mounting, would have carried 61 charges. X-4 would have been a Hedgehog combined with a conventional 5 in. gun mount.
37. Typical World War II Hedgehog range was 188 yds. According to the data package prepared for the Hartwell scientists in 1949, the maximum Hedgehog range was 221 yds, and it was being increased to about 330 yds. A typical postwar figure for the trainable Mk 15 was that it created a 279 ft circle of charges at a range (to its center) of 265 yds. The fixed Mk 10 was credited with an average range of 283 yds, creating an ellipse 168 ft deep in range.
38. According to the British 1947 report of a visit to U.S. torpedo development facilities (ADM 292/214) "torpedo countermeasures have practically the same priority as A/S measures in America." These countermeasures embraced both direct destruction (using explosive streamers alongside and astern of a ship) and noisemakers. As described

in ADM 292/214, Project General Type 1 was a shipboard rocket launcher. Type 3 was a towed detector, 200 ft abeam, wired to the Project General Type 1 launcher. Type 4 employed paravanes towed from various parts of a ship, each with its own detecting gear, firing its own towed charge. The U.S. Navy was also developing noise-making decoys (projected by rocket from surface ships) and a towed decoy (Harp).

39. The British hoped that at a range of 200 yds Ruler could destroy an incoming torpedo with 75 percent probability. Initially it was proposed for escorts, but by 1949 it was considered too heavy; the British were considering adapting Limbo as an antitorpedo weapon.
40. From the NATO publication, *NATO Strategy Documents 1949–1969*, p. 65, appendix A of the 1 April 1950 report of the NATO Military Committee. No details were given. These numbers were revised after the 1952 Lisbon Conference, and again by SACLANT in January 1953. Specific requirements were levied on the different countries.
41. France had 15 destroyers (including the *Fantasque* class, rated as cruisers) and 12 frigates transferred by the U.S. and British governments in wartime. Two ex-Italian light cruisers were later converted into ASW leaders, but that was well in the future in 1950. The Dutch had 7 destroyers, Norway had 5 modern destroyers, and Portugal had 5 destroyers, for a total force of 32 destroyers and 12 frigates.
42. Transcript of 1948 Undersea Warfare Conference, 118, remark by Rear Admiral Briscoe.
43. The *Buckley*-class destroyer escorts working initially with *Mindoro* were *Coolbaugh* (DE 217) and *Greenwood* (DE 679). Both had been upgraded with 5-in. guns, hence had No. 2 position available for a Hedgehog in place of a twin 40 mm gun. They had QHB, integrated attack systems, and trainable Hedgehogs, but they lacked the equipment intended to train the Hedgehogs remotely through the attack recorders; the Hedgehog train orders were telephoned to the mounts, based on estimates.
44. To meet the 27 kt speed figure, in November 1948 Engineer in Chief was authorized to develop a 30,000 SHP steam plant (Y.100). The Y.100 plant was simpler than the YEAD.1 being designed for higher-performance destroyers, but it was still considerably more compact and more efficient than its predecessors. Because it did not embody radical new technology, a prototype could be included in the 1949–50 program. Compared with previous practice, the new plant was considered 30 percent lighter, 25 percent more compact, and 10 percent less expensive. In effect it made the new frigate possible. Work was accelerated because in 1949 the Canadians were designing their own new frigate. They had been convinced to specify Y.100 rather than a U.S. plant in the interest of standardizing within the Commonwealth (the first Canadian plant would be ordered from Britain).
45. The second-rate idea was suggested in August 1949 by ACNS RADM R. S. Edwards. To avoid calling ships first- or second-rate, Edwards invented the numerical designation system the Royal Navy has used ever since. ASW ships were numbered in a 10 series, beginning with Type 12 for the ASW frigate (presumably 11 was not used to avoid confusion with the Roman number II). The second-rate frigate became Type 14. Destroyers converted as first-rate ASW frigates became Type 15, their second-rate version becoming Type 16. A third-rate frigate conceived in 1950 (but never built) became Type 17, and a more elaborate destroyer conversion (short of Type 15 standards) became Type 18. Neither Type 17 nor Type 18 ever materialized. Later ASW frigates had higher numbers: the abortive Type 19, Type 21 (Amazon class), Type 22, Type 23 (Duke class), and abortive Types 24 and 25. The AA frigate became Type 41, the aircraft direction ship Type 61.
46. ADM 239/489 and ADM 239/490.
47. Transcript of 1948 Undersea Warfare Conference, 153.
48. Transcript of 1949 Undersea Warfare Conference, 120.
49. The standard war-built type was the ZPK, modernization of which was supposed to begin in December 1948. BuAer bought a prototype of a much larger ZPN, and in 1947 it planned to install an APS-20 radar on board one. In 1948 BuAer began procurement of a prototype ASW airship with a 14-man crew, 112 hours' endurance, and a maximum speed of 75 kts. It would be capable of refueling, reballasting, revictualing, and rearming at sea, with a payload of 5,500 lbs of detection and attack equipment, airship sonar. One problem was that it was difficult for an airship to hover at low altitude. In 1949 a sonar towed by a blimp at 30 kts showed the same self-noise as a ship at 15 kts.

CHAPTER 6. THE CRISIS AND ITS SOLUTION

1. Low did not know that the first Whiskey had just been laid down. Presumably he deduced construction numbers from the number of yards in the program, of which he knew of two in Leningrad (Baltic and Admiralty), Gorkiy (Sudomekh), and Nikolaev on the Black Sea (U.S. intelligence probably did not yet know about Komsomolsk-on-Amur in the Far East). Gorkiy had been built specifically to mass-produce submarines, and it built 113 of the 236 Whiskeys. Low may have lumped the four ex-German Type XXIs with any new Soviet submarines to get the total of eight B types. The Soviets' other five long-range U-boats were one Type IXC and four Type VIICs, all of which had snorkels.
2. Mk 30 was the unsuccessful lightweight alternative to the wartime Mk 24. This account is from DCNO (Air) papers responding to the Low Report, NARA II.

3. Volume 1 of the report is ADA 800012, and volume 2 is ADA 800013; both have been declassified in full. A summary delivered at the 1951 Undersea Warfare Symposium is AD 103991.
4. These examples are from the declassified transcript of the 1959 Undersea Warfare Conference, NHHC.
5. Data on nuclear depth-bomb ancestry is from Chuck Hansen, *U.S. Nuclear Weapons: The Secret History* (Arlington, TX: Aerofax, 1988). Hansen mentions several canceled Navy weapons. The early W-5 warhead was associated with nuclear depth bombs ("Candy" and "Chris"), a guided surface-to-underwater rocket ("Clara"), and torpedoes ("Claudia" and "Clifford"). The abortive W-12 weapon was associated with nuclear depth bombs ("Alice" and "Alvin"), a surface-to-underwater rocket ("Anna"), and with nuclear torpedoes ("Adeline" and "Andy"). According to Hansen, 207, ASTOR and the Mk 101 Lulu depth bomb had the same W-34 nuclear element. Little Lulu seems to have been an abortive 1955 depth charge version of the Mk 105 Hotpoint bomb. Instead, a new lightweight depth bomb was developed beginning in 1957 under the AEC designation Mk 57 (applied in 1959); it was also used as an Air Force tactical bomb. Nuclear yields are from NHHC declassified Double Zero files, 1978 Box 116, Folder 2, which gives the yields of all the U.S. nuclear ASW weapons, all of which are now extinct. SUBROC yield was 130 kT. This paper gives Betty yield as 28 kT, ASTOR as 9 kT, and ASROC as 10 kT. Yields are also given in a declassified Naval War College document, *Selected Readings in Tactics 1975 for the Technological Factors and Constraints in System Performance Study: Tactical Nuclear Weapons: Effects and Tactical Implications*. In addition to the SUBROC, ASTOR, and ASROC yields, it describes B57 (Mk 57) as a variable-yield weapon (0.4–10 kT).
6. Some hints in relevant directions had gone unrecognized. In 1946 Dr. Glenn Camp of OEG described the existence of discrete low-frequency components in diesel submarine spectra and proposed using the visual spectral analyzer developed by Dr. Ralph K. Potter of Bell Labs for classification of submarines, but excited no interest. Early in 1949 NRL reported that SOFAR hydrophones off Point Sur in California were detecting submarines at a range of 10–15 nm; by the end of the year they were being detected at hundreds of miles. In 1949 Woods Hole concluded that the components of diesel submarine signatures were very stable, so it would be possible to improve the signal-to-noise ratio of detection by narrowing filter bandwidth. SubDevGru2 had just announced submarine detection using one-eighth-octave filters.
7. Gary E. Weir, *An Ocean in Common: American Naval Officers, Scientists, and the Ocean Environment* (College Station: Texas A&M University Press, 2001), 301. The key point was that instead of spreading spherically (sound energy lost at the square of range), sound in the deep channel spread cylindrically (energy lost at a multiple of range). That offered a bonus of 60 dB at a range of 500 nm. Beyond that there would be additional gain due to the convergence of great-circle transmission paths. Hunt envisaged vertical receiving arrays at least 20 wavelengths long; SOSUS used horizontal ones twice as long. He argued that work should concentrate on sounds below 500 Hz.
8. Strictly speaking, the frequency components of a slice of speech lasting a short time, because frequency cannot be instantaneous.
9. Sandy Hook seems to have been a test of LOFAR itself, without reference to long-range propagation. It employed two hydrophones: one near the shore and one farther out in the sound; between the two the water deepened due to the flow of the Hudson River.
10. An article by Bruce Rule in *Cable* (dated 15 September 2013, on the IUSS Association website) describes the initial arrangements to make Project Bridge, the SOSUS station in Norway, operational on 1 June 1963. The Norwegian Defense Research Establishment had already been operating its own detection system (Lydia), which had a better bearing determination capability but not a useful classification capability, presumably because it operated at higher frequency. Bridge must have been very highly classified, because another article in *Cable* describes Keflavík as the first station looking up into the Arctic. In the Pacific, according to Cross Associates, *Sea-Based Airborne Antisubmarine Warfare 1940–1977*, report dated 17 February 1978, prepared for Op-095, declassified, vol. I (hereafter Cross Report), the new array off Adak could detect Soviet submarines as they emerged from Petropavlovsk and began their transits.
11. According to a winter 1997 article in *Cable* on the origin of the Keflavík station, the initiative to build a SOSUS station there arose at a 7 March 1965 meeting to discuss possible establishment of such a station in the Norwegian Sea (the Norwegian installation was not mentioned). According to the Cross Report, II, 75, as of 1966 the five-year defense plan called for installation of two additional arrays in mid-Pacific and five additional ones in the Philippine Sea during FY 68–70. New arrays were installed near Chichi Jima and north of Midway, their NAVFACs at Guam and Midway. They became operational in December 1968. Additional surveys were conducted during 1968 in the Philippine Sea north of Midway and in the Hawaiian Islands area, in both cases to support added surveillance.
12. A summary of issues for POM-78, issued as a student text by the Naval War College (and declassified), includes a

proposal for modest improvements to SOSUS, including coverage of the Gibraltar entrance to the Mediterranean and coverage of Yankee-class SSBN holding areas in the North Atlantic. SSPA size would shrink, and there would be a sixfold increase in target capacity. At present SOSUS signal processing added 8 dB to target signatures; projected improvements would add 14 dB. The improved system would cover virtually the whole of the North Atlantic. The proposed improvements would also cover U.S. SSBN egress from Guam, and upgrades to two cable ships (a full program would replace them altogether). The Naval War College also issued a copy of CNO guidance for FY 79–82 (28 November 1975) that included the intent to provide inter-array correlation in FY 79 to reduce SSPAs. It seems likely that there was already some sort of cover in the Strait of Gibraltar, but that this paper documents later installation of arrays outside the strait (frequent references to knowing when Soviet submarines entered or left the Mediterranean suggest this).

13. A map purporting to show SOSUS arrays and NAVFACs appeared in the February 1981 issue of *Scientific American*. It showed a NAVFAC in Norway associated with arrays between Norway and Bear Island, arrays off Gibraltar (associated with a NAVFAC there), an array in the Tyrrhenian Sea off Italy, an array in the mouth of the Aegean between Greece and Turkey, an array off Diego Garcia, arrays off the Indonesian coast in the Indian Ocean, arrays off the Philippines in the Philippine Sea, and arrays off Japan. Many of the arrays shown were doubtless fictitious given limited U.S. security arrangements with the countries involved. The use of T-AGOS ships in the Mediterranean suggests that reports of arrays there were fictitious.
14. As indicated by U.S. claims made about 1972 that a proposed array off the Irish coast could be monitored from the United States, albeit at a high cost. Ultimately the array was terminated in the United Kingdom, at Brawdy in Wales. In the 1965 SOSUS lecture given at the Naval War College, the speaker said that cable then in the development stage would make it possible to position hydrophones 3,000 nm from the readout point.
15. As stated in a declassified 8 December 1965 lecture on SOSUS delivered at the Naval War College by Robert D. McWethy, from the NWC Archive.
16. Cross Report, I, 170–72.
17. The 1970 ASW master plan described Colossus (FQQ-10) as an active acoustic barrier system with limited passive capability, intended to detect submerged transits of straits, harbor entrances, or other narrow waters. By this time one Colossus array had been specially configured for implantation across the Strait of Gibraltar, but it had not yet been deployed. That would depend on the results of experiments and concept formulation studies which would be completed in FY 74.
18. ADM 204/1041 is an account of a 10 November–23 December 1952 visit to the United States by R. Hogben of the Admiralty Research Laboratory Acoustics Group. He visited both Sandy Hook and Eleuthera and brought information on the new British Corsair project. It was clear that snorkeling submarines produced considerable recognizable noise, while submarines running at low speed on battery were essentially undetectable. At this time the Eleuthera array produced a single steerable beam 3 degrees wide, rather than the multiple beams of later installations.
19. ADM 285/15, "Corsair Project," is a paper by E. G. B. Megaw, a scientist at the British acoustic center at Portland, dated May 1952. He was responding to an April 1952 request by the chief of the RN Scientific Service for applications of VLF to specific British ASW problems. The 27 September 1952 introduction explained that this paper was the beginning of the Corsair project undertaken by the Admiralty Research Laboratory. ADM 204/2401 is an October 1959 report of a May 1959 trial against HMS *Porpoise*, the full-silenced prototype British postwar submarine. She was easily detected at a range of 8 nm while snorting at 10 kts—but it had to be admitted that at this speed she was not particularly quiet, her noise level being comparable to that of the French *Roland Morillot*, an ex-German Type XXI. Better ranges, up to about 70 nm, were achieved against noisier snorkeling submarines.
20. Each AN/GRD-6 receiving system employed two sets of vertical antennas, one for 2–8 MHz and one for 8–32 MHz. Both sets of antennas were arranged around two separate circles, respectively 100 ft and 25 ft in diameter. According to an Internet account of the Canadian site at Masset, off the coast of British Columbia, its GRD-6 (GRD-501 in Canadian parlance) was installed in 1956. Masset was later the site of an FRD-10 array. The Atlantic Canadian HF/DF net consisted of four stations linked to the U.S. part of the net (ten more stations): Coverdale, Frobisher Bay, Bermuda (1963–66), and Gloucester, Ontario. The EASTPAC net employed Skaggs Island, CA (which received its GRD-6 in 1956); Imperial Beach, CA; Marietta, WA; Masset; Cape Chimiak, AK; Adak, AK; Wahiawa, HI; and Midway. Midway was also part of the WestPac net, consisting of San Miguel, Philippines; Guam; Adak, AK; Misawa, Japan; Kamisaya, Japan; and Midway.
21. Wullenweber features in a 15 September 1945 U.S. Navy document titled *Electronics Research in the German Navy*, a series of lectures by Freg. Kapt. Dr. Ing. Schlicke on "Direction-Finding in U-Boat Warfare." Schlicke was on

board *U 234* when she was intercepted en route to Japan. He lectured at the Navy Department between 19 and 31 July 1945. A copy is in the series of wartime NSA documents in NARA II. A prototype built at Skisby, Denmark, was studied by the British before being destroyed in accordance with agreements on captured German technology. A new array built at Langenargen in Germany after the war for experiments was brought back to the United States as a prototype, reassembled at the University of Illinois, and used for HF/DF work in 1947–60 under a U.S. Navy contract. The German prototype had a diameter of 120 m (393.6 ft), but the experimental array built at the University of Illinois at Bondville had a diameter of more than 1,000 ft.

22. Cross Report, II, 18–19.
23. The contract to build a network of FRD-10s was awarded to ITT Federal Systems in 1959; the first was built at Hanza, Okinawa. Atlantic sites were at Gander, Newfoundland; Winter Harbor, ME; Chesapeake, VA; Homestead, FL; Toa Baja, Puerto Rico; Galeta Island, Panama; Edzell, Scotland; and Rota, Spain. Pacific sites were Wahiawa, HI; Imperial Beach, CA; Skaggs Island, CA; Marietta Alderwood, WA; Masset, British Columbia; Adak, AK; Hanza, Okinawa; and Dededo, Guam. Two FRD-10s at Sugar Grove were used as communication antennas, not as part of Bullseye. In addition to the U.S. systems, the Japanese government installed two large circularly disposed arrays at Chitose and Miho. The WestPac net was extended to the Indian Ocean in the 1970s. In December 1972 an HF/DF station was activated at Udorn, Thailand; it worked with an RAAF station at Pearce, Australia. A joint U.S./UK station at Diego Garcia began operating in August 1974, followed by a U.S. station at Masirah in the Arabian Sea opened in October 1974. This extension, to cover increasing Soviet naval activity in the Indian Ocean, was apparently not part of the original Classic Bullseye net; it is described in the Cross Report, II, 210, with the comment that further extensions were expected (they have not been described in public). The British Plessey company developed a smaller (about 400 ft diameter) Pusher, at least 25 of which were installed in many countries. The United States adopted it as FRD-13 or AX-10; these arrays were installed at Keflavík and at Diego Garcia. Some unofficial sources have identified the Indian Ocean net with Classic Wizard, a code designation otherwise associated with U.S. ocean surveillance satellites.
24. The test ships were *Mitscher* and *Josephus Daniels*; plans called for installations on board nine cruisers and nine *Spruance*-class destroyers. A more austere combat DF system was planned for wider installation.

CHAPTER 7. THE MOBILIZATION DECADE

1. There is an unsubstantiated report of an attack on some Soviet submarines that got too close to the fleets off Korea. According to the Cross Report, I, 108, during the first 8 months of the war there were 96 possible submarine contacts, of which 2 were positive and 8 probable. Over the next 8 months there were 88 (6 positive, all outside the operating area).
2. Cross Report, I. CinCPac estimated in June 1950 that the Communists had 78 submarines in a position to threaten UN forces on the Korean coasts.
3. The first two active carriers to be redesignated were *Leyte* and *Antietam*. The inactive ones were *Enterprise*, *Franklin*, and *Bunker Hill*; none was ever activated. The two additional active carriers were *Princeton* and *Valley Forge*.
4. ADM 1/22001, file TSD 214/1950, a 23 November 1950 paper by the Director of the Tactical and Staff Duties Division (DTSD) of the Naval Staff, responsible for framing ship requirements: "Minimum Performance Data for NATO Ship Types." DNC offered two sketches of ships that could be built in comparatively large numbers, in peace or war, "without noticeable strain on our productive or financial capacity."
5. Details of the NATO competition are from the papers of the French Conseil Supérieure de la Marine (CSM), 1945–58, consulted at the then Service Historique de la Marine (SHM) at Vincennes (now part of the Service Historique de la Défense [SHD], also at Vincennes). The CSM consisted of senior admirals, including the chief of staff, who advised the minister of marine. Since the Fourth Republic changed ministers so often, the CSM was in effect the governing body of the French navy. Its papers covered overall policy and also approval of projected ship designs. I have not found references to the 1953 competition in Admiralty papers.
6. Peter T. Hayden, "The Non-Utility of 'Utility' Warships: The Royal Canadian Navy and the Vancouver-Class Escort," *Canadian Naval Review* 3, no. 1 (spring 2007). The design was also described by LCDR J. R. Coulter RCN of the RCN Directorate of TAS Warfare at the British TAS Meeting of 1954 (ADM 189/237).
7. ADM 204/1379, a report on a British visit to the United States between 23 and 25 May 1960 to discuss ASW weapon systems. The ballistic error in ASROC was probably too great for the Mk 44 torpedo. When dropped within its nominal acquisition range, Mk 44's probability of acquiring a 20 kt target at 700–800 ft was about 25 percent; that had to be halved to give kill probability. Blind time at 10,000 yds was 67 seconds to the detonation of a nuclear depth bomb and 142 seconds to enabling a torpedo at 450 ft.

8. ADM 302/483.
9. ADM 189/237, 13th TAS Liaison Meeting, 30 June–2 July 1954.
10. This account is based partly on Wayne Mutza, *Lockheed P2V Neptune: An Illustrated History* (Atglen: Schiffer, 1996), supplemented by Standard Aircraft Characteristics Cards for P2V-1 (1 April 1947), P-2V-5, and P2V-7; by material in RG 72 VV series for 1949; NARA II; and by BuAer presentations at the 1948 and 1949 ASW conferences.
11. For details of the Shackleton project and alternatives, see Chris Gibson, *Nimrod's Genesis: RAF Maritime Patrol Projects and Weapons since 1945* (Manchester: Hikoki, 2015). Coastal Command representatives told the 1948 U.S. ASW Conference that initially, on 22 September 1945, the British Air Staff wanted to adapt the Lincoln bomber for long- and medium-range reconnaissance, but in February 1946 it was rejected in favor of a new airplane. The new internal layout was applied to the second prototype. The designated role was general reconnaissance and antisubmarine duties, apparently in that order. The U.S. officer was CDR J. A. Masterson; his report is in RG 72 VV Secret file for 1949, at NARA II. The 12 October 1949 BuAer memo is in the same file.
12. This account is based on BuAer presentations at the 1948 and 1949 ASW conferences and on Robert J. Kowalski, *Naval Fighters Number 20: Grumman AF Guardian* (Simi Valley, CA: Steve Ginter, 1991). Mr. Kowalski was a Guardian pilot in 1953–54.
13. APR-9 covered the full potential radar band, from 1 to 10 MHz, displaying signals on a CRT and in audio form. It used a spinning antenna supplemented by two wing-mounted antennas (one for 1–2.3 MHz and one for 2.1–4.5 MHz). APR-9 used six tuning heads to cover the full radar spectrum, but it could accommodate only one at a time. In flight the ECM operator tuned through the available frequency range. APA-70 covered the forward arc and was considered suitable for carrier aircraft. It was limited to 1–4 MHz and to right-left indications, measured by switching lobes. Output was displayed on the landing meter otherwise used to home on carrier landing beacons. When a signal was intercepted, the ECM operator called for a hard 90 degree turn in the indicated direction (a second intercept would resolve the 180 degree ambiguity). U.S. submariners, with whom ASW aircraft exercised, typically swept their radars only two or three times, so the airplane would determine only a sector in which they were. At least initially, the APS-20 radar interfered with the APR-9 when both were operating; there were also problems with the orientation of the APS-20 antenna. VS-871 had the improved APA-69A radar direction-finder, which offered a visual bearing indication or a signature, and it required only two tuning heads.
14. The account of the rationale for selection is from the BuAer VV Confidential file in RG 72, NARA II. This account is based mainly on Robert J. Kowalski and Tommy R. Thomason, *Grumman S2F/S-2 Tracker,* part 1: *Development, Testing, Variants, and Foreign Users*, Naval Fighters Number 101 (Simi Valley, CA: Steve Ginter, 2016).
15. Such operation was first tested in 1953 using the carrier *Valley Forge*. The first fast carriers were redesignated for the ASW role in 1954.
16. Initial S2F deployments included AD-4W or AF-2W aircraft for wide-area snorkel detection. They were eliminated when the APS-38 proved sufficient. AD-5W detachments were later reinstated for wide-area surveillance, communications relay, and, when CVSs were given fighters, their original role of fighter support.
17. Secret letter in VV file for 1949 in RG 72, NARA II, signed by W. G. Knapp of the BuAer Observation and Rotary-Wing Branch.
18. Much of this is from VV files for 1950 in RG 72, NARA II; the Cross Report; and transcripts of the 1947–49 Undersea Warfare conferences in NHHC, supplemented by Tommy H. Thomason, *The Forgotten Bell HSL: U.S. Navy's First All-Weather Anti-Submarine Warfare Helicopter,* Naval Fighters Number 70 (Simi Valley, CA: Steve Ginter, 2005). Thomason describes early Navy attempts to develop an ASW helicopter using a dipping sonar. Numbers of the initial operational ASW helicopters were limited. The Navy bought a total of 32 HUP-1s and 165 HUP-2s (another 50 transferred from the Army became HUP-3s) but retrofitted only 12 HUP-2s.
19. The S-58 purchase is generally dated 1952, but the card in the NHHC collection of descriptions of competitions is marked "1951 Interim ASW Helicopter" and (in pencil) HSS.
20. Shawn Cafferky, *Uncharted Waters: A History of the Canadian Helicopter-Carrying Destroyer* (Halifax, Canada: Centre for Foreign Policy Studies, Dalhousie University, 2005). Like the Royal Navy and the U.S. Navy, the RCN experimented with smaller helicopters operating from helipads on board frigates. Off Korea on 8 February 1952, an H-5 helicopter landed on HMCS *Sioux* to bring a surgeon on board; it was thought to have been the first helicopter landing on a destroyer.
21. Cafferky, *Uncharted Waters*, 97. Canadian officers gained experience with the U.S. Navy as it introduced ASW helicopters.
22. While characteristics were being discussed, Chief of Fleet Maintenance R. L. Hicks wrote (28 May 1948) that he strongly disagreed with plans to convert many FMRs

on the ground that at the outbreak of war they would be needed as ocean ASW escorts; he argued that with the installation of scanning sonar (QHB) they could cope effectively with Guppy-type submarines, the perceived threat standard. He strongly preferred the slower DETs, 48 of which were earmarked for transfer to other governments. Hicks lost this fight.

23. The conversion file is in RG 19 Code 440 documents in NARA II.
24. Details from a Secret (now declassified) paper signed by Secretary of Defense Robert McNamara for Senator John Stennis of the Senate Preparedness Committee, stamped 3 November 1961, responding to a 5 October 1961 query about U.S. early warning systems (including SOSUS). Declassified paper released by National Security Archive.
25. JIC (57) 65, dated 17 June 1957, in CAB 158/29; I assume that British intelligence analyses took account of contemporary U.S. information. The 23 September 1958 JIC summary of Soviet forces noted, "there is evidence to show that at least one submarine in each of the Northern and Pacific fleets has been converted to carry some form of guided missile" (CAB 158/31). This was presumably a reference to initial Whiskey-class conversions. According to an April 1958 report on Soviet R & D in 1957 (in CAB 158/32), these weapons were probably cruise missiles. According to the 22 October 1959 summary of Soviet forces (in CAB 158/35), a Northern Fleet submarine had been sighted that had a hangar suitable for stowing a cruise missile (this was presumably the prototype Whiskey cruise missile submarine), and three Zulu class in the same fleet had been given "an enlarged fin structure which could possibly have been designed to accommodate two missiles. It is considered that the missile for the Z class would probably be a tactical anti-ship weapon of short range, rather than a strategic weapon." According to the 20 October 1960 summary of Soviet bloc forces (in CAB 158/39), the weapon in the Z class was probably a ballistic missile, and "it is believed [correctly] that the G class are similarly equipped." The missiles were credited with a maximum range of 300 nm. "It is estimated that the Z Class conversions, though having operational capabilities, were intended primarily for trials and evaluation of a submarine ballistic missile system with which a new class of submarine, namely the G class, was to be equipped." Design range was 350 nm.
26. Development of the SS-N-5 system, D-5, was authorized in March 1959, the first successful launch being made in October 1960. A modified Golf-class submarine (*K 102*) fired her first underwater-launched missile on 24 February 1962. Design range of the initial version was 755 nm (1,400 km); range of a modified version was 862 nm (1,600 km).
27. Cross Report, I, 172; the Pacific Fleet was then averaging 125–159 submarine candidates each year, and the Atlantic Fleet a little over twice that many.
28. By 1969, the number of identified out-of-area long-range patrols by attack submarines had grown from fewer than 40 in 1963 to 145. Patrol-days at sea grew sevenfold. In the Atlantic, in 1963 there were 500 patrol-days, compared with 4,700 in 1969. The Mediterranean was the next most active area, patrol-days growing from 200 in 1963 to 4,100 in 1969. It was the area of greatest change, as a result of the creation of a Soviet Mediterranean squadron after the 1967 Arab-Israeli War. Patrol-days in the Pacific grew from 600 in 1963 to 1,700 in 1969. CIA analysts traced growth in activity in the Atlantic and the Pacific to anti-Polaris operations and to tracking carrier strike groups. Data from CIA paper SR-IM 71-11-8, *The Soviet Attack Submarine Force: Evolution and Operations* (September 1971), in declassified Naval War College *Selected Readings on Antisubmarine Warfare 1972–73*, vol. 1.
29. Naval Message, July 1969 to JCS J-3, enclosed in a 7 August 1969 memo for CNO on SLBM warning/attrition capability, from the deputy director of ASW programs. It mentions the current Blue Mail study and notes that a previous assessment of effectiveness against Yankees had been conducted in the Blue Steel study. Documents released by the National Security Archive.
30. Kuzin and Nikol'skiy, *The Soviet Navy 1945–1991*, identify the submarines that introduced types of computer. The impact of a navigational computer is evident from U.S. experience with Polaris; the submarine not only had to know where it was, it had to compute the commands for the missiles' ballistic computers based on that changing position. In early U.S. Polaris submarines, the task was so difficult that this computation was done largely ashore, and the submarines carried computer tapes to sea. Providing for multiple targets complicated the process.
31. According to the oral history of VADM Edwin P. Hooper, declassified pages of which have been published by the National Security Archive.
32. Bruce Rule, "Faulty Intelligence Nearly 'Sank' SOSUS during the Cuban Missile Crisis," from the November 2012 issue of *Cable*, the newsletter of the IUSS Alumni Association.
33. According to the Cross Report, II, 40, during 1962 CINCLANT received 401 submarine reports, of which 139 were evaluated as U.S. or friendly, and another 169 as "positive submarine." Of those, 149 were from surveillance of a Soviet Norwegian Sea exercise in July; 6 were detected during the Cuban quarantine. Of the remaining submarine contacts, four were probables and five were possibles. Submarines were detected not only by ASW patrol aircraft but also by carrier-based E-1B (WF-2)

airborne early warning aircraft. It is not clear whether these aircraft, like their APS-20-equipped forebears, were detecting periscopes and snorkels. The sudden appearance of the Soviet oiler *Terek* in the western Atlantic on 18 October was considered an indicator of likely Soviet submarine activity, since it would be used to fuel submarines.

34. DoD Report to National Security Council on "Status of U.S. Military Programs as of 30 June 1960," NSC 6013, redacted document from National Security Archive, 83. The demonstrated ability of HMS *Porpoise* was described as "disquieting." This report did not directly connect the *Porpoise* experience with attempts to develop active sonars (which it also described). Details of the systems are from the transcript of the 1959 Undersea Warfare Conference. Details of Artemis are from published reports.
35. NRL Memorandum Report 1299 of 8 March 1962 (released as ADA 035001) describes the sound source, which operated at 400 Hz (100 Hz bandwidth) and generated 1.3 kW of acoustic power (source level 152 dB) with pulse lengths between 10 milliseconds and 60 seconds. Beams were 20 degrees wide and 12.5 degrees deep. Power was provided by an 8 kW gas turbine generator (maximum power transmitted to the array was 5.2 kW). The array structure, containing 1,440 transducers, was 44.5 ft wide, 54 ft high, and 22.5 ft deep at the lower end, tilted up 11 degrees. Initial plans called for emplacing the sound source on the sea bottom near a suitable receiving site. No suitable site was found, so it was decided to place the source on a ship so that location could be varied (later it would be mounted on the bottom). The shipboard installation made it possible to lower the source to a depth of 1,200 ft. The vertical-array receiver was on Plantagenet Bank southwest of Bermuda.

CHAPTER 8. THE NUCLEAR REVOLUTION

1. Based largely on the redacted Nobska report in the National Security Archive. On 18 October 1955, shortly after taking office as CNO, Admiral Burke wrote to the Chief of Naval Research to ask whether the Committee on Undersea Warfare could undertake an ASW weapons study. In a meeting in December with bureau chiefs and with the president of the National Academy of Sciences and the Committee on Undersea Warfare, Admiral Burke expressed the Navy's growing concern over the potential threat posed by enemy nuclear submarines. The study, which began on 18 June with briefings, was conducted at Woods Hole; general conclusions were discussed at a 6 September 1956 meeting.
2. See D. A. Rosenberg, "The Origins of Overkill," *International Security* 7 (spring 1983): 4. The realization that Polaris could be a key deterrent seems to have come when President Eisenhower had the Army and the Navy review an Air Force application to expand the national nuclear stockpile; it was soon obvious that the Air Force requirement was grossly excessive.
3. At this time the Naval Ordnance Lab at White Oak envisaged dead-reckoning the target position, the torpedo being wire-guided. Because the target might change bearing, White Oak planned to provide ASTOR with a short-range (500 yds) pinger to confirm target proximity; it would be fired on command. The passive PUFFS fire control system, which could triangulate the target based on noise it emitted, seems to have eliminated the need for the short-range pinger, which would have alerted the target. At this time Stinger (which became SUBROC) was envisaged as a 3,000–20,000 yd missile carrying an 8 kT nuclear depth bomb.
4. Developed under a CNO order to BuShips, 8 October 1964, to establish the ASWSCCS project. The prototype system was installed on board the ASW carrier *Wasp* and two destroyer escorts (*Voge* and *Koelsch*, DE 1047 and DE 1049). Further installations were precluded by Vietnam War budget cuts.
5. The exception is natural circulation, in which the temperature gradient of water in a reactor causes circulation without pumps. This technique was tested in the U.S. submarine *Narwhal*, and later used in *Ohio*-class strategic submarines. It requires considerably more hull volume than a pumped reactor but is inherently quiet. The British produced their own natural circulation reactor (PWR 2) for their Trident strategic submarines and, presumably, for the *Astute* class. No Soviet submarine appears to have used natural circulation.
6. Naval War College pamphlet, "Antisubmarine Warfare in the Atlantic," part of a longer lecture, Naval War College archives, declassified document. In the 1965 Naval War College lecture on SOSUS, students were told that *Nautilus* had been held by one of the Cape Hatteras arrays at a distance of 160 nm. They were shown the relevant lofargram, which showed a 28.5 Hz line (due to blade rate), two turbine lines at 105.5 and 106.2 Hz, and the main coolant pump at 139.8 Hz. The turbine-to-blade ratio of 3.71:1 just fitted *Nautilus*; no other known ship showed the same ratio. The lofargram covered the range from 5 to 150 Hz, the normal SOSUS search coverage. On selected bearings the upper limit could be extended to 280 Hz. *Nautilus* tracking in the Pacific in 1958 is described in John Byron, "SOSUS—The Early Days on the West Coast," pt. 2, *Cable* (winter 2009). She had been tracked up from Panama to the Seattle area and then as she left Pearl Harbor. SOSUS detected her turn north in her secret run toward the North Pole. She was lost as she entered the

Aleutians. U.S. first-generation nuclear submarines were so noisy that sometimes they could be tracked all the way from the West Coast to Pearl Harbor. The Naval War College lecturer gave other examples of SSN detection. As she returned to the United States in October 1964, running at high speed, USS *Scorpion* was detected by stations at Antigua and Barbados at a range of about 3,000 nm. The submarine CO thought he could not be detected until he reached longitude 50° west, at which point he slowed, went shallow, and was lost. Of the 58 hours and 30 minutes from initial to final contact, he was held for 44 hours and 30 minutes; his final plotted position was within 22 nm of his actual position. Similarly, SOSUS tracked HMS *Dreadnought*, the first British nuclear submarine, on her track from Gibraltar to Norfolk. She was detected northeast of the Azores and was held nearly continuously until reaching a position inshore of the 1,000-fathom curve off Norfolk. The worst location error was 58 nm, the average was 28 nm, and the least error was 7 nm.

7. According to the formerly classified System Log for SOSUS on the Integrated Undersea Surveillance System (IUSS) Association website, *George Washington* was tracked from the western Atlantic to the United Kingdom in 1961. A program to test SOSUS by having every U.S. SSN and SSBN report her path in detail showed that very few detection opportunities were being missed. The latest (quietest) U.S. SSNs would likely be detected at any speed over 17 kts, and SSBNs would be detected over about 13 kts.
8. Late in 1964 Soviet submarines began out-of-area operations in the Pacific. One of them was held intermittently by the Pacific system from 8 to 22 October in an area halfway between Kodiak and Pearl Harbor. In the Atlantic, in September 1964 a Soviet SSBN (Hotel class) was first detected at a range of over 3,400 nm and held for 58 hours while making 13–22 kts. It appeared that the submarine was detected after she passed through the relatively shallow water of the GIUK Gap, and was detected soon after crossing the 1,000-fathom curve and reaching deep water. In his article on acoustic intelligence problems involved in identifying the Foxtrots during the Cuban Missile Crisis, Bruce Rule mentions that a November was detected by the Barbados array in the Norwegian Sea in July 1962 at a range of about 3,200 nm making 24 kts in the GIUK Gap, between Iceland and the Faroes.
9. In 1960 it was stated that installation of the new DELTIC time-compression signal processor would probably increase range by 50 percent. Lewes was the first station backfitted with digital spectrum analysis equipment (operational 20 June 1962). The next step was Vernier and super-Vernier analysis, which enabled much finer-grained analysis. The first station backfitted seems to have been Lewes, in May 1966. The development of SOSUS signal processing is described in an article by Jim Weinel in the spring 2003 issue of the *Cable*, the newsletter of the IUSS Caesar alumni association. Initially streams of hydrophone data were amplified, arranged into beams using delay-line beamformers, then subjected to time compression and passed through banks of frequency filters to form the spectra seen on "grams." Caesar Phase II substituted magnetic for delay-line beam-formers, considerably reducing the number of electronic cabinets. Magnetic delay lines employed the then-new technology of rotating-drum magnetic memory. Phase III substituted a single multiplexed coaxial cable for the 40 wires previously connected one by one to the hydrophones. Phase IV substituted digital spectrum analysis for the previous time-compression technique. It was introduced beginning in September 1963. Vernier processing added in 1966–67 made it possible to examine part of the spectrum in greater detail. A new coaxial cable, which could be much longer, was introduced in 1972. By that time all processing was digital, including beam-forming based on sampling of what the hydrophones detected. Initially these processes were hardwired, but later they were embodied in standard programmable digital computers. The initial sonar system was designated FQQ-1(V); when magnetic delay line beam-forming was introduced it became FQQ-2. The addition of digital spectrum analysis was indicated by changes to FQQ-1A and -2A. The system with the new SD-C cable was FQQ-11. The array was FQQ-7. Standard military designators were abandoned in 1976–78 with further digitization.
10. According to the chronology published by the IUSS Association.
11. According to the Cross Report, II, 73, in 1966 Pacific SOSUS coverage allowed only limited coverage of transiting Golf-class strategic submarines. Coverage from the East Coast was not reliable enough to track submarines in the mid-Pacific holding areas. As a stopgap, P-3s were using sonobuoys for wide-area surveillance from high altitude. When the transiter approached the limits of Adak coverage, P-3s were deployed, using square 9-buoy patterns, the buoys spaced 75 nm apart, the airplane monitoring the field from 20,000 ft. The airplane could monitor an area of about 50,000 square miles, and predicted probability of detecting a conventional submarine at least once while transiting was 95 percent. Data from Cross Repot, II, 74.
12. Desmond Ball and Richard Tanter, *The Tools of Owatatsumi: Japan's Ocean Surveillance and Coastal Defense Capabilities* (Canberra: Australian National University, 2015), 81, quoting a former U.S. Navy intelligence officer, Thomas S. Burns, to the effect that in the mid-1960s the United States offered SOSUS to friendly countries.

13. Statement by RADM Horacio Rivero Jr. as director of the Long-Range Objectives Group, August 1960, quoted in Cross Report, II, 14.
14. DEFE 69/526. This document mentions American interest in Mediterranean arrays but includes no details. An initial British evaluation of SOSUS seems not to have taken into account the extent to which SOSUS could benefit from the library the system built up of detailed spectra of particular submarines and surface ships. Not having used a narrowband system, the British had no such library. The last British report on SOSUS, made in 1963, was concerned largely with snorting conventional submarines, mainly American ones with distinctive noise characteristics. "We have little information on which to assess current classification performance, though it is understood that nuclear submarines have identifiable noise spectra." According to this report, bearing accuracy was usually 3 degrees, and cross-fixes were generally within a 100-nm square. This time British acoustics specialists said that detection range depended on bottom topography, on conditions (e.g., season, water depth, and change of temperature along the propagation path), and on background noise level. It also appeared that long ranges were achieved by looking from warm to cold water, and vice versa. The "cold" region meant cold isothermal water (e.g., north of Iceland). Vagaries of propagation often resulted in poor detection at short ranges but good detection at long ranges. This was most significant for moderate sound levels (e.g., made by snorting conventionals). SSN detection could be similarly affected by the slope of the sea bottom near the submarine. Noise reduction would be an effective countermeasure, but blade rate would not easily be reduced except by reducing speed. Propagation was best in the deep channel, so detection range would be reduced if the submarine stayed shallow. Going ahead with the project gained the British access to the U.S. lofargram library. NAVFAC Brawdy began operations in 1974.
15. Ball and Tanter describe an upward-looking chokepoint system called LQQ-3. In the 1970s the Japanese began development of a longer-range system designated LQQ-4, which seems to correspond to SOSUS, with a follow-on LQQ-5. They mention contracts for recording paper for LQQ-4 stations, which would seem to indicate the use of lofargrams. A 1980 article in a European magazine claimed that there were three Japanese SOSUS stations: at Wakkanai, at Tsushima, and in the Ryukyu Islands. Scuba divers have described cables leading from Honshu to arrays in the Sea of Japan. Japan installed systems in shallow coastal areas.
16. Cross Report, II, 78.
17. Cross Report, II, 93. After the 1967 deployments, the commander of the Sixth Fleet ASW force pointed out that conditions in the Mediterranean so badly degraded the performance of acoustic sensors (particularly LOFAR) that visual and radar sensors accounted for about two-thirds of initial detections on evasive diesel submarines. Of recorded initial detections of conventional Soviet submarines in the Mediterranean at this time, 51 percent were visual, 23 percent were by radar, 12 percent were by LOFAR, 11 percent by sonar, and 3 percent by other sensors. In the Mediterranean, LOFAR meant sonobuoys, as there were no SOSUS arrays. Of only 36 possible contacts on Soviet nuclear submarines, 22 were LOFAR or submarine sonar, six were visual, and two were active sonar contacts by destroyers. The other six were assessed nuclear based on other information, presumably SIGINT. LOFAR accounted for about 80 percent of initial detections in the Atlantic and the Pacific.
18. ADM 1/25319, a British file on a proposed British LOFAR buoy, includes two of the Bell Labs LOFAR progress reports, dated 1 October 1953 and 1 January 1954, the eleventh and twelfth in the series of quarterly reports. The 1953 report included successful direction of an airplane (based on LOFAR) to a submarine, the airplane employing modified sonobuoys to locate the submarine more accurately. The submarine was at an unknown bearing, 105 nm from the experimental Bermuda SOSUS array, using Vernier LOFAR. Even 105 nm from the array, the submarine signals were still several decibels above the detection threshold. The target was USS *Tigrone*; the airplane was a P5M Marlin. This was apparently the first time a submarine whose position was unknown was located using a combination of LOFAR information from a fixed array and aircraft-monitored low-frequency buoys. The second report described tests of the new Bermuda SOSUS array conducted against the U.S. submarine *Guavina* in November 1953, which was detected at ranges of 350–400 nm. Tests against HMS *Auriga* confirmed previous indications that British submarines had their strongest signature components at about 50 Hz, compared with 100 Hz for U.S. submarines. *Auriga* proved quieter than her sister *Alcide*, previously tested. *Auriga* was slightly quieter than the U.S. submarine *Bergall*, although at a range of 290 nm their strongest lines (respectively, at about 64 and 107 Hz) were about equally evident on the lofargram. The strongest British lines were due to their propellers, but the strongest U.S. lines were due to their diesels, meaning that on batteries *Auriga* could be detected at substantially greater range than the U.S. submarine. It seemed, therefore, that for a given speed, direct-drive submarines should be detected at comparable ranges whether on batteries or snorkeling. It is not clear whether those writing the report realized that all Soviet diesel submarines operated on direct drive.

19. Wayne Mutza, *Lockheed P2V Neptune: An Illustrated History* (Atglen: Schiffer, 1996), 103. Jezebel P2V-5 and P2V-7 aircraft became SP-2E and SP-2H under the revised designation system introduced in 1962 (initially they had "S" suffixes, as P2V-5S and P2V-7S).
20. Minutes of 1959 Undersea Warfare Conference, 2–22 (document in NHHC).
21. Engineers working on the explosive echo-ranging concept named it after Julie Gibson, a stripper who could "turn a passive boy [pronounced like buoy] active." Kowalski and Thomason, *Tracker*, 69, shows Julie with the contemporary motto "what makes a passive buoy active?" A diagram in the proceedings of the 1959 Undersea Warfare Conference showed a recorder trace with a peak for the initial explosion, a second peak for the bottom return, and a third for the target echo. A related technique, Gilda, used the blimp-towed AQS-2 as the receiver.
22. This account is based on a paper presented by VX-1, which was conducting initial tests, given at the ABC (American-British-Canadian) ASW Evaluation Symposium at Canadian Forces Base Summerside, PEI, 4–10 June 1967. Transcript in AIR 65/408.
23. A Gyrodyne Helicopter Historical Foundation account of the QH-50D program includes a 2 July 1969 photo of a DesJez DASH launching from USS *Moale* (DD 693). According to Gyrodyne, this program included the use of a DASH as an antimissile decoy for aircraft carriers. *Moale* was apparently the DesJez test ship between September and October 1969, achieving the 50-nm range.
24. This NATO designation covered several Soviet missiles, the first being RPK-2 Vyuga-53 (82R). Its payload was a nuclear depth bomb. It entered service in 1969, equipping Victor I/II (Project 671 and 671RT) and Alfa (Project 705) class submarines. RPK-2 was inspired by SUBROC (the design of which was compromised in 1964); it was developed by the Novator OKB under the same 1960 decree as the Shkval torpedo and other anti-Polaris weapons. Range was originally 8–10 km. The later RPK-6 could deliver either a homing torpedo or a nuclear depth bomb (respectively, 83R Vodopod and 84R Vodopei). RPK-7 was a 65 cm equivalent carrying a 53 cm torpedo. NATO missed the distinction between two generations of systems, both of which existed in both 53 cm and 65 cm versions, instead identifying all 53 cm weapons as SS-N-15s and all 65 cm as SS-N-16s. There were also surface-launched versions.
25. DEFE 58/164, a Jezebel analysis manual, unfortunately not dated, but probably compiled in the early 1970s (because it refers to the Yankee power plant as Type III, whereas later it would be lumped with the Type II plants of the Charlie and Victor classes).
26. According to Bruce Rule, longtime chief of U.S. Naval Intelligence acousticians, for many years the U.S. Navy nuclear power establishment refused to believe the performance of these submarines. Initially the HENs were described as "unconventional" because Rickover's office in charge of naval reactors (usually called simply Naval Reactors), refused to believe that the Soviets could successfully design, build, and operate nuclear submarines. Once they had clearly demonstrated nuclear performance, ONI was directed to estimate a maximum speed of 22 kts because it was assumed that the Soviets could not package more than one reactor in the available space. In 1965 the ONI Scientific and Technical Intelligence Center (NAVSTIC) released a message claiming near 30 kt capability, attracting a furious response from Naval Reactors. The 30 kt figure seems not to have been widely accepted for some time after that. The low quoted speeds seem to have been based on the assumption that Novembers were powered by a single reactor producing about 70 kW (translated into 15,500 SHP by the turbine). Rule's comments are on the website of the IUSS Association.
27. Change 5 (September 1963) to the British handbook of Soviet warships (ADM 239/624) credited the submarine with "probably" a single pressurized-water reactor and turboelectric drive, for a maximum of 15,000 SHP and a submerged speed of 22 kts. Normal diving depth was given as 800 ft and maximum diving depth as 1,200 ft. Submerged displacement was given as 4,300 tons (3,750 surfaced). Presumably turboelectric drive was assumed because it seemed that the nuclear icebreaker *Lenin*, which used turboelectric drive, was the prototype of the November-class power plant. Change 11 (August 1965, ADM 239/670) credited her with much the same performance (15,500 SHP) but deeper diving depth (900 ft normal, 1,300 ft maximum) and considerably greater displacement (6,300 tons submerged, 5,400 tons surfaced). The British incorrectly credited the submarine with two turboelectric motors rather than the geared turbines she actually had. At about this time, presumably reflecting U.S. Naval Intelligence data, the CIA credited the November with a maximum speed of 25 kts. In Change 18 (October 1968, ADM 239/830) the submarine was finally credited with 25 kts (but still with a single reactor), but she was correctly credited with direct drive above 10 kts (and probably turboelectric below that). She was still credited with 22,500 SHP, but maximum submerged speed was now given as 25–26 kts. Normal operating depth was given as 1,350 ft. and emergency operating depth as 1,600 ft. Displacement was given as 4,550 tons surfaced and 5,225 tons submerged. A reliable Russian source (Yu. V. Apalkov, *Podvodnye Lodkyi Sovetskogo Flota,*

1945–1991 [Moscow, 2009]) credits the November with a displacement of 4,075 metric tons submerged (3,075 surfaced) and a diving depth of 320 m (400 m emergency), equating to 1,050/1,312 ft. She had two 17,500 SHP turbines, giving a speed of about 28 kts on 80 percent power and about 30 on 100 percent power. Armament was eight bow torpedo tubes (the British listed six bow and four stern). The change in understanding of the November class power plant seems to reflect British access to U.S. data collected using SOSUS.

28. OEG's Haystack analysis, "Passive Defense Aspects of Dispersed Formation Operations under EMCON," was reproduced (with notes) in a declassified Naval War College compendium for students on electronic warfare produced about 1974. The reprinted research contribution referred to recent (meaning ca. 1974) proposals for dispersed fleet formations intended to deceive airborne radars. During the 1980s the U.S. Navy adopted large-scale dispersed formations that required screening ships to be effective far from the units they were protecting.
29. The carriers *America* (CVA 66, FY 61) and *John F. Kennedy* (CVA 67, FY 63) were both to have had SQS-23 sonars, but the one planned for *Kennedy* was not installed. Similar reasoning led to installation of SQS-23 on board *Cleveland*-class missile cruisers. Several *Essex*-class ASW carriers (CVS) were also given SQS-23 in their bows, a modification evident in their revised anchor arrangements.
30. Captain Groverman (Op-71C), quoted in the Cross Report, II, 31.
31. Cross Report, II, 70.
32. Cross Report, II, 70. In a 17 June 2015 post on the IUSS Association website, Bruce Rule (the longtime ONI acoustics analyst) stated that the November involved was limited to 25–26 kts by her particular bow shape. At ONI's suggestion, the Navy directed the carrier to increase speed; the submarine gave up the chase about 60 nm from her. Rule was dubious that the submarine was ever chased off, which would contradict the claim that *Truxtun* was able to coach a P-3 into position to surprise the submarine on the surface. For Rule, this was the first concrete evidence of a November's actual speed, but as early as 1966 NAVSTIC had correctly credited the November with a maximum speed of about 30 kts, based on system data collected in April 1964.
33. This incident was apparently declassified in 1998 or 1999. It was the subject of a Naval Submarine League seminar. The CO of USS *Guardfish*, CAPT David C. Minton III, provided a detailed account on pp. 248–51 of the Naval Submarine League's *United States Submarines*, ed. D. R. Hinkle, H. H. Caldwell, and A. C. Johnson (Waterford: Sonalysts, 2002).
34. Robert Angevine, "Innovation and Experimentation in the U.S. Navy: The UPTIDE Antisubmarine Warfare Experiments, 1969–72," *Journal of Strategic Studies* (February 2005).
35. The 12 April 1968 report on submarine-versus-submarine exercises was declassified (with some redactions) on 31 December 2013. An interim report was dated 1 July 1966. Foster's initial memo was dated 30 November 1966. By this time UPTIDE was apparently considered significant enough that the 16 January 1973 report of UPTIDE 3B was included in the (now declassified) Naval War College tactics material. The Blue carrier *Ticonderoga* was screened by the large destroyer (DLG) *Sterett* (with a LAMPS), the large missile destroyer *Mahan* (DLG), the missile destroyer *Buchanan*, the FRAM I destroyer *Higbee*, and the destroyer escorts *Roark* and *Bagley*. The carrier was supported directly by the fast underway replenishment ship *Camden* with the Coast Guard cutter *Midgett*. The oceanographic ship USNS *Lee* (T-AG-192) towed a surveillance array. *Buchanan*, *Higbee*, and *Mahan* all had towed acoustic deception devices, and the carrier had eight helicopter-deployable deception devices. In addition, *Mahan* had a radar simulator (Blue Blinker) that she could operate in conjunction with her towed acoustic device. The exercise area off the Southern California coast was a 400-by-380-nm box with 25-nm boundaries, so that the Blue force had to make 300-nm transits. *Ticonderoga* was allowed up to 30 hours per transit. The submarines were to simulate either Charlie or Echo II missile submarines employing known Soviet tactics. Simulation of an Echo II required the submarine to operate a sound projector to simulate that submarine's higher noise level (155 dB). Initial submarine positions were randomly chosen. The Blue force operated in HF silence except outside a 30 nm radius of the high-value unit. The same applied to aircraft radar and IFF and to active sonar. At night ships simulated merchant ship lighting. The carrier E-1B early warning airplane directed helicopter planting of deception devices.
36. From a lecture at the Naval War College, in declassified archives.

CHAPTER 9. FIGHTING A NEW SOVIET SUBMARINE GENERATION

1. According to the Cross Report, II, 152, the first major deployments of ASW helicopters on board attack carriers were on two Atlantic Fleet carriers (*Forrestal* and then *Independence*) deployed to the Mediterranean in 1969–70. *Independence* reported particular success, and CinCLantFleet ordered that every other attack carrier going to the Mediterranean should carry ASW helicopters. Each of the two initial Atlantic Fleet carriers had half an HS squadron, eight SH-3Ds, on board.

2. As described in the Minutes of the British 1973 Anti-Submarine Conference, DEFE 69/727.
3. See, for example, a January 1975 CIA intelligence report: "Soviet Naval Strategy Concepts and Forces for Theater War against NATO," approved for release on 14 June 2017. It was based largely on classified Soviet papers.
4. Evidence that the SLOC issue had been dormant can be found in comments in the DEFE 69/208 file. One says that this is an important contribution to the current debate on a previously somewhat neglected field of warfare. Another is SACLANT's (Vice Admiral Kidd's) very warm reaction to the initiation of the 1978 study. His 17 November 1977 letter mentions planned new conferences (Anchor and Seahawk, the latter planned for 7–9 February 1978) focused on shipping defense, reflecting his personal interest. Shipping protection was currently (1977) a hot potato in NATO. At this time the most recent NATO study was an August 1977 effort, which had never received formal approval from either SACLANT or NATO headquarters in Brussels. In 1977 SACLANTCEN and SACLANT were just beginning a smaller report on attrition at sea. A report on fast convoys based on the Teamwork exercise was beginning. One aim was to update standard NATO naval formations (ATP-1B). The introductory talk for the study mentioned Admiral Kidd's new strategy, which gave reinforcement shipping "new relevance." In 1977 he saw it as his primary task to ensure the safe and timely arrival of reinforcements "upon which the fate of a European land battle will depend." He was determined that these reinforcements would sail at the earliest possible moment, noting "that when hostilities start they may have to be fought across every inch of the way. Our ability to do this is in itself one of the major planks of the NATO deterrent." He asked whether a NATO army could last beyond seven days. Admiral Kidd was SACLANT in 1975–78, so his new interest as stated late in 1977 suggests that he was responding to a major strategic shift in U.S. policy.
5. DEFE 69/643 and DEFE 69/644, the final draft (and comments) being in DEFE 69/644. The study was conducted in 1977–78 and was intended to cover the period 1980–89. According to the paper, the Soviets recognized "the necessity to break [the Atlantic connection] in any war with the West. . . . They have also noted how nearly this was achieved in World War II by a relatively small number of submarines. They have declared their conclusion that the submarine lost the last Battle of the Atlantic only through insufficient support and counter ASW effort. The destruction of reinforcement and resupply shipping might therefore become a major Soviet preoccupation in a conflict short of nuclear war and the powerful modern nuclear submarine fleet, which forms the spearhead of Soviet Sea Power, endows them with the ability to sever the connection." Presumably this argument was intended to justify an ASW-oriented Royal Navy, which had just lost its last major strike asset, its carrier.
6. Comment on the paper by ACNS (Operational Requirements).
7. DEFE 69/208. The Maritime Tactical School seems to have been the prime British center for developing tactics. Its earlier studies had been concerned with nuclear depth bombs (1968), Exocet (1970), an underwater-launched antiship missile (1971), the tactical employment of helicopters (1972), ASW in the 1980s (1973), antiship missiles (1974), tactical communications (1974), the SOSS (1976), and towed array tactics (1977). Possible tactics for 1977 were problems of operating an attack carrier in an Irish Sea sanctuary; countering the AS-4 missile as carried by Backfire bombers; IFF ("preserve ourselves from our friends"); and command/control. The study period was 15 March–7 April 1978. The study group comprised 6 Royal Navy captains, 13 RN commanders, 16 RN lieutenant commanders, plus officers already on the Maritime Tactical School staff. There would also be two U.S. and two Canadian officers, and one RAF group captain and two RAF wing commanders. The study coordinator, CAPT A. R. Barnden, RN, had visited the United States during the Seahawk ASW conference. This file includes numerous comments on the study, but not the study itself. The comments include the evidence that the Soviets were not particularly interested in SLOC interdiction because France was not part of the NATO military structure at this time; the assumed destination ports were Amsterdam, Antwerp, Rotterdam, and Zeebrugge. For purposes of the study, assumed convoy size was 40–60 ships, convoys departing from North America every 3 or 4 days. There was little agreed NATO guidance on shipping defense. Threat input to the Maritime Tactical School study was that Britain and NATO acknowledged that the Soviets could ascribe high priority to attacking reinforcement ships, but it would be less important to them than "contributing to the defense of the Soviet Union and its allies against maritime strategic nuclear strikes and to provide an offensive/retaliatory Soviet maritime strategic strike capability." Secondary roles would be contributing to defending the USSR against all other maritime threats and providing support to ground forces as required, and conducting ocean maritime warfare on conventional and nuclear basis. Open-ocean forces were deployed and armed for the top-priority job. There was some flexibility, which could be exploited using the efficient SOSS. The Soviets would surely be able to decide when they could attack opportune targets of lesser importance (reinforcement shipping) while remaining in position to counter the prime threat (such as the Strike Fleet).

8. Estimates of Soviet loadouts had recently been drastically reduced. In 1974, the Victor class submarine was credited with a total of 33 torpedoes, 8 in its tubes and another 25 stowed. These figures are from CAB 186/17, the 1974 JIC papers, which includes a summary of Soviet bloc war potential dated 16 April 1974. A November was credited with eight 533 mm tubes and four 406 mm tubes, with a total of 37 torpedoes. In fact a Victor carried 18 weapons and a November carried 20 torpedoes (including 8 in her tubes; she had no 406 mm tubes). Estimates of torpedo capacity of diesel attack submarines were much more accurate; the report credited the Foxtrot and Tango with 22 torpedoes each, the Zulu with 18; the shorter-range Romeo was credited with 12, and the older Whiskey with 14. In fact, a Tango carried 24 torpedoes, a Foxtrot and a Zulu both 22. A Romeo carried 14 torpedoes, a Whiskey 12.
9. NIE 11-14-79, Warsaw Pact Forces Opposite NATO, dated 31January 1979, declassified 16 August 2010.
10. These amounted to 22 torpedoes in Zulu and Foxtrot class submarines and 24 in Tangos. It may not yet have been known that Victor-class submarines carried only 18 torpedoes. The comments in the NIE do not address the question of Soviet submarine tactics, which was discussed at some length in connection with the Maritime Tactical School paper.
11. V. Zharkov and CAPT 1st Rank B. Tyurin, "Project 632 and 648 Mine-Laying Submarines," *Morskoi Sbornik* no. 3 (1996). Project 632 was a submarine minelayer with a secondary ability to transport aviation fuel and troops (in the mine spaces), but Project 648 was intended primarily to resupply submarines operating in oceans and along SLOCs. The requirement (TTZ) was published in May 1957. The project was approved on 10 July 1958; at the same time Project 632 was abandoned and the remaining submarine assigned an alternative minelaying role. Two submarines and a *Skoriy*-class destroyer were assigned to carry out tests of the proposed underway replenishment equipment and technique. The project was formally approved in August 1959 and the building project formally endorsed in June 1960, builders and equipment suppliers being assigned. The Severodvinsk yard apparently tried to shut down the project because it wanted to build series of nuclear submarines rather than the single Project 648. Initially, plans changed to provide a mixed power plant incorporating two small reactors, but then Project 648 was superseded by the nuclear Project 664. It is described by the same two authors in "Large Transport-Minelayer SSN of Design Project 664," *Morskoi Sbornik* no. 10 (1997). Parallel projects (648 and 664) for diesel and nuclear transport-minelayers were included in the Seven-Year Plan (1959–65). These submarines would provide submarines at sea with weapons and fuel. They could also transport small assault forces, evacuate casualties, and lay mines. Project 648 was canceled on 21 June 1961. Formal specifications for the nuclear-powered Project 664 were approved on 1 March 1960. Construction was assigned to Severodvinsk (Yard No. 402) at the end of 1962. The submarine-to-submarine fuel transfer system was tested on board a Romeo and a Zulu. The ammunition transfer system (including cruise missiles) was tested using a Zulu-class submarine and the transport *Kholer*. The complexity of the project drastically slowed it down, and it was overtaken by the high-priority Yankee (Project 667A) project. Although about 600 tons of metal had been worked, in June 1965 construction of the Project 664 submarine was stopped. A further Project 717 for a supply submarine was later pursued, but again, not built.
12. Kuzin and Nikol'skiy, *The Soviet Navy 1945–1991*, describe the evolution of Soviet naval doctrine during this period. Their account is credible because both were captains in the Soviet navy, but it does not comport with Western understanding of how Soviet strategy evolved. Some of the chronology is suspect; for example, the authors date rough nuclear parity with the United States to the end of the 1970s rather than their beginning. They rank neutralizing enemy naval nuclear weapons as the second naval mission of the 1970s, after maintaining the ability to strike enemy targets ashore. According to them, the defensive mission raised awareness of the Soviet navy within the political leadership. This mission comprised attacks on both carriers and NATO SSBNs; a special "combat service" was created to track both. Soviet studies found that neutralization was possible only if the Soviets preempted. Kuzin and Nikol'skiy do not address the failure to track SSBNs. Instead, they present a table showing the growth of Soviet ASW forces—of input, in effect, rather than output (expected destruction of SSBNs). Thus they list force ratios: ultimately 2.8 large ASW ships, 1.8 ASW nuclear submarines, and 3 long-range ASW aircraft per U.S. SSBN, "which in the opinion of specialists was adequate for the given level of destruction of these SSBNs." They write of two major naval elements, the strategic nuclear force (SSBNs) and an anticarrier force (SSGNs); they refer to Tomahawk missiles but assume they were non-nuclear. They write that the advent of the Tomahawk made it necessary to provide cover for the main navy strategic force and for the main anticarrier force. The submarines would require ASW cover. The other main anticarrier force, the missile bombers, required fighter cover, hence Soviet carriers. Kuzin and Nikol'skiy do not mention the bastions and the forces needed to protect them.

13. CAPT Peter M. Swartz, USN (Ret.), "Understanding an Adversary's Strategic and Operational Calculus: A Late Cold War Case Study with 21st Century Applicability: U.S. Views on Soviet Navy Strategy and Operations," CNA paper COP-2013-U-005622 Final (August 2013). Captain Swartz was deeply involved in the development of the Maritime Strategy of the 1980s and was, in effect, its historian. The Office of Naval Intelligence (and DIA) concentrated on Soviet naval capabilities, as reflected by Soviet ships and systems, rather than on their strategy. The Navy set up a special operational analysis unit in ONI to work out the implications of the new information, which, apparently for the first time, enabled U.S. officers to understand how the Soviets saw the naval situation. See also Christopher A. Ford and David A. Rosenberg, "The Naval Intelligence Underpinnings of Reagan's Maritime Strategy," *Journal of Strategic Studies* 28, no. 2 (April 2005). They do not date the new intelligence, as Captain Swartz does.
14. The idea of fully "scientific" planning was central to Soviet ideology. By the 1970s Western defense establishments also invested in apparently similar forms of analysis, but always subject to pressure to avoid excessive spending. The figures that Kuzin and Nikol'skiy give for the resources required to fight a new Battle of the Atlantic, already cited, are a case in point; they would have stunned German and Allied participants in the Battle of the Atlantic. Kuzin and Nikol'skiy claim that through the early 1960s Zhukov's ideas of SLOC attack continued to guide Soviet naval development, even though the expected rapid army offensive ashore would have rendered attacks on NATO shipping pointless. They claim that at the end of the 1970s the three Soviet naval priorities (other than maintaining the SSBN deterrent) were, in order, destroying NATO nuclear attack ships, attacking NATO naval lines of communication, and assisting the army.
15. The British notes are in DEFE 69/635. The proposed reassignment entirely ignored the need for carriers to counter the most flexible Soviet antishipping threat, the mass of Soviet naval aviation (SNA) bombers armed with antiship missiles.
16. This account of NARAC-G and SEAMIX I is based on the declassified draft final report of the SEAMIX I exercise in the Naval War College archives. NARAC-G was a Navy analysis of general-purpose naval forces called "Navy Sea Control and TACAIR." The draft report, prepared in April 1973 by the OpNav System Analysis Division comparing NARAC-G and SEAMIX I results, was dated 10 May 1973. It was part of a larger analysis of Navy general purpose forces in sea control missions.
17. According to the declassified SOSUS log produced by the IUSS Alumni Association, a SOSUS modernization program began in 1975. Inter-array correlation was available in FY 81, and there was improved real-time secure communication between SOSUS and tactical forces. All stations received a major computer upgrade in 1981–84 that incorporated automatic detection of acoustic signature components (Bell Labs Computer Alerted Surveillance System [COMPASS]). At the same time vacuum tube hydrophone shore amplifiers were replaced by solid-state ones. Standard analog magnetic recorders were installed. The paper recorders remained. In the early 1980s data processing began to be consolidated at regional processing centers. A fourth generation of sonar sets used the new Lightweight Under Sea Components (LUSCs) beginning in 1984. The paper gram printers were finally replaced by standard workstations in the early 1990s. Articles by Jim Weinel in the spring 2003 and summer 2004 issues of *Cable*, the official newsletter of the IUSS Caesar Alumni Association. The 1970 Surface ASW Plan, in the declassified volume 1 of *Selected Readings for Antisubmarine Warfare for the 1972–73 Academic Year*, described near- and far-term possibilities as understood at that time. They included the Improved Processing System (IPS, 5–8 dB), Automatic Processing of Jezebel Information/Multibeam (APOJI/M, 5–8 dB with an improved spectrum analyzer), and adaptive beam-forming. Potential array improvements were longer arrays (2 dB), arrays with vertical directivity, multidimensional arrays, arrays suspended above the bottom, and distributed arrays of individual or small numbers of hydrophones linked together. The last could be used in areas of poor propagation. The hope was that suspending arrays well above the ocean floor near the depth of the deep channel would gain up to 5 dB. Longer arrays might gain 2 dB. Adaptive beam-forming could gain 1–10 dB. It is not clear to what extent the new kinds of arrays were ever deployed. Another 1970 proposal was an active adjunct, the existing SOSUS arrays being used as receivers; the active transmitter would operate at 250–400 Hz.
18. The ships were to have been designated AGOS. In a quirk of the 1980 presidential campaign, Ronald Reagan promised the Seafarers' Union that the vessels would be civilian-manned (i.e., by the union's members), so the ships were assigned to the civilian-manned Military Sealift Command, becoming T-AGOS. The first ship, USNS *Stalwart*, arrived at Little Creek to begin test and evaluation of its SURTASS system, commissioning on 9 April 1984. With the addition of these ships, SOSUS became the Integrated Undersea Surveillance System (IUSS). Three ships had been commissioned by October that year, followed by four in 1985, two in 1986, one in 1987, one in 1988, six in 1989, and one more, the last in the initial class, in 1990. *Victorious*, the first SWATH ship, followed in 1991. Ships

were designed to steam 1,500 nm to a patrol area, where they towed their arrays at 3 kts. The original arrays were 8,575 ft long, towed at a depth of 500–1,500 ft; the tow cable was reportedly 6,000 ft long. There were also two (of a planned five) Japanese surveillance ships.

19. The lanes are shown as an example of barrier ASW in the declassified Naval War College publication *Selected Readings in Tactics 1975: Projection Study*, enclosing a January 1974 sea control analysis by David K. Pack and Dale Bryson of the Naval Weapons Center (China Lake) Weapons Planning Group.
20. JIC(68) Memoranda 1–8 in PRO. I have used British intelligence documents as proxies for their U.S. counterparts, on the theory that both countries shared their technical intelligence; comparable U.S. files are not available.
21. SOSUS log, from the website of the IUSS Association, a formerly classified document. It is not clear whether submarines had previously obtained the relevant signatures.
22. JIC(A)(69) 1 of 1 April 1969, a periodic intelligence survey for NATO commands, in CAB 186/1 and JIC(A)(70) (3) (Final) of 15 January 1970, on the Soviet bloc war potential, in CAB 186/5.
23. JIC(A)(72)3 of 25 January 1972 in CAB 186/11. This seems to have been the last year for which JIC papers included detailed analysis of Soviet forces (the 1971 papers do not include any such analysis).
24. DEFE 58/164, a handbook for Jezebel analysis, undated, but probably from about 1970. Soviet submarine signatures are discussed in detail in a declassified Naval War College publication, Volume III-1 of the 1976–77 *Selected Readings of the Engagement Analysis Study*, the section on SSN versus SSN dated 1977. Part I describes Soviet submarine characteristics and weapons as then understood. Propeller and hull design had eliminated broadband propeller cavitation as a major source of noise, but significant low-frequency tonals, such as the 50 Hz line, remained. The strongest tonal was at 300 Hz (the harmonic of the 50 Hz line), followed by a 580 Hz line due to the 49 Hz main coolant pumps, and then the 50 Hz line itself. In order of decreasing strength were lines at 600 Hz (harmonic of 50 Hz), 246 Hz, 294 Hz (main coolant pump vane), and 250 Hz (harmonic of 50 Hz). Line levels of Soviet SSBNs (then called Type III) showed an effort to silence. The strongest line was now 50 Hz, followed by the 300 Hz line. That was followed by the 580 Hz harmonic of the main coolant pump (still 49 Hz), and then by the 147 Hz of the main coolant pump vane. By this time U.S. SSNs were trailing Soviet SSBNs; the chart for these ships extended through 11 lines, compared with the 7 listed for Type II SSNs.
25. V. Bildin, "Design 661 Nuclear Attack Submarine," *Morskoi Sbornik* no. 4 (April 1993).
26. L. Samarkin, "Multipurpose Nuclear Submarines of Project 671," *Morskoi Sbornik* no. 2 (1995).
27. This account is based mainly on the special issue of the Russian magazine *Taifun* on the Alfa, published in 2002. The contributors were involved in the Alfa project. The remark about a submarine as fast as a torpedo is from another Russian article. It added that Malakhit assigned the project to an engineer recently released from the Gulag. The engineer who had conceived the project was so angry that he changed jobs, abandoning submarine design altogether in favor of oceanography.
28. The concept design Malakhit selected for development would have displaced 1,620 metric tons surfaced and 2,260 submerged; it would have made 43–45 kts. At the technical design stage, displacement was 2,000/2,883 tons and full speed was 40–41 kts. The adjusted technical design showed a displacement of 2,250/3,890 metric tons and full speed of 37.5–38.5 kts. Dimensions of the concept design were 58.4 by 8.0 m, but the adjusted technical design showed 78.8 by 9.5 m.
29. The antiship version was Project 672, armed with five 53 cm tubes and four 65 cm tubes; it would have had a 60,000 SHP engine and a diving depth of 500 m. Project 673 was the SSBN version, competing designs being produced by both Rubin and by TsKB-112, which developed the Charlie class. The Rubin version had an unusually high reserve buoyancy, 42 percent rather than the 10 percent of the Malakhit designs. There was also a high reserve buoyancy version of the basic Malakhit design.
30. Alternatives were conventional twin-loop water-cooled, single-loop water-cooled (i.e., a boiling water reactor), single-loop gas-cooled, and twin-loop liquid metal-cooled. Little work had been done on the two single-loop alternatives, so the choice was either a conventional water-cooled reactor or a metal-cooled reactor. By the time submarines were being built, two alternative liquid-metal reactors were under development; they equipped submarines of Project 705 and Project 705K. According to the design history of the Victor class (Project 671), Malakhit had gained experience of an AC turbo-generator in developing the design of the abortive Project 639 SSBN. Presumably Project 645 came somewhat later.
31. Gerhardt Thamm, "The Alfa SSN: Challenging Paradigms, Finding New Truths, 1969–79," *Studies in Intelligence* 52, no. 3 (2008), the CIA internal magazine (unclassified). The article initially appeared in the fall 1993 issue and was republished in slightly redacted form.
32. This account of the development of U.S. towed arrays is based largely but not entirely on Stanley G. Lemon, "Towed Array History, 1917–2003," *IEEE Journal of Oceanic Engineering* 29, no. 2 (April 2004). Lemon does not explain the TUBA connection, for example. The account

of the 1975 exercise is from a declassified Naval War College publication, volume III-1 of the 1976–77 *Selected Readings of the Engagement Analysis Study*, the section on SSN versus SSN dated 1977.

33. According to Lemon, TUBA meant "to be announced," a cover name.
34. AIR 8/3282, "Anti-Submarine Warfare." An enclosed Royal Navy paper includes a silencing chronology that includes the three Yankees but makes no mention of Victor III. It includes a reference to construction of a sound range beginning in 1977, but the earlier date of the first of the silenced Yankees suggests that the Soviets were measuring narrowband noise well before that. The main subject of the file was a fight between the two services, the Royal Navy arguing that silencing would soon make sonobuoys (hence RAF Nimrods) useless.
35. This progression is evident from the in-house history of the Okeanpribor design bureau, which was responsible for nearly all Soviet submarine sonars.
36. Shipping noise in the Mediterranean blanked out the 50 Hz line used by SOSUS, so the Mediterranean towed arrays used the 300 Hz harmonic, which was detectable. It appears that initially the destroyer escorts sought the 50 Hz line, and thus detected nothing until they were advised to look at 300 Hz.
37. APL memorandum 16 June 1971 to CO of NAVORD, referencing their June 1971 report on the *Barney* tests. APL had issued a preliminary report on 27 May 1971.
38. A detailed analysis of Squeeze Play by the Operational Evaluation Group (OEG Study 755, September 1971) is included in volume IV of *Selected Readings for Antisubmarine Warfare for the Academic Year 1972–73* (Naval War College), declassified. This paper was released by CNO on 15 March 1972. The sound characteristics of a Yankee were considered comparable to the U.S. *Skate* class, which was used in exercises to simulate a Yankee. Yankees could be used either as a relatively invulnerable deterrent, or to spearhead a first strike from close to shore, offering minimum warning time. Warning time might be reduced further by firing on depressed trajectories, then and later a considerable U.S. preoccupation. Because no point in the continental United States was more than 900 nm from the sea, Yankees could cover the whole country with 1,300 nm missiles. According to the OEG report, the submarine could launch missiles in salvoes of two, 30 seconds between salvoes.
39. For an account of British trailing experiences, see David Parry, *Perisher: 100 Years of the Submarine Command Course* (Gloucester: Choir Press, 2022). Parry, a former Royal Navy submarine commander, explains that he collided with an Echo II because he was unable to use Doppler to sense her maneuvers.

CHAPTER 10. END-GAME

1. For a description of some of the NATO-oriented exercise scenarios, see Geoffrey Till, "Holding the Bridge in Troubled Times: The Cold War and the Navies of Europe," *Journal of Strategic Studies* 28, no. 2 (April 2005, special issue, *The Cold War at Sea*). Till reports that after a lapse, a major convoy exercise was held in 1968 (Silver Tower). It seems to have been the first by NATO to take into account the prospect of attacks using antiship missiles, both by submarines and by smaller missile craft. Two of the carriers and several underway replenishment ships had been sunk by the end of the second day of Strike Fleet operations at the cost of only two enemy submarines. According to Till, a 1969 SACLANT study (Brosnio) projected inability to handle the Soviet navy in northern waters by the mid-1970s. Till also mentions the 1972 exercise Strong Express, involving the Strike Fleet in an entirely non-nuclear scenario, supporting the defense of northern Norway against Soviet submarines, aircraft, and missile-armed surface ships. It was the largest NATO sea-land-air exercise since Silver Tower. A new series of Ocean Safari exercises began in 1975, focused on resupply shipping before and during hostilities. Concepts included support by the Strike Fleet, which might operate near or south of the GIUK Gap. The Northern Wedding series begun in the late 1970s tested capability in the Norwegian Sea. Till cites a dismal prognosis for operations north of the GIUK Gap leading to an emphasis on pre-D-day reinforcement of northern Norway. Ocean Safari and Northern Wedding were amalgamated in the first U.S.-led exercise (1981) reflecting ideas of the Maritime Strategy.
2. CAPT. Robert Bathurst, "The Lemming Complex: Ritual Death in the Norwegian Sea," *Naval War College Review* 27, no. 8 (1974). Bathurst was then an intelligence officer at the Naval War College. He pointed to the ongoing series of twice-yearly carrier battle group exercises in the Norwegian Sea, which he considered pointless. They were intended to demonstrate that NATO could reinforce northern Norway in a war, but the United States lacked the necessary shipping, and the carriers were far too vulnerable. In explaining how revolutionary the far north exercises of the 1980s were, John Lehman, *Oceans Ventured: Winning the Cold War at Sea* (New York: Norton, 2018), points to the earlier prohibition of such exercises as too provocative. The combination of his comments and the Bathurst article suggests that the provocation issue was raised by the Carter administration, which was generally interested in arms control agreements that would restrict naval operations.
3. Lehman, *Oceans Ventured*, 140.
4. V. N. Burov, *Otchestvennoye Voennoye Korablestroeniye* (St. Petersburg: Sudostroyeniye, 1995), 461–62.

5. The British wrote that FOSM considered GCHQ (communications intelligence) by far the most important source on Argentine naval forces and an invaluable contribution to the success of the submarine campaign. There was almost no human intelligence. FOSM wrote that a single reliable source in Puerto Belgrano could have clarified the whereabouts of the Argentine submarines and thus "might have permitted a more relaxed ASW posture by British units." My account of British submarine operations is based on my book, *British Submarines in the Cold War Era* (London: Seaforth; and Annapolis: Naval Institute Press, 2021), which is based on declassified British official documents.
6. For the Argentine side of the submarine war, I have relied on Mariano Sciaroni with Andy Smith, *"Go Find Him and Bring Back His Hat": The Royal Navy's Anti-Submarine Campaign in the Malvinas/Falklands War* (Warwick: Helion, 2020). It seems to be the only attempt to combine British and Argentine archival material on the submarine war.
7. According to Sciaroni, *"Go Find Him,"* 42, the submarine's HF message was intercepted by HMS *Endurance*, transmitted to the UK, and decrypted there. The message included the submarine's estimated time of arrival at South Georgia.
8. Sciaroni, 29–30. This incident does not appear in British submarine force summary reports.
9. According to a 10 September 1982 paper by the ASW section of the Admiralty Underwater Weapons Establishment (AUWE) in DEFE 69/1086, a compilation of Falklands ASW lessons including a supplemental AUWE Portland analysis dated 23 October 1983. The file was a reply to a request for information by the Operational Evaluation Group.
10. In the P-3C Orion, Update I (introduced into the fleet in April 1975) added 392,000 words of memory, as well as Omega navigation, a tactical display for the sensor operators, improved acoustic processing, and significantly better reliability and maintainability. Update II (first delivered January 1977) doubled the capacity of the acoustic tape recorder and added a sonobuoy reference system (SRS) offering an accurate tactical plot of the positions of sonobuoy fields in the water, a FLIR, the Harpoon antiship missile, and an improved acoustic tape recorder. Using the SRS, the airplane no longer had to overfly the buoys to locate them. This system was developed for the S-3A Viking. Update III was an extensive acoustic redesign using the Single Advanced Signal Processor (UYS-1) to replace the hardwired AQA-7 of earlier versions. A 99-channel buoy system replaced the earlier 31-buoy system. Update IV introduced a new distributed architecture developed by Boeing for the New Zealand Rigel I P-3B modernization. Compared to Update III, it had 2.5 times the acoustic data throughput, 8 times the bulk memory, 30 times the display memory, and 5 times the MTBF. The program began in 1984, and Boeing was awarded a full-scale engineering development contract in 1987.
11. AD-A 231952 is a December 1990 report by J. A. Rice of the Naval Ocean Systems Center in San Diego describing theoretical and experimental (1986–90) work on arrays of free-floating sonobuoys up to 3,000 m across. Their number was limited by current buoy uplinks to 99. The concept was called Array Element Localization, and a prototype buoy was made for tests.
12. Declassified NIE 11–15–85/D, *Soviet Naval Strategy and Programs through the 1990s*, based on information available through 21 December 1984. No publication date is given, but it was probably early in 1985. Note the general agreement that there would be a limited SLOC offensive. There was no longer any discussion of the possible activation of Soviet submarines in reserve. According to the NIE, current active Soviet submarine strength was as follows. Northern Fleet: 41 ballistic missiles, 32 cruise missiles, 92 torpedoes; Pacific Fleet: 32 ballistic missiles, 26 cruise missiles, 67 torpedoes; Baltic Fleet: 6 ballistic missiles, 4 cruise missiles, 23 torpedoes; Black Sea Fleet: 1 ballistic missile, 2 cruise missiles, 24 torpedoes.
13. This account is based on Lehman, *Oceans Ventured*, 68–85.
14. Lehman, 83, writes that the towed arrays were effective.
15. The most obvious example was the cancellation of the first Soviet nuclear carrier, *Ulyanovsk*, in 1985. Note too the drastic decline in numbers of Soviet attack submarines. Eighteen Victor I/II were followed by 26 Victor IIIs, but there were only 14 Project 971s (the Victor follow-on) and only 4 of the titanium-hulled equivalent (Project 945).

APPENDIX A.
SURFACE SHIP AND AIRCRAFT SONARS

1. SQS-1 was a modernized QHB. It entered production in November 1952, and 236 were made. It operated at 20–28 kHz. SQS-2 was redesignated as a variant of SQS-1. SQS-3 became the SQG-2 attack sonar. SQS-10 and SQS-11 were modernized QHBs (SQS-10 was initially QHB-3). As QHB-3, SQS-10 (20 kHz) predated SQS-4, and it seems to have been the first scanning sonar developed by Sangamo Electric. It appeared in 1950. It was a higher-frequency (20 kHz) SQS-4. It was installed on board many recommissioned destroyers, as well as the new Canadian *St. Laurent*–class frigates (DDEs). It had a source level of 208–18 dB (with about eight times the input power of QHBa), Range scales were 1,000, 3,000, and 6,000 yds. It had variable pulse lengths (6, 30, and 80 msec). SQS-11 operated at QHB frequency (25.5 kHz).

The A versions of these sonars had higher power. In 1956 estimated SQS-11A range was 1,600 yds in the layer and 800 below it, roughly what would have been expected of a wartime 25.5 kHz sonar. These figures were for 0.25 probability of detection at 15 kts. Initially these sonars were specified for U.S. destroyers and frigates in the mid-1950s, but they ended up with SQS-4 instead.

2. Thaddeus G. Bell, *Probing the Ocean for Submarines: A History of the AN/SQS-26 Long Range Echo-Ranging Sonar* (Newport Beach: Peninsula Publishing, 2011), 130. In May 1968 the Sixth Fleet requested NUSL's assistance with convergence zone detection in the Mediterranean; controlled convergence zone trials compared an SQS-26(BX) with an SQS-23. The SQS-23 performed substantially worse than expected due to chemical differences in the Mediterranean. For example, it performed poorly against submarines at nonbeam aspects—that is, not broadside to the sonar beam.
3. Bell, *Probing the Ocean*, 14. I have relied heavily on Bell's book in my account of SQS-26 development. He was a key participant. For example, he was the one who realized that a combined bottom-bounce/convergence zone sonar was practicable. He was also responsible for many other innovations incorporated in the new sonar. Bell (103) points out that through the late 1940s "deep ocean" was defined as water so deep that the bottom was irrelevant to sonar propagation, which was understood to mean water more than 100 fathoms deep. According to Bell, in 1947 Robert Young of NEL presented the first experimental evidence in the open literature that bottom reflection might be important for near-surface points in the deep ocean. Discussions of sonar operation in the Navy's Secret *Undersea Warfare* magazine include no references to bottom bounce, but they do include the convergence zone, which was mentioned at the Undersea Warfare Conferences.
4. Bell, *Probing the Ocean*, 130.
5. Bell, 108.
6. Data from declassified Naval War College instructional material, 1976, for an "Engagement Analysis Study of the ASW Surface Ship in the Escort Role vs Submarine."
7. Prairie-Masker was an air bubble screen shielding the sonar from propeller and machinery noise.
8. According to an account of proposed modernization (to deal with nuclear submarines, ADM 302/42), the Staff Requirement issued in 1949 called for both greater range and the ability to sweep 360 degrees as quickly as possible. The prototype was to be ready in 1955. Controller approved the Staff Requirement on 21 October 1949. Range and other details were not included. However, the Way Ahead requirement stated in 1948 had already set a minimum range of 8,000 to 10,000 yds. According to Hackmann, *Seek & Strike*, 348, alternatives considered at the time were a long keel array, a big cylindrical array, and a flat transducer using a combination of electronic and mechanical scanning. The latter was chosen, the first transducer being ready for tests in 1951. The transducer employed scaled-up versions of the Type 170 (fire control) transducer. Data are from the 1966 U.S. handbook of foreign sonars (STIC-CS-05-4-66, declassified copy courtesy of Chris Carlson), also from the British sonar handbook (ADM 239/767 of 1964) and from a study of Type 177 modernization to deal with nuclear submarines (ADM 302/52). Source level was 138 dB. The modernized Type 177M had two rather than three pulse lengths (the 5 msec pulse was eliminated) to increase the Doppler range to ± 40 kts. Echoes were displayed on two CRTs. All target bearing was displayed as true bearing. More pings were provided on the range recorder to improve visual integration.
9. ADM 1/24104, "Near Future Developments in Shipborne Asdics," dated 22 April 1952.
10. Peter Marland, "Postwar Sonar Systems in the Royal Navy," in John Jordan, ed., *Warship 2021* (London: Osprey, 2021).
11. Types 177 and 170 typically carried a Type 176 torpedo detection sonar (introduced in 1952) on their shafts, its display a PPI showing the "noise spoke" of a torpedo. It operated at two frequencies, 39 and 12.5 or 15 kHz, but it was considered ineffective, unable to detect even acoustic torpedoes in time.
12. The logic of the Type 184 design is described in ADM 259/531, dated November 1958. Some of this description is based on ADM 302/198, "Introduction to Asdic Type 184," dated December 1961.
13. As evaluated by the Defence Operational Requirements Staff in 1968 (DEFE 24/134). New requirements are discussed in DEFE 69/726, the minutes of the 1970 Anti-submarine Warfare Conference.
14. DEFE 69/727, in connection with a discussion of helicopters in ASW.
15. Description based partly on Marland, "Postwar Sonar Systems in the Royal Navy."
16. Data from declassified U.S. Navy publication, *Sonar Equipment—Foreign* (STIC-CS=05-4-66), published 31 May 1966.
17. Data from declassified Naval War College instructional material, 1976.
18. ADM 1/24148. Details of the AQS-4 are in the 1953 edition of *Progress in Underwater Warfare*, ADM 239/684. There is also a discussion of the trials in the papers of the 1953 TAS Conference, ADM 189/236.
19. Notes about the initial tests are from the Admiralty *Progress in Underwater Warfare* for 1955. Sonar ranges and

the comments about helicopter capability are from a 12 March 1958 ASW R & D report by a DRPC panel in DEFE 7/1396. I have also drawn on a lecture on ASW helicopters in the papers of the May 1958 TAS Conference (ADM 189/239). Details of versions of Wessex and the initial requirement for the Sea King are from DEFE 10/482.

20. Details from Roger A. Holler, "The Evolution of the Sonobuoy from World War II to the Cold War," *U.S. Navy Journal of Underwater Acoustics* (January 2014) (ADA 597432).
21. The solution, proposed in the spring of 1957 and tested in principle in 1959, was a signal compressor that reduced Jezebel sonobuoy data to transmittable radio form. The initial test employed telemetry equipment and four transmission sub-bands. An airplane flying at 5,000 ft successfully transmitted data to an AQA-3 up to 90 nm away. A VS/CVS relay system was successfully tested in 1962. It could handle 16 signals, 4 from each of 4 aircraft, which would be processed on board an ASW carrier (CVS). The redesigned and enlarged S2F-3 (S-2D/E) could be fitted with the initial Jezebel airborne processor (AQA-3). *Naval Weapons Bulletin* 4-62, formerly Confidential, available as ADC 960224.
22. DEFE 10/941 and DEFE 10/954, papers of the Operational Requirements Committee, 1968.
23. AIR 2/11090, a file on decisions on the initial Shackleton design. The idea that the airplane should be handled like a ship comes from comments made by a Coastal Command officer at the 1948 U.S. Undersea Warfare Conference, transcript in NHHC Operational Archives.
24. This was not a radically new perception. The Air Staff wrote and approved a Staff Requirement (ASR 805) for a new sonobuoy system in 1954, defining an improved capability needed by 1966. It was assigned a low priority due to the lack of R&D resources but raised to Priority 2 in 1960. The British had to adopt Jezebel to meet an urgent requirement for long-range detection. An Air Staff Target (AST) for an improved Sonics Mk 1C was endorsed in April 1965.
25. DEFE 10/944, Operational Requirements Committee for 1969, analysis of requirements for the Phase 2 Nimrod, November 1969.
26. DEFE 10/951, Operational Requirements Committee 1971, 2 December 1971.
27. At this time, about 1971, the CEP of an initial CODAR plant against a stopped submarine was about 5,000 yds. It would worsen as the target moved. With additional CODAR plants CEP on a stopped submarine could be refined to less than 1,000 yds, but the time required was too great to contain a fast SSN or a diesel submarine snorting over a period of 40 minutes or less. Automatic hyperbolic fixing and Doppler fixing (which required a digital processor and were implemented in AQA-7) would not meet the basic requirement for reliable and timely localization to within attack sensor range.

APPENDIX B. SUBMARINE SONARS

1. John Merrill and Lionel D. Wyld, *Meeting the Submarine Challenge: A Short History of the Naval Underwater Systems Center* (NUSC, 1997, PB97–202402).
2. The submarine LOFAR processor was already being developed. Feasibility studies began in January 1954. An Operational Requirement was issued in May 1954, and a contract in September 1960. The processor in the BQQ-1/2 system was designated BQQ-3.
3. ADM 204/3153, notes of a visit to the Naval Ordnance Laboratory, February 1964.
4. ADM 204/3327.
5. This section is based largely on a post–Cold War in-house history of the Morfizpribor design bureau, which was responsible for all Soviet submarine sonars except for the Kerch type used in Charlie-class submarines (which were designed by a bureau in Ukraine).
6. Data on Soviet tactics and underwater communication from Engagement Analysis Study 1976–77 from the College of Naval Warfare (selected readings), declassified.

APPENDIX C. TORPEDOES

1. In 1946 the U.S. Navy's Bureau of Ordnance announced a six-torpedo program, Mks 35 through 40, of which it was already working on Mk 35. They were soon reduced to three (Mks 35, 37, and 39) because money was not available for new antiship torpedoes. Mk 39 was a wire-guidance testbed converted from a Mk 27 Mod 4 homing torpedo. Mk 37 was another ASW homing torpedo (initially also an anti-escort torpedo). According to the report of a March–April 1947 visit by the British underwater weapons team to the United States at U.S. invitation (ADM 292/214), "the highest priority is being given in America to the development of homing torpedoes as a counter to the fast submarine. A really prodigious effort involving about 120 scientists and engineers is being put into this development, which is far in advance of anything being done in this country." As during World War II, the British visitors reported that the scale of the U.S. effort made it possible to support competing projects, which the British would have considered wasteful.
2. Mk 41 was also proposed as the payload of the ship-launched Grebe (originally Kingfisher E). guided missile, developed in response to a request by the first Undersea Warfare Conference (1946) for a means of launching a Mk 35 torpedo from a surface ship at a distant target, up to 20 nm away. It was canceled in 1950, Admiral Low pointing out that its range exceeded that of existing sonars.

For a time its developers argued that it could be fired on the basis of data from another ship, a concept later taken up in the Australian-British Ikara program.

3. Mk 27 was essentially a stretched Mk 24 for submarines, powered by two batteries in tandem. It swam out of the tube under its own power. It was used late in the Pacific war as an anti-escort weapon, about 106 being fired to make 33 hits, sinking 24 ships and damaging another 9. Performance matched that of Mk 24 (12 kts), but range was greater (12 minutes' endurance, range over 5,000 yds). The postwar Mod 4 was also nominally an anti-escort weapon, adapted to be set using standard electrical umbilicals. Speed increased to 15.9 kts, and range to 6,200 yds (12 minutes). It was produced between 1946 and 1954 (about 3,000 were made).
4. Homing and some other details are from Frederick J. Milford, "U.S. Navy Torpedoes Part Five: Post–World War II Submarine-Launched/Heavyweight Torpedoes," *Submarine Review*, October 1997. Details of U.S. weapons and sensors as of 1949 are from a notebook prepared for Hartwell members, in RG 72 enclosures to Secret correspondence for that year, box 170.
5. BuAer files (RG 72 in NARA II) include a 24 March 1950 memo setting out desired characteristics of a lightweight helicopter or blimp torpedo, responding to a 6 February 1950 request by BuOrd as the basis of a study. Maximum permissible weight would be 250 lbs. A speed of 12 kts would be satisfactory, although 15 kts was desirable. Required range was 3,000 yds. The explosive charge should be 40–50 lbs, and the torpedo would attack down to 400 ft. The helicopter might drop the torpedo 500 yds from the desired drop point. Homing should be active acoustic, although passive homing might be acceptable for interim use. Launch speed would be under 40 kts, and launch altitude under 150 ft. Mk 43 Mod 0 weighed 370.4 lbs (12.75 in. diameter, length 88.25 in., warhead 60 lbs), with a range of 4,300 yds at 20 kts. About 500 were made. Mod 1, presumably based on the wartime Mk 30, weighed 260 lbs (10 in. diameter, length 91.5 in., 54 lb warhead, with tail fins 15.4 inches across). Range was the same, but at a speed of 15 kts. Because its tail fins extended beyond its body, Mk 43 Mod 1 could not be fired from a surface ship tube; instead it was tossed overboard. Five hundred were made.
6. According to a December 1954 NOTS report, EX-2 was envisaged as a 400 lb (450 lbs acceptable) torpedo 12 in. × 98.5 in., with 15 in. fins with a 75 lb HBX-3 warhead. Desired range was 6,000 yds at 30 kts (6 minutes), with an active homing range of 1,000 yds. NOTS began development in January 1954, with hoped-for evaluation in January 1956. This was the 430 lb EX-2A, which was to be electrically powered. EX-2B diameter was 12.5 (12.75 by October 1956) in., and weight was 425 lbs. Initially GE planned to use a monopropellant engine, but when work was suspended on it they had to rapidly develop a contra-rotating electric motor. Their EX-2B used only a contact exploder. Data from ADM 290/640, an account of a visit to the United States by the chief scientist and principal scientific advisor of the British Torpedo Experimental Establishment to the United States, 7–26 October 1956.
7. Situation in the fall of 1956 from ADM 204/969, an April 1957 report of a British visit to hydrodynamics laboratories in the United States in the fall of 1956 (24 September–26 October), including a report on U.S. ASW weapon projects. This document does *not* refer to the new lightweight torpedo that would become Mk 46.
8. ADM 302/372, an account of a 1962 visit to the United States to review the current state of U.S. torpedo research and development.
9. DEFE 69/609, the report of a British mission that visited the United States in October 1965 to learn about Mk 46 Mod 0 as an input for the new British torpedo project. The Specific Operational Requirement was 90 percent at a 95 percent confidence level: a fully operational torpedo, on being launched, should have a 90 percent probability of correctly completing a six-minute run. The 3 kt Doppler gate considerably alleviated the serious problem of a torpedo homing on its own wake. If the torpedo failed to acquire a target during its first turn, it would encounter its wake, which at the least would increase self-noise. Below 125 ft the Doppler gate was optional. Without it, the torpedo would home on its own wake after completing three-fourths of a search circle, unable to detect a real target before running through the wake and resuming its search.
10. NEARTIP program goals from a 10 July 1975 paper, "Notes on US Navy Lightweight Torpedo Programme," in DEFE 69/609 (papers E66 and E47/48, the latter a table comparing various lightweight torpedoes).
11. Data such as the target strength goal, the shallow depth requirement, and the search and attack speeds are from DEFE 69/610, the report of a British visit to the United States to decide whether to buy NEARTIP or the developmental NASR 7511 (Stingray). According to this report, NEARTIP would be effective against current Soviet submarines, both coated and uncoated, in a wide variety of scenarios. In some areas, however, NASR 7511 requirements would not be met fully; in others, information was still incomplete. A single NEARTIP could not search all the way down to the desired 760 m (2,493 ft). Maximum depth cover was 500 m (1,640 ft) against an uncoated (+8 dB) target, reduced to 300 m (984 ft) against a coated target. U.S. tactical policy was to drop two weapons

searching at different depths when target depth was uncertain. The British wanted the torpedo to pull out at 45 m (148 ft) rather than 60 m, but they thought that in practice NEARTIP could manage that. It might not pull out reliably at 37 m (121 ft) as required of NASR 7511. Another issue was acquisition of low-Doppler targets (less than about 4 kts); given existing dropping accuracies probability of success against targets making less than 5 kts would be significantly reduced, and there would be some effect at 8–10 kts. It was not clear how often targets would operate at slow speed, but it seemed that they might use low speed as a defensive tactic in shallow water, where it would be difficult to evade at high speed. At this stage the U.S. Navy was not prepared to discuss likely counter-countermeasures performance, though it would be significantly better than for Mod 2. The British mission commented that total countermeasures capability would still be limited because the torpedo was not collecting high-resolution sensor information. By way of contrast, NASR 7511 had a broadband coherent homing system that would provide that information except at very short range. The U.S. assessment was that given a hit, NEARTIP would have a 70–90 percent chance of rupturing the hull of a worse-case target (a Soviet SSBN at shallow depth), taking into account that hits on medium-speed targets would be distributed over the after halves of their pressure hulls, and over a wider area of slower targets. This met the UK requirement, but the British mission considered that performance would be worse against third-generation targets, which would be present in substantial numbers by the mid-1980s. The preferred hit distribution required by NASR 7511 could solve this problem. When the British mission visited the United States (1977) the passive surface attack mode had not yet been implemented. The British observed that the program hardly addressed operation in shallow depths (in which they were vitally interested) except for reducing plunge depth. U.S. officers emphasized that not all modifications would be accepted or put into all torpedoes; with money tight, the plunge depth modification would probably have the lowest priority.

12. The Naval Underwater Warfare Center at Newport produced a prototype deep experimental torpedo (DEXTOR) that could be converted into an operational torpedo if needed. It was launched several times by the deep-diving submersible *Dolphin*. Dextor was a full-size torpedo (21 × 230 in.) capable of diving to 4,000 ft, with a combination of wire and homing guidance. It used a motor from a Mk 45 torpedo and Mk 37 controls, with an acoustic projector enabling it to track a target. The fiberglass hull, with a honeycomb core, was considered a breakthrough, possibly enabling the torpedo to dive to 8,000 ft. At full power (38.2 kts) it had a range of 12,000 yds (16,000 yds at half power, 26.8 kts). It made nine shakedown runs at Newport, then operated at full depth at the AUTEC underwater range, at 4,000 ft, on 14 February 1968.
13. Data from ADM 302/372.
14. Mk 48 data from DEFE 69/609 (paper E72), a British file on the decision to develop what became the Spearfish torpedo instead of buying Mk 48 Mod 1. The discussion arose from a mid-July 1975 determination that not later than 1985 the Royal Navy needed a new heavy torpedo significantly better than the existing Mk 24 Mod 1. At that point the issue was whether to buy Mk 48 Mod 1 or to modify the existing Mk 24 Mod 1 to achieve increased depth, improved homing, and increased speed—which was impossible. According to British data, at 55 kts Mk 48 was 14 dB louder than the British Mk 24 running at 24 kts (10 dB louder when at 40 kts). At 40 kts it was 15 dB louder than a Y-class submarine at 10 kts (10 dB louder than the same submarine at 20 kts). At 55 kts it was 19 dB louder than a 10 kt Y-class submarine and 14 dB louder than a 20 kt Y-class submarine.
15. According to a March 1972 article by G. J. Kirby (a torpedo researcher at AUWE since 1966) on British postwar torpedoes in the *Journal of the Royal Naval Scientific Service* (U.S. AD 595842), work on an antisubmarine homing torpedo began in 1943. It was split into the passive antisubmarine Dealer and the active-homing Trumper.
16. Speed limitations from Llewellyn-Jones, *The Royal Navy and Anti-Submarine Warfare, 1917–1949*, 100.
17. Admiralty, *Progress in Underwater Warfare* for 1947, 83, ADM 239/421. The next year's edition stated that this torpedo was not going into production.
18. As described in the 1951 edition of the Admiralty *Progress in Underwater Warfare*.
19. *Progress in Underwater Warfare* 1953, ADM 239/684.
20. *Progress in Underwater Warfare* 1955, ADM 239/424. Kirby claimed that improvements planned for Mod 1 would have given it a hit probability better than that of any other torpedo.
21. Foreign Naval Publications collection, RG 38, NARA II.
22. As described in *Progress in Underwater Warfare* 1948, ADM 239/422, in the 1949 edition (ADM 239/274), and in the 1951 edition (ADM 239/275). There is also ADM 1/24435, "Pentane: Suitability for Use as an Air-Launched Weapon," paper by Director of Naval Air Warfare, 20 May 1952.
23. My account of British submarine torpedoes is based on the more extensive one in my *British Submarines in the Cold War Era* (Barnsley: Seaforth; and Annapolis: Naval Institute Press, 2021).
24. ADM 332/7 is a file on Naval/Air Staff Target NAST 1204 for the revised UK-made Mk 44 Mod 2. In addition

to the British requirement, the new torpedo would be a candidate for a large planned West German purchase of a torpedo with good shallow-water performance; it would compete with a Canadian Westinghouse proposal.

25. ADM 332/8. The origin of the Mk 31 project was presumably the Naval/Air Staff Target (NAST 1204) for Mk 44 Mod 2, endorsed by the Operational Requirements Committee on 24 May 1964.
26. The Deep Mobile Target had a redesigned Mk 44 afterbody with contra-rotating screws, a modified nose section (flat area less than two-thirds diameter), a modified homing nose, 7 in. more length, and an additional short transducer section (2 in. long) between the control section and the afterbody, carrying an LF transducer ring and Mk 24 transducers to radiate noise at 33 kHz. The homing nose had a ring of Mk 24 transducers around its periphery (echo repeat receiving and transmitting). The contra-rotating motor was adapted from a starter motor already used for a Rolls-Royce jet engine. According to the Admiralty's feasibility study for NSR 7709, the new motor would reduce self- and radiated noise. The torpedo would be further silenced by replacing the alternator in Mod 1 with static inverters (DC from the battery into AC). Eliminating gearbox losses would increase power and speed. The motor incorporated an extra slip ring to enable shunt field current to be changed (BHP would be increased 50 percent, from 30 to 45, for the burst of speed). Opening a solid-state switch would make it possible to weaken the field and increase power when range was down to a few hundred yards, increasing torpedo speed to the desired 35 kts. The battery was silver-zinc.
27. ADM 302/467, a book of preliminary (advance) information for Mk 31.
28. OR/P(66)13 in ADM 332/8, in the justification for the original staff target dated 24 May 1966. Required Mk 31 maximum depth was then 1,200 ft, compared with 1,500 for Mk 46. Mean search time (half of search cycle) was 10 seconds for Mk 31 (both passive and active) vs. 9 seconds active or 18 seconds passive for Mk 46. Mk 31 acquisition range was expected to be 700–1,000 yds in shallow water. For Mk 46, the corresponding figures were 200 yds for low-Doppler and 1,400 yds for high-Doppler targets. Run was 6,000 yds, compared with 10,000 for Mk 46. Mk 46 offered alternative search depths without reference to water depth: 125, 275, 500, 750, or 1,000 ft. In water up to 200 ft deep Mk 31 could search down to 20 ft above the bottom. In water 200–600 ft deep, it could search down to 100 ft from the bottom. In water more than 600 ft deep it would search down to 500 ft from the bottom. Both torpedoes were credited with 75 percent reliability, compared with 40 percent for the British Mk 44 Mod 1.
29. According to a February 1966 note comparing Mk 44 Mod 2 (later Mk 31) with Mk 46, estimated acquisition costs were very different: £9,000 for Mk 44 Mod 2 versus $65,000 for Mk 46. At an exchange rate of $2.80 to the pound, the estimated unit cost of a Mk 44 Mod 2 would have been about $25,200.
30. "Soviet War Potential 1955–59," JIC (55) of 24 March 1955.
31. JIC (56) 3 of 14 March 1956 in CAB 158/23G.
32. JIC (56)1/9 of 24 September 1956 in CAB 158/23.

BIBLIOGRAPHY

Note: documents referred to in the text with prefixes ADM, AIR, CAB, and DEFE are in the British national archives, formerly known as the Public Record Office (PRO). Ships' Covers for the Type 14, 15, and 16 frigates, which have been used as references for this book, are held by the Brass Foundry out-station of the British National Maritime Museum; the Cover for Type 12 seems to have been lost.

Apalkov, Yu. V. *Podvodniye Lodki*. 2 vols. St. Petersburg: Galeria Print, 2002–3. For characteristics of Soviet submarines; part of Apalkov's larger series on Soviet warships.

Ball, Desmond, and Richard Tanter. *The Tools of Owatatsumi: Japan's Ocean Surveillance and Coastal Defense Capabilities*. Canberra: Australian National University, 2015.

Bell, Thaddeus G. *Probing the Ocean for Submarines: A History of the AN/SQS-26 Long Range Echo-Ranging Sonar*. Newport Beach: Peninsula Publishing, 2011.

Burov, V. N. *Otchestvnnoye Voennoye Korablestroeniye*. St. Petersburg: Sudostroyeniye, 1995.

Cafferky, Shawn. *Uncharted Waters: A History of the Canadian Helicopter-Carrying Destroyer*. Halifax, Canada: Centre for Foreign Policy Studies, Dalhousie University, 2005.

Cave Brown, Anthony, ed. *Dropshot: The American Plan for World War III against Russia in 1957*. New York: Dial Press/James Wade, 1978.

Cote, Owen R., Jr. *The Third Battle: Innovation in the U.S. Navy's Silent Cold War Struggle with Soviet Submarines*. Newport: Naval War College, 2003.

Cross Associates. *Sea-Based Antisubmarine Warfare 1940–1977*. 2nd ed. 17 February 1978 [Cross Report]. Report prepared for the Office of Naval Research and later declassified from Secret. For many years a declassified copy was available in the Operational Archives of NHHC.

Ewing, M., and J. L. Worzel. "Long-Range Sound Transmission." In *Propagation of Sound in the Ocean. Geological Society of America Memoir* 27 (1948): 1–35.

Fagan, M. D., ed. *History of Engineering and Science in the Bell System*. Bell Laboratory, 1975.

Friedman, Norman. *British Destroyers and Frigates: The Second World War and After*. London: Chatham, 2006.

———. *British Submarines in the Cold War Era*. Barnsley: Seaforth; and Annapolis: Naval Institute Press, 2021.

———. *The Fifty-Year War: Conflict and Strategy in the Cold War*. Annapolis: Naval Institute Press, 2000.

———. *U.S. Destroyers: An Illustrated Design History*. Rev. ed. Annapolis: Naval Institute Press, 2004.

———. *U.S. Naval Weapons*. London: Conway Maritime Press; and Annapolis: Naval Institute Press, 1983.

———. *U.S. Submarines since 1945: An Illustrated Design History*. Rev. ed. Annapolis: Naval Institute Press, 2018.

———. *World Naval Weapons Systems*. Annapolis: Naval Institute Press, various editions, 1989–2006.

Gardner, W. J. R. *Anti-Submarine Warfare*. London: Brassey, 1996. Volume 11 in Brassey's Seapower series.

Gibson, Chris. *Nimrod's Genesis: RAF Maritime Patrol Projects and Weapons since 1945*. Manchester: Hikoki, 2015.

Grove, Eric J., ed. *The Defeat of the Enemy Attack on Shipping, 1939–45*. Aldershot: Ashgrove, 1997. Published version of the 1957 Admiralty Staff History (BR 1736), whose principal author was LCDR D. W. Waters. To some extent it is a British equivalent of OEG 51, but much more extensive and also strongly advocating the value of convoys.

Hackmann, Willem. *Seek & Strike: Sonar, Anti-Submarine Warfare, and the Royal Navy 1914–54*. London: HMSO, 1982.

Hansen, Chuck. *U.S. Nuclear Weapons: The Secret History*. Arlington, TX: Aerofax, 1988.

Hinkle, D. R., H. H. Caldwell, and A. C. Johnson, eds. *United States Submarines*. Naval Submarine League: Waterford: Sonalysts, 2002.

Holler, Roger A., Arthur W. Horbach, and James F. McEachem. *The Ears of Air ASW—A History of U.S. Navy Sonobuoys*. Warminster, PA: Navmar Applied Sciences Corp., 2008.

Holmes, W. J. *Double-Edged Secrets*. Annapolis: Naval Institute Press, 1978.

Jackson, James. *The Admiralty and the Helicopter: Royal Navy Helicopter Projects*. Blue Envoy Press, 2018.

Jolie, E. W. *A Brief History of U.S. Navy Torpedo Development*. NUSC Technical Document 5436, 15 September 1978.

Kowalski, Robert J. *Grumman AF Guardian*. Naval Fighters Number 20. Simi Valley, CA: Steve Ginter, 1991.

Kowalski, Robert J., and Tommy R. Thomason. *Grumman S2F/S-2 Tracker*. Part 1: *Development, Testing, Variants, and Foreign Users*. Naval Fighters Number 101. Simi Valley, CA: Steve Ginter, 2016.

Kuzin, V. P., and V. I. Nikol'skiy. *The Soviet Navy 1945–1991*. St. Petersburg: Historical Naval Society.

Lehman, John. *Oceans Ventured: Winning the Cold War at Sea*. New York: Norton, 2018.

Llewelyn-Jones, Malcolm. *The Royal Navy and Anti-Submarine Warfare, 1917–1949*. London: Frank Cass, 2006.

Mayne, Richard O. *The Annapolis Riddle: Ship Design and the Canadian Navy's Force Structure Crisis, 1957–1965*. Dissertation, Queen's University, Kingston, Ont., 2008.

Merrill, John, and Lionel D. Wyld. *Meeting the Submarine Challenge: A Short History of the Naval Underwater Systems Center*. NUSC, 1997, PB97-202402.

Mutza, Wayne. *Lockheed P2V Neptune: An Illustrated History*. Atglen: Schiffer, 1996.

Parry, David. *Perisher: 100 Years of the Submarine Command Course*. Gloucester: Choir Press, 2022.

Pavlov, A. S. *Warships of the USSR and Russia 1945–1995*. Annapolis: Naval Institute, 1997.

Radley, Clive. *Sonobuoy History from a UK Perspective: RAE Farnborough's Role in Airborne Anti-Submarine Warfare*. Camberley, Surrey: Clive Radley, 2016.

Rodgaard, John. *Tailships: The Hunt for Soviet Submarines in the Mediterranean, 1970–1973*. Warwick: Helion, 2023.

Rowher, Juergen, and Mikhail S. Monakov. *Stalin's Ocean-Going Fleet*. London: Frank Cass, 2001.

Sciaroni, Mariano, with Andy Smith. *"Go Find Him and Bring Back His Hat": The Royal Navy's Anti-Submarine Campaign in the Malvinas/Falklands War*. Warwick: Helion, 2020.

Simonov, Nikolai. *Voenno-promyshlenniy complex SSSR v 1920–1950-e gody*. Moscow: Rosspen, 1996. Trans. Steve Zaloga as *Military-Industrial Complex of the USSR 1920–1950*.

Solomon, Louis P. *Transparent Oceans: The Defeat of the Soviet Submarine Force*. Thriller Tales, 2003 (the story of LRAPP).

Sternhell, C. M., and A. M. Thorndike. *A Summary of Antisubmarine Warfare Operations in World War II*. Summary Technical Report of Division 6. Washington, DC: NDRC, 1946; originally issued as a classified report. It is often referred to as OEG [Operation Evaluation Group] 51.

Swartz, CAPT. Peter M., USN (Ret). "Understanding an Adversary's Strategic and Operational Calculus: A Late Cold War Case Study with 21st Century Applicability: U.S. Views on Soviet Navy Strategy and Operations." CNA paper COP-2013-U-005622 Final, August 2013.

Thomason, Tommy H. *Naval Fighters No. 70: The Forgotten Bell HSL: U.S. Navy's First All-Weather Anti-Submarine Warfare Helicopter*. Simi Valley, CA: Steve Ginter, 2005.

Weir, Gary E. *An Ocean in Common: American Naval Officers, Scientists, and the Ocean Environment*. College Station: Texas A&M University Press, 2001.

INDEX

Italicized *page numbers* indicate photographs.

ABC Partnership, 3–4
acoustic fences, 273
acoustic intelligence program (AIDA), *189*
acoustic torpedoes, 94, 95
acoustics, underwater, 28–32. *See also* noise
ADAWS (Action Data Automated Weapons System), British, 166, 168
ADCAP (advanced capability) redesign, Mk 48 torpedo and, 345, 351
Admiralty: as surrendered-U-boats' caretaker, 83. *See also* Britain and British; Royal Navy
Admiralty Research Laboratory (ARL), 41–42
Admiralty Underwater Weapons Establishment (AUWE), 269, 301, 391n9
Advanced Light Weight Torpedo (ALWT), 235, 243, 345–6
Advanced Lightweight Torpedo (EX-11), 338
Advanced Weapons Unit (AWU), Key West Operational Development Force, 335
Aegis antiair system, 255–6, 278
Aegis combat system, 277
AF-3S aircraft, 173
Afghanistan, 3, 15, 255
afternoon effect, calm weather and sun and, 29, 366–7n14
AGI (intelligence-gathering ships), Soviet, 211
AGSSN (experimental submarine), U.S., 194
ahead-thrown weapons, 44–45, 72, 76–77. *See also* Hedgehog; Weapon A
Air Force, U.S.: land-based AWACS aircraft, 281; national nuclear stockpile and, 381n2; nuclear deterrence and, 11; on Soviet non-acoustic submarine finding, 20; on strategic ASW, 180; Western Europe defense and, 13
air reconnaissance, 37–39, 105
air threats, postwar ASW warfare and, 121, 122
air-capable ships, 220. *See also* Sea Control Ships, U.S.
aircraft, AEW (airborne early warning), 172, *175*, 175, 178, 380–1n33
aircraft, Argentine, 263
aircraft, ASW: ASW simulation battle and, 235; British development of, 46–47; British exercise experience on, 105; Cuban Missile Crisis and, 380–1n33; early radar detection of submarines by, 131; Fido (homing torpedo) and, *52*, 52–53; fixed-wing, Hartwell group on torpedoes for, 141; German, 37–39, 369n23; German snorkeling U-boats and, 67–70; Korean War and, 152–3, 154; LOFAR installations, 200; mobilization decade and, 169–78, 379n13; nuclear submarines and, 186; postwar, 129–32; quieter Soviet submarines and, 219; Royal Canadian Navy on, 177; shipbuilding and, 53–54; sniffers, 33; sonobuoys, 51, *51*, 305–10, 393n23; SOSUS and, 143, 145; torpedoes for, 335; types, *49*; U-boat radio traffic and, 50; U.S. ASW force (1980s) and, 278; U.S. development of, 48–49; U.S. long-range planning (1950s) and, 158; U.S. patrol squadrons designated as, 132; U.S. postwar warfare exercises (1949) and, 106; World War II and, 36, 54–66. *See also* blimps; helicopters
aircraft, land-based patrol, 201–5. *See also* P-3 Orion, U.S.
aircraft, maritime patrol: ASW simulation battle and, 236–7; long-range detection and, 274–5; World War II ASW and, 4
Air-Deployed Active Receiver (ADAR), 275
Akkord computer, Soviet Okean (integrated sonar system) and, 325, 326
Akula attack submarine, Soviet, 259, *260*, 260, 327
Albacore (nuclear submarine), 134, 185, 244
Alfa-class (Project 705) submarine, Soviet: characteristics, *241*, 241–3, 389nn27–28; Mk 48 torpedo and, 345; as puzzle to the West, 242, 389n31; rafts in, 247; sonar prototype tested in, 326
Algerian war for independence (1956), 8–9
ALICE (adaptive line-canceler and enhancer), 251
ALIEX study, 206
Ametist (SS-N-7) cruise missiles, Soviet, 325
Amfora bow array, Soviet, 330
AMOS (Acoustic, Meteorological, Oceanographic Survey), 290
Andreyev, Vladimir, 16, 365n29
Andromeda torpedoes, Soviet, 358
anechoic coatings: Soviet submarines and, *225*, 259, 295; U.S. torpedoes and, 235, 340, 341, 344. *See also* echoes
antiship missiles, 4, 15, 259, 260, 261, 350
antisubmarine attack plotter (ASAP), U.S., 45–46
antisubmarine craft, 60. *See also* ships, ASW; submarines
Aphrodite radar decoy, German, 65
APR-9 radar, 173, 379n13
APS-20 radar, 107, 129, *129*, 169–70, *170*, 379n13
AQA-1 airborne buoy processor, 306
AQS-13 helicopter sonar, *303*, 303–4
Arabian Sea, ASW simulation battle and, 238
Arctic: British operations in, 257; Doenitz on diversion to, 61; German World War II operations in, 59, 61–62; navigation difficulties in, 128; sonar use in, 29; Soviet operations in, 255, 257; U.S. operations in, 158, 257, 280–1

Argentina. *See* Falklands Conflict
Argus (Canadian maritime patrol aircraft), 204, *204*
Arktika fire control sonars, Soviet, 322–3
Army, Soviet, 256
Army, U.S., 13, 48–49, 223
Army Air Corps, U.S., 51
Army Air Force, U.S., 59
Artemis (submarine detection system), 183, 381n35
A/S Attack Plotters, 112
ASCL (Advanced Sonobuoy Communications Link), 235
asdic (British submarine finding tool): bottom-profiling, development of, 69; Churchill and acronym for, 365n1; Corsair, 146, 318–9, 377n19; development of, 21, 22–23, 366n2; improvements to, 73–76; Type 184, as torpedo detector, 116. *See also* Admiralty Research Laboratory
ASPECT (Acoustic Short-Pulse Echo Classification Technique), for sonars, 289
ASROC (rocket-boosted depth charges and torpedoes): development, 140, 141; Extended-Range, 206; mobilization decade and, 159, *160*, *161*, 161, 378n7; SQS-23 sonar and, 289; torpedo choices for, *339*
associated support for submarines, 194–5
ASTOR ASW torpedo. *See* Mk 45 ASW torpedo, U.S.
ASW Classification and Analysis Centers (ASCACs), 220
ASW Conference (1948), U.S. *See* Undersea Warfare Conferences, U.S. Navy
ASW FORPAC, U.S., 212, 215
ASW legacy: aircraft, 47–53; organization for development of, 41–42; shipbuilding 1920–1950s, 53–66; the snorkel, 66–70; surface attack tactics, 42–47. *See also* Cold War; World War II
ASW Operations Research Group (ASWORG), U.S., 42
ASW Ship Command and Control System (ASWSCCS), U.S., 189, 381n4
ASW task groups, U.S., 193–4, 362n13
Atlantic, Battle of the: Americans examining captured Type XXIs and, 87; British codebreaking in, 57; British Committee on (1943), 63; British lessons from, 105; new, British on fast frigates and, 122; new, British–U.S. concepts paper on Soviets and, 224; new, Soviet intentions (1955) and, 108; new, Soviet Whiskey program and, 99; Royal Navy's ASW and, 2–3; U-boat failure in, 368–9nn15–16, 369n18; Western Cold War assumptions and, 36
Atlantic aircraft project, European, *203*, 203–4
Atlantic Antisubmarine Development Detachment, U.S., 63
Atlantic ASW simulation battle: LAMPS helicopters and, 235, 237; maritime patrol aircraft and, 236–7; merchant ships and, 232–3, 237; SOSUS and, *234*; Soviet Backfire bombers and, 233, 237–8; Soviet radar surveillance satellites and, 236; Walker spy ring, SOSUS array and, 234–5, 236
Atlantic Ocean: deep channel in, 367n15; HF/DF net in, 147, 377n20; SOSUS contacts in, 199; substandard postwar U.S. Navy performance in, 100; supply lines between U.S. and Europe and, 1–2, *2*; U-boat war in, 59. *See also* convoys, Cold War; convoys, World War II; North Atlantic
Atlantis (submarine detection system), 183
atomic bombs, 104, 136, 393n11. *See also* depth bombs, nuclear
Atrina, Operation (March 1987), Soviet, 283
attack at source, 104, 135
Attack Director Mk 4 fire control, for Squid, U.S., 76–77
Aurand, E. P., 212–3, 215
Australia and Australians, 3–4, 166, *166*, 205, 310. *See also* Royal Australian Navy
automatic gain control (AGC), U.S. Navy submarine finding and, 24
automatic line integration (ALI), sonobuoys and, 307
Avengers, AEW, British, 172
Avrora towed array versions, Soviet, 329

B-29 Superfortress, U.S., 169
Backfire bombers, Soviet, 233, 237–8
Backscratch arrays, 215
Balaklava submarine sonar upgrade, 325
Balkon (passive sonar array), Soviet, 85, 127, 128, 313, 322
ballistic missile submarines: British sonars and, 320; flash reporting via SLBM voice reporting net, 181, 380n29; nuclear, 186–7; U.S. development of, 9–10, *9*, *10*, 222. *See also* SSBNs
Barnden, A. R., 386n7
Barra buoy, 205, 307, 310
barriers: Burke on, 159; coastal U.S., 144, 145; nuclear submarines and, 191; quieter Soviet submarines and, 217, 218; snorkeling U-boats and, 69; sonobuoy, British ship trade convoys and, 104–5, 137; Soviet saturation and, 193; U.S., against Soviets, 158
bastions, Soviet: in Barents and White seas, 225; British or U.S. Navy awareness of, 15; Cold War end-game and, *19*, *254*, 255, 257; naval exercises demonstrating interest in, 282–3; new generation of submarines and, 230; Ocean Venture '81 exercise and, 281–2; readiness issues (1980s) and, 280; *Seawolf*-class submarines and, 258; as Soviet priority, 232, 252; Soviets on Western priority for, 231; U.S. carrier strike forces as threats to, 18
Batchelder, I., 366–7n14
Bathurst, Robert, 390n2
Bauer, Hermann, 367n4
Baya (experimental submarine), *315*
B-dienst, 58. *See also* codebreaking
BdU (*Befehlshaber der Unterseebooten*), 37, 38–39
Beach, Edward L., Jr., 101, 102, 362n10
Bear radar reconnaissance aircraft, Soviet, 212
bearing deviation indicator (BDI), U.S. Navy submarine finding and, 24
bearing recorder, British submarine finding and, 23
Belgium, 55, 203

Bell Telephone Laboratories: BQR-16 arrays and, 245, 246; COMPASS and, 388n17; Fido homing torpedo and, 331; helicopter production, 176; interwar ASW development and, 42; LOFAR sonobuoys and, 199–200, 383n18; Mobile Listening and, 244; Project Hartwell study and, 136; sound recorders (BQH-2), 244; submarine sound spectrograph and, 111, 142–3, 144
Belyea, James, 113
Bennett, Fred C., 215
Berlin, West Germany, 5, 12
Betty (Mk 90) depth bomb, 140, 185, 376n5
Bidder (Mk 20) passive large-caliber homing torpedo, 348–9
bioluminescence, Falklands Conflict and, 264
bistatic submarine sonar systems, submarine explosive echo-ranging and, 188
Blackett, P. M. S., 42, 51
blade rate: detecting Soviet Novembers and, 208; of fast single-screw nuclear submarines, 196–7, 381n6; Soviet nuclear submarines and, 240, 329; Soviet reductions in, 259; SSN detection and, 383n14
blimp torpedoes, 394n5
blimps: as ASW aircraft, 49, 132, 135, 375n49; Korean War and U.S. mobilization of, 152–3; non-acoustic sensors for, 33; sonars, 303; U.S. strategic air defense and, 178
Blohm, Rudolf, 93
Blohm & Voss shipbuilders, 91, 92–93
Blue Orchid Doppler navigator, Wessex helicopters and, 304–5
Bolide antiship missile, 261
bombs: B-57 (Mk 57), 140, 277; conventional, depth charges and, 368n13; hydrogen, Khrushchev on Soviet testing, 16; thermonuclear , U.S. and Soviet testing, 10. *See also* atomic bombs; depth bombs, nuclear; retro-bombs
boost-phase tactic, 252
Boresight, as *Kurier* solution, 148
Borey 1 and 2 (Projects 955 and 935) strategic submarines, Soviet, 262
bottom arrays, 188, 329
bottom-bounce ping, from seabed, *29*
bottom-bounce sonar, 186, 291–3, 295, 300–301, 314–5
bottom-bounce sound propagation, 31–32, 290
bottom-profiling, British and, 69, 76
BQH-2 intelligence-gathering analyzer, 320
BQH-4 sound analysis module, 244–5, 317
BQQ-1 submarine sonar, 316
BQQ-2 submarine sonar, 206, 316
BQQ-5 towed arrays, 246, 317
BQR-7 passive sonar, 316
BQR-15 submarine array, 245, 246, 321
BQR-25 towed array, 246
BQS-6 spherical active sonar, 315–6
BQS-13 spherical active sonar, 317
Brackett, *40*
BRASS (Bottom Reflected Active Sonar System), 314–5
Bravo (Project 690) submarine, Soviet, 243
Brazenface (underwater communication signal), Soviet, 324
Breadfruit (Soviet voice and Morse system), 324
Brezhnev, Leonid, 17, 212, 260, 364–5n28, 365n31
Britain and British: ASW aircraft, 46–47; ASW weapons in World War II of, 40; Autolycus project, submarine exhaust and, 33; convoy code changes and, 63; defending Germany against Soviet assault possibility, 102; deterrence strategy and, 10–11; on electro-U-boats, 87; escorts for Korean War and, 154, 156; German electro-U-boats and, 79–80, *80*, 82–83; helicopter program, *165*, 165–6; homing torpedo testing by, 331; interwar ASW development in, 42; LOFAR buoy results for, 200, 383n18; LOFAR units sent to, 144; Luftwaffe air reconnaissance and, 39; maritime patrol aircraft, *171*, *172*, 172, 263; Mediterranean crisis (1936) and, 37; on Mk 48 torpedo, 344; narrowband noise analysis and, 244; on NATO's Mainbrace exercise, 7; postwar ASW improvements, 73–76; postwar ASW warfare and, 100, 113; postwar ship trade convoy games, 104–5; postwar submarine hunter-killers of, 127; scanning sonars, 27–28; silencing machinery by, 195; skeptic on U.S.–British concepts paper, 228; snorkeling Type XXI misapprehension by, 131; sonobuoys, 308–10, 393n23; SOSUS offered to, 199, 383n14; SOSUS technology and, 146, 181, 377n18; Soviet army end-game and, 256; on Soviet November submarines, 384–5n27; submarine sonars, 318–21; tactical towed arrays, 310–2; torpedoes, 346–57; Type XXIs modifications by, 89; ultimate Cold War weapons' research by, 5; Undersea Warfare Conferences and, 3, 101; U.S. concepts paper on Soviets and, 223–5, 226–7, 386n5; U.S. on attacking at source versus, 104; U.S. torpedo development and, 394nn7–9; Walter boats and, 93; World War I submarine finding and, 21–25; World War II shipbuilding, 54. *See also* Falklands Conflict; postwar ASW warfare; Royal Air Force; Royal Navy
British Chiefs of Staff, 10–11, 363n18
British Joint Intelligence Committee (JIC): British–U.S. concepts paper on Soviets and, 227–8; on Cold War year of maximum danger, 5; detailed Soviet forces analysis by, 389n22; on postwar Soviet submarine fleet, 108–9, 374n22; Soviet fighting-readiness estimates by, 2; on Soviet potential behavior (1970s), 223; on Type XXIs captured by the USSR, 88
British Maritime Tactical School, 225–6, 386n7
Brown, Harold, 223
bubble detection, Hartwell study of, 138
Buckley-class destroyer escorts, 375n43
Bullseye sites, 148, 149, 378n23
Bureau of Aeronautics, U.S. (BuAer): ASW aircraft, 172, *173*, 173–4, 175; ASW blimps, 132, 375n49; design contest for single-package plane, 173–4; helicopters, 52, 135–6, 176; Regulus missiles and, 372–3n6; seaplanes, *170*, 171; on ship versus aircraft ASW, 121; side-looking radar and,

139; on snorkel-detection radars, 129–30. *See also* LOFAR; Navy, U.S.; torpedoes; United States
Bureau of Standards, U.S., 42
Burke, Arleigh: on backup barrier, 159; on long-range planning, 158; on nuclear submarines, 186, 190; on nuclear SUBROC, 206–7; on strategic ASW, 180, 381n1
Butoma, B. Ye., 241, 242

Cam Ranh Bay, Vietnam, Soviet naval base in, 283
Camp, Glenn, 376n6
Campbell, Isabel, 361n7
Camrose (antitorpedo torpedo), British, 116
Canada and Canadians: FRD-10 antenna array built in, 149; maritime patrol aircraft, 204; postwar ASW warfare and, 113; sonobuoys, 310; SOSUS and, 3; SQS-505 sonar, 296. *See also* Royal Canadian Air Force; Royal Canadian Navy
Cann, Jerry, 250
CAPTAS towed-array receiver, French, 276
Captor rising mines, 207, 235, 237, 282
carrier task forces, U.S., 158
carriers: converted for ASW, 174, 379n15; countermeasures to Soviet threats, 212–3; Falklands Conflict and, 263, 265; helicopter, convoy air support and, 220; Soviet (1980s), 280; U.S. ASW force and, 277. *See also* escort carriers; fleet carriers, U.S.; helicopter/VSTOL carriers
Carter, Jimmy and administration: defense budgets under, 3; on NATO war in Europe, 14, 223; on naval exercises as provocative, 390n2; Navy's reaction to budget-cutting by, 255, 256; Soviets on submarine sanctuary areas and, 253
CASS buoys, *163*, 307, 310
Cauliflower tactic, 104
cavitation sounds: blade rate and, 196–7; bombardier monitoring sonobuoys for, 173; *Nautilus* exercises and, 185; propeller noise and, 196–7, 308; single propellers and, 182; Soviet nuclear submarines and, 240; Type XXIs' and, 81. *See also* noise
Cedar (Project 957) attack submarine, Soviet, 262
centimetric radar, 50, 56
Channel, English, 55, 68–69
Charlies, Soviet tactical missile submarine, *238*, 238–9, 324, 325
Chelomei, Vladimir, 235–6
chemical range recorder, British and U.S. Navy submarine finding and, 22–23
Chernyavskii, Sergei, 364–5n28
Chesapeake Instrument Company, 244, 245, 246, 248
chin arrays, sonar, 317
China and Chinese, 16–17, 207, 233
CHLOE, 139
Churchill, Winston, 365n1
CIA (Central Intelligence Agency), 20
cigarette lighters, 371–2n30
Clarinet Bullseye, 148. *See also* Bullseye sites
Classic Outboard, 149, 378n24
Clevite, 343–4
close convoys, 237. *See also* convoys, Cold War
closed-cycle oxygen torpedo, 95
closed-cycle propulsion: German development of, 90; German interest in, 71, 72, 73, 369n2; Soviet development of, 93, 371–2n30. *See also* Walter boats
Cluster Bay, Soviet RVM rocket as, 207
CNA (Center for Naval Analyses), 230
CO Tactical Plot (COTP), 113
Coast Guard, U.S., 51–52, 151, 279
Coastal Command, RAF: aircraft, radar on, 49; Anglo-American active buoy and, 308–9; ASW aircraft, World War II, 55, 59, 64, 69, 368n13; bombing U-boats, 368n13; on offensive air ASW, 4; postwar ASW aircraft, *171*, 171–2, 379n11, 393n23; quieter Soviet submarines and, 274–5; radar floodlighting by, 137; shore-based reconnaissance aircraft of, 48; sniffers of, 33; sonobuoy deployment, 51; torpedoes, 346–7, 355; Type XXI electro-U-boats and, 131; U.S. maritime patrol force and, 153. *See also* Royal Air Force
coastal waters, electro-U-boats for, 39, 83, 85
CODAR (correlation detection), 200, 306, 310, 393n27
codebreaking (radio intercepts): Anglo-U.S. Naval Cipher, 60–61; British, against Germany, 59, 64, 65–66, 368n12; Combined Anglo-U.S. Naval Cipher No. 5, 64–65, 367n4; electro-U-boats and, 79; Falklands Conflict and, 263, 264, 266, 268, 391nn5, 7; German, against Britain, 56–57, 58, 64–65; German, postwar study on, 367n4; German Enigma and, 39, 62; German ocean surveillance and, 37
Cold War: ASW overview, 1–4; burden of limited war, 11–12; Low's ASW readiness report (1950), 134–6, 151, 375n1; rethinking nuclear war, 10–11; Soviet thinking, 15–19; Soviets on Western hostility and, 4–10; strategic shift in, 12–15. *See also* ASW legacy; end-game; mobilization decade; nuclear submarines; postwar ASW warfare; Project Hartwell study, U.S.; sound in the sea; submarines; Type XXI electro-U-boats
collapse depth, for submarines, 72, 351
Colossus (shallow-water SOSUS complement), 144–6, 377n17
Commencement Bay-class escort carriers, 118–9, 172–4
Committee on Undersea Warfare (CUW), 100, 136, 381n1
communications: among U.S. destroyer escorts, 113; British, Falklands Conflict and, 273; British tactical submarine hunting trials and, 105; German, with submarines, 96–98, 372n35; Soviet HF radio with submarines, *226*; underwater, Soviet versus U.S. systems for, 324; U.S. underwater telephone, 323. *See also* codebreaking
commutator sonar, 25–27, 366n11
computer-aided detection, 110, 272
computers: Akkord, Soviet Okean and, 325, 326; British combat direction system and, 268; Nimrod and, *202*, 203; Soviet problem with, 276, 380n30. *See also* microcomputers
Conqueror (British), *262*, 264, 265–6, 268, 321
contact-fuzed depth charges, 44

containment strategy, Cold War, 5
continental shelf off Argentina, 263–4
convergence zone: British sonars and, 299; characteristics, 31–32, 367nn15–16; first demonstration of, 30; research on, 290–1; sonar, bottom-bounce sonar and, 186, 291–3, 295; sonar propagation paths and, *29*
convoys, Cold War: Aegis antiair system and, 255–6; as ASW measure, 7; British Cauliflower tactic for, 104; British ships for, 121–5, *122–4*; dispersed, Haystack tactic of, 210, 281, 282, 385n28; escort ships and, 11, 12; France abandons, 9; Hartwell study of, 139; helicopters and, 220; Kidd on, 14, 363n14; limited-war scenario and, 13; logic applicability (mid-1960s), 211; off Newfoundland (1942), *2*; quieter Soviet submarines and, 217, 219; SACLANT and, 8, 10; screening ships for, 102, 103; SEAMIX I exercises and close or expanded, 237; Soviet naval exercises and, 227, 387n8; Soviet radar satellites and, 225–6, 235–6; surface ships and, 4, 219; U.S. ASW force (1980s) and, 278; U.S. on Soviets attacking SSBNs instead of, 232. *See also* merchant ships
convoys, SSN, U.S. Navy and, 194
convoys, World War II: air ASW and, 60; British assumption on, 54; emergency turns by, 81, 82, 87; escort strengthening for, 57–58; German use of codebreaking and, 39, 56–57; long-range German attacks on, 58; Luftwaffe air reconnaissance and, 38–39; snorkeling U-boats and, 67–70; U-boats' night surface attacks against, 55–56. *See also* Doenitz, Karl
Corsair (British long-range passive asdic), 146, 318–9, 377n19
Courageous (British), 264, 272
creeping barrage tactic, for depth charges, 43
Crowe, William P., 232, 256
CRTs (cathode ray tubes): APR-9 radar and, 379n13; British directional sonobuoys and, 308; British FH4 wide-open detector and, 372n35; British Type 177 asdic and, 297; British Type 184 asdic and, 272; British Type 2016 asdic and, 299; German use of, 50, 98; GRD-6 system and, 147; LOFAR analyses and, 307; paper recorder in place of, 294; photographing, 319; as plotter, 306; Soviet flicker-free displays on, 328; Soviet wire-guided torpedoes and, 327; U.S. Navy's PUFFS and, 317; U.S. SQS-23 sonar and, *289*, 289–90; visual indication of submarines, 24–25
cruise missiles: German researchers on, 101–2; nuclear, U.S. plans (1950s) for, 158; Soviet, as threat to U.S., 180, 380n25; Soviet, SS-N-7, 207; test launch by U.S., *101*, 372–3n6
Cuban Missile Crisis (1962), 150, 181–2, 380n31
currents, ocean layers' distribution and, 29–30
Curtis, Tom, 312
Cusk, test cruise missile launched by, 101, 372–3n6
CVE. *See* escort carriers
cylindrical sonar intercept array sonar, Soviet, 325
Cyprus War (1974), 358
Czechoslovakia, 5, 13
Dabbler (Soviet underwater communication signal), 324
Dachsund (Soviet Morse code signal system), 324
DASH (drone antisubmarine helicopter): ASW task group Charlie (1958) and, 362n13; buoy fields laid by destroyers, 201, 384n23; development of, 161–2, *162*, *163*; LORELI and, 33; mobilization decade and, 159
data links, UPTIDE exercises and, 214, 385n35
Davies, Thomas D., 220
Dealers (British homing torpedo), 346–7, 395n17
Dealey or *Dealey*-class destroyer escorts, 114, *150*
deception exercises and tactics, 210, 213–4, 281–2, 385n35
decoys, 25, 116
deep channel: in Mediterranean, 30–31, *31*; Nobska on nuclear submarines and, 190; sonobuoys, 307; sound propagation and, *29*, 30, 142, 367n16, 376n7
deep experimental torpedo (DEXTOR), 395n12
Deep Mobile Target, British, 354, 396n26
DEER (destroyer explosive echo-ranging), 188
defended lanes, to protect shipping, 226, 237. *See also* SLOCs
defense budgets, mid-1950s, ASW and, 3
Defense Science Board (DSB), U.S., 191–3
delousing, as Cold War operation, 195
Delta-class SSBNs, Soviet, 230, 253
DELTIC time-compression signal processor, 307, 382n9
DEMON (noise demodulator), 197, 320, 327–8, 329
density layer, convergence zone and, 30, 367nn15–16
depth bombs, nuclear: Hartwell study of, *139*, 140; Lulu (Mk 101), *139*, 188, 189, 277, 376n5; mobilization decade and, 160; types, 376n5
depth bombs, Western, 277
depth charges: ASW tactics and, 25; characteristics and tactics for, 42–43, *44*; conventional bombs and, 368n13; Doenitz on electro-U-boats and, 86; Falklands Conflict and, 271–2; fast-sinking, U.S. development of, 76, 114–5; high-speed U-boats and, 72; homing torpedoes as alternative to, 52–53; improvements, 44; lethal range, 367–8n6; postwar ASW warfare and, *115*, 115–6; shallow waters, Falklands Conflict and, 270; SOFAR (SOund Fixing And Ranging) channel, 30, 376n6; timing determinations, 22–23; types and storage on board, *43*; U-boat diving depths and, 366n8; UEPs as proximity sensors for, 33
depth recorder, British deep-diving submarines and, 75
depth-determining sonar (QDA), 24, *71*, 77
Deschimag shipbuilding yard, Brennan, *73*, 79, 81
Destroyer Division (DesDiv) 81, 112, 118
destroyer escort radar pickets (DERs), 178–9, *179*
destroyer escorts (DDEs): ASW upgrades to, 152, *153*; Battle of the Atlantic and, 64; British on Cold War need for, 105; British spacing, for convoys, 56; building, 53–54; Cold War need for, 7, 361nn4–5; European NATO navies and, 125; French withdrawal from NATO military and, 9; as Korean War problem, 154–8, *157*; Lend-Lease plan and production of, 60; new, cut from 1949 U.S. budget, 6; as postwar ASW

convoy components, 118; SACLANT's convoy strategy and, 8; surveillance and, *248–9*, 249; for World War II, *40*, 40–41, *41*; as World War II survivors, *3*
Destroyer Jezebel (DesJez), 201
destroyer killers (DDKs), 118, 120–1
destroyers, 7, *123*, 185, 335
DF (direction finding) transmissions, German use of, 96–97
DICASS (directional command-activated) buoys, 307
diesel engines, for frigates, Korean War and, 155
diesel submarines: Khrushchev on construction of, 17; low frequency adjunct to SURTASS for tracking, 274; PUFFS and nuclear torpedo on, 318; Soviet, active searchlight sonar on, 322; Soviet Varshavyanka, 260–1. *See also* submarines
DIFAR (directional) buoys, 201, 205, 275, 307
digital signal processing, Soviet, 328–9
digital spectrum analysis, SOSUS data and, 196
DIMUS (digital multibeam sonars), 28, 244, 314
dipping sonars: anticarrier exercise (1971) and, 324; British helicopters and, 265; deep channel and, 31; Falklands Conflict and, 269, 270; French Lynx helicopters and, *168*, 169; helicopter lifting power and, 175; helicopters as pickets and, 103; LOFAR compared with, 221; U.S., Low on, 134; U.S. adaptation from blimps to helicopters, 51; U.S. development of, 111; U.S. helicopters in Mediterranean and, 219; U.S. types, 176, 178, *204*, *303*, 303–4, 305. *See also* helicopters; sonars; Type 195 dipping sonar, British; VDS fish
directional buoys, 51, 305, 308
Director of Naval Construction (DNC), British, 154
Director of the Admiralty Anti-Submarine Warfare Division (DA/SW), 74
Director of the Tactical and Staff Duties Division, (DTSD) British, 123, 154
Director of Torpedoes and ASW (DTASW), British, 154
Directorate of Miscellaneous Weapons Development (DMWD), 42
Dismukes, Brad, 230
distance-measuring equipment (DME), Soviet, 239, 323–4
diving depth: deep experimental torpedo and, 395n12; Hedgehog and, 71–72; *Komsomolets* and, 279; Mk 48 torpedo and, 345; proximity fuzes and, 75, 369n4; Soviet, British torpedoes and, 350, 351; Soviet, towed arrays and, 243–51; *Thresher*, 338; U-boats, 24, 71–72, 366nn8, 13. *See also* submarines
Doenitz, Karl: on all-acoustic torpedo shots, 96; antiradar and antiaircraft weapon for U-boats and, 65; Atlantic convoy attacks (1942) and, 61–64; on attacking convoys, 54, 56; codebreaking reliance by, 39, 56–57; communications with U-boats, 96; on electro-U-boat production timing, 81; failures for, 59; German submarine strategy and, 36–37; snorkels and offensive by, 66–70; on torpedo use, 94; Type XXI instructions, 84–86; on Walter confessing fast-U-boat details, 93; Walter on open-ocean HTP submarine and, 91–92; withdraws U-boats from U.S. East Coast, 368–9nn15–16. *See also* Germany
Dolan, J. H., 169
Doppler gate, homing torpedoes and, 339, 394n9
Doppler sonar benefits, 137–8
Doppler-controlled gain (DCG), 24
Downes, William A., 291
draft, peacetime U.S., 5
Draft Staff Requirements, British (1951), 155
Dreadnought nuclear attack submarine, British, 319–20, 350
drones. *See* DASH
dual purpose echo sounder, British, 76
Dutch. *See* Netherlands and the Dutch
DUUX system, French passive ranging, 269, 318
Dvorkin, Vladimir, 282

Echo II–class submarines, Soviet, 210, 324, 325
echoes: audible low-frequency sonic, 20; classifying, 289; due to wakes, 22, 25; quality changes, 138. *See also* anechoic coatings; explosive echo-ranging; noise
echo-ranging array sonar, Soviet, 325
echo-ranging submarine finders, 21, 306. *See also* sonobuoys
economic recovery, postwar Europe and, 5
Eden, Anthony, 9
Edwards, R. S., 375n45
Eilat (Israeli destroyer), Egyptians sink, 211
Eisenhower, Dwight D.: as first SACEUR, 7, 256; Gaither Committee and, 180; national nuclear stockpile and, 381n2; on nuclear deterrence, 11; on nuclear war, 254
Eisenhower Strike Fleet, 281
electric torpedoes, 94, 95, 357, 358–9
electro-U-boats. *See* Type XXI electro-U-boats
Enactment (Soviet voice, Morse, and tone system), 324
end-game: ASW force, 277–9; drivers of war, 279–84; elements in, 254–9; Falklands Conflict and western ASW, 262–73; modern conditions compared with, 284; new Soviet submarines and, 259–62; replacing SOSUS, 273–6; weapons, 276–7
Enhanced Echo-Ranging (E2R), sonobuoys and, 275
Enigma, Germany's, 39, 62
Enterprise, 211
ERAPS (Expendable Reliable Acoustic Path Sonobuoy, SSQ-75), 275, 307
escort carriers: ASW aircraft and, *49*; Battle of the Atlantic and, 63, 64; Korean War roles, 153, 154; nighttime guppy search and scrappers from, 173; postwar HUK groups and, 118–20, *120*. *See also* carriers
Escort Towed Array Surveillance System (ETASS), 250
escorts. *See under* convoys; destroyer escorts
Ewing, Maurice, 367n16
EX-2 torpedoes, U.S., 336, 394n6
EX-4 torpedo, U.S., 335
exhaust trail detection, Hartwell study of, 138

Exocet missiles, Argentine, 263, 265
explosive echo-ranging, 188, 200–201
Extended-Range ASROC (ERA), 206, 276

Fairey Gannet, British aircraft, *175*, 175, 347–8
Fairey Seamew, British aircraft, 175
Falke torpedoes, 95
Falklands Conflict: Argentine fleet/forces and tools, 264, 265–7; ASW without MAD in, 33; Britain and late 1970s crisis in, 3, 15; British ordnance expended during, 270–1; British tracking Argentine submarines, 268–9, 272–3, 391nn5, 7; environmental conditions, 263–4; false targets during, 270; possible Argentine torpedo attack and, 269–70; Royal Navy fleet/forces and tools, 264–5, 266–8; U.S. Navy and Royal Navy on, 262
Fanfare (towed decoy), U.S., 116
Far East: Hayward's Sea Strike concept for, 256. *See also* China and Chinese; Japan; Korean War; Vietnam War
FAT pattern-running torpedoes, German, 65, 94
Feniks (low-frequency sonars), Soviet, 322, 325
Ferranti Company, 113
fiber optics, towed arrays and, 321
Fido (homing torpedo), U.S., *52*, 52–53, *53*, 331–2, 346
Fife, James R., Jr., 127
50 Hz line, 240, 241, 242, 246, 247, 389n24; Soviet awareness of, 259
53-51 torpedo, Soviet, 358
53-56VA acoustic homing torpedo, Soviet, 358
53-61 torpedo, Soviet, 358
53-65 torpedo, Soviet, 358
fire control, 94–96, 111–3, 343
fire control center (FCS), Soviet, 325
Firefly AS Mk 5 to Mk7 sonobuoys, *174*, 308
Fishbowl (submarine detection system), 183
Fisk, J. B., 136
Five-Year Defense Plan (FYDP), U.S. (1981), 233
Fixed Distributed System (FDS), U.S., 196, *271*, 273–4
flaming datum (a sinking ship), 131, 258
flash signals, 147, 148
fleet air defense, ASW and, 212
fleet carriers, U.S., 119, 154, 378n3
fleet destroyers, U.S., 118, 121
Fletcher-class destroyers, U.S., *115*, 121, *134–5*, 151
flow noise: blade rate and, 197; British solution for, *23*, 23; homing torpedoes, 331; quieter submarines and, 276, 320; Soviets on single-hull submarine and, 241; submarine signatures and, 196; towed arrays and, 244, 245, 247, 390n35. *See also* DEMON
FMRs, 379–80n22. *See also* destroyer escorts
Forrestal battle group, U.S., 281
Foster, John S., 191, 385n35
four-square transducers, British, 28
Foxtrot submarines, Soviet, *1*, 110, 181–2, 322–3
FQQ-1A and -2A sonar systems, U.S., 382n9
FRAM (Fleet Rehabilitation and Modernization), U.S., 159, 218, *331*, 334
France and French: Algeria's war for independence and, 8–9; Atlantic ASW simulation battle and, 233; Bay of Biscay radar attacks on U-boats by, 50; Cold War ASW and, 4; DUBV 43 fish, *31*; European Atlantic aircraft project, 203; German U-boats based in, 55, *55*; Korean War escorts and, 156; LOFAR destroyer of, 146; postwar ASW warfare and, *125*, *126*, 375n41; Type XXIs' postwar transfers to, 88–89; U.S. Navy and navy reconstruction for, 7, 361n5; U.S. Navy ASW aircraft for, 171; Vietnam War and, 12; World War I submarine finding and, 21
FRD-10 (Navy's circularly disposed antenna array), 149, 377n20, 378n23
Freedom Torpedoes, U.S., for sale abroad, 337
French Union, in Africa, 12
frequency, sonar, 21–22, 102, 104, 111. *See also* 50 Hz line
frigates: Britain's naval staff on Korean War and, 154–5; British, with torpedo tubes, *348*, 348; British postwar ASW warfare and, 121–3; British towed array, 312, *312*; Canadian postwar ASW warfare and, *125*, 156–8, *177*, 177; NATO design competition for, 378n5; U.S. ASW force (1980s), 278

Gaither Committee, U.S., 180
Gearing-class destroyers: converted to DDEs, 112–3, 118; FRAM and, 159; postwar ASW warfare and, *120*, 120–1, *121*, 151; RAT installation on, 161
Geber transmitter, 97–98
General Belgrano cruiser, Argentina, 263, 265–6
Germania shipbuilders, 91
Germany: Atlantic ASW simulation battle and, 233; bubble decoy (*Pillenwerfer*), 24; Enigma (Ultra machine), 39, 62; on failures against convoys, 368–9nn15–16; GHG (*Gruppenhorchgerät* or group listening device), 314; sonars, U.S. adoption of, 313–8; on Soviet submarines and equipment, 107; Soviets' enmity of, 5; specialists working in Soviet Union from, 357; submarine strategy, 36–41; submarine technology, 2; U-boats' night surface attacks, 55–56; undersea missile-launching barge and, 101; Walter's high-speed U-boats and, 91. *See also* electro-U-boats; U-boats; West Germany; World War II
Gibraltar, Strait of, 32–33, 58, 69
Gilda technique, sonobuoys and, 384n20
GIUK (Greenland-Iceland-United Kingdom) Gap: British–U.S. concepts paper on Soviets and, 223, 224; Falklands Conflict water management and, 264; SLOC protection mission and, 255; SOSUS contacts in, 199, 382n8; Soviet potential war (1980s) and, 281; Soviet threat in North Atlantic and, 7, 361n8; submarine barrier across, 143
Glover (experimental escort ship), 188
Gnat torpedoes, 95
Goliath transmitter, 96, 226
Gorbachev, Mikhail, 282, 363n22

Gordon, A. P., 300–301
Gorshkov, Sergei: on Alfa-class submarine, 241, 242; on Allied force to defeat U-boats, 2; modern Soviet navy and, 18; Soviet military industrial base and, 17; as Soviet naval chief admiral, 16, 364–5n28; on Soviet sea power, 223–4; on SSBNs as effective deterrents, 230; Western intelligence on statements by, 227
Granay (Project 885) cruise missile submarine, 262
GRD-6 (as early HF/DF), 147, 148, 377n20
Greece, 5
Guardians (AF), U.S. aircraft, *172*, 172–3
Guppy (Greater Underwater Propulsive Power) program, U.S., 89, *89*, 127, 169, 371n25
Gyrodyne Corporation, 161, 162

Hagenuk radar search receiver, German, 65
Haines, R. G. C., 104
Hallett, C. F., 362n11
Hartwell study. *See* Project Hartwell study, U.S.
Harvard Underwater Sound Lab (HUSL), 24, 25
Haystack deception tactic, U.S., 210, 385n28
Hayward, Thomas B., 256, 257
Hedgehog: Argentine, Falklands Conflict and, 273; ASAP and, 45–46; Attack Director Mk 4 and, 112; British development of, 44–45; characteristics, 368n7; deep-diving submarines and, 71–72; explosions due to, *45*; lacking postinvention support, 42; loading on board, *44*; Mortar C as British replacement for, 74; projectiles, determining timing for, 22–23; trainable, 114, 374n36; U.S. improvements to, 77–78, 374n37; Weapon A and, 113
helicopters: ASW, early U.S., 132; ASW, fast cargo ships and, 139; ASW, Hartwell group on benefits of, *140*, 140–1; ASW and, 51–52, 135–6, 375n2; ASW simulation battle (1981) and, 237; British development of, *165*, 165–6; British surface strike and reconnaissance by, *168*, 168–9; Canadian ASW warfare and, 176–8, *177*, 379n20; deception against Soviet air attacks and, 210; Falklands Conflict and, *265*, 265–7, *267*; Korean War and U.S. mobilization of, 153; as LAAV, ASW and, 162–3; as LAMPS, ASW and, *163*, 163–5, *164*, 235; light, torpedoes dropped from, 394n5; nuclear submarines and, 186; sonars, 303–5; updated versions (late 1960s), *205–6*; U.S. first deployments, 385n1; U.S. naval air forces (1980s) and, 278–9; U.S. on Korean War ASW role of, 175–6, 379n18. *See also* DASH
helicopter/VSTOL carriers (CVHGs), 158, 220–1, *221*
HEN nuclear submarines, Soviet, 207, 384n26
HF/DF (high-frequency direction-finding): as antisubmarine sensor, 103, 373n9; British development of, 56; Doenitz and, 39; Hartwell study of, 146–9, *148*, 377n20; Soviet-built arrays (Krugs), 16
Hicks, R. L., 379–80n22
hindsight contract, for towed arrays, 245, 320–1
Hitler, Adolf, 37–38
homing torpedoes: ASW, of U.S. Navy, 116; British, 308, 309, 346–57, 395n15; German, against Britain, 65; German World War II use of, 94–95; Hartwell study of, 140, 141–2; helicopter or blimp dropping, 394n5; high-speed U-boats and, 72, 80; mobilization decade and, 159–60; *Nautilus* exercises and, 185; Nobska on types of, 189; snorkeling U-boats and, 68; Soviet blade rate reductions and, 259; Soviet development of, 357; speed of, 331; for U.S. ASW aircraft, 169; U.S. development of, 331–46; U.S. helicopters and, 175–6; U.S. postwar ASW and, 104. *See also* Mk 48 torpedo, U.S.
Horizontal Line Array (HLA) sonobuoys, 275
horizontal scanning sonars, 77
Horton, Max, 93
Hotel (Project 658) ballistic missile submarines, Soviet, 207–8, *208*
HTP (high-test hydrogen peroxide), German high-speed U-boats and, 90, 91, 92
HTP (hydrogen peroxide) torpedoes: German development of, 93, 95, 371n27; Soviet, 110, 357, 358
Hughes Aircraft, 245, 246, 248, 274, 345
HUK (hunter-killer) groups: aircraft, non-acoustic sensors for, 33; destroyer escorts and, 375n43; exercises, UPTIDE and, 213, 385n35; experimental, creation of (1946), 103, 118; as guppy and scrapper, 173; Low on HF/DF net and, 373n9; *Nautilus* exercises and, 184–5; postwar ASW warfare and, *120*, *121*; snorkel detection and, 129–30, *130*; sonar scouts and, 291; U.S. integrating submarines with, 169; U.S. postwar ASW warfare exercises (1949) and, 105–6; western Pacific, Korean War and, 153–4
hull arrays, Soviet, 329
hull sonars: dipping sonars with, *31*, 289; frigate self-defense and, 296; LOFAR processing and, 248; SQS-43 sonar, 288–9; standard U.S. postwar, 27; whales detected by, 272. *See also* sonars; Type 184 asdic, British; VDS fish
Hungary, 5
Hunt, Frederick, 142
hydrofoils, 185–6
hydromotors, Soviet underwater rockets and, 360
hydrophones: acoustic torpedoes and, 95; deep channel sounds and, 143; electro-U-boats' use of, 84, 85, 86; FDS and, 273–4; German, in World War II, 39; line, PUFFS passive ranging sonar and, 317; postwar ASW warfare and, 128–9; scanning sonar and, 366n12; World War I use of, 20. *See also* passive array sonars; sonars; sonobuoys
Hyulloise (Canada), *38*

Iceland, 59, 69, 233
IDD (initial detection display), British sonars and, 319–20
IFF (identification friend or foe), 163, 258, 264, 273
Ikara missile, 166, *166*, 167–8
Indonesia, independence from France, 8
infrared (IR) jammers, 278
infrared (IR) submarine detector, 34, 138

integrated support for submarines, 194–5
Integrated Undersea Surveillance System (IUSS), *271*, 274, 382n7
intelligence: acoustic, towed arrays and, 248; British, as proxies for U.S. sources, 389n20; British–U.S. concepts paper on Soviets and, 227–8; on enemy intentions versus enemy capabilities, 230; Falklands Conflict and, 268, 391n5; on Soviet SSN sonars, 323; on Soviet submarines, 389n20; on Soviet versus American thinking, 231, 232, 388n13; on Soviet wartime intentions, 258–9; UPTIDE exercises and, 214, 385n35; U.S. agencies on Soviet SLOCs campaign, 228–9. *See also* codebreaking
Interim Escort Towed Array Surveillance System (IETASS), 250, 251
Interim Towed Array Surveillance System (ITASS, SQR-14), 248
inverse Loran technique, Hartwell on HF/DF and, 147–8
Iran, 3, 5, 15, 255
Iraq, 9
Irtysh-Amfora (MGK-600) sonar, Soviet, 262
Iselin, Columbus O., 100, 366–7n14
Israel, 9, 211
Italy and Italians: closed-cycle submarine, 369n2; Cold War ASW and, 4; European Atlantic aircraft project, 203; Mediterranean crisis (1936) and, 37; postwar ASW warfare and, 375n41; Trackers certified to deliver U.S. weapons, 277; U-boats requested by, 61–62; U.S. Navy and navy reconstruction for, 7

Jane's, British knowledge of Argentine force and, 263
Japan: convoys and, 217; DASH effectiveness for, 162; maritime patrol aircraft and, 171; SOSUS offered to, 199; SOSUS stations in, 383n15; Soviet submarine warfare against, 258–9
Japanese Maritime Self-Defense Force, 276
Jezebel sonobuoys: British and, 309; frequency analysis, 244, 306, 393n21; LOFAR and, 144; lofargrams of, 182; progress report (1954), 200; signatures catalog, 189, 389n24
Joint Chiefs of Staff, U.S., 117
Julie sonobuoys, 182, 200–201, 306, 384n20
Juliett-class submarines, Soviet, 322

K2 torpedo, French, 125
Kahn, Louis, 125
Keflavík (sound surveillance system), 145, 198, 215, 376nn10–11; detecting quieter Soviet submarines, 239, 252, 389n21
Kelly, M. J., 136
Kennan, George, 5
Kennedy, John F., 12, 362n16
Kerch submarine sonar, Soviet, 324, 325, 326
Key West Operational Development Force, Advanced Weapons Unit (AWU), 335
Khrushchev, Nikita: deposed for military revolution demands, 231; Gorshkov and, 364–5n28; radar satellites and, 235; Revolution in Military Affairs under, 16–17; ship and submarine building under, 110; on submarine-based missiles, 180; Suez crisis (1956) and, 9
Kidd, Isaac "Ike," 14, 236, 363n19, 386n4
Kidd or *Kidd*-class destroyers, *26*, 278
Kilo, Soviet diesel submarine, 260–1
King, Ernest J., 51–52, 63
Kit passive ranging sonar, Soviet, 330
KNOUT, British submarine alternative to SOSUS, 319
knuckle in water, 24, 25
Kodak, for recharging Soviet submarine batteries, 261
Kola Peninsula, Soviet base complex on, 255, 257, 281
Komsomolets (Soviet submarine), 260, 279
Korean War: Communist submarines and, 378n2; NATO escorts inadequate for, 154; naval war, ASW in, 153–4; onset, Project Hartwell and, 149; outbreak of, as Western shock, 2; Skyraiders and Guardians and, 172; Soviet submarines and, 378n1; Stalin's approval of North Korea instigating, 6; U.S. helicopters for ASW in, 176; U.S. long-range planning after, 158–9; U.S. Navy budgets and, 150–3
Krasny Kavkaz, 16
Kretschmer, Otto, 58
Krivak (Project 1135) class frigates, Soviet, 187
Kurier (transmitters): convoy attack warnings and, 82; electro-U-boats and, 87; expected Soviet use of, 146–7; Germans compressing messages with, 39, 97, 98; Navy solutions to, 148; Soviet adoption concerns, 103
Kursk, Soviet loss of, 358
Kuznetzov, Nikolai, 16, 364–5n28

LAAV (Light Airborne ASW Vehicle), 162–3
Lamont Geophysical Laboratory, 244
LAMPS (Light Airborne Multipurpose System) helicopters, *163*, 163–5, *164*, 220, 235
lanes. *See* SLOCs
Lasky, Marvin, 244
LCS (low-cost sonobuoy), 275
Leanders, British, 167–8
Lehman, John, 256–7, 273, 282, 391n14
Leigh light, ASW aircraft and, 49
Lemon, Stanley G., 389–90n32
Leningrad fire control system, 322
Lerche (wire-guided acoustic homing torpedo), 95
Liberty ships, 60
LIDAR (laser radar), 34
Limbo (ASW mortar): British development of, *74*, 74–75, *75*, 369n6; British postwar ASW warfare and, *122*, 122, *123*; Canadian postwar ASW warfare and, *124*; shallow waters, Falklands Conflict and, 270, 273
Link 11, Canadian development of, 113
liquid-metal reactors, for Soviet nuclear submarines, 207, 242, 389n30

littoral ASW, 1. *See also* shallow waters
LOFAR (low-frequency spectral analysis): for ASW aircraft and submarines, 145; concealing meaning of, 143–4; Cuban Missile Crisis and, 182; for European Atlantic aircraft, 204; extremely secret status of, 146; Mediterranean detections, 383n17; PAIR (SQQ-23) sonar with, 290; processor for, 306, 393n2 (Appendix B); proposed British buoy with, 383n18; Soviet detection of, 327–8; Soviet Novembers detection and, 208; Soviet nuclear submarines and, 329; submarine sonars and, 316; tactical, nuclear submarines and, 199–201; towed arrays and, 244, 245
Long Range Objectives Group (LRO), U.S., 158, 217
Lorad (submarine detection system), 183
LORELI, 33, 206
Lotus (British), *36*
Low, F. S., 121, 373nn9, 11, 393n2 (Appendix C); ASW readiness report (1950), 134–6, 151, 375n1
Low Frequency Adjunct (LFA), or SURTASS ships, 274
LQQ-3 (upward-looking chokepoint system), 383n15
Luftwaffe, 37–39, 55, 84
LUT pattern-running torpedoes, German, 94

MacArthur, Douglas, 48
Mackle (Mk 22) torpedo, wire-guided, 349
MAD (magnetic anomaly detection): false targets during Falklands Conflict and, 270; for finding submarines, 20; Hartwell study on, 138; *Nautilus* exercises and, 185; as non-acoustic sensor, 32–33; of ships beyond the horizon, 163
Magic Sword exercise (1980s), 282
magnetic mines, 142
magnetometer, as non-acoustic sensor, 33
magnetorestriction transducers, 22, 24, 28, 366n3
Mahan, Alfred Thayer, 256
Mainbrace (NATO naval exercise), 7
maintenance of true bearing (MTB), U.S. Navy submarine finding and, 24
Malakhit design bureau, Soviet, 240, 241–3, 327, 389nn27–28, 389n29
Marine Physical Laboratory, University of California, 290
Maritime Defence (British magazine), 357
Maritime Strategy: development, 256–7; success of, 283–4; U.S. Navy (1980s), 14, 222, 232, 254–5, 277–9
Marlin study, Nobska's submarine rocket and, 187
Mars passive array, German, 321–2
Marshall Plan, 5, 6
MATCH (Medium Range Torpedo-Carrying Helicopter), British, 165, *165*, 168, 349
MCAR (Multi-Channel Aural Reception), for sonars, 289
M-class submarines, Soviet, 363–4nn24–25
McNamara, Robert, 198
Mediterranean Sea: ASW simulation battle and, 238; British codebreaking in, 39; degraded acoustic sensors in, 383n17; Interim Towed Array Surveillance System, 248–9, 250; NATO exercises (1980s) in, 282; sonar use in, 30–32, 290, 367n15; SOSUS arrays in, 377n13, 383n14, 390n36; SOSUS contacts in, 199; Soviet naval exercises in, 283; U-boat demands in, 62; U.S. on domination there (late 1970s), 232
merchant ships: Atlantic ASW simulation battle and, 232–3; containers and RO-ROs for, 225; postwar ASW warfare planning and, 125; quieter Soviet submarines and, 218; reinforcement shipping, Soviet SSNs and, 225, 386n4; SEAMIX I and NARAC-G exercises and, 236; shipbuilding versus World War II losses for, 59–60; Soviet naval nuclear tasks and protection of, 255; Soviet SSNs available for harassing, 227–8; U.S. postwar defense of, 102; World War II losses, 63; World War II threats against, 41. *See also* convoys, Cold War; convoys, World War II
MEZ (Maritime Exclusion Zone), Falklands Conflict and, 264–5
Michael (oceanographic analysis codename), 144
Michelsen, Andreas, 58
microcomputers, 14–15, 223, 351. *See also* computers
Middle East, 12
midget submarines, 110, 374nn22, 30, 393n12
Midway, 7, 361n6
Military Assistance Program, United Kingdom and, 171
Military Sealift Command, 388–9n18
Miller, J. D., 206
mine detection arrays , Soviet, 325
minelayers, Soviet submarines as, 387n11
mines: antisubmarine, development of, 104; ASW simulation battle and, 235; Captor, 207, 235; German, World War II, 4, 55; Hartwell study of, 142; postwar ASW warfare and, 105; postwar blockades by, 103; snorkeling U-boats and, 68; Undersea Warfare Conferences on, *346*; underwater electric potential sensor for, 33. *See also* Mk 24 torpedo, U.S.
minesweepers, 4, 41, 55, 249
Minton, David C., III, 385n33
missile destroyers, 251–2
missile submarines, Soviet, 258
missiles: nuclear, ASW warfare and, 373n6; as postwar threat, 101–2; solid-fuel torpedo tube bombardment, 206; Soviet, 4, 207–8, *208*, 210, 215, 384n24; Soviet sea-launched, as threat to U.S., 179–81, 380nn29–30; Soviet torpedo-tube, 261; submarine rocket, 187. *See also* ballistic missile submarines; cruise missiles; Polaris ballistic missile submarines; Regulus missiles, U.S.; SSBNs
Mitscher-class destroyers, 120
Mk 23 torpedo, British (Grog), 349
Mk 24 torpedo, British (Ongar), 348–52, 354, 355
Mk 24 torpedo, U.S. (Fido), 331–2, 333
Mk 27 torpedo, U.S., 333, 394n3
Mk 30 torpedo, British, 347, 395n20
Mk 31 torpedo, British, 354–6, 396n28

Mk 32 surface ship torpedo, U.S., 333, *333*, 335
Mk 34 Mod 1 torpedoes, U.S., 333–4
Mk 35 homing torpedo, U.S., 332, 393n1 (Appendix C)
Mk 37 homing torpedo, U.S., *334*, 334–5, 337, 393n1 (Appendix C)
Mk 39 torpedo, U.S., 393n1 (Appendix C)
Mk 41 torpedo, U.S., 332, 393–4n2 (Appendix C)
Mk 43 torpedo, 333, 335–6, *336*, 347
Mk 44 torpedo: British, 354, 355, 395–6n24, 396n26; U.S. development of, 336–7, *337*; U.S. replacement of, 337–8
Mk 45 ASW torpedo, U.S. (ASTOR), 140, 187, *188*, 252, 337, 381n3
Mk 45F Freedom Torpedo, U.S., 358
Mk 46 torpedo, British, 354, 355, 356, 396n29
Mk 46 torpedo, U.S., 259, 338–42, *339*
Mk 48 torpedo, U.S.: ADCAP redesign, 345, 351; characteristics, 191, 193, 244; development of, 335, 342–5; Mk 35 as predecessor to, 142; Mk 54 compared with, *189*; Nobska study on, 189; pump-jets propulsor of, *342*; speed of, 252, 256; on surface ships, *343*
Mk 50 lightweight torpedo, U.S., 276, 277, *346*
Mk 50 torpedo, U.S. (Barracuda), 345–6
Mk 54 torpedo, U.S., 346
Mk 101 (regenerative underwater fire control system), U.S., 112
Mk 102 (regenerative underwater fire control systems), U.S., 112
Mobile Listening, 244
mobilization decade: ASW aircraft, 169–78; Cuban Missile Crisis and, 181–2; Korean War and, 150–8; sonar and weapons, 159–69; strategic ASW, 178–81; submarine silencing, SOSUS and, 182–3, 381n34; submarines as direct support, 169; U.S. long-range planning after Korean War, 158–9
Momsen, Charles B., 111, 136
Moore's law, 275
Morse, Professor at MIT, 100
Mountbatten, Louis, 11, 165–6
Mousetrap projectiles, 22–23, 368n7
MSS (Moored Surveillance System), 235, 274
Musical Tones (Soviet underwater communication signal), 324
Mutual Defense Assistance Program, European, 6
Mutual Weapons Development Program, 204

NARAC-G (Atlantic ASW simulation battle), 232–3, 388n16
narrowband acoustics, 208, 319, 321
narrowband noise analysis: British submarine sonars and, 217; British Type 184 asdic and, 298; British use of, 244; Falklands Conflict and, 264, 272; Soviet submarines and, *247*; SQS-53A sonar and, 295–6; U.S. sonobuoys and, 306; by Western powers, 247. *See also* LOFAR
NASR 7511 torpedo, British (Stingray). *See* Stingray lightweight torpedo, British
Nasser, Gamel Abdel, 9
National Academy of Sciences, U.S., 100
National Defense Research Committee (NDRC), U.S., 42, 331
National Physical Laboratory (NPL), Britain, 42
National Resources Defense Council (NDRC), 100
National Security Industrial Association (NSIA), 244
NATO (North Atlantic Treaty Organization): ASW exercises, 390n1; ASW frigates, 155–6; becomes military alliance, 7; British–U.S. concepts paper on Soviets and, 223–4; on broken-backed war, 8; on Canadian shipboard helicopters, 178; creation and initial purpose, 6; disparate European and U.S. views on, 361n3; escorts inadequate for Korean War, 154; European Atlantic aircraft project, *203*, 203–4; frigate design competition, 156, 378n5; Kennedy and strategic shift for, 12, 13, 362n15, 362–3n17; Korean War and, 149; LOFAR processors released to, 307; Neptune as standard for, *170*, 171; on postwar ASW convoy components, 116–7; quieter Soviet submarines and, 217; Silver Tower convoy exercise (1968), 211; SLOC issue and, 386n4; on Soviet potential behavior (1970s), 223; Soviet potential war (1980s) and, 280, 281; Soviet-style torpedo designations of, 357; Suez crisis (1956) and, 9; Supreme Allied Commander for the Atlantic (SACLANT) and, 2; war, expectations for, 262–3
Nautilus, 184–5, 190–1, 195, 381–2n6
Naval Air Development Center, 200, 203
Naval Electronics Laboratory (NEL), 290–1
Naval Ordnance Laboratory, White Oak, 142
Naval Ordnance Test Station (NOTS), 160, 336
Naval Research Laboratory (NRL), U.S., 33, 42, 111, 287–8, 290, 366n4
Naval Tactical Data System (NTDS), 189
Naval War College, 101
NAVFACs (Naval Facilities), 144, 145, 197–8, 383n14
Navy, U.S.: aircraft-carrying non-acoustic sensors and, 33; ASW aircraft of, 49; ASW experimental task groups (1958), 362n13; ASW force (1980s), 277–9; ASW history and, 2–3; attack-at-source strategy for carriers and, 7; budget cuts (1948–1949) and, 100, 372n3; Carter's budget-cutting and power projection by, 255; closed-cycle engine research, 94; Cold War ASW and, 4; on continental U.S. defense, 362n11; countermeasures to Soviet threats, 210, 212–3; cruise missile test launch, 101; Cuban Missile Crisis and, 181–2; East Coast convoys (1942), 60; German electro-U-boat lessons for, 87; Hedgehog adopted by, 45; helicopter sonars, 303–4; Korean War and budgets of, *150*, 150–1; limited-war scenario of, 11; NATO's strategic shift and, 13, 14; nuclear air firepower (1948), 373n11; operator-friendly sonars, 25; postwar ASW convoy components, 117–8; postwar ASW warfare and, 99–101, 116–21; QHB scanning sonar, 26–27, 28; *Seapower 2000*, 256–7; silencing machinery by, 195; sniffers for ASW aircraft of, 33–34; sonar development, 21, 287–96, 366nn2, 4, 391–2n1; Soviet submarines tracked by (1950s), 13;

Soviets on power and competence (late 1960s) of, 231, 387n12; SSKs (ASW submarines) of, 127; strategic ASW and, 9–10, 178–9; on submarines and surface ship coordination, 193–4; torpedo development by, 331–46; towed arrays adopted by, 244–5; variable-depth sonars and, *302*, 302–3; weapons (1980s), 276; on Western weapons in Falklands Conflict, 273; World War II sonars of, 22–25. *See also* homing torpedoes; torpedoes
Navy Mine Warfare Test Station, 333
Navy Underwater Sound Laboratory (NUSL) U.S., 290, 291, 314, 315, 392n2
NEARTIP (near-term improvement program), U.S., 259, 341, 342, 345–6, 394–5nn10–11
Neptune (Navy ASW aircraft), U.S., 131
Netherlands and the Dutch, 156, 203, 277, 375n41
New Threat Upgrade (NTU) air defense systems, U.S., 278
New Zealand, 3–4
Newfoundland, 59, 69
Nibelung active sonar, 85, 86
NIDAR (radar phenomenon). *See* non-acoustic sensors
NIE (National Intelligence Estimate), 207, 230, 279–80, 391n11
Nightshirt, reducing ship self-noise and, 111
Nimitz, Chester, 99, 100–101
Nimrod, British aircraft: computerized data system on, *202*, 203; for NATO, *203*; nuclear depth bombs of, 277; on patrol, *204*, 263; sonobuoys, 275, 309; specifications, 33–34, 204–5
Nixon, Richard M., 211, 212, 219
Nobska study, 186–90, 337–8
noise: Falklands Conflict and, 268; 50 Hz line, 240, 241, 242, 246, 247, 389n24; Hartwell study of sonar and, 137; in Mediterranean, 249, 390n36; natural circulation reactors and, 381n5; sea, long arrays and, 143; ship self-noise, 111, 244, 272, 316; Soviet awareness of 50 Hz line, 259; Soviet awareness of sources of, 195–6; Soviet nuclear submarines and, 239–42; steam versus gas turbine ships, 312; of submarine heard during war, 371n22; supporting submarines and, 195, 381n6; torpedo self-noise, 141–2, 331, 342–4, 350; towed array self-noise, 245, 332. *See also* blade rate; cavitation sounds; flow noise; narrowband noise analysis; sound in the sea; towed arrays
noisemakers, 65, 72, 80, 374–5n38
non-acoustic sensors: environmental sensors, 35; future of, 284; Hartwell study of, 138–9; LIDAR (laser radar), 34; MAD (magnetic anomaly detection), 32–33; magnetometers, 33; sniffers, 33–34; Soviet experiments with, 34–35, *330*; thermal wake detection, 34; underwater electric potential, 33
Norfolk, 120
North Africa, 62
North American ASW command, 180
North Atlantic: convoy attacks (1942–1943) in, 61–63; shore-based aircraft and, 48; Soviet threat in, 361nn7–8; surface layer in, 29. *See also* Atlantic Ocean
Northern Wedding '86 exercise, 282, 390n1
Norway, 55, 282. *See also* Keflavík
Norwegian Sea: ASW simulation battle and, 233, 238; British–U.S. concepts paper on Soviets and, 224–5, 228, 229; Northern Wedding '86 exercise, 390n1; postwar ASW warfare planning and, 125; Soviet nuclear submarine transiting, 199, 215; U.S. Maritime Strategy and, 257. *See also* Keflavík
Nott, Hugh, 223
Nott, John, 15
November-class nuclear attack submarine, Soviet: early design issues, 209; intelligence reports on, 384–5nn26–27; Kit passive ranging sonar and, 330; Pacific operations, 382n8; speed of, *184*, 253, 385n32; U.S. tracking, 215–6
nuclear missiles, Soviet submarines and, 261
nuclear striking power, 4, 158
nuclear submarines (SSNs): in action (1973), 215–6; airplanes and, 201–5; ASW function of, 190; British, Falklands Conflict and, 264; closed-cycle propulsion and, 94; diesel submarines compared with, 184–6; direct support for, 193–5; early, self-noise of, 381–2n6; Khrushchev on, 17; land-attack force, Soviet, 17; liquid-metal reactors and, 207, 242, 243; other nuclear submarines versus, 190–3; Project Nobska study, 186–90; quieter, Soviet introduction of, 238–43; sonar development, 314–5; SOSUS and, 195–9, 252, 383n14; Soviet, *184*, 207–9; Soviet, SQS-26 sonar and, 292; Soviet fast carriers and, 209–12; tactical countermeasures, 212–5; tactical LOFAR and, 199–201; weapons, 206–7; Western fast carriers and, 209–12. *See also* SSBNs
nuclear war, 254
nuclear weapons: British research on, 5; as cruise missiles, potential of, 101; era of nuclear plenty for, 6–7; Hartwell study of, 139–42; Khrushchev on, 17; postwar ASW warfare and, 393n11; for Regulus, 373n6

Ocean Safari '81 exercise, U.S., 282, 390n1
ocean shipping. *See* merchant ships
Ocean Venture '81 exercise, U.S., 281–2
ODN (own-Doppler nullifier), 24
ODT (omnidirectional transmission), 25, 274
off-frequency transmissions, German, 96
Office of Naval Intelligence (ONI), 108, 279, 374n22
Office of Naval Research (ONR), 100
Office of Scientific Research and Development (OSRD), U.S., 42
Off-Shore Purchases (OSP) program, U.S., 6, 7, 156, 361n5
Ogarkov, Nikolai V., 14–15, 223
Okean (integrated Soviet sonar system), 325–6
Okean 70 (Soviet exercise), 210, 231
Okean 75 (Soviet exercise), 231, 282
omnidirectional buoys, 51, 164, 305, 306
Ongar (Mk 24) torpedo, British, 348–51
Oniks missile, Soviet, 262

Operational Development Force (OpDevFor): dipping sonars on helicopters and, 176; experimental HUK groups and, 103; on Fido (homing torpedo), 53; formation of, 42; MAD bird from a blimp and, 33; postwar ASW warfare and, 100, 118; trainable ahead-thrown weapon for, 113–4
Operational Evaluation Group (OEG), U.S., 42, 102, 105, 372n4
Ordnance Research Lab, Penn State, noise studies, 142
Oscar antiship cruise missile submarines, Soviet, 259, 260, 261
OSD (Office of the Secretary of Defense), U.S., 13, 223, 278
Otto fuel, 343, 351, 356
Outer Air Battle, SEAMIX I report advocating, 238
oxygen-kerosene engines, for Soviet torpedoes, 358

P2V Neptune, U.S., 169–71, *170*, 172, 200–201, 384n19
P-3 Orion, U.S.: characteristics, *202*, 203, *203*; LOFAR processors and, 307; sonobuoys and, *201*, 275, *307*; updates, 391n10
P4M Mercator, U.S., 169
P5M Marlin, U.S., 131, *170*, 171
Pacific Ocean: HF/DF nets in, 147, 377n20; Soviet out-of-area operations (1964), 382n8; substandard postwar U.S. Navy performance in, 100; U.S. codebreaking in, 39; U.S. countermeasures to Soviet threats, 212–3. *See also* Far East
PADLOC (Passive-Active Detection and Localization), U.S. Navy, 290
PAIR (Performance and Integration Refit), U.S. Navy, 238, 290
Papa (Project 661) submarine, Soviet, 240, 243, 323
Parry, David, 390n39
passive array sonars: German (GHG), 116, 313, 367n3; German, Soviets at World War II end and, 321; Hartwell study of, 138; Soviet cylindrical, 325; submarine rocket and, 187; U.S. Navy testing, 111; Western, Soviet awareness of, 261. *See also* hydrophones
passive homing torpedoes, 331
passive range-finding, 317. *See also* PUFFS
passive surface torpedo attack mode, 395n11
pattern-running torpedoes, 65, 94, 116, 358. *See also* homing torpedoes
PB-4Y-1 Liberator, U.S., *48*
PBM Mariners, U.S., 131, 132, 171
Pelmida towed array, Soviet, 329
Penkovskiy, Oleg, 17
Pentane homing torpedo, British, 175, 308–9, 347–8
Permit-class nuclear submarines, 192
Perry-class frigates, *221*, 222
Philippines, Kennedy on army division in, 12
photography, British sonars and, 319, 320
Pike Jaw (NATO nickname for Artika fire control sonars), 322
pinging: buoys marking wrecks and, 68; convergence zone and, 30; Doppler shift and, 24; radar, 21; sonar, 20; U.S. sonobuoys, 306; wakes and, 93
Pirie, Robert, 180
plotting, automatic, 309
Plutoniy active searchlight sonar, Soviet, 322
Polaris ballistic missile submarines: British nuclear striking power and, 4; as key deterrent, 381n2; navigational computer on, 380n30; smaller missiles and, 187; Soviet hunting for, 17; U.S. Long Range Objectives on, 158, 159
Polaris missiles, range, 225
POMCUS (Prepositioning of Materiel Configured in Unit Sets) warehouses, Europe's, 12–13, 223, 362n16
Porpoise or *Porpoise*-class submarines (British), 146, 183, 381n34
Portugal, 375n41
Poseidon missiles, U.S., 225
postwar ASW warfare: aircraft, 129–32; estimated threat for, 107–10, 374n22; estimating the situation, 102–7; fire control, 111–3; new missile threat, 101–2; new shipboard weapons, 113–6; ships, 116–27; sonars, 110–1; submarines, 127–9; U.S. Navy and navy reconstruction for, 99–101. *See also* Project Hartwell study, U.S.
Potsdam Conference, 87, 88, 371n23
Potter, Ralph K., 376n6
pouncers, British convoy, 104
PPI (plan position indicator): audio search compared with, 291; gain control for, 24; scanning sonar and, 25–27, 366n12
Prairie-Masker, reducing ship self-noise and, 111, 392n7
Pratt, William V., 48
preformed-beam sonars: British development of, 138, 296–301, 319–20, 392nn7, 11; British use of, 28; Soviet use of, 324; U.S. development of, 302–3
Prien, Günther, 58
Prinz Eugen (cruiser), 116, 313, *314*
Privateer (ASW aircraft), U.S. Navy, 131, 132
Problem of China, The (Mahan), 256
Project 833 (antisubmarine submarine), Soviet, 262
Project Bridge, Norway, 376n10
Project Caesar, U.S., 143. *See also* SOSUS
Project Hartwell study, U.S.: agenda for 1950s and, 136–7; convoys, 139; HF/DF submarine locator, 146–9; recommendations, 149; sonar, 137–9, 291; sonobuoys, 305; SOSUS beginnings, 142–6; weapons, 139–42
Project Nobska study, U.S., 186–90, 337–8
Project Trident, British, 123
propeller reshaping, 196–7. *See also* blade rate
propulsion: electro-U-boats and, 89–90; high-speed German, projecting scenarios for, 72–73, *73*; Mk 48 torpedo, 343. *See also* Walter boats
proximity fuzes, 75, 115, 344, 357, 369n4
PUFFS (Passive Underwater Fire Control Feasibility Study), *188*, 317, 337, 381n3
PUFFS passive ranging sonar, 316
PUFFS sonar dome, *314*, 317
pump-jets, reducing ship self-noise and, 111, 197

Q attachment, British, 24, 366n8
QC-1 sonar, U.S., *20*
QGB sonar console, U.S., 25, 110
QGB sonars, U.S., 287
QH-50D sonar, U.S., 201, 384n23
QHB scanning sonar, U.S.: Attack Director Mk 4 compared with, 112; British sonars compared with, 28; development of, 26–27; Navy on searchlight sonars versus, 110–1; operational tests, 77; torpedo evasion and, 116
quartz transducers, 22, 366n3
Quebec class submarines, Soviet, 110

radar: air ASW against U-boats and, 63; as airborne sensor, 49–50; for aircraft, 56; APS-20, postwar ASW warfare and, 107, 129, *129*; Argentine, Falklands Conflict and, 263; back echo problems for escorts, 59; on electro-U-boats, 86; for escorts, 56, 368n11; Hartwell participants on, 136–7; radio traffic and, 50; snorkel detection and, 129–31, 361n8; substandard postwar U.S. Navy performance with, 100; U.S. Navy development of, 21
radar picket forces, 178–9, *179*
radar satellites (RORSATs), Soviet, 235–6
radio, for Doenitz to U-boat communications, 96
radio fingerprinting (RFP), 56
radio intercepts. *See* codebreaking
RAFOS (Ranging and SOund Fixing), 30
rafting, to silence machinery, 195, 236, 247
range recorder, British deep-diving submarines and, 75
RAPLOC (rapid passive localization), U.S., diesel-submarine sonar and, 318
RAT (rocket-assisted torpedo), U.S., 141, *160*, 160–1
RDSS (Rapidly Deployable Surveillance System), U.S., 235, 274
RDT (rotationally directed transmission), for U.S. Navy sonar, 288, 291
Reagan, Ronald and administration: Aegis program and, 278; on arms control, 253; arms control under, 277; blade rate secrecy and, 256; on ship-towed SOSUS array equivalents, 388n18; stresses on Soviet Union and, 283, 391n15
REFORGER exercises, U.S., 12
refracted surface reflection, 243
refraction and reflection, asdic beams in water and, 28–29, 366–7n14
Regulus missiles, U.S., *9*, 102, 158, 362n10
reinforcement shipping, Soviet SSNs and, 386n7
Reitlinger, Otto W., 343
reliable acoustic path (RAP), for towed arrays, 243, 244
RETORC (Research TORpedo Configuration) program, U.S., 337–8, 342
retro-bombs, 32–33
retrospective direction-finding, 148
Rickover, Hyman, 94, 194, 278, 384n26
Robbins, T. H., Jr., 373n10
Roberts, G. H., 84, 87
Rochelle salt transducers, 22
rockets, for attacking surface submarines, 169
Romeo submarines, Soviet, 110, *192*, 322, 323, 330
rotating active planar array sonar, Soviet, 325
Rotoscope (scanning sonar), 25–26, 366n11
Royal Air Force: on ASW aircraft, 47–48; Cold War ASW and, 4; sniffers for ASW aircraft of, 33–34; very-long-range aircraft for ASW, 59. *See also* Britain and British; Coastal Command, RAF
Royal Australian Navy, *174*, *205*, *289*
Royal Canadian Air Force (RCAF), 4
Royal Canadian Navy: adopts twin 4.6 in. gun, 122; Cold War ASW and, 3; destroyer escorts, 59; helicopters operating from destroyers by, 176–7, *177*, 379n20; Korean War escorts and, 156–8, 378n6; postwar ASW convoy components, 117–8; postwar ASW warfare and, *124*, *125*; Soviet threat in North Atlantic and, 361n7; variable-depth sonars and, 301–2, *301–2*. *See also* Canada and Canadians
Royal Navy: asdic submarine finding and, 21; ASW history and, 2–3; on broken-backed war, 8; Classic Outboard and, 149; Cold War ASW and, 4; Defence Reviews (1966 and 1967), 166–7; on depth-charge limitations, 43–44; destroyer escorts, 11; Falklands Conflict limitations, 263; Firefly fighter, *174*; helicopters, *165*, 165–6; Korean War and, 6; Low Report upgrade equivalent, *135*; Mk 46 torpedo and, 339; NATO and, 13, 15; numerical designation system for, 375n45; postwar ASW convoy components, 117–8; postwar ASW warfare and, 116, 121–5, *122–4*, 375n44; on preformed-beam sonars, 297; on Soviet potential behavior (1970s), 223; Soviet submarines and, 181; Soviet threat in North Atlantic and, 361n7; on Soviets silencing submarines, 390n34; on submarines and surface ship coordination, 194; tactical submarine hunting trials by, 105; towed-array receivers and, 276; Type XXIs' postwar transfers to, 87–88; weapons (1980s), 277; Wessex helicopter, *206*; World War II sonars of, 22–24. *See also* Admiralty Research Laboratory; Britain and British
Royal Navy Torpedo Factory, 346–7
Royal Netherlands Navy, 4, 7. *See also* Netherlands and the Dutch
RPK-2 Vyuga-53 (82R) missile, Soviet, 384n24
RPK-6 missile, Soviet, 384n24
RPK-7 missile, Soviet, 384n24
RT (rocket torpedo), Soviet, 261
rubber domes, sonars and, 23, 238, 290, 295
rubber sound window, in steel sonar dome, 294, 295
Rubikon submarine sonar, Soviet, 326–7
Rubin submarine sonar, Soviet, 323, 324, 326
Rule, Bruce, 382n8, 384n26, 385n31
Ruler (mortar-firing proximity-fuzed bomb), British, 116, 375n39
rules of engagement, British, Falklands Conflict and, 273

Russia, 55, 61–62, 69–70. *See also* Soviet Union
RVM rocket, Soviet, 207

S2F Tracker, Grumman, 130, *173*, 173–5, *174*, *306*, 379n16
S-3 Viking ASW airplane, U.S., *219*, 219, 307
SACEUR (Supreme Allied Commander Europe), NATO's, 7, 224
SACLANT (Supreme Allied Commander Atlantic), 7, 8, 149, 155–6, 223, 361n8
SAET-50 antiship homing torpedo, Soviet, 358–9
SAET-53 ASW homing torpedo, Soviet, 359
SAET-60 antiship homing torpedo, Soviet, 359
SAET-65 ASW homing torpedo, Soviet, 359
salinity, sound propagation and, 28–29, 366n13
San Luis (Argentine submarine), 263–4, 266, 268–9, 270, 272–3
Sandy Hook, New Jersy, pilot underwater array at, 144, 376n9
Sandys, Duncan, 11
Sangamo Electric, 287, 391n1
Sapfir unified homing system, Soviet torpedoes and, 359
Sarsfield (ASW destroyer), 185
satellites, 225–6, 235–6, 263, 264, 270, 284
scanning sonar, 25–27, 366nn11–12
Schepke, Joachim, 58
Schlicke, Ing., 377–8n21
Scotsman (British electro-submarine), *83*, 84, 146
Scripps Institution of Oceanography, California, 290
SDT (steered directional transmission), 288
Sea Control Ships, U.S., 220–1
Sea King helicopters, *205*, 270, 305. *See also* SH-3 helicopters, U.S.
Sea Lance (standoff weapon), U.S., 276
sea power, Soviet, U.S. naval technology and, 255–6. *See also* Maritime Strategy
Sea Sparrow point-defense missile system, U.S., 212
Sea Strike concept, Hayward's, 256
seabed: bottom-bounce ping from, *29*; distinguishing from submarine near, 353. *See also* bottom-bounce sonar; shallow waters
SEAMIX I (Atlantic ASW simulation battle), 233, 237, 238, 388n16
seaplanes, Pacific Fleet, 134, 152–3
Seapower 2000, 256
Seapower of the State (Gorshkov), 223–4
search receivers, airborne, 50
searchlight sonars, 76, 77, 111, 308, 322
Seawolf-class submarines, 258
Sebastien, G., 248
seekers: active (echo-ranging) torpedo, 333; British Mk 23 homing torpedo and, 349; Mk 37 homing torpedo, 334; passive, British versus U.S., 347; passive, Mk 35 and, 332; torpedo, water noise and, 332
SEER (submarine explosive echo-ranging), 188
semiautomatic dead reckoner, German, 96, 372n33
semi-closed-cycle oxygen torpedoes, 110
semiconductors, in Soviet sonar, 326
Seraph (British), 79–80, *80*, 370nn11–12
SESCO (secure underwater communications system), 316
SH-3 helicopters, U.S., 235
Shackleton, maritime patrol airplane, British, 171, *171*, 308–9, 379n11, 393n23
shallow waters: British Mk 44 Mod 2 torpedo and, 354–5; British sonars and, 299; British Stingray torpedo and, 356–7; Colossus and, 144–6, 377n17; Limbo, Falklands Conflict and, 270, 273; Mk 46 torpedo and, 339, 354, 355, 356; quiet diesel-electric submarines in, 275; submarine speeds in, 342
Shark Gill (Soviet submarine sonar), 327–8
Shark Teeth (NATO nickname for Feniks-M diesel submarine sonars), 322
Shifting Tones (Soviet underwater communication signal), 324
Ship Characteristics Board (SCB), U.S. Navy, 152, 178
Ship Design Policy Committee, British, 155
shipping: independent routing to protect, 237. *See also* merchant ships
ships, anti-air warfare (AAW), Royal Navy and, 167
ships, ASW: Canada's need for, 177; as convoy escorts, 116–7; Nobska committee on, 188–9; postwar ASW warfare and, 113–6; ship self-noise and, 111; U.S., as radar pickets, 178, *179*; U.S. Navy, Korean War and, *150*, 150–2, *151*, *152*, 378n4
ship's inertial navigation system (SINS), U.S.-style, Soviets and, 181, 380n30
Shkval, Soviet rocket torpedo, 359–60
Short Seamew airplane, British, 348
Sidewinder missiles, U.S., 278
Sierra attack submarine, Soviet, 259, *259*, 260, 327
Sikorsky HO4S helicopters, U.S., 132, 176
Sikorsky S-58 helicopters, 176, 177, 379n19
Single Integrated [Strategic] Operational Plan (SIOP), U.S., 210
Sistema Obnaruzheniya Kil'vaternogo Sleda (Soviet non-acoustic sensors, SOKS), 34–35
SITTAR (Submarine Integrated Tactical Towed Array), U.S., 245–6
Six-Day War, 199, 211
65-73 torpedo, Soviet, 358
65-76 torpedo, Soviet, 358
Skat submarine sonar, Soviet, 327–9
Skipjack (fast attack nuclear submarine), 185, 190–1
Skyraider, Douglas, 120, 172, 175
SLBMs (ballistic missiles), Soviet D-35 system and, 262
SLOCs (sea lines of communication): as barrier ASW, 389n19; British on Soviet sea control goal and, 18–19; British–U.S. concepts paper on Soviets and, 223–4, 225, 226–7; defended lanes, British suggest, 226; defended lanes, to protect shipping, 237; as dormant issue, 386n4; future of,

284; as low Soviet priority (1980s–1990s), 280; NATO's strategic shift and, 13–14, 15; Soviet attacks against, 16; Soviet Foxtrot-class submarine and, *191*; Soviet minelaying submarines and, 387n11; Soviet naval aviation bombers and, 388n14; Soviet warfare intentions and, 259; Soviets protecting its SSBNs and, 230; Soviets unlikely to attack, 258; U.S. intelligence agencies on Soviet campaign against, 229; U.S. on Soviets attacking SSBNs instead of, 232

SLQ-48 mine-hunting device, 207

sniffers (chemical analyzer of submarine exhausts), 33–34

snorkel-detecting radar, SACLANT's Emergency Defense Plan on, 361n8

snorkels: British summarize experiences for U.S. with, 369n25; German, radar search receivers on, 50; German introduction of, 36, 66–70; indiscretion time and, 89; noise due to, 127; radar detection of, 129–31, 361n8; submarine approaching convoy and, 105; Walter's claim on inventing, 91

SOFAR (SOund Fixing And Ranging) channel, 30, 376n6

Sokolov, A. N., 364n26, 365n30

Solberg, T. A., 136, 138

sonar scout ship, 291. *See also* HUK (hunter-killer) groups

sonar signal acoustic processors (SSAPs), 272

sonars: bell-ringer feature, 195; BQQ-2 submarine, 206; British, Argentine submarines and, 272; British, Falklands Conflict and, 267–8; British and preformed beams, 28, 138, 296–301, 319–20, 392nn7, 11; Falklands Conflict and, 270; German low-frequency, 313; helicopter, 303–5; long range active, 187–8; low-frequency, Soviet postwar use of, 102; low-frequency, U.S. postwar use of, 104, 111; mobilization decade and, 159–69; nuclear submarines and, 185; passive, World War I use of, 20; postwar ASW warfare and, 110–1; postwar training and effectiveness of, 99–100; Project Hartwell study of, 137–9; propagation paths, *29*; quieter Soviet submarines and, 217; sea life around Falkland Islands and, 263; shipboard, quiet Soviet submarines and, 276; on Soviet Alfa-class submarine, 241; submarine support by, 193; on submarines versus surface ships, 127–8; surface ship and aircraft, 287; torpedo-detecting, 116; on U.S. carriers, *209*, 210; U.S. Navy development of, 21, 287–96, 366nn2, 4, 391–2n1. *See also* asdic; depth-determining sonar; searchlight sonars; sonobuoys; transducers, sonar; VDS fish

Sonics Mk 1 sonobuoys, 308

sonobuoy receivers, early Western, 131, 132

sonobuoys: Army's ASW units and, 49; British development of, 307–10, 393n23; characteristics and examples, *51*, 368n8; Cuban Missile Crisis and, 182; Falklands Conflict and, *271*; free-floating, 391n11; LAMPS helicopters and, 163; marking U-boat that dived, 51; power limitations, 200–201; silencing and improvements to, 274–5; for submarine detection, 131; substandard postwar U.S. Navy performance with, 100; U.S. development of, 305–7, 308

SOSS (Soviet Ocean Surveillance System), 209–10, 211, 386n7

SOSUS (sound surveillance system): Atlantic ASW simulation battle and, *234*; Canada and, 3; classified documents and analysis of, 273; as continental U.S. defense, 362n11; Cuban Missile Crisis and, 182; development of, *142*, 142–6; Falklands Conflict and, *271*; FDS and, 274; international installations, 145, *196*, 197–8, 199, 376nn10–11, 377n13; LOFAR data and, 200; Mediterranean countries and, 4; modernizations/improvements to, 388n17; Nobska committee on, 187–8; nuclear submarines and, 195–9, 382nn7, 11; planned phased expansion of, 198; quieter Soviet submarines and, 219, 254; signal processing, 196, 382n9; SURTASS as ship-towed equivalents to, 234, 388–9n18; tracking Soviet submarines, 13, 215–6; warning of submarine-launched missiles by, 180. *See also* Keflavík

SOSUS probability area (SSPA), 143, 377n12

sound in the sea: early ASW efforts, 20–25; German submarines and early Western sonars, 25–28; non-acoustic sensors, 32–35; underwater acoustics, *26*, *27*, 28–32, *29*, 290, 366n13, 367n16. *See also* blade rate; cavitation sounds; noise; noisemakers; submarine sound signatures

Soviet Union: Atlantic ASW simulation battle and, 232–8; blade rate issues and, 197; British on submarine fleet war plans of, 222; Cold War ASW and, 4; countermeasures to missile submarines of, 212–5; Defense Science Board predictions on, 193; economy, military element in, 365nn31–32; estimates of postwar submarine program of, 108, 109–10, 374n22; Falklands Conflict and, 270; fast carriers and, 209–12; German U-boat technology postwar and, 102–7; Khrushchev's disruption of economic planning in, 17; missile submarines, SOSUS and, 145, 180–1, 380nn27–28; missile-launching submarines, 159; naval exercises, convoy groups and, 227, 387n8; navy, 8, 18, 230, 365n34; new submarines, *259*, 259–62; non-acoustic sensor testing by, 34–35; on nuclear balance, 280; nuclear submarines, *184*, 186–90, 207–9, *208–9*; nuclear weapons, NATO and deterrence of, 6; nuclear weapons obsession of, 254–5; Pacific submarine fleet, Korean War and, 153; postwar ASW warfare against, 393n11; prioritizing new construction over maintenance (1980s), 279; quieter submarines, 238–43; rules of engagement and, 273; on "scientific" analysis, 231, 388n14; shipbuilding 1920–1950s, 365n30, 375n1; submarine design organizations, 243; submarine fleet, 2, *133*, 363–4n24; submarine sonars, 321–30; submarine-launched missiles' threat of, 180, 380n25; surface force survival and, 283; torpedoes, 357–60; tracking U.S. and British nuclear weapons, 258; Type XXIs' postwar transfers to, 88, *99*; U.S. Maritime Strategy and, 257; Walker spy ring and, 236; Walter boats and, 93, 371nn27, 29; on Western power and competence (late 1960s), 231, 387n12; Wullenweber technology captured by, 149. *See also* postwar ASW warfare; Russia; submarines, quieter Soviet

Spanish Civil War, 41
Spartan (British), 264, 265, 268
Spearfish torpedo, British, 350, 351, *352*, 352, 395n13
speech, frequency components of, 142–3, 376n8
Speer, Albert, 81
Spinne (T X) electric torpedo, 95
Splendid (British submarine), 263, 264, 266
split-beam sonar, Limbo and, 76
SpringEx-84 (Soviet exercise), 282
Spruance-class destroyers, 222, 277
Sputnik, 180
SQG-1 attack sonar, U.S., 111
SQQ-28 sonobuoy processor and tracker, U.S., 296
SQQ-89 towed array, U.S., 251, 296
SQR-14 towed array, U.S., 249–50
SQR-15 towed array, U.S., 249, 250
SQR-19 towed array, U.S., 250, 251
SQS-4 sonar, U.S., 24, 111, 159, 287–8
SQS-23 sonar, U.S.: on carriers, *209*, 210, 385n29; development of, 289–91, 392n2; elements of, *288–9*; SQQ-23 active-passive version, 238, 290
SQS-26 sonar, U.S.: bell-ringer feature, 195; bottom-bounce and, 300–301; convergence zone and, *186*, 186; countermeasures to Soviet threats, 213, 215; development of, 291–5, *292*, 392n3; nuclear submarines and, 206; prototype, *287*
SQS-29 sonar, U.S., 288
SQS-43 sonar, U.S., 288–9
SQS-53A, SQS-53B, SQS-53C sonar, U.S., 295–6
SQS-56 hull sonar, U.S., 296
SQS-504 variable-depth sonar, U.S., *301*, 301–2
SQS-505 variable-depth sonar, U.S., 301–2
Squeeze Play, Operation, 252, 390n38
Squid (ahead-firing weapon), British: British postwar ASW warfare and, 124–5; Cold War need for, 105; deep U-boats and, 71–72; postwar fitting on destroyers, 76; trials, 45–46, *46*; U.S. improvements, 76; Weapon A compared with, 114
Squid Arm (integrated Soviet sonar system), 325–6
SSBNs (nuclear-powered ballistic missile submarines): British–U.S. concepts paper on Soviets and, 224–5; *George Washington*, *18*; Security Program, 34; SOSUS probability areas and, 377n12; Soviet, Northern Fleet bastion, 281; Soviet concerns on protection of, 18; as Soviet national deterrent, 284; trailing enemy SSBns, 230–1; U.S. development of, 9–10, 362n10. *See also* ballistic missile submarines
SSKs (ASW submarines): as continental U.S. defense, 362n11; development of, 127; as direct support for surface ships, 169; exercises testing, 127–8; mass production of, 104; U.S. example, *128*
SS-N-3 missiles, Soviet, 210
SS-N-6 missiles, Soviet, 384n24
SS-N-7 cruise missile, Soviet, 207, 215, 240
SS-N-15 missiles, Soviet, 261, 384n24
SS-N-19 missiles, Soviet, 261
SS-N-32 Bulava missile, Soviet, 262
SSNs. *See* nuclear submarines
SSQ-2 omnidirectional buoys, U.S., 172, 305
SSQ-53 DIFAR buoys, U.S., 201
Stalin, Joseph, 5, 6, 8, 15, 110, 363–4n24, 364n26
standoff (magnetic) exploders, Soviet torpedoes and, 261
Standoff Weapon (SOW), U.S., 276
standoff weapons: all-weather, British, 166; lower-frequency sonars and, 165; rocket-propelled, mobilization decade and, *160*, 160; rocket-propelled, Soviet, as threat to U.S., 8
STASS (Submarine Towed Array System), 246–7, 249
steam plants, for frigates, Korean War and, 155
steam torpedoes, German World War II use of, 94
Steinbarsch (HTP) torpedoes, 95
Stinger study, U.S., 187, 381n3
Stingray lightweight torpedo, British: British consider, 394–5nn10–11; characteristics, 355–7; at Euronaval 2006, *356*; helicopters and, 305; Ikara and, 167; Lynx helicopter and, 168; Nimrods and, 309; Tigerfish torpedo and, 351
Strategic Air Command (SAC), 181
Stratton, J. A., 136
Strike Fleet Atlantic, NATO's, 7, 361n7
Strike Fleet Mediterranean, NATO's, 7–8
Strong Express exercise (1972), 390n1
STWS-2 torpedo launcher, British, *268*
Sub-Harpoon, British, 193
Submarine Operating Authority, 194
submarine rocket (SUBROC), ASW, 187, 206, 252
submarine sonars: British, 318–21; Soviet, 321–8, 393n5; U.S., 313–8, *314*
submarine sound signatures: catalog/library of, 189, 307, 389n24; flow noise and, 196; Soviet, 142–3, 145, 189, 376n6
submarine-launched one-way transmission (SLOT) buoys, 195
submarines: Argentine, Falklands Conflict and, 263, 266–7; ASW, postwar Soviet threat and, 103; Cold War ASW and, 4; communications, 96–98, 372n35; destroying enemy submarines in transit, 104; as direct support, 169; German, British detection of, 39; German deep-diving, 24, 71–72, 366nn8, 13; German high-speed, 72–73, *73*; German improvements, 25; navigation, strategic ASW and, 181; nuclear material for, 7; nuclear submarines compared with, 184–6; postwar ASW warfare and, 127–9, 314; postwar British ASW improvements, 73–76; postwar U. S. ASW improvements, 76–78; silencing, by British, 146; Soviet (1939), 363–4n24; Soviet, at sea, 1980s limits on, 279–80; Soviet building plans for, 364nn26–27; Soviet land-attack-missiles on, 9–10, *10*; Soviet sound signatures, 383n14; Soviet W-class and Z-class, 15–16; strategic, with long-range weapons, 18, *18*; threats, changes since Cold War in, 1; torpedoes and fire control for, 94–96;

Type XXI postwar distribution, 87–90, 371n23; Type XXI tactics, 84–87; Type XXI U-boat, 78–84; Type XXVI, 90–94; U.S., trailing Soviet SSBNs, 230–1; U.S. ASW force (1980s) and, 277, 278; U.S. postwar ASW warfare exercises (1949) and, 106; Whiskey project submarines, 88, *99*, 109–10. *See also* ballistic missile submarines; cruise missiles; diesel submarines; diving depth; electro-U-boats; nuclear submarines; postwar ASW warfare; SOSUS; SSKs; torpedoes

submarines, quieter Soviet (1960s–1970s): ASW strategy for, 222–32, 387n12; development of, 238–43; fighting an ASW war, 232–8; strategic ASW and, 251–3; towed arrays and, 243–51; U.S. prospect of, 217–22

Suez crisis (1956), 9

Sumner-class destroyers, U.S., 151, 159

supersonic submarine finders, 20–21

surface ASW tactics, 42–47, 235

surface bounce. *See* refracted surface reflection

surface capture: avoiding, submarines and, 192–3, 339; torpedoes and, 353, 354

surface layer: in Mediterranean, 30–31, *31*; *Nautilus* exercises and, 185; sound velocity and path in, *29*, 29–30

surface ship radar, 56

Surface Ship Sonar Modification Program (SSMP), U.S., 295–6

surface ships, British, 263, 310–2

surface ships, U.S.: ASW equipment on, 78; *Nautilus* exercises and, 185; Soviet Yankee submarines and, 252; submarines as direct support for, 169; towed arrays and, 247–51, *248–51*. *See also specific ship types*

surface warships, Soviet, 4

SURTASS (Surveillance Towed Array Sensor System), *234*, 234, 250, 274, 388–9n18

SWATH ships, LFA for, 274

Sweden, 89

synchros, U.S. Navy's, 112, 374n35

T IV torpedoes, German, 95

T V (*Zaunkoenig*) torpedoes, German, 95

T X (*Spinne*) electric torpedo, German, 95

T XI torpedo, German, 95–96

T-5 nuclear torpedo, Soviet, 358

TACCO (tactical coordinator), U.S., 203

Tactical Support Centers (TSCs), U.S., 219

tactical surveillance buoy, U.S., 275–6

T-AGOS surveillance ship, U.S., *271*, 274, 388n18

Tango submarine, Soviet, *225*

tankers, German, for refueling, 61, 64, 66

target motion analysis (TMA), 128, 312, 324

TB-16 array, U.S., *246*, 246–7

Teller, Edward, 187

temperature, sound propagation and, 28–29, 366n13

Tenth Fleet, U.S., 42, 63

TEST-68 wire-guided torpedo, Soviet, 359

TEST-71 wire-guided torpedo, Soviet, 359

Thatcher, Margaret, 265, 351

thermal wake detection, 34

thermocline, 29, 30

Thermopylae (British), 318

Third World war: Western weapons and systems and, 263, 273. *See also* end-game

Thresher or *Thresher*-class nuclear submarines, U.S., 190, *190*, 193, *313*, 316, 338

Tigerfish torpedo, British, *349*, 351, 352

Till, Geoffrey, 390n1

titanium, Soviet submarines and, 240, 241, 242, 260

Tomahawk cruise missiles, 258, 277–8

torpedoes: Advanced Light Weight, 235, 243; ASW, air-launched, 333; British development of, 346–57; cavitation by, wake due to, 269–70; drone version, 161; Falklands Conflict and, 271; Hartwell group on helicopters and, *141*, 141; HTP (high-test hydrogen peroxide), 93; interwar ASW development and, 42; long-range, warning needed for, 297, 392n11; long-range pattern-running, postwar Soviets and, 102; new German submarines and, 94–96, 372nn32–34; nuclear, features of, *189*, 189; postwar ASW warfare and, *115*, 129; RAT (rocket-assisted), *160*, 160; Soviet antiship, 261; Soviet development of, 357–60; Soviet naval exercises and, 229, 387nn8, 10; Soviet postwar projects, 110, 374n32; SSN versus SSN and, 190–3; U.S. and shallow-water performance of, 273; U.S. countermeasures to, 116, 374–5n38; U.S. development of, 331–46. *See also* homing torpedoes; *specific types*

Torquay-class fast frigate, British, 348

Total Exclusion Zone, Falklands Conflict and, 265

towed arrays: ASW simulation battle (1981) and, 237; benefits, 165; British, Falklands Conflict and, 264, 272; British, for Type 23 frigate, 305; British Duke class and, *227*; British equivalent to, 320–1; characteristics, 32; detecting quieter Soviet submarines, 240; free-floating sonobuoys and, 391n11; nuclear submarine detection by, 256; plans for, 164; quiet, deep-diving Soviet submarines and, 243–51; Soviet Project 885 and, 262; Soviet Victor III submarines, 329; Soviets on defended SLOCs and, 226; SQR-19, *164*; surface ships and, 247–51, *248–51*, 390n35; tactical, ASW simulation battle and, 235, 238; tactical, British development of, 310–2; tactical countermeasures to missile submarines and, 213; U.S. naval air forces (1980s) and, 278–9. *See also* passive array sonars; STASS; SURTASS

towed decoys, 116, 272

towed-array receivers, French multiline, 276

Towflex (towed array), 244

trailing, as Cold War operation, 195, 252, 253, 255, 390n39

transducers, sonar: British retractable dome, *23*, 23; example, *21*; lower sound frequency and, *26*; nuclear submarines and, 314–5; paired, 290; size of, 21; sound absorption and, *27*; Soviet Skat submarine sonar and, 327; Soviet tilting active, 325; types developed, 22, 299

transmitters (*Kurier*), Germans compressing messages with, 39
Trident C-4 missile, 258
Trident system, 196
Tripartite Naval Commission, 87, 88, 371n29
Triton cruise missiles, 206, 362n10
Trout Cheek (Feniks-M nuclear submarine sonars), 322
Truman, Harry S., 5, 6
Tu-16 "Badgers," Soviet, 8
TUBA program, 244–5, 390n33
Turner, Standfield, 363n20
Type 144 asdic, British, 23
Type 147 asdic, British, 24
Type 177 asdic, British: development, 138; fast submarines and, 297–8; as first preformed-beam sonar, 28, 296–7, 392n11; helicopters and, *165*; replacement of, 168; ships installed on, *122*, *123*; whales detected by, 272. *See also* DIMUS
Type 184 asdic, British, 116, 168, 269, 272, 298–9
Type 195 dipping sonar, British, 266, *267*, 304, 305
Type 2001 asdic, British, 272, 319, 320
Type 2007 passive submarine flank array, British, 321
Type 2016 sonar, British, 272, 299–300, *300*
Type 2020 asdic, British, 320
Type 2023 towed array, British, 321
Type 2026 towed array, British and Dutch, 321
Type 2032 asdic, British, 320
Type 2046 towed array, British, 321
Type 2050 sonar, British, 300
Type 2052 clip-on towed array, British, 321
Type XVIIB (high-speed U-boat), German, 91, 92
Type XVIII (open-ocean HTP submarine), German, 91–92
Type XXI electro-U-boats: British misapprehension of snorkeling by, 131; British response to, 79–81, *80*, 82–83, 371n17; British testing, 83; captured, postwar distribution of, 87–90, 371n23; Fido's lack of speed and, 332; German introduction of, 78–79, *79*, 369–70nn8–10; German production problems, 81, 370nn14–15; as postwar threat, 103; *Scotsman* modified as, *83*, 84; tactics, 84–87, 371n21; Walter's development of, 92; wolf packs and, 81–82. *See also* Guppy program
Type XXVI, Walter design for, 92, 371n28
Typhoon strategic submarines, Soviet, 259, 260

U-boats: British on possible modifications to, 71–73, *73*; central control for, 96; new, electro-U-boat, 78–83; postwar transfers to Soviets, 88. *See also* submarines; Type XXI electro-U-boats
Ultra machine, Germany's, 39, 62
Undersea Warfare Conferences, U.S. Navy: on ASW training needs, 134; British officers attending, 3; on destroyer escorts' transfer to Europe (1948), 6, 361n2; estimating postwar situation, 102–3; on experimental HUK groups, 103; on hunter-killer groups, 118; on mass installation of QHB sonars, 287; on mines, *346*; Nimitz establishes (1946), 99; on ship self-noise reductions, 111; sniffers, ASW aircraft and (1959), 33
underwater acoustics, 28–32, *29*
Underwater Battery Plots, 113, 114
underwater communications systems, Soviet SSN sonars and, 323
Underwater Detection Establishment (UDE), British, 319
underwater electric potential (UEP), as non-acoustic sensor, 33, 139
underwater rockets, Soviet interest in, 360
underwater-launched antiship missiles (USGW), Royal Navy on, 15
United Kingdom, 100–101, 145, 303, 304–5, 377n14. *See also* Britain and British; Royal Air Force; Royal Navy
United States: ASW aircraft, 46–47; battle of convoy routes along, 60; British concepts paper on Soviets and, 223–5, 226–7, 386n5; British on attacking at source versus, 104; budget cut effects (1949), 133–4; Cuban Missile Crisis (1962) and, 181–2; dropping torpedoes when target depth uncertain, 394–5nn10–11; East Coast barrier, 143; frigates for postwar ASW warfare, 158; German Type XXIs and, 87, 88, 89, 94; intelligence community, on resupply shipping attacks, 228–9; interwar ASW development in, 42; Korean War and, 6; long-range planning (1950s), 158–9; Low report on ASW readiness, 134–6; postwar ASW upgrade, 76–78, 369n7; postwar ASW warfare exercises (1949), 105–6; postwar torpedo program, 393n1 (Appendix C); sonobuoys of, 305–8; SOSUS offered to friendly countries by, 199, 382n12; ultimate Cold War weapons research by, 5; Walter boats and, 93; World War I submarine finding and, 21; World War II shipbuilding, 54. *See also* Air Force, U.S.; intelligence; mobilization decade; Navy, U.S.; postwar ASW warfare; Project Hartwell study, U.S.; SOSUS; submarines, quieter Soviet
United States, 6, 103–4
UPTIDE (United Pacific Fleet Project for Tactical Improvement and Data Extraction), 212–5, 385n35
uranium, stockpiling, 6–7
Ustinov, Dmitry, 260
U.S.-UK-Canada Tripartite Technical Cooperation Committee, Working Panel on Surface Systems, 178
utility helicopters (UTTAS), U.S., 164, 165
UYS-1 (Advanced [Acoustic] Processing System), *219*, 310

V-1 cruise missile, Soviet, 101–2, 109
V-2 ballistic missile, Soviet, 101–2
Valiant (British towed array submarine), 264, 265, 268–9
Valley Forge, U.S., 362n13
Varshavyanka, Soviet diesel submarines, 260–1
VDS (variable-depth sonar) fish: characteristics, 301–3, *302*; FRAM-1 and, 159; French, *31*; towed arrays and, 221, 251
Vernier processors and analysis, *163*, 198, 382n9
vertical linear array (VLA), for towed arrays, 243

Vertically Launched ASROC (VLA), 276–7
Victor (Project 671) class submarines, Soviet: characteristics, *217*, 238–9, 261; sonar for, 323; strategic ASW and, 253; with towed array, *247*, 247; towed array pod of, *329*, 329
Vietnam War, 12, 17, 198, *211*, 211–2, 218, 385n33
Virginia-class nuclear-powered missile cruisers, 278
visual integration, of SOSUS images, 144
VLAD (Vertical Line Array DIFAR, SSQ-77), 275, 307
Voenny Mysl' (Military Thought), Soviet, 17
VTOL Support Ship (VSS), 221

wake-following guidance, Soviet torpedoes and, 261
wakes: detection of, 34, 35, 138; echoes due to, 22, 25; electric torpedoes and, 94; following, homing torpedoes and, 357–8; homing torpedoes and, 141; pinging interference and, 93; submarine, reducing, 34; underwater, 79
Walker, John and spy ring, 39, 234–5, 236, 252
Walter, Helmut, 90–92, 93, 94, 371n27
Walter boats (submarine prototypes), 88, 90, *90*, 91–92, 109
war. *See* end-game
Warsaw Pact, 224
water, vertical structure of, *29*, 30
water pressure, sound propagation and, 28–29, 30, 366n13
Way Ahead Working Party, British Mk 24 torpedo, 351
W-boats, 90–91, 371n26
Weapon A: on British ASW destroyer, *45*; Limbo as British equivalent to, *74*; nuclear weapons compared with, 140; as trainable weapon, 47; on U.S. destroyer escorts, *115*; U.S. development of, 78, *78*, 113
Weapon B, 78
weather: Falklands Conflict and, 264; sound propagation/detection and, 29, 366–7n14
Weinel, Jim, 382n9
Wessex helicopter, *205*, *206*, 304–5
West Germany: Berlin, 5, 12; Cold War defense of, 101; European Atlantic aircraft project, 203; land-based Gannets operated by, 347; NATO European war study on, 223; U.S. combat army deployed to, 7
Western Approaches, Battle of the, 56
Western ASW: superiority, Soviets' perceptions and, 255. *See also specific countries*
Western European Union, 6
Westinghouse, 343
whales, Falklands Conflict and, 263, 269, 270, 272
Whirlwind helicopters, 304
Whiskey project submarines, Soviet: cruise missiles launched from, 180, 380n25; Feniks sonars on, 322; first mention of, 374n27; Low on ASW readiness by U.S. and, 375n1; mass production of, *99*, 109–10; maximum speed of, 118; Plutoniy active searchlight sonar on, 322; Type XXI power plant and, 88
Whitby-class fast frigate, British, *122*, 122, 348
White Sea, Soviet bastion in, 18, 225, 257
Whitworth, Jerry, 236
Wide Aperture Array (WAA), diesel-submarine sonar and, 318
wideband receivers, Hartwell on HF/DF and, 147–8
Windrow (AM voice system), Soviet, 324
wire guidance system, for torpedoes, 334–5, 343, 349, 350, 352, 359
wire-guided acoustic homing torpedo (*Lerche*), 95
wolf packs, German, 37, 58–59, 61, 81–82, 87
Woods Hole Oceanographic Institution (WHOI), 42, 100, 142, 290
Woodward, John, 265, 266
World War I, hydrophones during, 20
World War II: ASW aircraft, 47–53; ASW and course of, 54–66; ASW during, 4; British response to U-boat improvements, 73–76; German communications with U-boats during, 96; German submarine strategy during, 36–41; organization for ASW development during, 41–42; shipbuilding, naval and merchant, 53–54; snorkels and, 66–70; sonar development during, 21–25; Soviet alliance with U.S. and U.K. in, 5; surface ASW tactics, 41–47; U.S. postwar ASW upgrade, 76–78, 369n7. *See also* destroyer escorts; submarines
WQC-2 underwater telephone, U.S., 323
Wullenweber (German circularly disposed antenna array), 148, *148*, 149, 377–8n21

Yankee-class ballistic missile submarines, Soviet, 181, *222*, *224*, 238–9, 243, 252–3, 390n38
Yasen (Project 885) cruise missile submarine, 262
year of maximum danger, 5, 6
Yegorova, Natalia I., 363–4n24
Yenisey submarine sonar, Soviet, 324, 325, 326

Zacharias, J. R., 136
Zhukov, Georgi, 16, 136, 388n14
Zulu submarines, Soviet: ballistic missiles launched from, 180, 380nn25–26; characteristics, 110, 118; Feniks sonars on, 322; first mention of, 374n27; ocean mission of, *99*; as Project 611, 109; sonar prototype tested in, 326
Zumwalt, Elmo R., Jr., 163, 219, 220, 221–2, 277
Zusma transmitter, 97

ABOUT THE AUTHOR

NORMAN FRIEDMAN is a prominent defense analyst and historian specializing in the intersection between policy, strategy, and technology, mainly in a naval context. He has published more than forty books. He is the recipient of numerous awards, including the Commodore Dudley W. Knox medal by the Naval Historical Foundation, the Samuel Eliot Morison award by the Naval Order of the United States, the Westminster Prize awarded by the Royal United Service Institute, and the Anderson Medal of the British Society for Nautical Research. He has twice received the John Lyman Award from the North American Society for Oceanic History. He lives in New York.

THE NAVAL INSTITUTE PRESS is the book-publishing arm of the U.S. Naval Institute, a private, nonprofit, membership society for sea service professionals and others who share an interest in naval and maritime affairs. Established in 1873 at the U.S. Naval Academy in Annapolis, Maryland, where its offices remain today, the Naval Institute has members worldwide.

Members of the Naval Institute support the education programs of the society and receive the influential monthly magazine *Proceedings* or the colorful bimonthly magazine *Naval History* and discounts on fine nautical prints and on ship and aircraft photos. They also have access to the transcripts of the Institute's Oral History Program and get discounted admission to any of the Institute-sponsored seminars offered around the country.

The Naval Institute's book-publishing program, begun in 1898 with basic guides to naval practices, has broadened its scope to include books of more general interest. Now the Naval Institute Press publishes about seventy titles each year, ranging from how-to books on boating and navigation to battle histories, biographies, ship and aircraft guides, and novels. Institute members receive significant discounts on the Press' more than eight hundred books in print.

Full-time students are eligible for special half-price membership rates. Life memberships are also available.

For more information about Naval Institute Press books that are currently available, visit www.usni.org/press/books. To learn about joining the U.S. Naval Institute, please write to:

Member Services
U.S. NAVAL INSTITUTE
291 Wood Road
Annapolis, MD 21402-5034

Telephone: (800) 233-8764
Fax: (410) 571-1703

Web address: www.usni.org